Protective Relays: their Theory and Practice

VOLUME TWO

Protective Relays

THEIR THEORY AND PRACTICE

VOLUME TWO

by

A.R. van C. WARRINGTON

A.C.G.I., B.Sc. (Lond.), C.Eng.,
F.I.E.E., F.I.E.E.E.

LONDON
CHAPMAN AND HALL

First published 1969
by Chapman and Hall Ltd
11 New Fetter Lane, London EC4P 4EE
Second edition 1974

Printed in Great Britain by
Fletcher & Son Ltd, Norwich

SBN 412 12100 X

Library of Congress Catalog Card Number 70–385616

Distributed in the U.S.A.
by Halsted Press, a Division
of John Wiley & Sons, Inc.
New York

Foreword

The addition of this Volume II by Mr. Warrington provides, in the two volumes together, an exhaustive study of the development and present status of protective relaying, particularly in relation to the transition from electromagnetic to static relays using semiconductors. Volume II attempts, among other things, to explain their theory in a manner understandable to non-electronic engineers, as well as dealing more thoroughly with some of the subjects touched on in Volume I.

It gives me much pleasure to recommend this book and its counterpart Volume I to all interested in protective relaying, as they have been written by one of the country's outstanding specialists in this particular field. As Chief Engineer of the English Electric Company in the years 1948–1954, I was closely connected with Mr. Warrington's Forward Development Programme and I am particularly pleased, from his success in this field, that my confidence in his forward judgment has been amply justified. I am sure that the industry will greatly benefit from his putting down in writing the product of his many years of experience in the design and application of protective relaying.

J. T. Moore, B. Sc., A.R.T.C., M.I.C.E., M.I.E.E.,
M.I. Mech. E., A.M.I. Mar. E.,

Chairman, Ewbank & Partners Ltd.,
Consulting Engineers.

Author's Preface

In modern protective relay design transistor circuits are replacing electromagnetic movements. Since this is a new technique for many protective relay engineers it was felt that an explanation of static protective relays, written by a relay engineer for relay engineers, would be useful.

This Volume also supplements Volume I by providing additional information on c.t's and p.t's, fault incidence, transients and sources of relay error.

For those interested only in static relays, an attempt has been made to make this Volume self-contained by summarizing the basic principles of protective relaying in Chapter 1.

As in Volume I, the space devoted to each subject was determined by its importance and novelty.

A. R. van C. Warrington

ACKNOWLEDGEMENTS

I wish to thank the English Electric Co. Ltd. for their permission to publish this book, my colleagues in the Relay Department for engineering assistance, Dr. W. D. Humpage of the M.C.O.S.T. for editorial and technical checking of the manuscript, Mr. F. H. Birch of the C.E.G.B. for his assistance in forecasting future trends of protective relaying and Dr. J. Rushton of the C.E.G.B. for contributing Chapter 15 which summarizes recent work in the U.K. on Pilot Differential Systems.

A. R. van C. Warrington

Contents

List of Symbols

A	area; amperes
B	susceptance; magnetic flux density
C	capacitance
D	discrimination factor; diameter
E	e.m.f. (usually at power source)
F	force
G	conductance
H	magnetizing force
I	current
J	angular moment of inertia
K	a constant
L	self-inductance
M	mutual inductance; numeric ratio or constant
N	number of turns; numeric ratio or constant
O	origin of a graph
P	point on a graph; general constant
Q	steady-state amplitude of charge q; general constant
R	resistance; ratio
S	spacing or displacement
T	temperature
V	voltage
W	power
X	reactance
Y	admittance
Z	impedance

A and B are also used as unspecified quantities or ratios, real or complex

a	a $\underline{/120^\circ}$ operator $\left(-\frac{1}{2}+\mathrm{j}\frac{\sqrt{3}}{2}\right)$
b	susceptance per mile
c	capacitance per mile
d	diameter
e	instantaneous value of potential difference
f	frequency
g	conductance per mile
h	height

i	instantaneous value of current, unit vector
j	a $\underline{/90°}$ operator
k	a constant
l	length
m	mass; unspecified number
n	an unspecified number
p	in-phase component
q	quadrature component or electric charge
r	resistance per mile
s	modulus of attenuation
t	time
v	velocity
x	unknown quantity or reactance/mile
y	admittance per mile
z	impedance per mile
α	an angle
β	an angle
γ	attenuation factor (complex)
δ	an increment
ε	base of Naperian logarithms
η	efficiency
θ	characteristic angle; angle between source e.m.f's
λ	an angle
μ	permeability or prefix micro
π	radians in 180°
ϱ	resistivity
$\mathscr{R}$	reluctance
σ	conductivity
θ	phase angle of a relay characteristic
Φ	magnetic flux
ϕ	phase angle, generally the angle by which the current lags the voltage in a protected circuit
Ψ	an angle
ω	frequency in radians/sec; ohms
Σ	summation
Ω	ohms
$\overline{\vert 60°}$	lagged 60°
$\underline{\vert 60°}$	led 60°
$\vert V \vert$	scalar value
$\hat{V}$	peak value

α is also used as the complex ratio of two currents and β their inverse ratio. ϕ and G on circuit diagrams refer to phase and ground relays respectively.

List of Subscripts

A, B, C	the terminals of a protected line
a, b, c	the three phases
d	difference; direct axis
e	general suffix
f, F	fault
g, G	ground
h, i, j	general suffixes
l	load
L	line
m	magnetizing
n	neutral; nominal
o	a basic value; operating quantity
p	in phase component; primary; polarising
r	replica; restraint
R	relay; relay (to distinguish in the case of a secondary quantity); receiving end.
s, S	source; secondary; sending end.
t	suffix denoting quantity variable with time
res	residual
max	maximum
min	minimum
1	positive sequence
2	negative sequence
0	zero sequence

Abbreviations

B.S.S.	British Standard Specification (put out by the British Standards Institution)
C.E.G.B.	Central Electricity Generating Board of Great Britain

1

Basic Principles of Protective Relays

Operating characteristics and equations–Types of protection Level detectors, timing circuits and comparators–Duality of phase and amplitude comparators

VOLUME I of this book was a self-contained treatise on electromagnetic protective relays. This second volume deals with semiconductor protective relays and contains further information on the following subjects: power system faults, transient overvoltages, c.t's and p.t's, steady-state and transient sources of error in relay measurement and some new principles of protection.

Because some readers may be interested only in static relays, an attempt has been made to make Vol. II self-contained also. In this first chapter the general principles of protective relays have been summarized and thus duplicates some of the material in Vol. I. To those who possess Vol. I the author offers apologies and suggests that they now turn to Chapter 2.

1.1. THE ROLE OF PROTECTIVE RELAYS

In order to generate electric power and transmit it to customers a vast amount of money must be spent on equipment [36], so that it is important to run it at peak efficiency and protect it from accidents. Unfortunately, a certain number of accidents are inevitable as insulation deteriorates or unforeseen things occur, such as strokes of lightning or the entry of birds or animals into the equipment.

Insulation breakdowns are called 'faults' by relay engineers. When one occurs it is liable to be very expensive because of the damage that can be done by the tremendous amount of electrical energy in modern power systems; furthermore there is a loss of revenue due to the shutdown of the damaged circuit or equipment.

Protective relays minimize this damage and expense by locating the fault immediately and opening the correct switches to isolate the faulted circuit. Hence it is obvious that *reliability*, *speed* and *selectivity* are the most desirable qualities of a protective relay.

Figure 1.1 shows how the relays are arranged to trip only the breakers which isolate the faulted circuit and yet to overlap their zones of operation so as to leave no unprotected spots. Figure 1.2 shows how the speed of clearing faults affects the stability of a typical power system. The curves show the maximum time permissible for clearing each type of fault (relay plus breaker) versus system loading prior to the fault; the system will become unstable if these faults clearing times are exceeded.

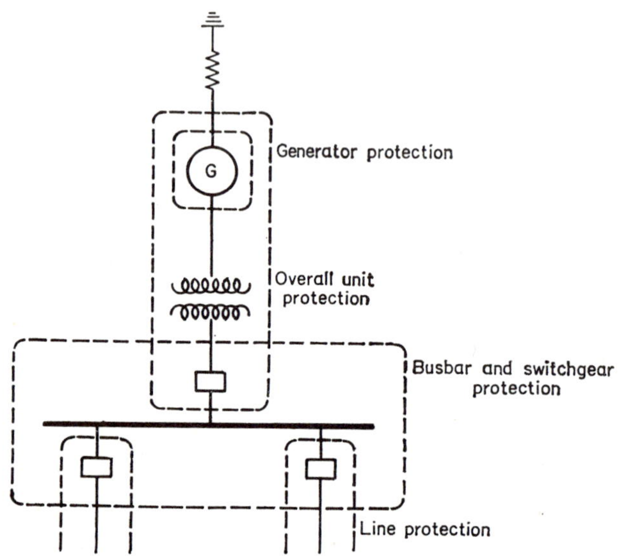

Fig. 1.1. Zones of protection

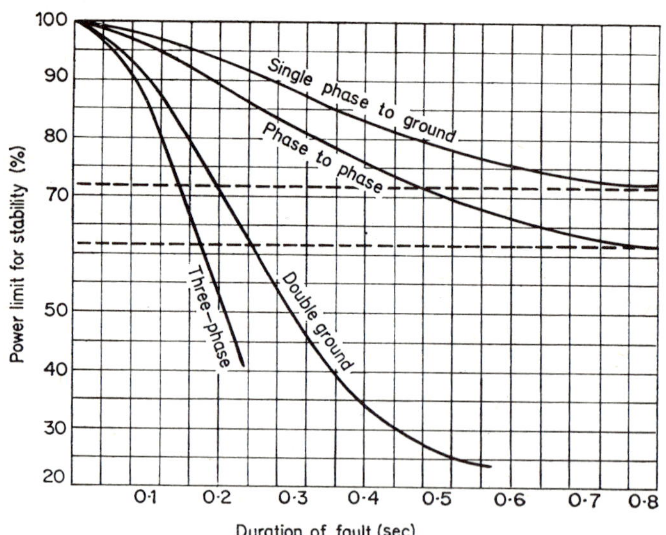

Fig. 1.2. Fast clearing of faults preserves stability

Relays recognize and locate faults by constantly measuring electrical quantities of the system, which are different during normal and abnormal conditions. The basic electrical quantities which may change when a fault occurs are current, voltage, phase-angle (direction) and frequency. It is generally necessary to provide relays responding to more than one of these conditions because, for instance, the current during a fault with minimum generation may be less than load current during maximum generation and power-factor may be as low during a power swing as during a fault.

Two sets of relays are used, main and reserve (back-up) [25]. The main relays clear faults in the protected section as fast as possible. The back-up relays operate if the others fail and usually protect not only the local section but the adjoining section also; they usually have a time delay long enough to permit the main relays to operate if they can.

1.2. LEVEL DETECTORS AND COMPARATORS

A relay operates when the measured quantity changes, either from its normal value or in relation to another quantity. The operating quantity in most protective relays is the current entering the protected circuit. The relay may operate on current level against a standard bias or restraint, or it may compare the current with another quantity of the circuit such as the bus voltage or the current leaving the protected circuit.

In a simple electromagnetic relay, used as a level detector, gravity or a spring can provide the fixed bias or reference quantity, opposing the force produced by the operating current in an electromagnet. The spring is thus a means of calibration of the relay pick-up. In static relays the equivalent is a d.c. voltage bias, as shown in Fig. 1.3a.

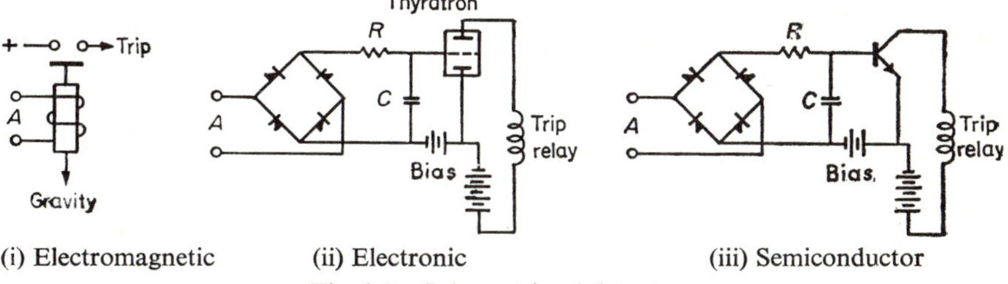

(i) Electromagnetic (ii) Electronic (iii) Semiconductor

Fig. 1.3a. Inherent level detectors

Since the fault current level changes with generating conditions, it is seldom possible to obtain selectivity on the basis of current magnitude alone. Usually a time function is added so that the relay nearest the fault, which sees the most current, will trip before relays in the unfaulted circuits.

It is difficult to obtain selectivity by measuring one quantity such as current, potential, phase-angle, etc., without using time delay. Hence most high-speed relays measure a derived quantity which is a combination of several simple quantities; for example, impedance, current-ratio, etc., in which two simple quantities are compared in magnitude and/or phase relation.

Figure 1.3b shows, in a very much simplified form, inherent amplitude comparators of the electromagnetic, electronic and semiconductor types; Fig. 1.3c shows inherent phase comparators. These comparators are discussed in more detail in Chapter 4.

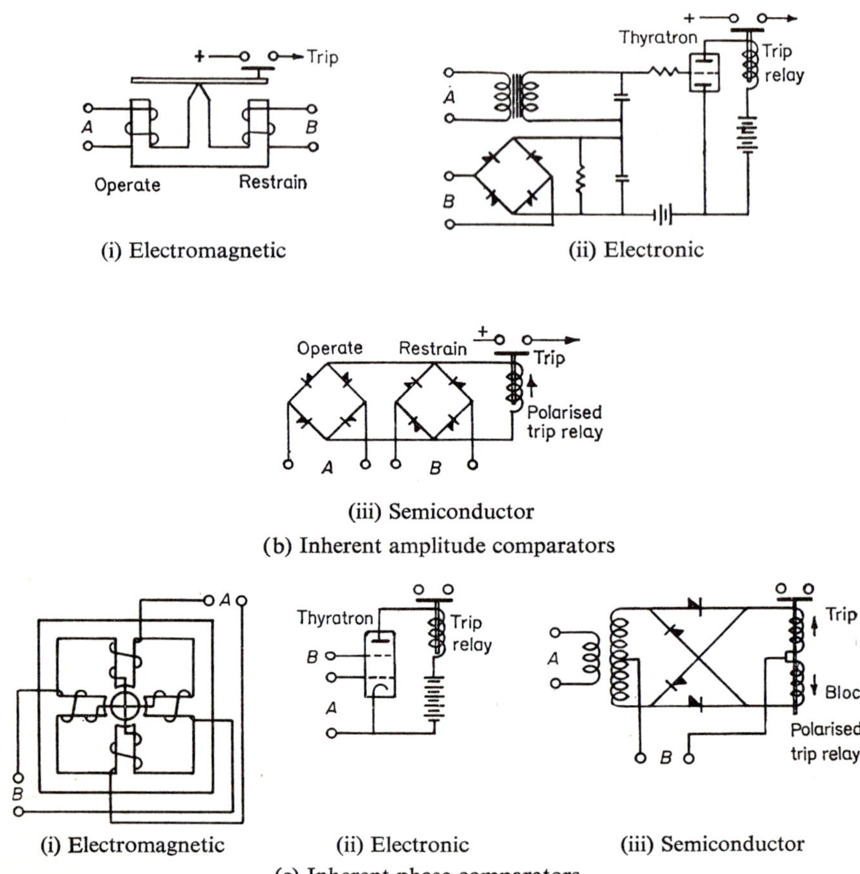

(i) Electromagnetic (ii) Electronic

(iii) Semiconductor

(b) Inherent amplitude comparators

(i) Electromagnetic (ii) Electronic (iii) Semiconductor

(c) Inherent phase comparators

Fig. 1.3. Basic electromagnetic and static relays

1.3. OPERATING CHARACTERISTICS [22, 41]

The most important operating characteristic of a single-input relay (level detector) is the relation between the input magnitude and the operating time, e.g. the time-current curve of a time-current relay.

Modern amplitude and phase comparator relays are virtually instantaneous but a curve of time versus the ratio of the inputs is of interest even though the time scale is in milliseconds; for example, a time-impedance curve of a distance relay. In such relays the most important characteristic is the ratio of the two input quantities at the threshold of operation for varying phase between them.

This operating characteristic is plotted on a polar graph whose ordinates are the real and imaginary components of A/B or B/A where A and B are the two quantities compared. The ordinates of the graph are $\left|\frac{A}{B}\right| \cos \phi$ and $j\left|\frac{A}{B}\right| \sin \phi$, where ϕ is the angle by which A leads B; this can be abbreviated as $\left|\frac{A}{B}\right| p$ and $\left|\frac{A}{B}\right| q$.

An example is the distance relay, where A is voltage and B is current, so that the ordinates of its operating characteristics are $\left|\frac{V}{I}\right| \cos \phi = R$ and $\left|\frac{V}{I}\right| \sin \phi = X$. This is generally referred to as the $R - X$ diagram or the impedance diagram. Similarly, the components of I/V give an admittance (G versus jB) diagram.

Since there are no terms like impedance for the general case of A/B where A and B are, for example, two currents, the A/B diagram is referred to as the α-plane and the B/A diagram as the β-plane diagram [22]. For two-input comparators these characteristics have second-order equations and hence are straight lines, circles or sectors of circles. They represent the threshold of operation where the comparator has zero output; hence the output is positive (tripping) on one side of the characteristic and negative (blocking) on the other side.

It is interesting to note that, if the characteristic of the relay is a circle going through the origin when plotted on the α-plane, it is a straight line outside the origin when plotted on the β-plane and vice versa as shown in Fig. 1.4 . Also, circular characteristics not passing through the origin are orthogonal in the two planes.

1.3.1. General Equation of Operating Characteristics

If not more than two input quantities are involved, they can produce effects individually and by co-operation. Hence the equation of the operating

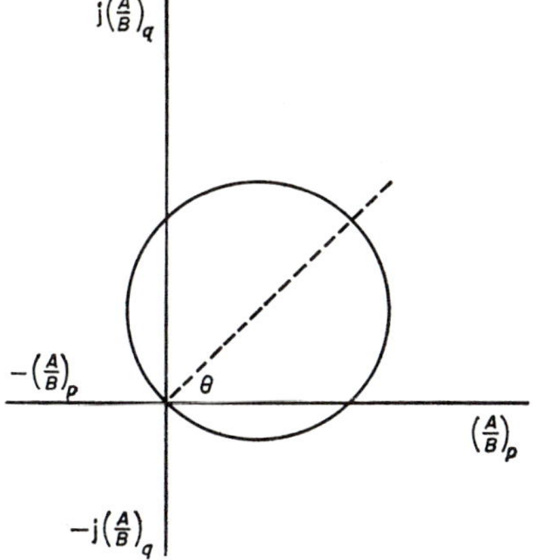

Fig. 1.4a. Typical relay characteristic plotted on α-plane

characteristic will [22] be of the form

$$K|A|^2 - K'|B|^2 + |A||B| \cos (\phi - \theta) = K'' \tag{1.1}$$

where A and B are the two electrical quantities being compared, K and K′ are scalar constants, K″ is a constant representing a bias which would take the form of a mechanical restraint in an electromagnetic relay, ϕ is the phase angle between A and B and θ is the phase angle of the characteristic. θ is usually the value of ϕ which provides maximum relay torque in an electromagnetic relay or maximum output in a static relay.

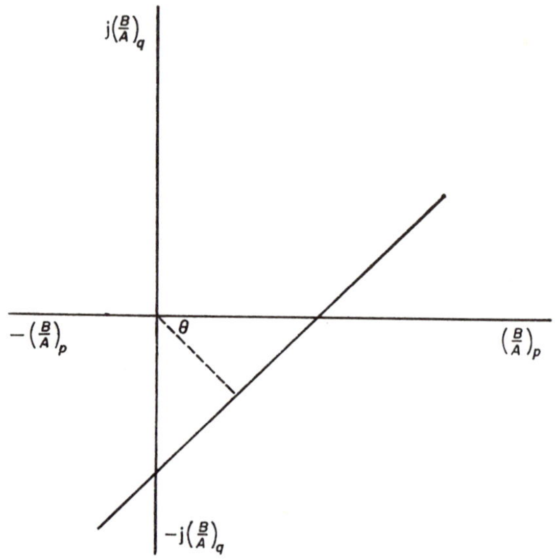

Fig. 1.4b. Typical relay characteristic plotted on β-plane

The Eq. (1.1) represents all the linear and circular characteristics which can be obtained from two-input relays. This equation is applicable to most of the common types of relay and simplifies the explanation of their operation and characteristic curves (as shown in Fig. 1.5, columns 3 and 4).

In Eq. (1.1) K″ is finite only in single-quantity relays where it is used as a level indicator; it is made substantially zero in relays that compare two input quantities and in this case the equation represents a circle or a straight line on a complex plane (polar diagram). This can be demonstrated by rewriting the equation (1.1) making K″ = 0 and dividing through by $K'A^2$:

$$\frac{K}{K'} - \left|\frac{B}{A}\right|^2 + \left|\frac{B}{A}\right| \frac{\cos (\phi - \theta)}{K'} = 0 \tag{1.2}$$

Moving the K/K', term to the right-hand side and adding $\left(\frac{1}{2K'}\right)^2$ to each side, Eq. (1.2) becomes:

$$\left|\frac{B}{A}\right|^2 - \left|\frac{B}{A}\right| \frac{\cos (\phi - \theta)}{K'} + \left|\frac{1}{2K'}\right|^2 = \frac{K}{K'} + \left|\frac{1}{2K'}\right|^2 \tag{1.3}$$

Type relay	Conditions	Resulting equation	Relay pick-up	Simplified relay static	Relay electro-magnetic	Polar diagrams: Admittance or current	Polar diagrams: Impedance or potential
Overcurrent	No potential windings hence no V^2 or VI terms	$KI^2 = K''$	$I > \sqrt{\frac{K''}{K}}$	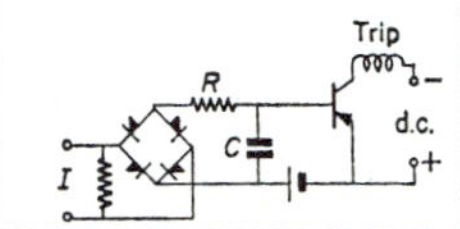	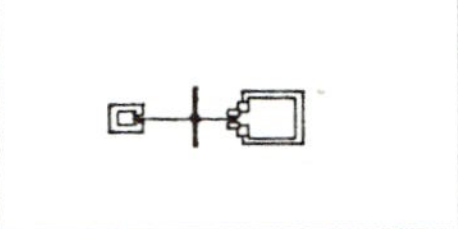	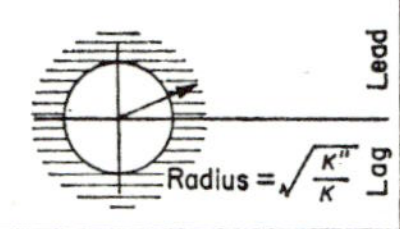 Radius $= \sqrt{\frac{K''}{K}}$	
Undervoltage	No current windings hence no I^2 or VI terms K'' is neg.	$-K'V^2 = K''$	$V < \sqrt{\frac{K''}{-K'}}$	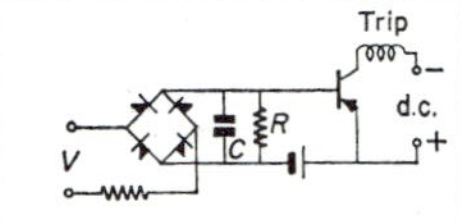	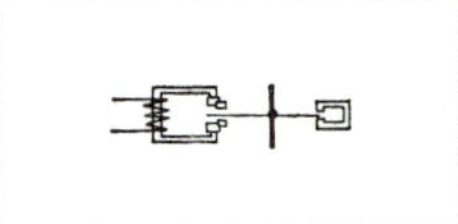		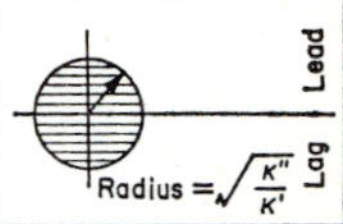 Radius $= \sqrt{\frac{K''}{K'}}$
Directional	$K = K' = 0$	$VI \cos(\phi-\theta) = K''$	$VI \cos(\phi-\theta) > K''$	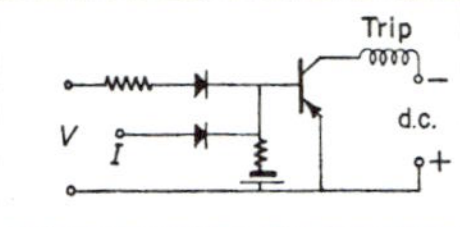	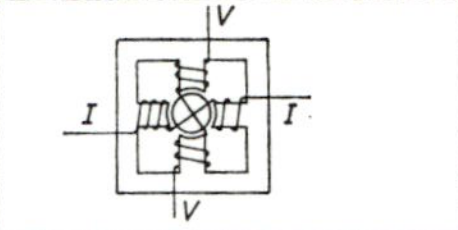	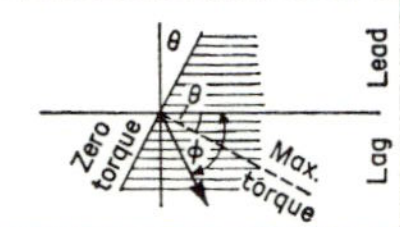	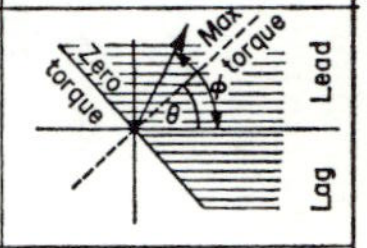
Reactance (ohm unit)	$K' = K'' = 0$ $\theta = 90°$	$KI^2 = VI \cos(\phi-\theta) = VI \sin\phi$	$Z \sin\phi < K$ i.e. $X < K$	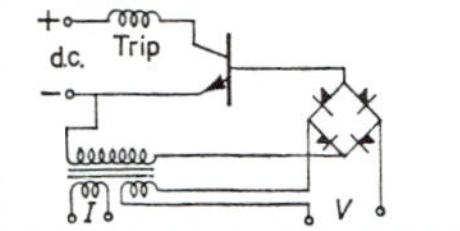	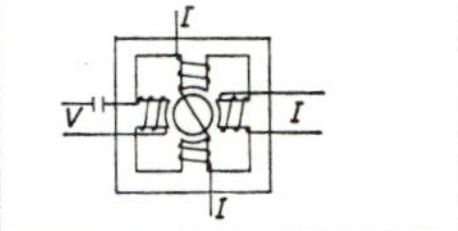	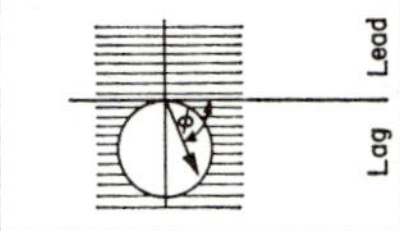	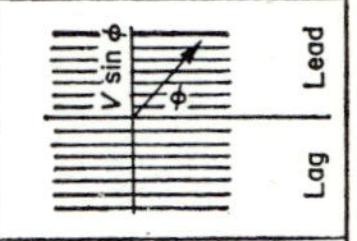
Directional impedance (mho unit)	$K = K'' = 0$ $\theta = 75°$	$-VI \cos(\phi-75°) = K'V^2$	$\frac{Z}{\cos(\phi-\theta)} < \sqrt{\frac{1}{K'}}$	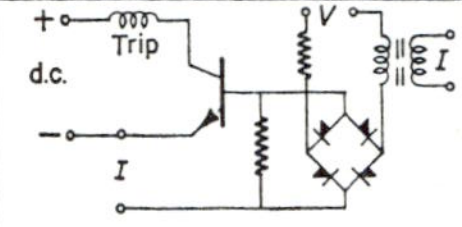	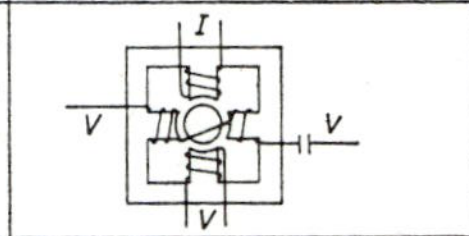	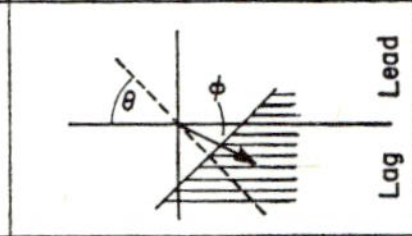	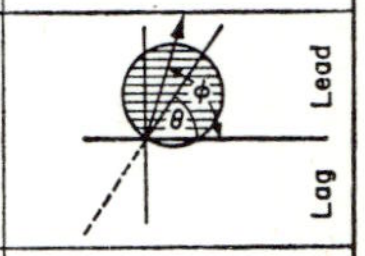
Impedance (ohm unit)	V and I scalar or separate so no VI term	$KI^2 = K'V^2$	$Z < \sqrt{\frac{K}{K'}}$	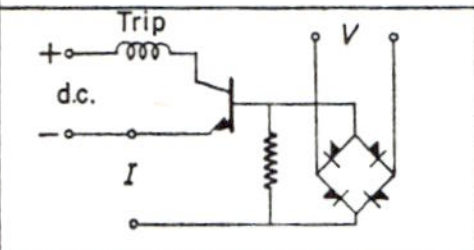	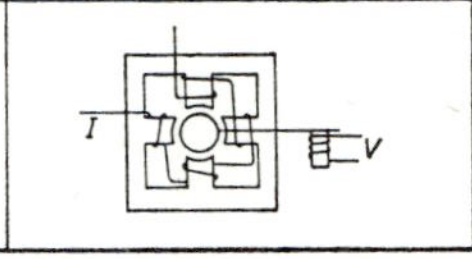	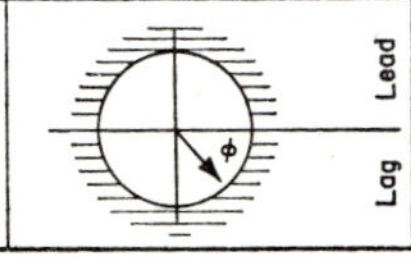	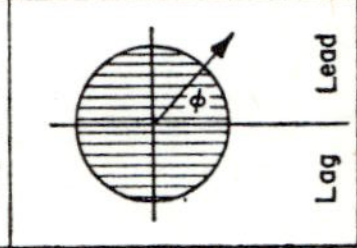

General relay equation $= KI^2 - K'V^2 + VI \cos(\phi-\theta) = K''$ (where all K's are torque constant)

All characteristics are loci of V or I for zero torque. In static relay column integration and level detection assumed incorporated in trip device

Fig. 1.5. Analogy of static and electromagnetic relay units

Equation (1.3) represents a circle on the β-plane (Fig. 1.6) whose radius is $\{(1 + 4KK')/2K\}$ and whose centre is at $1/2K'$ from the origin at an angle of θ from the reference axis $\left|\frac{B}{A}\right|p$. Similarly, it can be shown that the same characteristic on the α-plane would be a circle of radius $\{(1 + 4KK')/2K\}$ and whose centre is at $1/2K \underline{/-\theta}$ from the origin.

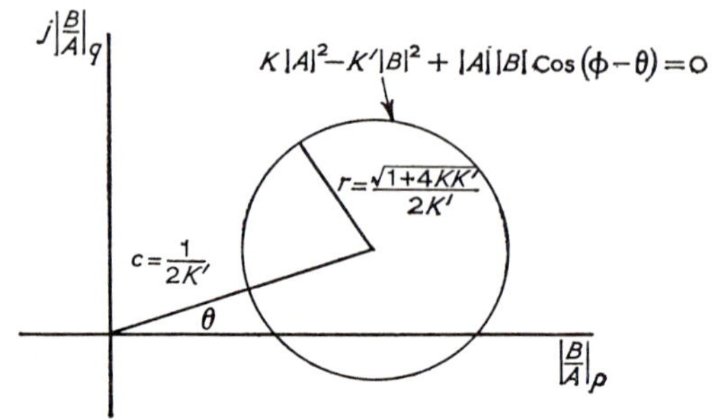

Fig. 1.6. General equation for a relay comparing two quantities $A + B$
$$K|A|^2 - K'|B|^2 + |A|\,|B|\cos(\phi - \theta) = 0$$

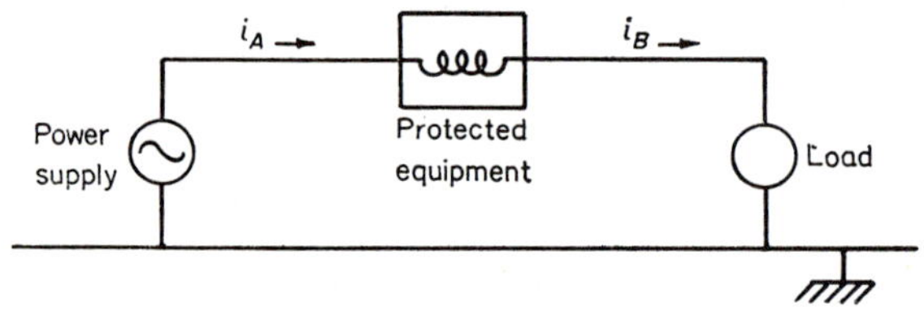

(a) Normal conditions ($i_A = i_B$)

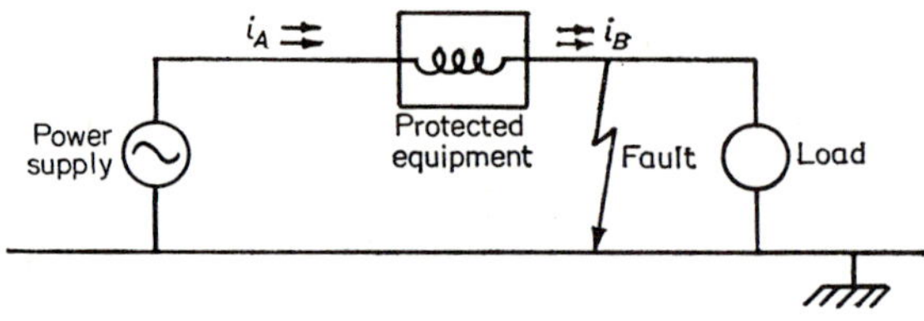

(b) External fault ($i_A = i_B$)

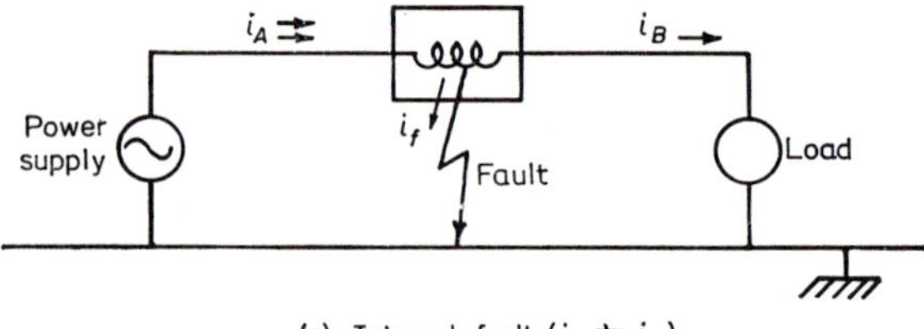

(c) Internal fault ($i_A \neq i_B$)

Fig. 1.7. Operating principle of differential current relay

1.4. TYPES OF PROTECTION [40]

There are hundreds of types of protective relays but the most important ones can be grouped under the headings differential current, distance, over-current and directional.

The most obvious way of detecting and locating a fault is for the protective relay to compare the current going into each piece of equipment with the current coming out of it. The two currents should be identical unless some of the current is diverted through a short-circuit, as shown in Fig. 1.7. This method is used wherever possible and such a relay is called a differential current relay.

In the following diagrams that illustrate the various forms of protection, electromagnetic relays will often be shown for the sake of simplicity because the electrical circuitry of the equivalent static relay is usually much more complicated. The operation and application of these relays will be very briefly explained, but Ref. (40) is recommended as an admirable introduction to the subject. This article is remarkable for its simplicity and clarity.

1.4.1. Unbiased Differential Relays

This system merely requires an instantaneous relay to be connected across the paralleled secondaries of the current transformers (c.t's) on each side of the equipment so that the relay receives the differential or spill current; if the c.t. ratios are equal, spill current should exist only if there is an internal fault (see Fig. 1.8a). This principle is very effective with generator or transformer winding protection and bus protection. In the latter case all the c.t's are connected in parallel so that the operating current is the vector sum of all the currents which, under Kirchhoff's law, should be zero under normal conditions but will have a resultant to trip the relay for an internal fault.

A differential current relay cannot be given a very sensitive setting if the c.t's are not perfectly matched (Fig. 1.9) because there would be a risk of tripping on heavy external faults. This is overcome usually by connecting a stabilizing resistance in series with the relay coil.

The value of this resistance depends upon the degree of saturation of the c.t's. In bus protection one c.t. can be completely saturated and act as if its secondary were short-circuited. In this case the stabilizing resistance required would be $R_r = (I_B R_B / I_r)$ where I_r is the pick-up of the relay, I_B is the maximum through fault current (secondary) and R_B is the resistance of the leads from the saturated c.t. to the relay. In generator or transformer protection, where much less c.t. saturation is possible, only about one third of this value is required.

Even a relatively high value of R_r should not prevent the relay from operating on the minimum internal fault, but in an extreme case (such as bus protection with long c.t. leads) part of the stabilizing resistance can be in the form of increased turns on the relay coil, which is in series with it, and this restores the sensitivity of the relay.

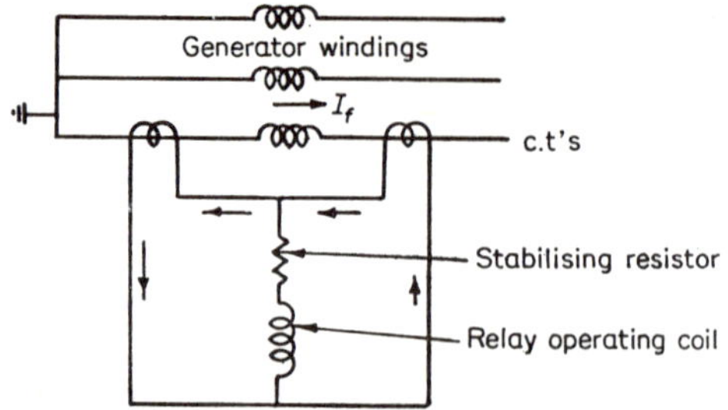

(a) CondititION during load or external fault: no current through relay (unbiased)

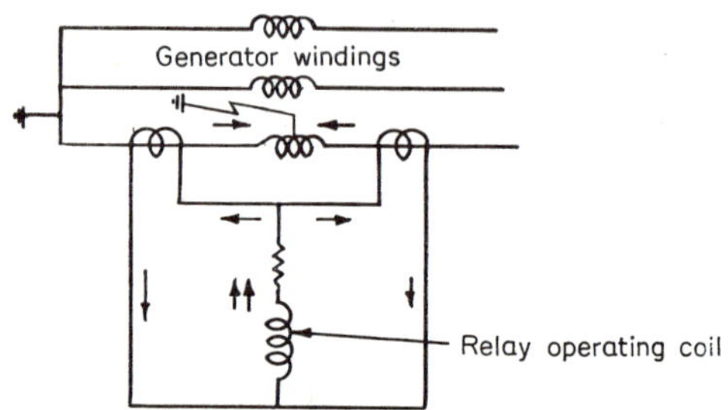

(b) Condition during internal fault relay energised to trip (unbiased)

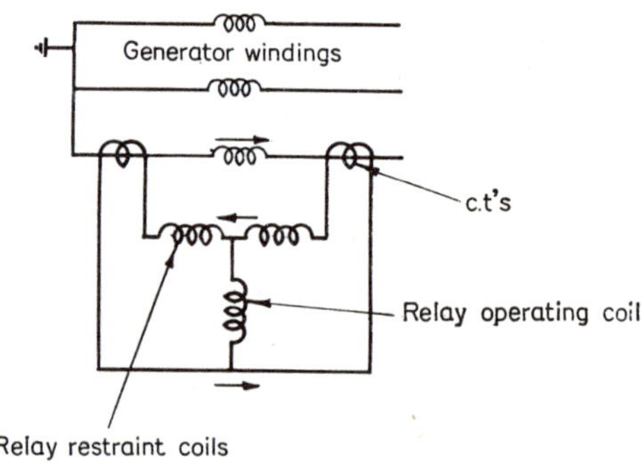

(c) Only restraint coils energised during normal conditions or an external fault (biased)

Fig. 1.8. Generator differential relays with and without through-current bias

1.4.2. Biased Differential Relays

An alternative method of preventing a differential relay from tripping undesirably on spurious differential current due to c.t. inequalities is to add a restraining winding connected as shown in Fig. 1.8c.

Current normally circulates between the c.t's through the restraining winding and keeps the relay inoperative (Fig. 1.8c). The ratio of the turns

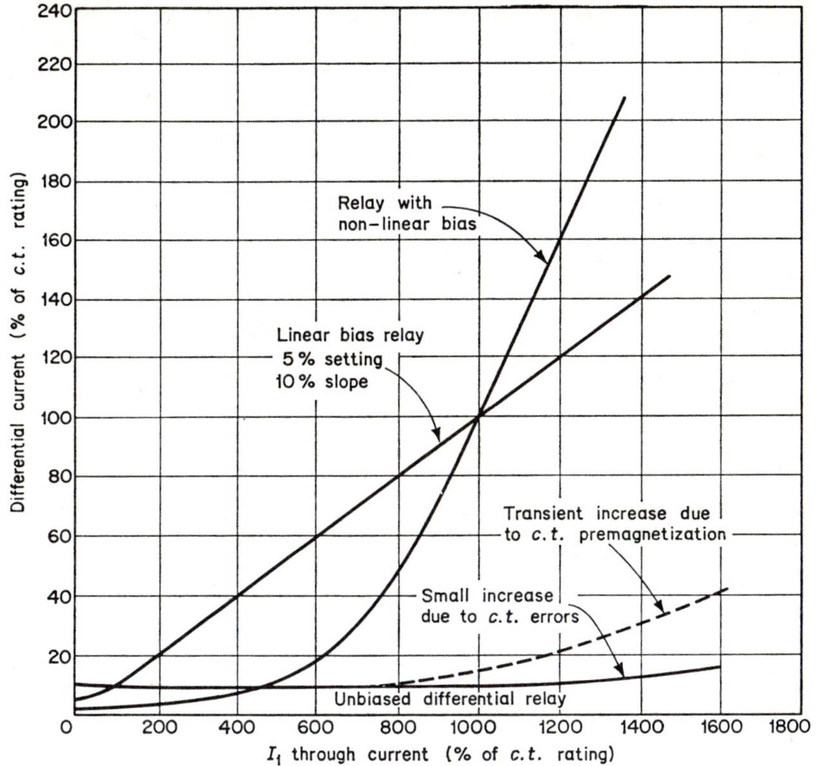

Fig. 1.9. Operating characteristics of biased & unbiased differential relays for generator protection

of the operating and restraining coils controls the difference current required to operate the relay as a fixed percentage of the through current. The operating coil receives $|I_A - I_B|$ and the restraining coils $|I_A + I_B|$.

The relay operates when

$$|I_A - I_B| > S|I_A + I_B| \tag{1.4}$$

and it will be seen that the relay operates when I_B is negative or when I_A is greater than I_B by an amount which is small at low currents and large at heavy currents, thus allowing for c.t. errors increasing at high currents. S is the ratio of the restraining input per ampere divided by the operating input per ampere and is of the order of 0.05 for protecting a generator, so that the slope of the characteristic in Fig. 1.9 is 5%.

In the case of a circuit with more than two ends, such as a 3-winding power-transformer or a multi-circuit bus (Fig. 1.10), the operating input to the comparator is the vector sum of the individual currents while the restraining input is their scalar sum; hence the relay operates when

$$|I_A + I_B + I_C \cdots + I_N| > S(|I_A| + |I_B| + |I_C| \cdots + |I_N|) \quad (1.5)$$

In overall protection of a power transformer, an additional restraining winding or blocking relay is provided to prevent the relay from tripping on magnetizing inrush current when the transformer is switched on, since

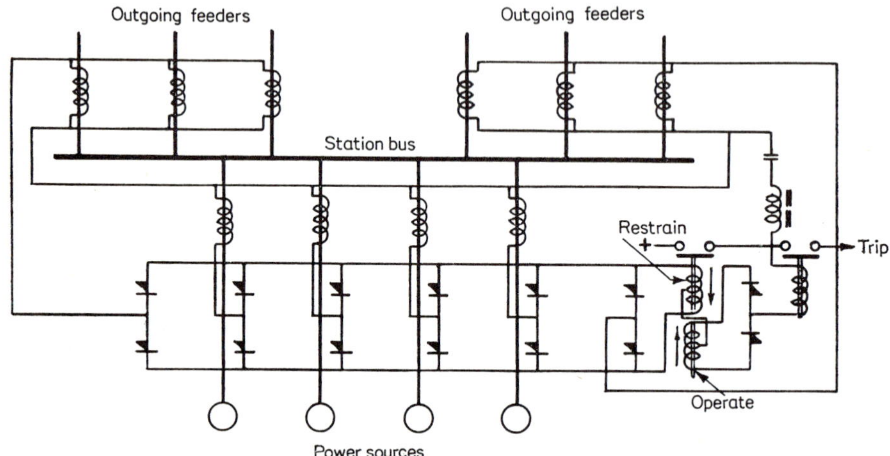

Fig. 1.10. Multi-input differential relay
Arrows on relay refer to pulls on armature

this current exists only on the supply side. The second harmonic component of the current is usually used for providing this blocking feature since it is the dominant component of the inrush current and does not appear normally [22].

Stabilizing Resistor versus Restraining Winding. The stabilizing resistor is used where the c.t's can be conveniently paralleled at one location, such as in generator or bus protection, and can be reasonably matched.

In overall transformer protection, where the c.t's are at different voltages and hence are more difficult to match, or in pilot protection where the c.t's are separated by long leads, the restraining winding is essential.

1.4.3. Pilot Differential Relays

For feeders a pilot wire is used to bring the currents together for comparison; on long overhead lines a high-frequency carrier channel is used. It would be too costly to provide pilot wires for each phase, so the three phase currents are combined into one composite current (Fig. 1.11) and also reduced in magnitude so that the c.t. lead burden will not be excessive. The local current and the pilot wire current are compared in amplitude by one of two methods.

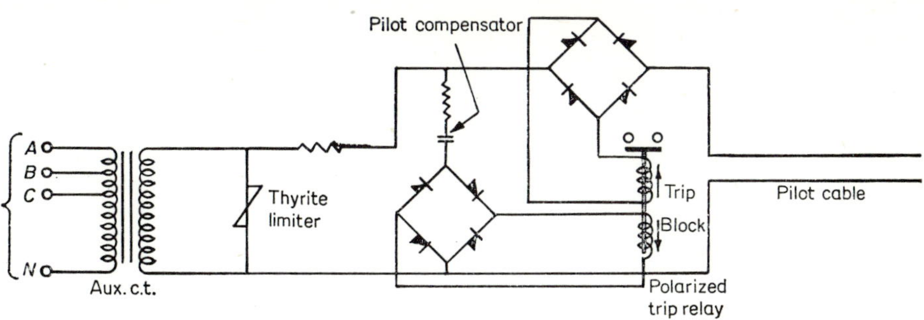

Fig. 1.11. Pilot wire differential relay (balanced voltage scheme)

In the circulating current scheme the through current flows in the pilot wires and in the restraining circuit; any difference between the currents flows in the operating circuit of the relay. In the opposed voltage scheme (Fig. 1.11) the currents are turned into voltages which are arranged to be opposed normally, so that no current flows in the pilot wires or in the operating circuit which is in series with them; during an internal fault the voltages are not balanced and current flows in the operating circuit of the relay.

In the case of very long lines the pilot wires are replaced by a carrier current channel and the currents at the two ends are compared in a phase-angle comparator. The latter has certain advantages over the amplitude comparator and has been applied to pilot-wire relaying both directly and through the use of directional relays; contacts of the latter are interconnected through the pilot wire so as to distinguish between an internal fault (where both close) and external fault (where only one closes).

The subject of pilot differential protection was discussed in some detail in Volume I, Chapters 3 and 8. Further discussion is provided in Chapter 9 of the present volume and a rigorous analysis of its limitations is given in Chapter 15.

1.4.4. Distance Relays

An alternative to the use of a pilot channel is to compare the local current with the local voltage instead of the far current. Since $V/I = Z$ the relay is known as an impedance relay and, since the impedance of the line is proportional to its length, it is called a distance relay.

Figure 1.12 shows how a relay, set for impedance Z_L, will trip if the fault is within that setting because $V < IZ_L$, whereas faults beyond that setting will not cause tripping because $V > IZ_L$. Since this relay ignores the phase angle between V and I, the characteristic is the circle in Fig. 1.13. It will be seen that the effect of fault resistance is to shorten the reach of the relay.

Here again there are refinements. Instead of comparing I with V it is sometimes necessary to compare I with a component of V, or a component

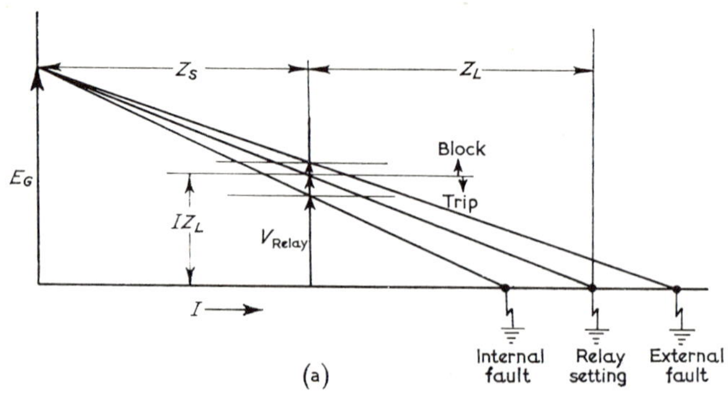

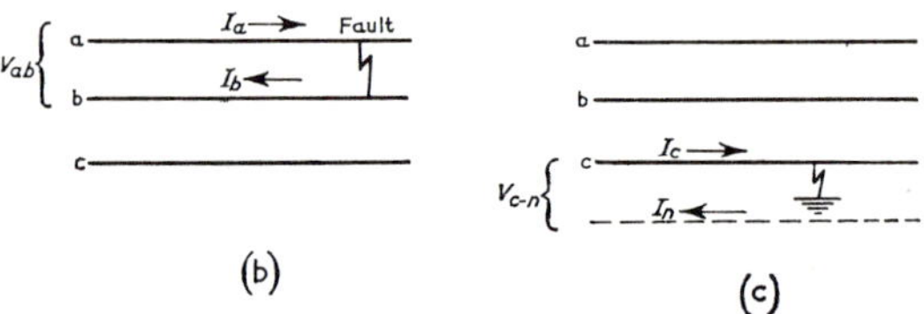

Fig. 1.12. Operating principle of distance relay

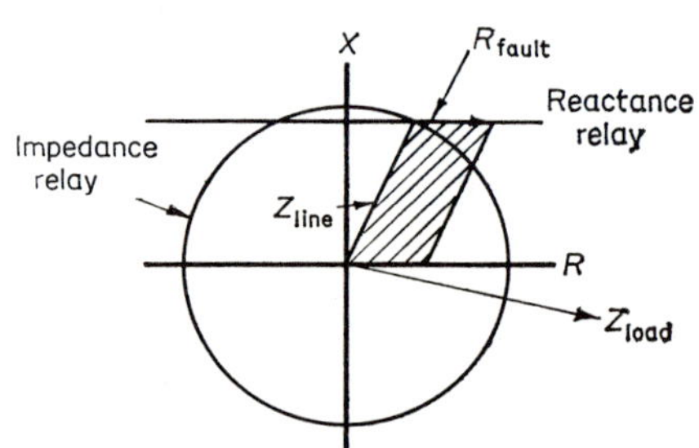

Fig. 1.13. Impedance and reactance characteristics

of I with V [29]. For example, one type of distance relay operates when $I > V \sin \phi$, where ϕ is the angle between V and I. The ratio $(V \sin \phi)/I = X$, so that the relay measures the reactance of the line (Fig. 1.13) instead of its impedance and hence its distance measurement is not affected by fault resistance. Such a relay is used for very short lines and for ground faults where the resistance of the fault path through the earth may be appreciable.

On the other hand, comparing V with a component of I makes the relay more suitable for longer lines where the impedance of the line is close to that of a heavy load (Fig. 1.14) or a power swing, except for the phase angle. In this instance the relay measures the ratio $V/[I \cos(\phi - \theta)]$ or $Z \sec (\phi - \theta)$, where θ is the value of ϕ for maximum sensitivity and hence it is the angle of the impedance circle relative to the R axis (Fig. 1.15). This is called a mho relay because it operates on a constant value of admittance, viz. $Y\cos(\phi - \theta)$, which is the angle-admittance $Y\underline{/\theta}$.

It will be noticed that the mho circle fits snugly round the fault area and hence is very selective between internal faults and any other conditions.

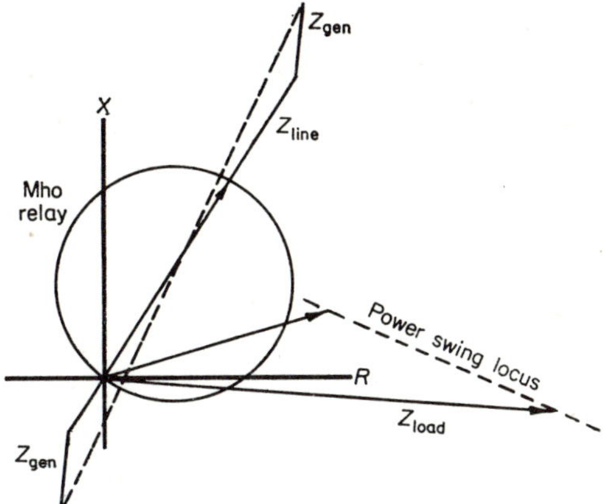

Fig. 1.14. Effect of power swing on mho relay

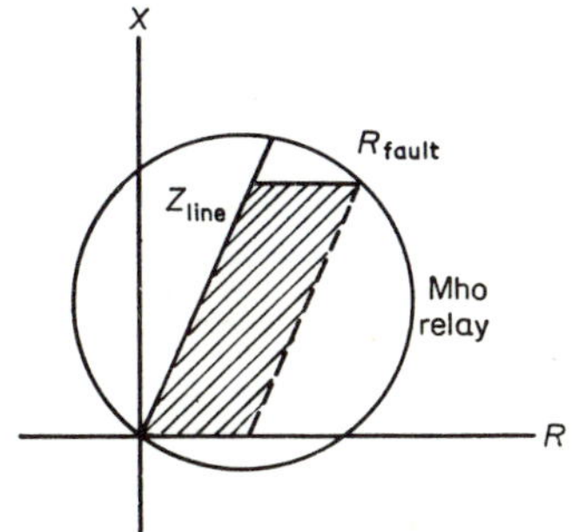

Fig. 1.15. Mho relay characteristic snugly fits fault area

1.4.5. Overcurrent Relays

On less important circuits, such as distribution systems, only the cheapest relaying is justified, so that the local current is merely checked for magnitude only and the relay is simple and cheap.

This method is possible in distribution systems because the short-circuit current is large compared with the load current and because the faulted section has the most current since it is fed from all the neighbouring unfaulted circuits. Its selectivity is improved by adding a delaying means so that the operating time is inversely proportional to the current; with this arrangement the relay in the faulted section tends to trip first and clear the fault before the others can trip. Such relays are also used for back-up protection on transmission lines in case the more sophisticated primary relays should fail.

The time-current curve of an inverse-time relay of the electromagnetic type used to be a direct function of the B-H magnetization curve of the iron core of an electromagnet. This made it very difficult to obtain consistency between the characteristics of individual relays and to obtain a time-current characteristic which had a simple mathematical equation. In the static relay the time-current relationship is based on an R-C circuit which can be precisely controlled and can be arranged to give any desired relationship between current and operating time; this is explained in Chapter 8.

The desirable form of equation is

$$\text{time } t = \frac{A}{m^n - 1} + B \tag{1.6}$$

where A and B are adjustable constants and m is the multiple of pick-up current. The integer n is 0 for a definite time relay, 1 for an inverse time relay and 2 for an extremely inverse time relay (Fig. 1.16).

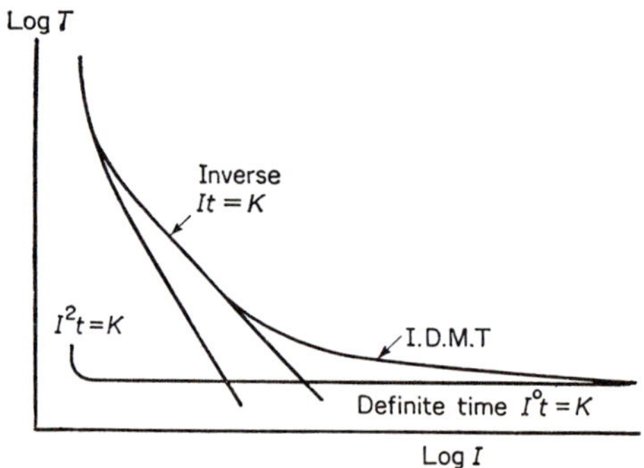

Fig. 1.16. Inverse, definite and I.D.M.T. time-current relay characteristics

1.4.6. Directional Relays

In many locations the magnitude of the current is not sufficient information for locating a fault. For instance, at a substation with power sources on both sides of it, the fault current could be about the same for a fault on either side of the station. Here a directional relay is required to monitor the over-current relay.

For phase faults the directional relay usually compares the phase relation of the current in one phase with the voltage between the two other phases because they will be less affected by a fault involving the first phase; this is called the quadrature connection; other connections are analysed in Volume I.

For ground faults the residual current ($3I_0$) of the c.t's is compared in phase angle with the residual voltage ($3V_0$) because it is known that V is a maximum at the fault and the current flows from the fault to the nearest grounded neutral. Alternatively, the residual current may be compared with the neutral current of a grounding transformer [2] (volume I section 4.5.2).

1.5. DUALITY BETWEEN AMPLITUDE AND PHASE COMPARATORS

It will be seen from the foregoing that some relays are inherently amplitude comparators and others phase comparators. For instance, in an impedance relay the current tends to make the relay operate and the voltage tends to restrain it so that the relay operates when $K|I|$ exceeds $|V|$ or when $|Z| < K$ irrespective of the phase angle between I and V; hence an impedance function is an amplitude comparison. On the other hand, a directional relay compares I and V in phase angle irrespective of their magnitudes, and this is a phase comparison.

This is only part of the matter, however, since an inherent amplitude comparator becomes a phase comparator and vice versa if the input quantities are changed to the sum and difference of the original two input quantities.

Consider a relay which operates when $|A| > |B|$, i.e. an amplitude comparator. If the input quantities are changed so that it operates when $|A + B| > |A - B|$ it is now a phase comparator because A and B must have the same sense or polarity to satisfy the equation and for the relay to operate. This is illustrated in Fig. 1.18.

Similarly, a directional relay whose torque is proportional to a vectorial product of A and B is a phase comparator which operates when A and B have the same direction. If, however, the input quantities are changed to $(A + B)$ and $(A - B)$, as in Fig. 1.18, the relay becomes an amplitude com-

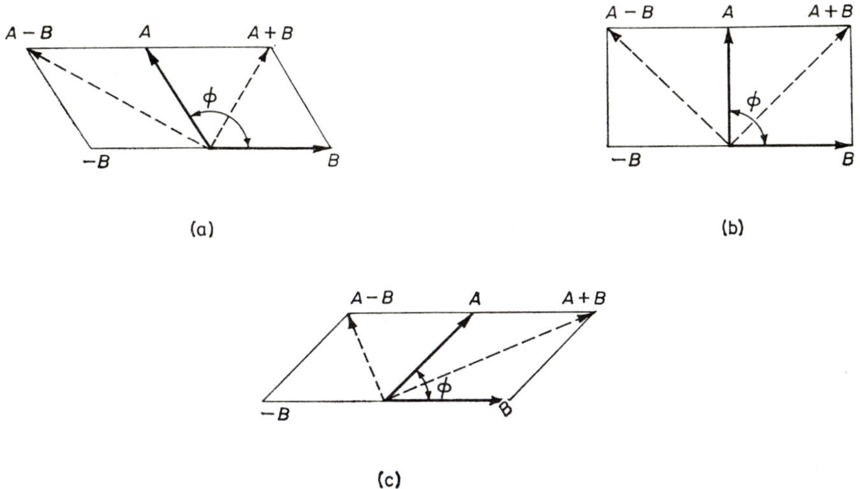

Fig. 1.17. Amplitude comparator used for phase comparison

	(a) $\phi > 90$	(b) $\phi = 90$	(c) $\phi < 90$
Comparison	$A + B < A - B$	$A + B = A - B$	$A + B > A - B$

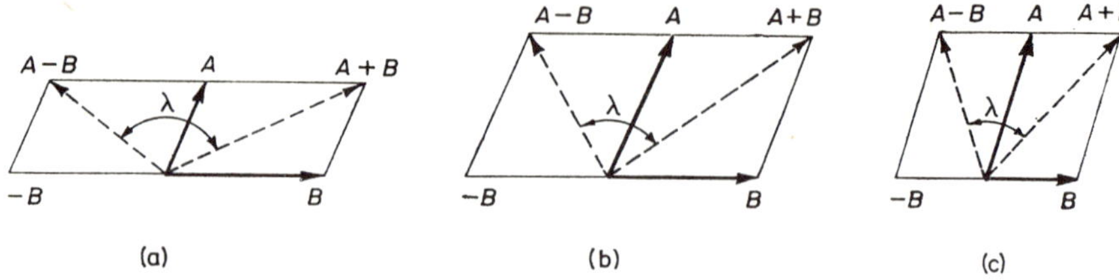

Fig. 1.18. Phase comparator used for amplitude comparison
(a) $A < B$ Comparison $\lambda > 90°$ (b) $A = B$ $\lambda = 90°$ (c) $A > B$ $\lambda < 90°$

parator because $(A + B)$ and $(A - B)$ have the same polarity only if $|A| > |B|$.

This can also be proved algebraically by taking specific cases of relays. For instance, a balanced beam relay operates when the pull of the operating magnet at one end of the beam exceeds that of the restraining magnet at the other end, i.e. when $|A|^2 > |B|^2$. If we change the input quantities as discussed above, the relay operates when $|A + B|^2 > |A - B|^2$, i.e. when

$$|A^2 + B^2 + 2AB \cos(\phi - \theta)| > |A^2 + B^2 - 2AB \cos(\phi - \theta)|$$

where ϕ is the angle between A and B and θ is a design angle; this is when $4AB \cos(\phi - \theta) > 0$, i.e. when $(\theta + 90°) > \phi > (\theta - 90°)$.

Similarly, in an induction cup relay, the torque is proportional to the vector product $|A|\,|B| \cos(\phi - \theta)$ and the relay operates when

$$(\theta + 90°) > \phi > (\theta - 90°).$$

If we change the input quantities as before, the torque $\propto |A + B|\,|A - B| \sin\alpha$ where α is the angle between $(A + B)$ and $(A - B)$ which must be 90° for maximum torque. Thus the torque $\propto |A|^2 - |B|^2$ and the phase comparator has become an amplitude comparator which trips when $A > B$.

1.6. GENERAL EQUATIONS FOR COMPARATORS

The two input quantities, A and B, can be compared directly or they can be combined so that the two input quantities are functions of both A and B, as mentioned above. For example, in the differential current relay (Eq. 1.4), one input is the sum and the other the difference of A and B. The general expressions for these input signals are therefore $S_1 = (K_1A + K_2B)$ and $S_2 = (K_3A + K_4B)$.

If A is taken as the vector of reference, they can be written

$$S_1 = K_1|A| + K_2|B| [\cos(\phi - \theta) + j \sin(\phi - \theta)] \tag{1.7}$$

and

$$S_2 = K_3|A| + K_4|B| [\cos(\phi - \theta) + j \sin(\phi - \theta)] \tag{1.8}$$

PLATES

Plate 1. Printed circuit modules (B. B. Cie.)

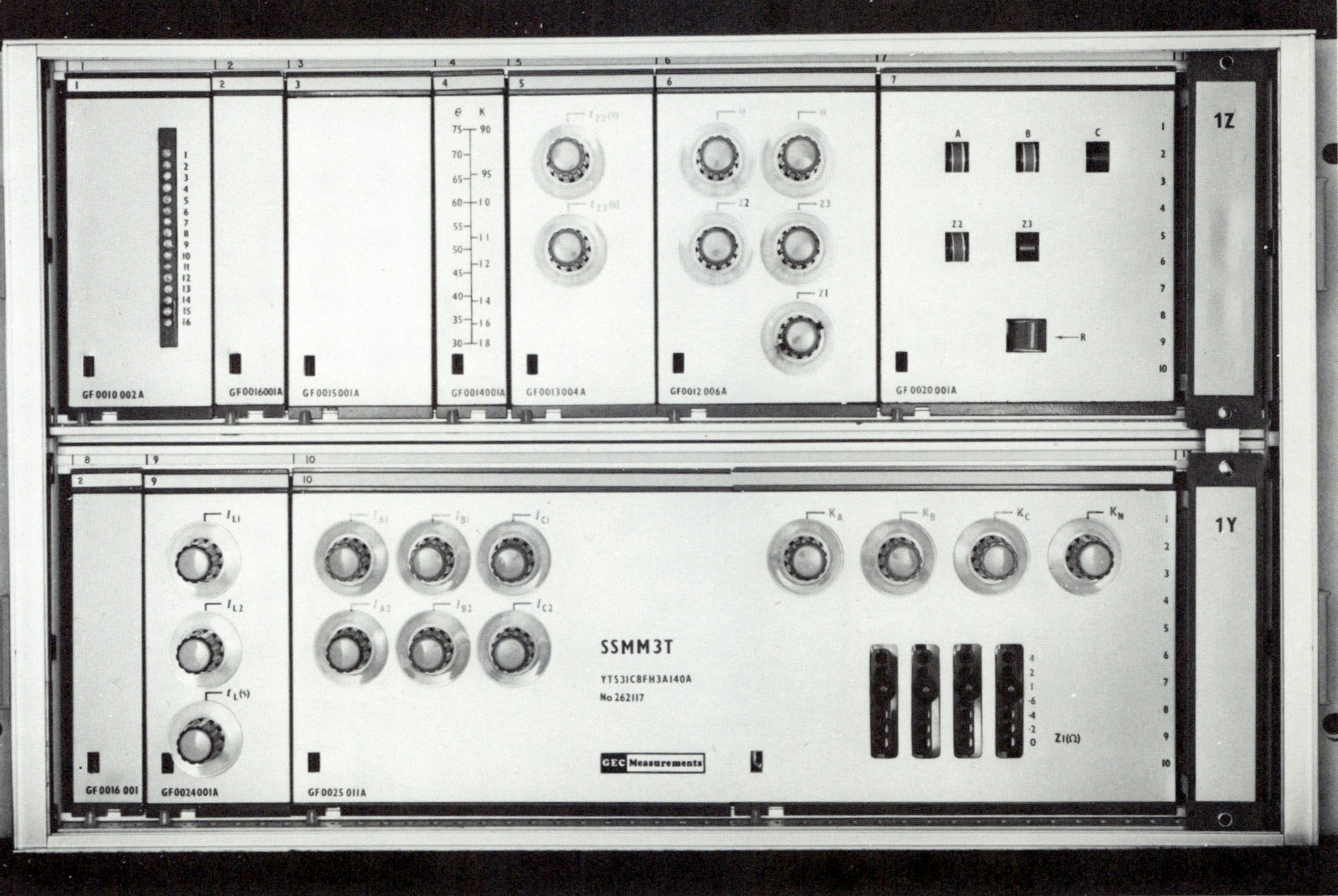

Plate 2. Static Switched Distance Relay for all faults

Plate 3 Electromagnetic versus static times-current relays (E. E. Co.) chassis of static relay ($\frac{3}{4}$ front view)

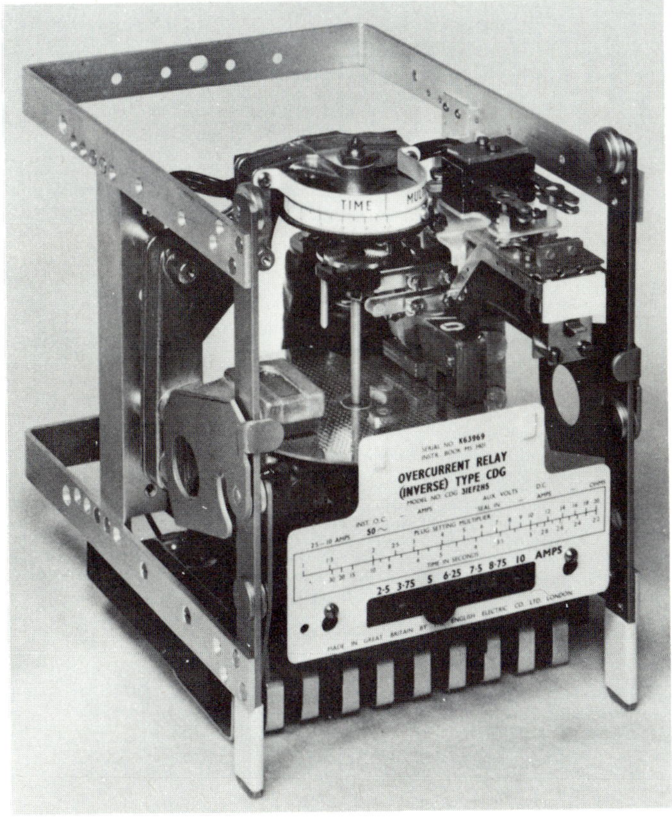

Plate 4 Electromagnetic versus static time-current relays (E. E. Co.)
Upper chassis of static relay (¾ rear view)
Lower chassis of induction relay (¾ front view)

Plate 5. 3-step, 3-phase static distance relay (E. E. Co.)

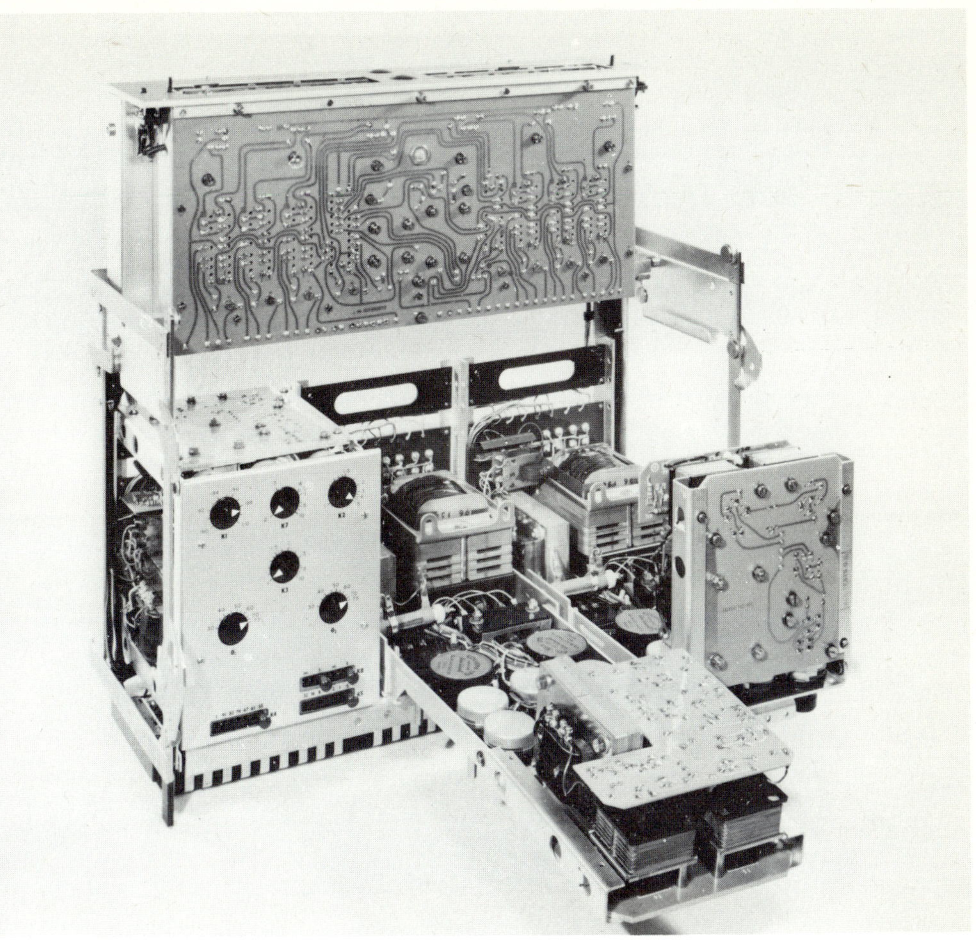

Plate 6. Fold-out construction of static distance relay (E. E. Co.)

This is shown vectorially in Fig. 1.19, K_1, K_2, K_3 and K_4 are design constants. In most relays at least one of them is zero and two of them are often equal. This makes the practical case relatively simple [22].

It will be shown in the following sections, 1.6.1 and 1.6.2, that, for a given characteristic (Fig. 1.20), the equations for the amplitude comparator and for the phase comparator are of the same form (Eq. 1.10) but with different values for K_1, K_2, K_3 and K_4. On the other hand, if the same values of K are used for both comparators, their characteristic circles will be orthogonal.

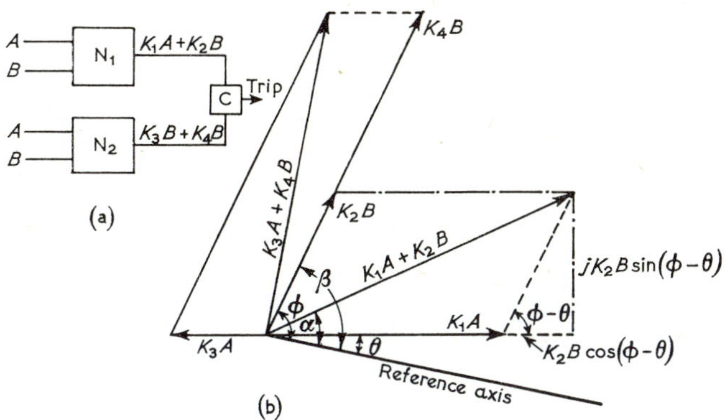

Fig. 1.19.

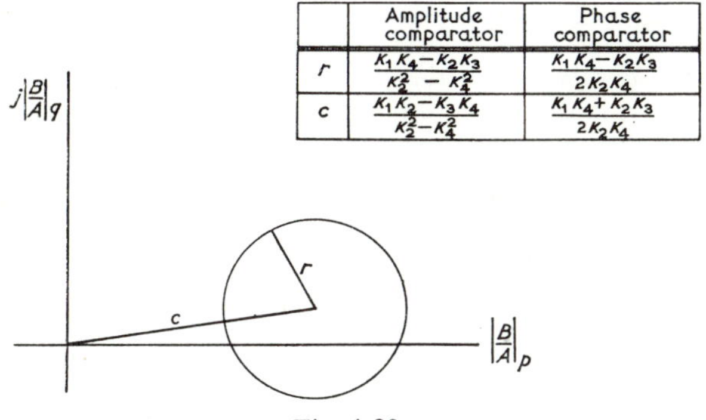

	Amplitude comparator	Phase comparator
r	$\frac{K_1K_4-K_2K_3}{K_2^2-K_4^2}$	$\frac{K_1K_4-K_2K_3}{2K_2K_4}$
c	$\frac{K_1K_2-K_3K_4}{K_2^2-K_4^2}$	$\frac{K_1K_4+K_2K_3}{2K_2K_4}$

Fig. 1.20.

1.6.1. Amplitude Comparator

In this case the two quantities are opposed and their modulii will be equal at the threshold of operation for any phase angle between them, the locus of which is the relay characteristic. Equating the modulii of the expressions (1.7) and (1.8) we have $S_1 = S_2$ or

$$[K_1|A| + K_2|B| \cos(\phi - \theta)]^2 + [K_2|B| \sin(\phi - \theta)]^2$$
$$= [K_3|A| + K_4|B| \cos(\phi - \theta)]^2 + [K_4|B| \sin(\phi - \theta)]^2 \qquad (1.9)$$

Rearranging these terms

$$(K_1^2 - K_3^2)\,|A|^2 + 2(K_1K_2 - K_3K_4)\,|A|\,|B|\cos(\phi - \theta) + (K_2^2 - K_4^2)\,B = 0.$$

Dividing through by $(K_2{}^2 - K_4{}^2)\,|A|^2$ for plotting in the β-plane,

$$\left|\frac{B}{A}\right|^2 + 2\left(\frac{K_1K_2 - K_3K_4}{K_2^2 - K_4^2}\right)\left|\frac{B}{A}\right|\cos(\phi - \theta) + \frac{K_1^2 - K_3^2}{K_2^2 - K_4^2} = 0 \quad (1.10)$$

The last term can be rearranged in a form indicating the location of the characteristic circle in the β-plane, viz.

$$\left|\frac{B}{A}\right|^2 + 2C\left|\frac{B}{A}\right|\cos(\phi - \theta) + C^2 = r^2. \quad (1.11)$$

This can be shown to be the equation of a circle of radius $(K_1K_4 - K_2K_3)/(K_2^2 - K_4^2)$ whose centre is at

$$-\left(\frac{K_1K_2 - K_3K_4}{K_2^2 - K_4^2}\right)\underline{/\theta}.$$

The same characteristic drawn on the α-plane is obtained by dividing the original Eq. (1.9) through by $(K_1{}^2 - K_3{}^2)\,|B|^2$ which gives a circle of radius $(K_1K_4 - K_2K_3)/(K_2 - K_4)$ whose centre is at

$$-\left(\frac{K_1K_2 - K_3K_4}{K_1^2 - K_3^2}\right)\underline{/-\theta}.$$

1.6.2. Phase Comparator

Here the relay operates when the product of the two input signals S_1 and S_2 is positive. If α is the phase angle of one input and β is the phase angle of the other, the threshold of operation is when $\alpha - \beta = \pm 90°$ because their product is greatest when the two signals are in phase.

$$\text{Hence } \tan(\alpha - \beta) = \pm\infty$$

$$\text{i.e. when } \frac{\tan\alpha + \tan\beta}{1 + \tan\alpha\tan\beta} = \pm\infty$$

$$\text{i.e. when } 1 + \tan\alpha\tan\beta = 0 \quad (1.12)$$

The values of $\tan\alpha$ and $\tan\beta$ can be seen from Fig. 1.17 and, if these are substituted in Eq. (1.12) the equation for the characteristics on the α-plane and β-plane can be obtained as before. Table 1.1 summarizes the values for the β-plane. The values for the α-plane are obtained by substituting K_1 for K_2 and K_3 for K_4 in their *denominators* only.

By substituting the values for K_1, K_2, K_3, K_4 the radius and the centre of the characteristics can be obtained for the various types of relays. Table 1.1 shows the summarized results for the common types of relays comparing one current with a voltage or another current. It will be seen that there are

TABLE 1.1.

β-Plane Characteristics for General Equation

Relay	Comparator	Quantities to be Compared in Amplitude or Phase		K_1	K_2	K_3	K_4	r	c
Impedance	Amplitude	$\lvert KI\rvert$	$\lvert V\rvert$	K	0	0	1	K	0
	Phase	$KI - V$	$KI + V$	K	-1	K	1		
Mho	Amplitude	$\lvert KI\rvert$	$\lvert -KI + 2V\rvert$	K	0	$-K$	2	$\frac{K}{2}$	$\frac{K}{2}$
	Phase	$KI - V$	V	K	-1	0	1		
Ohm	Amplitude	$\lvert 2KI - V\rvert$	$\lvert V\rvert$	$2K$	-1	0	1	∞	∞
	Phase	KI	$KI - V$	K	0	K	-1		
Balanced current	Amplitude	$K\lvert I_1\rvert$	$\lvert I_2\rvert$	K	0	0	1	K	0
	Phase	$KI_1 - I_2$	$KI_1 + I_2$	K	-1	K	1		
Differential current	Amplitude	$K\lvert I_1 - I_2\rvert$	$\lvert I_1 + I_2\rvert$	K	$-K$	1	1	$\frac{2K}{K^2 - 1}$	$\frac{K^2+1}{K^2 - 1}$
	Phase	$(K-1)\,I_1 - (K+1)\,I_2$	$(K+1)\,I_1 + (K-1)\,I_2$	$K - 1$	$-(K+1)$	$K + 1$	$-(K-1)$		
Percentage differential current (Slope $= S$)	Amplitude	$\lvert I_1 - \rvert_2\rvert$	$\frac{S}{2}\lvert I_1 + I_2\rvert$	1	-1	$\frac{S}{2}$	$\frac{S}{2}$	$\frac{S}{1-\left(\frac{S}{2}\right)^2}$	$\frac{1+\left(\frac{S}{2}\right)^2}{1-\left(\frac{S}{2}\right)^2}$
	Phase	$\left(1 - \frac{S}{2}\right)I_1 - \left(1 + \frac{S}{2}\right)I_2$	$-\left(1 + \frac{S}{2}\right)I_1 + \left(1 - \frac{S}{2}\right)I_2$	$1 - \frac{S}{2}$	$1 + \frac{S}{2}$	$-\left(1+\frac{S}{2}\right)$	$1 - \frac{S}{2}$		
Ditto Scheme B and balanced voltage scheme	Amplitude	$\lvert I_A - \gamma I_B\rvert$	$K\lvert I_A\rvert$	1	$-\gamma$	K	0	$K\left\lvert\frac{1}{\gamma}\right\rvert$	$1\left\lvert\frac{1}{\gamma}\right\rvert\underline{/-\gamma}$
	Phase	$(1 - K)\,I_A - \gamma I_B$	$(1 + K)\,I_A - \gamma I_B$	$1 - K$	$-\gamma$	$(1 + K)$	$-\gamma$		

never more than two constants, so that the general Eq. (1.1) fits all normal types of protective relays.

Consideration of the constants calculated in this manner will indicate which type of comparator is preferable for a given relay characteristic. In general an inherent comparator is better than the converted type because, if one quantity is large compared with the other, a small error in the large quantity may cause an incorrect comparison when their sum and difference are supplied as inputs to the relay.

1.7. DUALITY IN ELECTRICAL CIRCUITS

In Section 1.5 the duality of phase and amplitude comparators was discussed. Duality and symmetry are two very useful properties which are very helpful in the development of relay circuitry; often duality suggests an alternative solution when the first attempt is blocked by a difficulty.

If we group equations for a number of varieties of a certain type of relay we notice a certain symmetry of the parameters; it is most noticeable in its inverse or reciprocal form which we call duality. This can be clearly seen in Table 10.3, Chapter 10, where the equations for the characteristics of distance relays are grouped.

1.7.1. Duality of Circuit Parameters

Many of the parameters of electric circuits are duals. Some examples are:

TABLE 1.2. *Duality Table of Electrical Quantities and Conditions*

Quantity	Dual
Series Resistance	Shunt Conductance
Inductance	Capacitance
Capacitance	Inductance
Shunt Resistance	Series Admittance
Inductance	Capacitance
Capacitance	Inductance
Constant potential source (Thevenin's theorem)	Constant current source (Norton's theorem
Series connected e.m.f. and impedance	Current supply and shunt admittance
Mesh current	Nodal potential
Amplitude comparator with single quantities	Phase comparator with sum and difference
Phase comparator with single quantities	Amplitude comparator with sum and difference
Electrons moving against the current	Holes moving with the current
Electron tube triode in voltage amplifier	Transistor in current amplifier
Normally open contacts in series	Normally closed contacts in parallel
Normally closed contacts in series	Normally open contacts in parallel
Smoothing one input in an amplitude comparator	Spiking one input in a phase comparator

TABLE 1.3.
Dualities of L and C circuits

	INDUCTANCE		CAPACITANCE		
	Series	Parallel	Series	Parallel	Time characteristic
Circuit	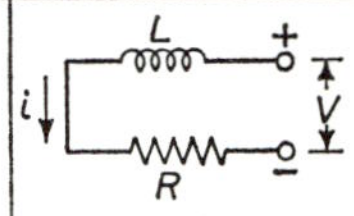 L, R, i, +, −, V	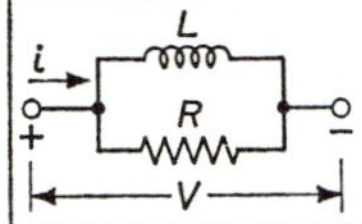 L, R, i, +, −, V	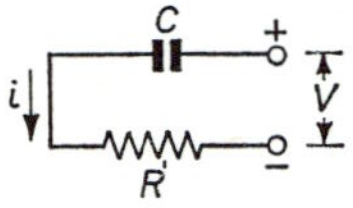 C, R, i, +, −, V	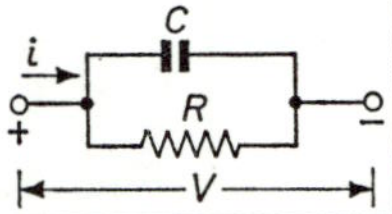 C, R, i, +, −, V	
Input	Constant V	Constant i	Constant V	Constant i	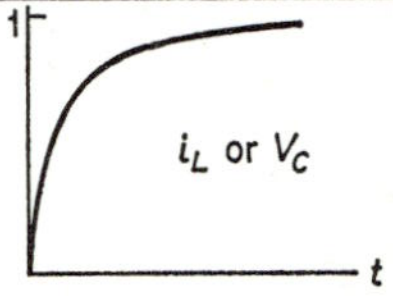 1, i_L or V_C, t
Rising value	$i_L = \frac{V}{R}(1 - e^{-\frac{R}{L}t})$	$i_L = i(1 - e^{-\frac{R}{L}t})$	$V_C = V(1 - e^{-\frac{t}{RC}})$	$V_C = iR(1 - e^{-\frac{t}{RC}})$	
Falling value	$V_L = V.e^{-\frac{R}{L}t}$	$V_L = iR.e^{-\frac{R}{L}t}$	$i_C = \frac{V}{R}e^{-\frac{t}{RC}}$	$i_C = i.e^{-\frac{t}{RC}}$	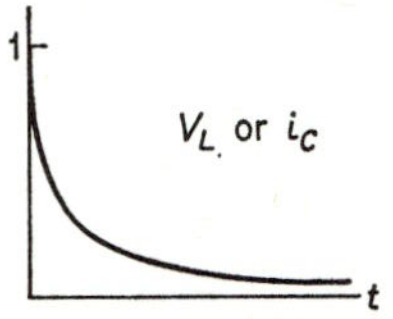 1, V_L or i_C, t

1.7.2. Application of Duality to Static Timing Circuits

In any electrical circuit many of the parameters, such as L and C, are duals. For example, any function of an L circuit can be obtained in the corresponding C circuit by substituting i for V or by substituting shunt connections for series, or vice versa.

For instance, with a pure capacitance, the voltage across it (V_C) increases linearly with time if a constant direct current is fed into it. Similarly, with a pure inductance, the current through it (i_L) increases linearly with time with a constant d.c. voltage across it. In both cases this is storage of energy (joules).

Series or parallel resistance causes the changes of current and voltage of L and C with time to be exponential instead of linear because its ohmic characteristics do not change with time.

Change of one parameter. Table 1.3 shows how the voltages and currents exchange roles when C is substituted for L in a series circuit. It also shows how the voltage and current also exchange roles when C is substituted for L in a shunt circuit. Furthermore, comparison of the columns shows how the voltage and current exchange roles in going from a series to a shunt circuit.

Change of odd number of parameters. The change of one parameter changes the characteristic curve; for example, the current rises with time in the series L circuit and falls with the series C circuit, with the other parameters left unchanged.

The change of three parameters has the same effect. For instance, the series RL circuit with constant voltage input has a rising current; the shunt RC circuit with constant current input has a falling current.

Change of even number of parameters. The change of two parameters causes no change in the characteristic. For example, the V_C time characteristic for a series RC circuit with a constant voltage input is the same as for a shunt RC circuit with a constant current input.

Similarly for four or six changes. For example, the time-current characteristic of the series RL circuit with a constant voltage input is the same as the time-voltage characteristic of the parallel RC circuit with constant current input.

If extra parameters are considered, such as open-circuiting or short-circuiting circuit components, six or more changes can be shown to cancel out their effect on the circuit characteristic.

2

Development of Static Relays

Early electronic relays. Semiconductor relays. Comparison of static and electro-magnetic relays–Reliability of Static relays and their components

Static protective relays are now coming into prominence because of their promise of better performance, lower input power (burdens), smaller size and freedom from maintenance. Like automation, it is difficult to say when they started. The date of the first static relay may be anywhere from 1931 to 1958, depending upon the definition of a static relay.

2.1. DEFINITIONS

To most engineers a static relay is one in which the measurement or comparison of electrical quantities is done in a static network which is designed to give an output signal in the tripping direction when a threshold condition is passed; the output signal operates a tripping device which may be electronic, semiconductor or electromagnetic. This definition excludes electromagnetic relays in which the measurement is derived from a balance of magnetic or mechanical forces within the relay, the electrical contacts operating as a result.

The B.S.I. definition refers to a relay with "no moving parts", which excludes static relays which trip through a slave device of the attracted armature type. Finally, there is the distinction between electronic relays using vacuum tubes and/or thyratrons and those using only semiconductors.

The first wholly static relay was developed by M.E. Bivens in 1930 (U.S. Patent 2,299,501) in the laboratories of the General Electric Company, U.S.A., but was not offered for sale because the comparator used a hot-cathode thyratron which at that time was not sufficiently consistent in operation to be practical.

The first wholly static relay offered for sale was the electronic mho relay produced in 1948 by the General Electric Company, U.S.A. [4]. In this relay a vacuum tube was used as the comparator and a thyratron used only as a tripping device. Unfortunately, no business resulted because at the time power system engineers did not have sufficient confidence in electronic components.

A more detailed account of the early work on static relays up to 1956 has been given by Dr. C. Adamson (Manchester College of Science & Technology) in his series of articles in the *Electrical Times* [114]. The present chapter will discuss in somewhat less detail the progress up to the present time.

2.2. EARLY ELECTRONIC RELAYS [118]

In the late twenties and early thirties much experimenting was going on with electronic tubes (valves) and thyratrons used as relays, but carrier relaying received the first serious attention because it was an adaption of carrier communication which was already in practice.

2.2.1. Carrier Relaying

The first electronic system to go into service in a power system was A.S. Fitzgerald's carrier pilot system in 1928 [12]. This was a phase-comparison system and its principle was much the same as that of present-day phase-comparison equipment except that overcurrent fault detectors were used instead of a static network to start carrier and control tripping (Fig. 2.1).

The transmitters at each end of the line produced a high-frequency carrier signal during each positive half-cycle of line current; these were coincident during an internal fault and alternate during an external fault. The receiver anode was connected to be positive during negative half-cycles of current. On an external fault there was a blocking signal every other half-cycle which prevented tripping but on an internal fault there was no blocking

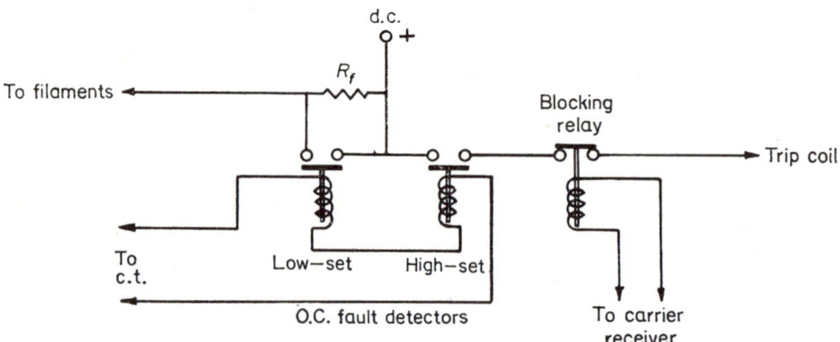

Fig. 2.1. Basic relays in Fitzgerald carrier scheme

signal. This was because the receiver anode was positive during the half-waves of signal received from the other end only during an external fault. Carrier was started by bringing the filaments of the thermionic tubes up to normal excitation (Fig. 2.1); the resistance R_f was used to keep the filaments warm so that carrier transmission would be started as quickly as possible.

The Fitzgerald circuit was relatively simple because it catered only for ground faults so that it used only zero sequence current and thus eliminated the complexity and time delay of sequence networks; it is interesting to note that this is the tendency in American practice today.

Two years later a directional comparison scheme was brought out [20] which became popular in the U.S.A. Here electromagnetic directional relays, by their joint action, discriminated between internal and external faults; electronic equipment was used only for the carrier channel. The first distance carrier scheme [26] entered service in 1942 and differed little from the blocking carrier schemes of today.

2.2.2. Thyratron Relays

From the point of view that Fitzgerald's carrier scheme used electronics only for the carrier channel, the earliest documented work on static relays was that of Rolf Wideröe who developed most of the standard types of protective relays in Germany in 1930–31 [1].

These circuits used thyratrons because (*a*) they could handle more current than 'hard' vacuum tubes and (*b*) when once triggered they would continue to pass trip coil current until the anode voltage was removed. Unfortunately, again this was an example of a good idea being ahead of its time because of the lack of suitable components to make it practical. In fact, the patents expired before reliable thyratrons became available.

Figure 2.2 shows an inverse time current relay which fired a thyratron causing tripping when the capacitor *C* had been charged to a voltage exceeding the negative bias from the battery. Figure 2.3 shows an instantaneous undervoltage relay which tripped when the negative bias from the rectified voltage fell below the positive bias from the battery; the capacitor *C* was provided to absorb the interference spikes which could cause undesirable tripping.

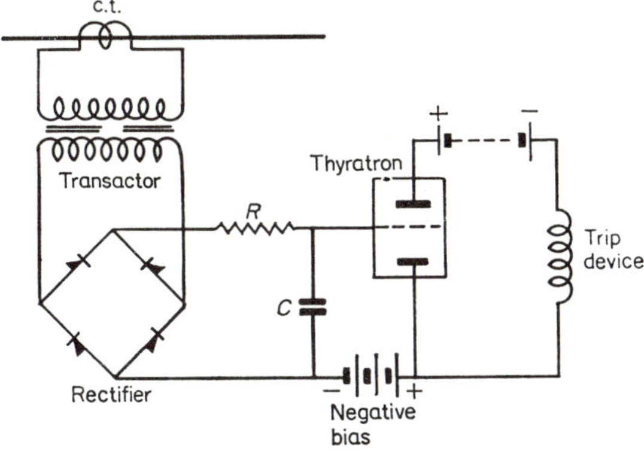

Fig. 2.2. Time-current relay (Wideröe)
R and *C* control the operating time

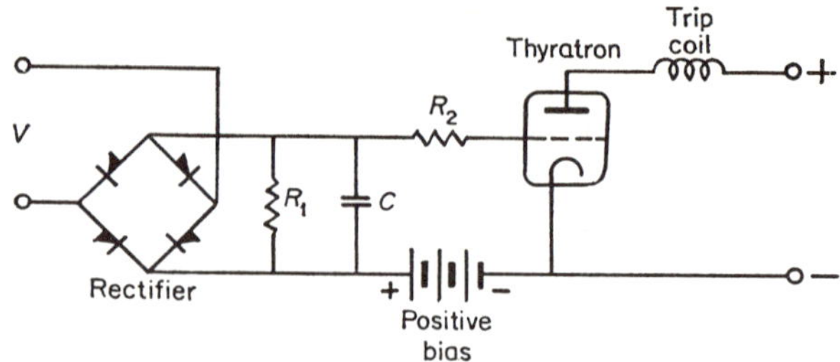

Fig. 2.3. Instantaneous undervoltage relay (Wideröe)
R_1 applies bias to grid;
R_2 limits grid current;
C is a smoothing capacitor

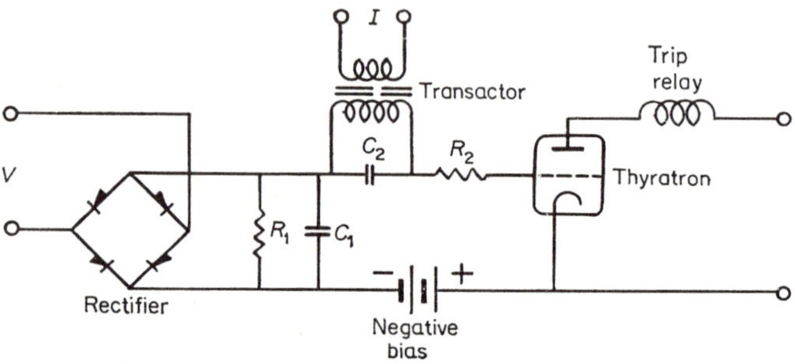

Fig. 2.4. Impedance relay (Wideröe)

In the impedance relay of Fig. 2.4 the current tripping signal on a positive half cycle had to overcome the negative bias from the rectified voltage [1].

2.2.3. Electronic Distance Relays

In 1948 the author, in conjunction with Hodges and Macpherson [4], developed a 1 cycle electronic distance relay (Fig. 2.5). This relay compared the current and voltage at the moment of voltage maximum, which gives a mho characteristic, (see Sections 4.3.1 and 4.3.2). The comparator was a pentode thermionic tube which triggered a thyratron for tripping purposes. A voltage $(IZ_L - V)$ was applied to the control grid of the pentode which was normally held non-conducting by a negative bias on its suppressor grid. This bias was counteracted by a positive impulse at the instant of voltage maximum, permitting a momentary measurement of $(IZ_L - V)$.

A three-phase three-step distance relay using this principle was put into service in 1950 and, although it performed excellently, it was not put into production because customers at the time still objected to the use of electronic tubes in relays as not being sufficiently reliable. This was mainly due to their experience with home radios, which did not use the specially selected tubes that were used in electronic relays and other important equipment.

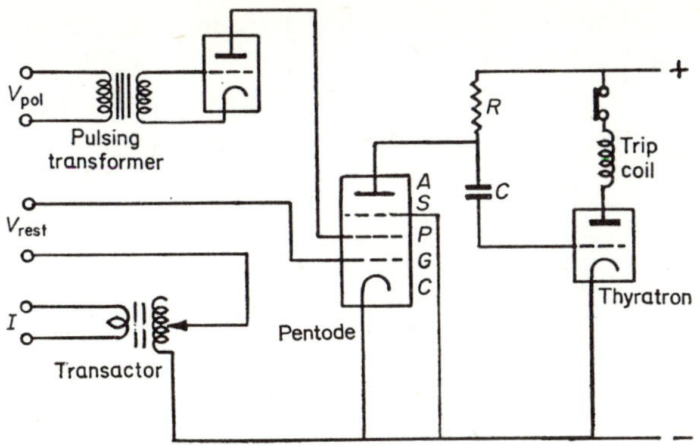

Fig. 2.5. Electronic distance relay
A = anode; S = screen grid; P = suppressor grid
G = control grid; C = cathode

Looking back from the present day it may seem odd that protective relays using electronic tubes had not been accepted by the industry on the grounds of unreliability whereas carrier equipment using these tubes had not only been accepted but used for the most important lines. The explanation of this paradox is that carrier relaying was the only practical means by which both ends of a long line could be tripped instantly, permitting immediate reclosure, and there was no alternative to electronics for the carrier channel except pilot wires which were uneconomical above 15 miles. Furthermore, the channel could be kept working by automatic checking equipment which could sound an alarm if any tube became defective, so that it could be replaced before the next fault occurred.

2.3. SEMICONDUCTOR RELAYS

Over the last ten years the major manufacturing companies have developed static equivalents of all the principal types of relays, but so far there has been a very conservative attitude towards them by the major power companies.

In 1959 Allis-Chalmers, U.S.A., brought out an inverse time-current relay [93]; since then many other manufacturers have developed similar relays but the induction disc type of inverse time-current relay still dominates the market. Somewhat more success has been achieved with static definite time-current relays, especially in the industrial field.

A number of differential current relays have appeared in the last seven years for the protection of generators [64, 101] and buses [115]. In 1961 an American company brought out a transistorized version of their electronic distance relay [15] and this was followed by others in England, Switzerland [88] and the U.S.A.

The performance of the more recent static distance relays has been so much superior to that of the electromagnetic type that the power companies have been forced to accept them. Thus they may be the thin edge of the wedge of general acceptance of static relays. This would never have been achieved by time-current relays because high performance is less important in their applications. Figure 2.6 shows the basic arrangement of differential or distance relays.

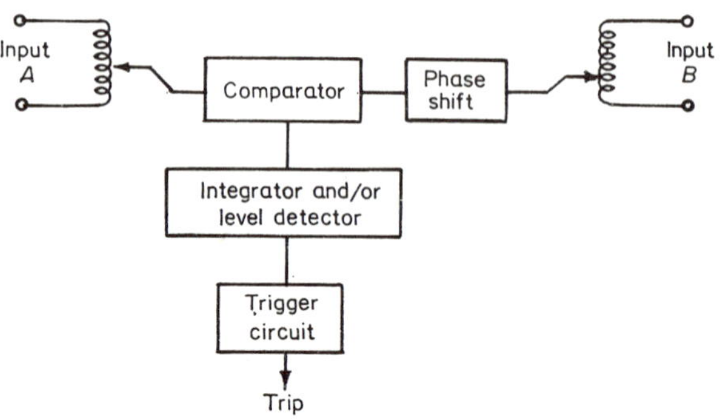

Fig. 2.6. Block diagram of a static relay of the comparator type

The high performance of distance relays has been achieved by advances in the design of phase and amplitude comparators. The most recent techniques include multi-input comparators [75, 78, 106] for obtaining quadrilateral impedance characteristics with a single comparator.

2.3.1. Rectifier Comparators

The earliest documented work on rectifier bridge comparators took place in 1947 in Norway [13] and Germany [2, 3]. Figure 2.7 shows the circuit used by the Jacobsen Company for a distance relay whose characteristics could be impedance, admittance (mho) or reactance, depending upon the taps chosen. For impedance, one coil is supplied with rectified current and the other with rectified voltage. For admittance, one coil is supplied with

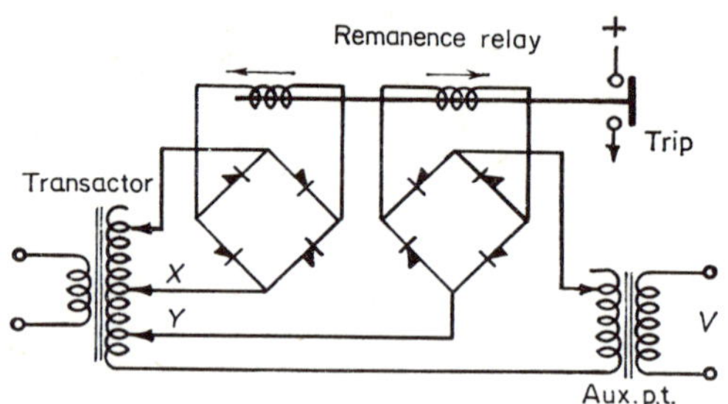

Fig. 2.7. Static distance relay with controllable characteristic

(a) with X and Y as shown, I^2 versus $(V - I)^2$: mho
(b) with X and Y on the bottom, I^2 versus V^2 : impedance
(c) with X and Y on the p.t., $(I - V)^2$ versus V^2 : reactance
(d) with third input to give I^2, V^2 and $(I - V)^2$: ellipse

the line current and the other with the line voltage minus a voltage proportional to the current, i.e. one current tap plug goes to a voltage tap. For reactance, one coil is supplied with voltage and the other with voltage minus current, i.e. one voltage tap plug goes to a current tap.

The German relays used a circulating current rectifier bridge comparator which had the advantage of great sensitivity at low inputs and a limiting action at high inputs. The output of this amplitude comparator (Fig. 4.4) operated an extremely sensitive moving coil relay [2] which was polarized by a permanent magnet so as to trip only for faults within the relay setting. The same principle was used in England in the Reyrolle H type distance relay [116] and in A.E.I. relays [119].

The modern tendency is towards phase comparators because they are better adapted to semiconductor logic changes. The English Electric type YTG distance relay uses a rectifier bridge phase comparator (Chapter 4, Fig. 4.29) which is the electrical dual of the amplitude comparator mentioned above but has the advantage of inherent directional action without the need for precisely balanced circuits.

2.3.2. Transistor Comparators

In 1949, soon after W. Shockley developed the transistor [91], J. D. Loving [8] developed a relay [Fig. 2.8) in which the current signal A was arranged to cause tripping unless it was diverted by transistor T when the signal B was negative. This was the first transistor phase comparator.

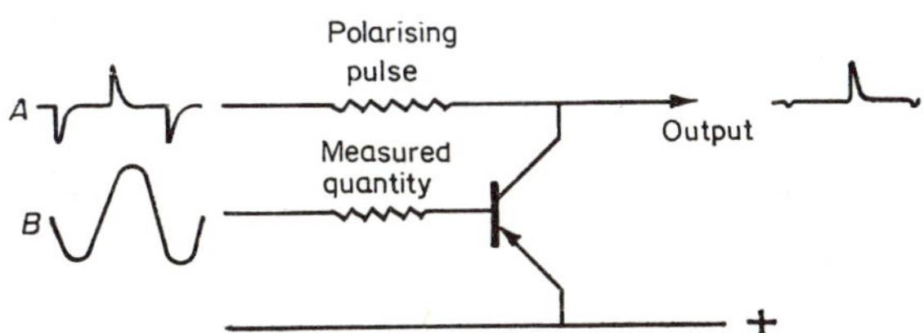

Fig. 2.8. Measuring unit of early ohm relay (Loving)
Negative pulse at A causes tripping output unless signal B is negative at that moment

In 1956 Bergseth [10] developed a mho distance relay using a phase comparator (Fig. 2.9) in which the two inputs were connected through half-wave rectifiers to the base-emitter circuit of a transistor. When the two inputs V and $(IZ - V)$ were in phase within $\pm 90°$ the transistor became conductive (during the negative halfcycles) long enough to pick up the tripping relay.

Neither Loving nor Bergseth were attached to manufacturing companies and their relays were never engineered for production, but they provided the incentive for subsequent work by relay manufacturers which culminated in fully transistorized distance relays.

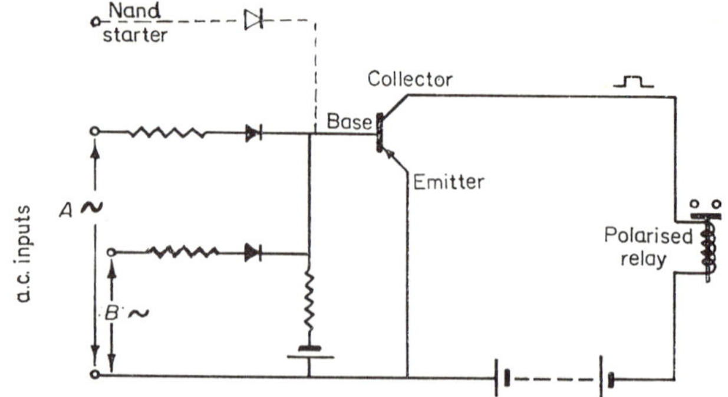

Fig. 2.9. Measuring circuit of early static mho relay (Bergseth)
Trips when neither input is positive

In 1957 Adamson and Wedepohl [9, 11] developed a transistor distance relay with a mho characteristic but, although it had very good characteristics, it had limitations which prevented it from being exploited commercially in its original form. A simplified diagram of the measuring circuit is shown in Fig. 2.10. Tripping occurs if the period during which the two transistors are both non-conducting exceeds 90°. Meanwhile, this was the first transistorized distance relay and its excellent characteristics caused the world's relay manufacturers to start work immediately on the high performance static distance relay.

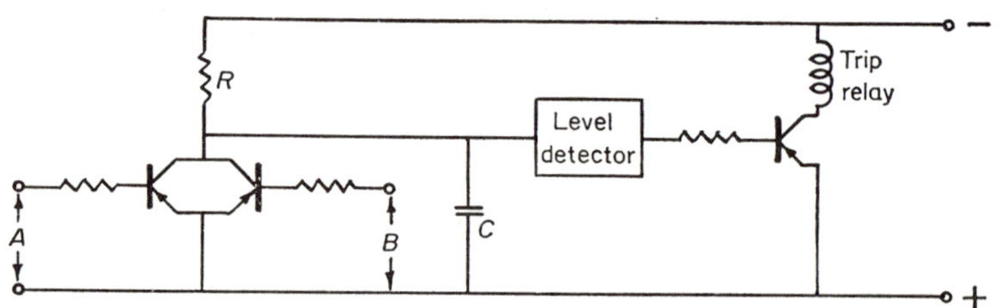

Fig. 2.10. Transistor comparator (simplified) with R-C integrator
T_1 and T_2 blocked with A and B both positive, causing full voltage across C
T_1 or T_2 conducting if either A or B negative; hence no voltage across C

The first wholly static distance relay [15] to be produced commercially was a transistorized version of the electronic mho relay with which the author was associated. This relay is capable of tripping in less than a half-cycle; the fast action is achieved by tripping through a thryistor (SCR) and thus eliminating the operating time of the tripping relay.

2.3.3. Multi-input Comparators

The Hoel-Bräten distance relay of 1949 (13) had a three-input amplitude comparator which produced an elliptical characteristic. More recently Humpage and Sabberwal (75) showed how to produce a quadrilateral characteristic with a three-input comparator. C.I.G.R.E. Committee No. 4 is investigating the application of such characteristics to distance relays.

2.4. COMPARISON OF STATIC AND ELECTROMAGNETIC RELAYS

Although transistor circuits offer many advantages over electromagnetic relay movement, they should be regarded as a valuable addition to the electrical tools from which protective relays can be made rather than a technique which must replace electromagnetic relays at all costs.

This can be graphically illustrated by considering first the reduced size and superior performance of static distance relays compared with induction cup or balanced beam distance relays; then one should consider the economy and simplicity of an electromagnetic multi-contact tripping relay compared with the equivalent group of silicon controlled rectifier circuits.

In most static relays available so far, the output or slave device has been an electromagnetic relay, chiefly for reasons of economy. A simple attracted armature relay can be manufactured for about £1; it will operate in 10 mS and carry a sufficient number of contacts to trip two breakers and control two auxiliary circuits. A very sensitive polarized d.c. relay with two contacts can be manufactured for a few pounds where the output of the static comparator is limited. In most cases a reed relay can be used for tripping.

The static equivalent of the output relay has the advantage of being faster, but it requires a semiconductor or thermionic tube (valve) circuit to replace each contact and would cost at present £ 20 ($ 50 U.S.) with the same contact arrangement, assuming that two of the semiconductors each have a rating efficiency sufficient to energise a circuit-breaker trip coil. A practical compromise is a combination of a thyristor (SCR) for one trip circuit and an attracted armature relay for the other contact functions of the relay.

The attracted armature relay can be of the telephone type or of the reed type. The former are well known and have been used for the last fifty years. Reed relays have only recently come into wide use where extremely fast operation is required.

2.4.1. Reed Relays

These relays have two accurately positioned flexible nickel-iron strips (reeds) sealed into a closed glass capsule in an inert atmosphere (Fig. 2.11). The inner ends of the reeds overlap and are separated by a gap of 0·01 inch or less. A coil surrounds one or more of the reed contact units and produces a magnetic field which makes the reeds come together, thus operating contacts consisting of wafers of royal metal welded to the inner ends of the reeds [82]. Reed relays have a very high degree of reliability and require no maintenance. Hence, from the service point of view, they can be classified as the equivalent of a static relay. Their performance relative to other switches is shown in Table 2.1.

The best but most expensive type of reed contacts are mercury-wetted; they are more suitable for normally-closed applications and are completely bounce-free in operation.

They are an alternative to thyristors as tripping relays (see section 5.1) and have the advantages of simplicity and low cost.

The best operating condition for most dry-reed relays is 2.5 times pick-up. This gives fast operation (1 mS with a low time-constant circuit), minimum bounce and good contact pressure. Correctly mounted, they are shock-proof up to 5 g but tend to resonate at about 900 Hz. Between coil and contact they will stand a 2 kV insulation test or a 5 kV 1/5 μS surge test and 500 V to 1 kV across open contacts. When set sensitively they require magnetic shielding.

TABLE 2.1.

Comparison of Typical Electromagnetic and Semiconductor Switching Relays

Function	Electromagnetic	Reed	Semi-conductor	Thyristor
Input	1 to 3 W	0·1 to 3 W	10 mW	20 mW
Switching Capacity	30 W	up to 20 W	50 W	600 W
Power Gain	10 to 30	7 to 200	5000	30,000
Continuous Carrying	5 A	0.1 A	1 A	2.5 A
Delay	10 mS	1 or 2 mS	20 μS	50 μS
Contacts	up to 6	up to 6	1	1
Operations	10^7	10^7	no limit	no limit
Ambient temp. Range	−5° to 70°C	−5° to 55°C	−20° to 100°C	−20° to 100°C
Affected by Vibration	Yes	Little	No	No
Affected by Corrosive Atmosphere	Yes	No	No	No

2.4.2. Advantages of Static Relays

The main advantages offered by semiconductor relays are low burden and superior performance, viz:

(*a*) Fast response, long life and high resistance to shock and vibration.

(*b*) Quick reset, a high resetting value and the absence of overshoot are easy to obtain in static relays because of the absence of mechanical inertia and thermal storage.

(*c*) With the absence of bearing friction and contact troubles (corrosion, bouncing and wear), better characteristics can be obtained and there is less necessity for maintenance.

(*d*) Very frequent operation causes no deterioration.

(*e*) The ease of providing amplification enables greater sensitivity to be obtained.

(*f*) The low energy levels in the measurement circuits permit miniaturization of equipment and minimize c.t. inaccuracy.

On the other hand, static relays have a number of limitations which must be compensated for:

(*a*) Variation of characteristics with temperature and age.

(*b*) Vulnerability to voltage spikes and high ambient temperatures.

(*c*) Dependence upon the reliability of a large number of small components and their electrical connections.

(*d*) The lack of life-test data due to rapid replacement by new designs.

(*e*) Low short-time overload capacity compared with electromagnetic relays.

The means for dealing with these limitations will now be mentioned briefly:

(*a*) Temperature error can be eliminated by appropriate use of thermistors. Ageing can be minimized by pre-soaking for a number of hours in a relatively high temperature.

(*b*) Protection against voltage spikes can be provided by filters and shielding. Silicon transistors can be used where high temperatures are involved.

(*c*) Modern methods of soldering, wire-wrapping, etc. and the selection of superior components can ensure a high degree of reliability.

(*d*) The manufacturers of transistors say that the experience gained with one transistor is embodied in its successor.

(*e*) Overload must be avoided by circuit design.

Transistors and electronic valves are, in many respects, duals, i.e. their characteristics are inverted.

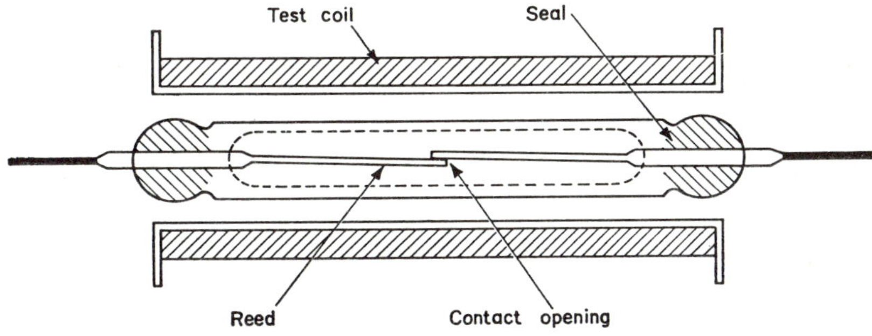

Fig. 2.11. Operating mechanism of make-action contact unit

Transistors are current-driven, they short-circuit with reversed d.c. supply and have solid material between electrodes. Electronic valves are voltage-driven, they open-circuit with reversed d.c. and have vacuum between electrodes.

2.4.3. Static and Electromagnetic Relay Equivalents

In order to produce characteristics which correspond to fairly complex equations, protective relays must be able to carry out most of the simpler mathematical processes such as addition, subtraction, multiplication, division, squaring and extracting the square root; the first four functions may be required in either scalar or vectorial form.

Table 2.2 shows how some of these processes are carried out in electromagnetic and in static relays. The wattmetric relay and the Hall generator are examples of vectorial multiplication. The Hall generator can also do scalar multiplication. Scalar division is done by an amplitude comparator. Vectorial addition and subtraction are done by the appropriate connections of the inputs. Scalar addition and subtraction result if the inputs are rectified before combination. Integrators, etc. were omitted for the sake of simplicity.

In the AND circuits it can be seen that all inputs have to be present to produce an output. Similarly, either input can produce an output in the OR circuits. The input B prevents the input A from producing an output in the NOT circuit.

2.5. EFFICIENCY AND SENSITIVITY

In one sense an electromagnetic relay is more efficient than a static relay because the energy required for moving the armature and contact does not affect the characteristic of the relay; in a static relay any power taken from the static network for operating the output device upsets the output and hence the relay characteristic.

This is shown in the early negative sequence relays which had to have an input of several hundred volt-amperes in order to operate a 2 VA relay with acceptable accuracy; on the other hand, the operating characteristic of an induction cup mho relay is negligibly affected by its burden or by the movement of the oup as the threshold of operation is passed.

This weakness of static relays has been overcome by two methods:

(*a*) the use of a supersensitive output device which can be a polarized d.c. relay designed to pick up at about 100 microwatts [2], or a thyratron [1], or a thyristor (S.C.R.) [14/15], which takes even less power;

(*b*) the use of a transistor amplifier between the measuring network and the output device, which can then be less sensitive, e.g. the ordinary attracted armature relay which requires about 1 watt to operate it.

With such aids to efficiency, static relays can be designed to have extremely low a.c. burdens and extremely high-grade performance. Furthermore,

TABLE 2.2. *Relay Tools*

FUNCTION	ELECTROMAGNETIC	SEMICONDUCTOR
Amplifier	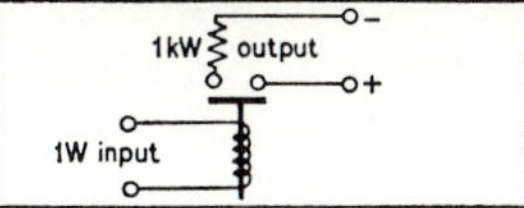 1kW output; 1W input; −; +	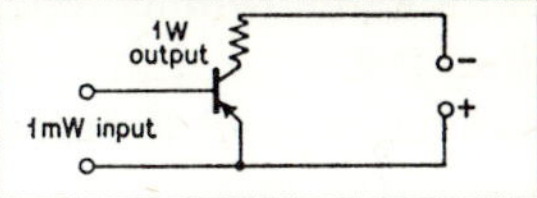 1W output; 1mW input; −; +
Level detector	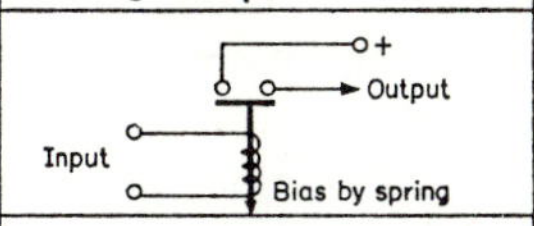 Input; Output; Bias by spring; +	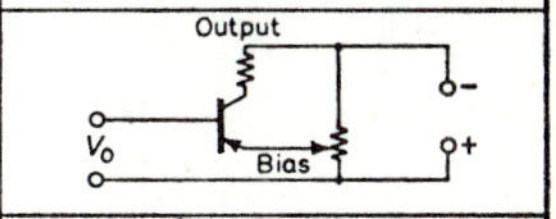 Output; V_0; Bias; −; +
Time delay A → 2 sec → O	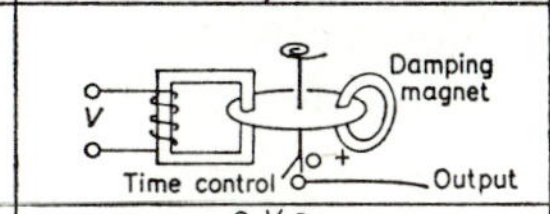 V; Damping magnet; Time control; Output; +	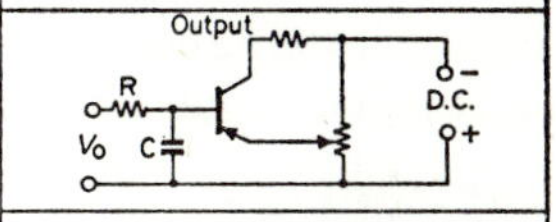 Output; R; V_0; C; D.C.; −; +
Vector product (wattmetric)	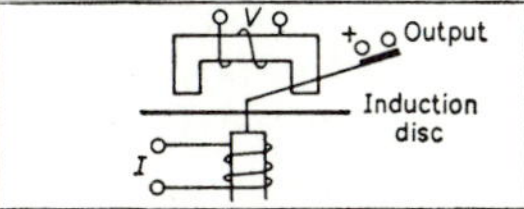 V; I; Output; Induction disc; +	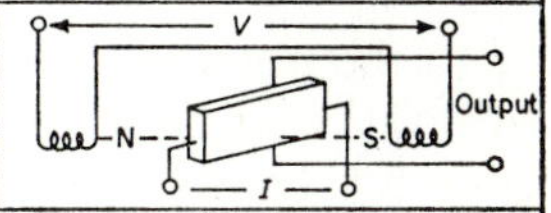 V; N; S; I; Output
Amplitude comparator (quotient relay)	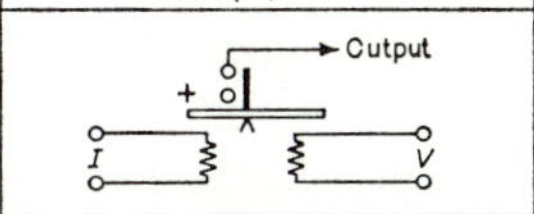 Output; I; V; +	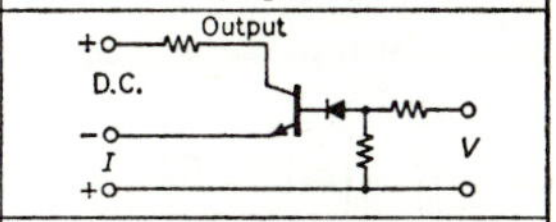 Output; D.C.; I; V; +; −
Phase comparator (directional relay)	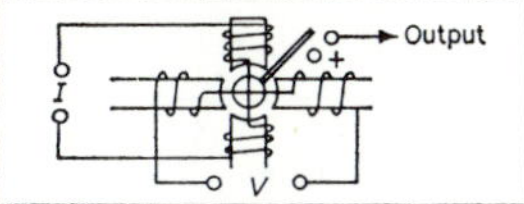 Output; I; V; +	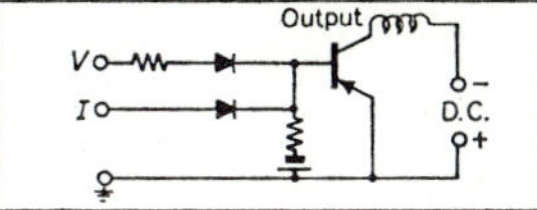 Output; V; I; D.C.; −; +
Snap-action (seal-in or trigger circuit)	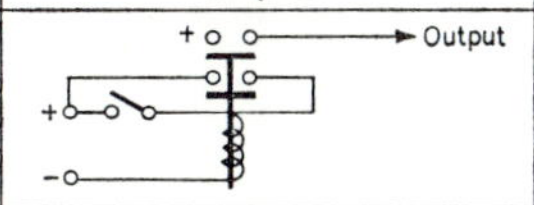 Output; +; −	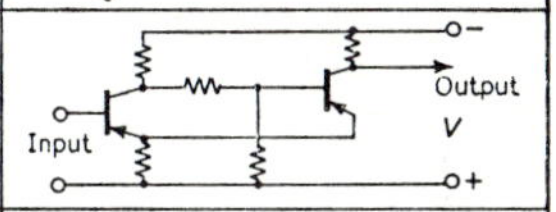 Input; Output; V; −; +
Ultra high-speed relay	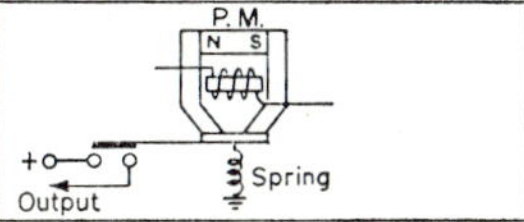 P.M.; N; S; Spring; Output; +	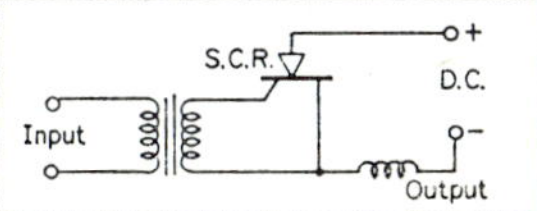 S.C.R.; Input; Output; D.C.; +; −
AND circuit A, B → AND →	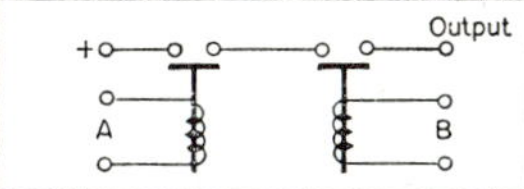 Output; A; B; +	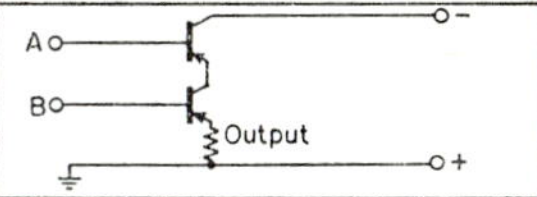 A; B; Output; −; +
OR circuit A, B → OR →	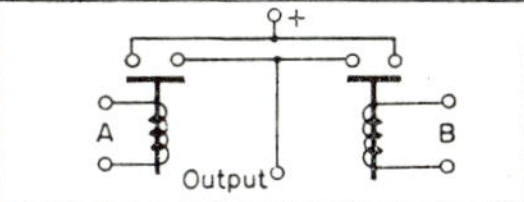 Output; A; B; +	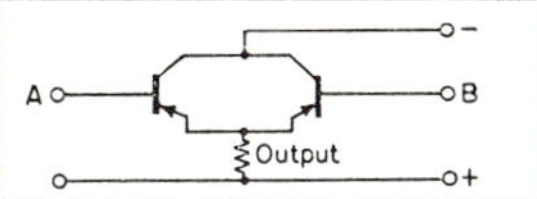 A; B; Output; −; +
NOT circuit A, B → I → NOT →	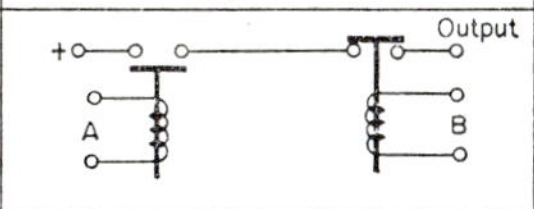 Output; A; B; +	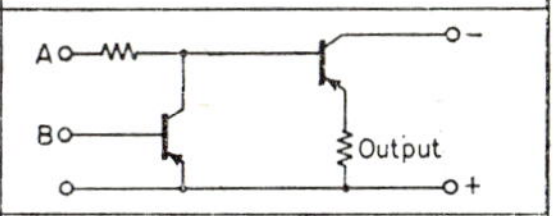 A; B; Output; −; +
Phase sequence relay (I_2)	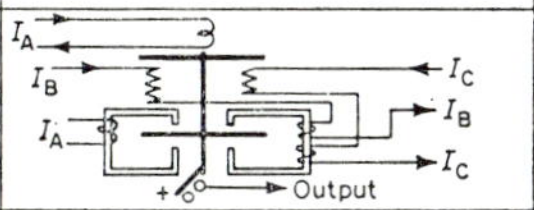 I_A; I_B; I_C; Output; +	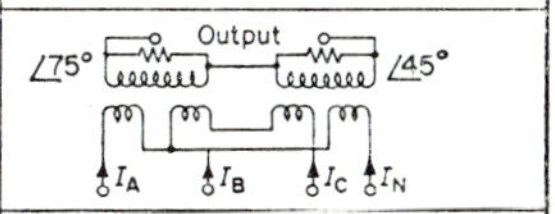 Output; ∠75°; ∠45°; I_A; I_B; I_C; I_N
Square wave generator	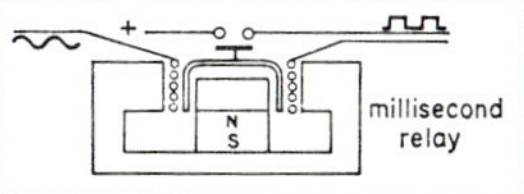 millisecond relay; N; S; +	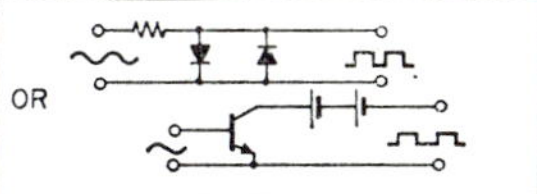 OR

NOTE:– The AND and OR circuits are interchanged when operated by removing the input instead of applying them. The semiconductor circuits operate with negative input signals

Integrators, level detectors and amplifiers omitted for simplicity.

inaccuracy due to magnetic saturation of iron-cored c.t's can be eliminated by air-cored c.t's which are practical only with low secondary burdens.

Reed relays tend to measure instantaneous values; induction and attracted armature relays measure root mean square values; dynamometer type relays measure the arithmetic mean; electronic devices inherently measure instantaneous values but their attendant circuitry can make them measure any of these values, and many others including power functions. On the other hand, their instantaneous operation necessitates the removal of the d.c. component from a.c. inputs and protection of the circuit from transient interference

2.6. RELIABILITY

Owing to the tremendous amount of energy in a modern power system, an electrical fault can do some very expensive damage if it is not cleared promptly. Table 2.3 shows the order of the cost of the damage that would be caused by uncleared faults in various types of equipment; in addition, there is the danger to personnel and the danger of starting a destructive fire. Quickly cleared faults, however, may cause little or no damage.

TABLE 2.3. *Cost of Uncleared Faults on a 30,000 MW Power System (1968)*

Equipment	Average Capital Cost p.u.	Faults p.a.	Unit Repair Cost p.a.	Total Repair Cost p.a.	Total Cost of Protection
Lines & Cables	£ 10,000/mile	230	£ 400	£ 92,000	£ 5.1 M.
Switchgear	£ 20,000 each	57	£ 3,000	£ 171,000	£ 0.75 M.
Transformers	£ 65,000 each	66	£60,000	£3,960,000	£ 2.8 M.
Alternators	£100,000* p.u.	42	£50,000	£2,100,000	£ 5.7 M.
Switch boards, etc.	£ 15,000 p.u.	157	£ 1,000	£ 157,000	£ 0.6 M.
TOTAL	£ 3.3 M. p.u.	—	—	£6,480,000	£15 M.

Notes on above Table

1. The "p. u." in Column 2 is per 100 MW.
2. *Generation costs about £ 1 M. per 100 MW including Turbines, Boilers, etc. Cost of Lines and Cables is also about £ 1 M. per 100 MW, Switchgear £ 0.3 M. and Transformers £ 0.75 M. per 100 MW of system.
3. Bus faults are included in Switchgear.

It is obvious from this that reliability is of paramount importance in a protective relay; but the very nature of a relay makes this difficult because it may remain inactive for years and deterioration due to time or adverse conditions may prevent the relay from operating when a fault finally does occur. Some of these adverse conditions against which the relay must be protected are:

(*a*) Corrosion of contacts or fine wires due to condensed moisture becoming acidified in industrial or tropical areas.
(*b*) Vibration and mechanical shock.
(*c*) Transient overvoltages and overcurrents due to interference or to circuit switching.
(*d*) Extremes of temperature.

Typical failure rates of the various components in static relay circuits are shown in Table 2.4. One failure per million relay hours is equivalent to one failure per year out of 125 components. These failure rates are for rated loading; the rates are generally much less for components run below rating, which is the practice of most relay manufacturers [5, 24, 37]. Furthermore the quality of components has improved since 1963.

TABLE 2.4

Catastrophic Failure Rate of Electronic Components (1963) at Rated Inputs

	Failure rate per million component hours
Diodes	4.5
Transistors	1.0
Connectors	0.8
Soldered connections	0.4
Switches	10·0
Resistors	0·1 to 1·4
Potentiometers	40.0
Capacitors (electrolytic)	0·84
Windings	5·0

The failure rate of semiconductors can be substantially reduced by heat-soaking, i.e. keeping the semiconductor near the maximum junction temperature (over 100°C for silicon transistors) for several hundred hours. This is an accelerated test which causes the failure of most of those which would ultimately fail in service.

The technique of heat-soaking of semiconductors in the manufacture of protective relays has been the subject of intensive development. There is at present some divergence of opinion as to the length of the soaking period and under what electrical conditions the test should be made, but it is hoped that eventually some sort of soak test can be evolved which will eliminate all components liable to subsequent failure and that it will include soldered joints and circuit components.

Electromagnetic protective relays have a known reliability which has been very good (less than 1·5% wrong operations due to relay failure or inaccuracy have been reported on major power systems), but annual testing and maintenance has always been thought necessary for them. It is hoped that the absence of moving parts in static relays will reduce these wrong operations to perhaps 0·5% and hence enable periodic maintenance to be simplified or eliminated. Whether this will be possible will depend upon the reliability of the components and electrical connections and the practice of certain design techniques such as the galvanic separation of the input circuit, the transistor circuit, the output (tripping) circuit and the d.c. supply (if derived from the station battery). This subject is dealt with in more detail in Chapter 5.

2.6.1. Electrical Connections [24]

Since milliamps and millivolts are used in most static circuits rather than amps or volts, it is necessary to take special precautions against corrosion because there is not enough voltage to break down corrosion and not enough current to keep the contact 'wetted'. All pressure contacts should be gold-plated and bifurcated tips should be used with at least 20 gm pressure.

Dry soldered joints can be avoided by proper precautions in soldering. A dry joint is one with insufficient contact area (poor bonding) which is later reduced by vibration, differential expansion due to temperature and corrosion due to leakage; this results in high resistance which may interfere with the operation of the relay. All soldered connections should be very carefully cleaned before tinning, especially those of components which have been in stock for more than a month; the joint should be sealed by varnish or encapsulation after soldering. Care should be taken to keep the soldering iron tip close to the correct temperature and to use the type and size appropriate to the job.

An alternative to soldering is wire wrapping (Fig. 2.12) which has a reliability factor somewhat higher than that of soldering. It is used mostly for connections to terminals where it is possible to use the wrapping tool.

2.6.2. Plug-in Modules and Connectors [5]

For test and replacement purposes relay units should be on modules which are easily removeable without disconnecting the wiring (plate 1). In the past the reliability of plug-in modules and contacts was not considered high enough for static relays, but now several reliable types are available, due to the improved quality and closer tolerances of printed board and contact materials.

Modern board material is physically stable and has almost no water absorbtion; modern gold plating for contacts is wear-resistant and non-porous. The final contribution to reliability is provided by bifurcated contacts, based on the premise that, if the possibility of failure is 1 in 1000 for a single contact, it is 1 in 1,000,000 for two in parallel.

The plating of the contact at present recommended is 0·00015 in. of gold on top of 0·0005 in. of nickel (Englehard process). This provides protection against the common sources of corrosion such as industrial atmospheres, but it is also important to wipe the plug-in contact clean before insertion since a small particle of dust or lint would keep the contact open at the low voltages employed in static relay circuits.

The main problem with plug-in contacts is that, if sufficient contact pressure is provided to ensure conductivity under all conditions, the withdrawal pressure of the module will be high and there will be a tendency to scrape the gold plating off.

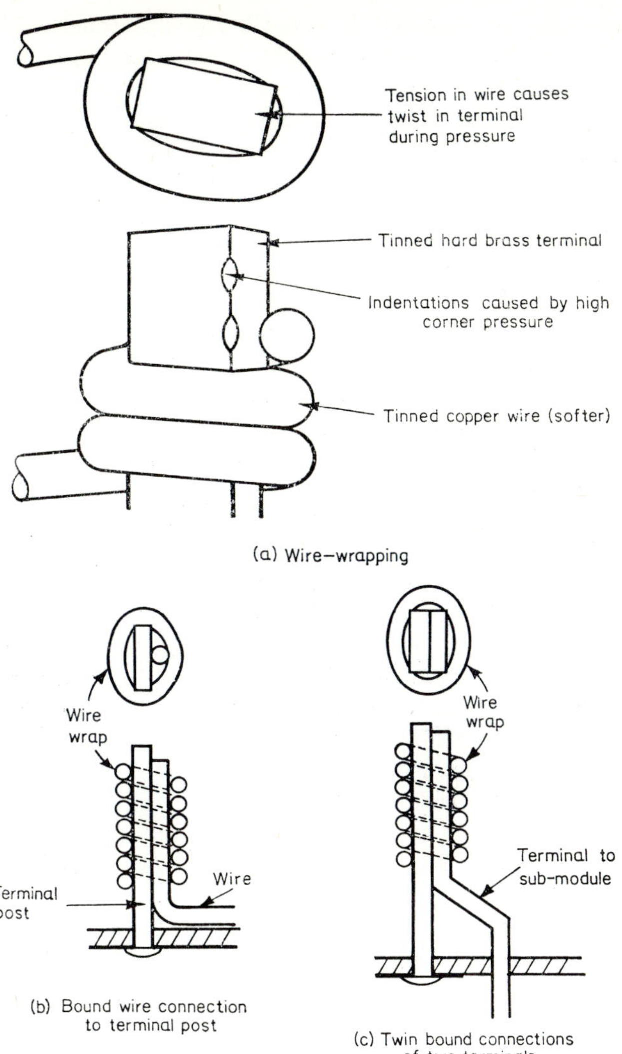

Fig. 2.12

An acceptable compromise is a contact pressure of at least 20 gm per contact with an incremental rate not more than 50 gm/in.; the withdrawal pressure should not exceed 170 gm per contact. This limits the practical number of contacts to about 20 per module.

One good design has a turn-screw which compresses all the contacts simultaneously after the module has been inserted; the contact pressure is released by the turn-screw before the module is withdrawn.

Another problem is the leakage between adjacent contacts. This is usually negligible where the temperature and humidity are controlled, as is generally the case with computers, but in some humid locations it is possible for moisture to condense on a cold module. In industrial atmospheres this can cause the creepage resistance to fall to a value comparable with resistances in the static circuit and hence to change the calibration of the relay. This can be avoided by air-conditioning or by increasing the creepage distance between contacts.

2.6.3. Semiconductors [37]

Semiconductors are inherently long-life devices but they are easily punctured by voltage spikes [16] and hence should be well protected by appropriate low-pass filters or traps. All circuits containing inductive or capacitative reactances should be carefully checked to see that circuit switching does not create any transient overvoltages across semiconductors. This subject is discussed in more detail in Chapter 5.

A high rate of change of voltage or heat may damage transistors, since it produces a high voltage or temperature gradient across their junctions. The duration is even more important than the steepness since it determines the amount of damage. For this reason an extremely short spike will not change the parameters of a transistor, even if strong enough to drive a hole through it. On the other hand, the energy involved in a spike of longer duration may change the characteristics of the transistor. Excessive heat may open-circuit internal connections of the transistor. Spikes may break down junctions, increasing leakage currents.

Transistors can also be damaged by high temperatures; hence the soldering of their leads should be done with great care, preferably using a heat-sink between the soldered connection and the transistor electrode. Furthermore, those semiconductor devices which are subject to continuous loading at or near their ratings should also have heat-sinks or fins. One alternative to soldering is micro-welding; this is used in space equipment but makes difficult the replacement of a component. A preferred alternative is flow-soldering.

The parametric failure rate of modern silicon transistors is about 0·05% per 1,000 hours, or perhaps 0 01% per 1,000 hours if run at less than half their rating. Parametric failure means going outside the specifications. The catastrophic failure rate is more like 0·001% per 1,000 hours (Table 2.4); this is the rate applying to properly designed protective relays in which transistors are used only for switching or output amplification and are under-rated. Modern diodes have even better failure statistics. Planar silicon diodes and transistors have a parametric failure rate of only 0·005% per 1,000 hours at rated load and many times better if under-rated.

Mechanical stress is another cause of damage. Transistor leads should not be bent sharply near the capsule unless the lead has been clamped between the bend and the capsule. The application of varnish to a printed circuit is no longer hazardous for transistors since they are now hermetically sealed.

All transistors should be operated well within their ratings [35, 38] because their life is inversely affected by their junction temperature. Heat soaking of semiconductors at a high temperature within 10°C of their junction failure temperature for several days before use offers two advantages:

(*a*) it eliminates those semiconductors which have manufacturing weaknesses;

(*b*) it minimizes any subsequent drift of their characteristics.

2.6.4. Resistors [24]

In all types of resistors those of high ohmic value are the most liable to failure. Running them at 20% of their rated watts decreases their failure liability many times but, in the case of wire-wound resistors, good humidity protection must be provided if the resistors are to be run cold. Metal oxide film and carbon composition resistors are the most reliable at present but the former are limited in ohmic value and are not of the precision type.

Wire-wound resistors are used up to about 3,000 ohms where stability (constant resistance) or high power is required; they are preferably coated with vitreous enamel to protect them against mechanical damage and electrolytic corrosion due to moisture in d.c. circuits. Metal film resistors are also subject to electrolytic corrosion but metal oxide films are very good in this respect.

Carbon resistors are satisfactory from 100 to 100,000 ohms where up to 25% increase in resistance due to ageing and a temperature coefficient of 0.17% per degree is permissible. Cracked carbon resistors are now available which have less than 2% increase in resistance due to ageing but they have a risk of open-circuiting unless given good protection against humidity because electromechanical corrosion of the spiralled carbon film can occur in the presence of d.c. and high humidity.

Potentiometers have the highest rate of failure of any component and hence should be very carefully chosen and checked. Wire-wound potentiometers are reliable only for low ohmic values below 3,000. Carbon track potentiometers are liable to open-circuit due to dust between the wiper and track, unless they are of the sealed type.

2.6.5. Capacitors [24]

Paper capacitors are used where an initial tolerance of ±20% in capacitance is permissible and 10% stability (long-term capacitance variation). Failure is generally due to the ingress of moisture but modern sealing methods are very effective and the use of hard wax and circular cans reduces the possibility of change of capacitance due to movement or distortion of the rolls of paper and foil.

Mica capacitors are used where lower power factor, higher voltage and greater stability are required; they have about 2% initial tolerance and 2% stability.

Plastic film capacitors are now available with high insulation resistance and stability of less than ½%. Their reliability depends upon the uniformity and porosity of the plastic film. The modern polystyrene type is considered very reliable.

Electrolytic capacitors use a liquid dielectric. They have had a bad reputation in the past, but reliable ones are now available. They are used in d.c. or low-frequency circuits where a high ratio of capacitance to volume is required. They have a large error at low temperatures and their initial leakage (after a period of de-energization) is high.

Tantalum capacitors are a very superior type of electrolytic capacitor which have similar but much better characteristics. They are practically leakproof and their ratio of capacitance to volume is an order higher than that of the ordinary electrolytic type, so that a 100 μF 30-volt capacitor is only 1 cm in diameter and 2 cm long. Their resistance is high, about 5/VC megohms (C is in μF). The latest solid tantalum capacitors have even better characteristics and reliability; also they cannot leak nor deform.

2.6.6. Reed Relays [82]

As with many other electronic components, reed relays were originally designed for applications requiring millions of operations with no long periods of inaction, whereas in protection the relay may be called upon to operate only once in ten years. The problem is aggravated by the fact that failure of a reed contact could have serious consequences such as the destruction of large electrical equipment, whereas in other applications, such as computers or telephone exchanges, an occasional failure is acceptable since it means only the loss of perhaps 0·1% of the capacity of the equipment.

The chief hazard with reed relays is failure of the seal. Reliability is theoretically ensured by making the capsule of glass which has nearly the same thermal coefficient of expansion as the nickel-iron contact members. Unfortunately glass is brittle and nickel-iron is hard, so that the seal can be damaged by a relatively small mechanical shock such as careless handling in the factory.

Most reed units are given a leak test before assembly into a protective relay and any units that have been dropped or bumped are either rejected or retested. The leak test is usually either a pressure test at normal temperature or immersion in boiling water at atmospheric pressure. A leak can be detected either by the insulation between contacts dropping below 100 megohms, by using a dye in the water, or by water vapour which shows up in colour inside the tube.

Simple reed units are given an operational test consisting of 10,000 operations at 50 V d.c. in an inductive circuit (L/R = 0 02) at a rate of 50 operations per second. The contact resistance after the test should not exceed 50 milliohms and the pick-up and drop-out currents should not have changed more than 10%.

Where reed relays are used for tripping they require additional tests for short-time current closing and carrying ability.

2.6.7. External Interference [18, 19, 71]

Transistors are replacing thermionic tubes because of their small size, because no filament supply is needed with transistors and because they are more robust. On the other hand, transistors are much more vulnerable to reversed voltage and voltage spikes. Whereas a thermionic tube requires only resistors in series with its grids, transistors require much more elaborate protection, as explained in Section 5.2, Chapter 5.

It will be obvious from Fig. 3.10 and 3.11 in Chapter 3 that reversed collector-emitter voltage could burn out the transistor junction; however, the transistor can be protected against accidental reversal by clear marking of terminals and by the use of protective diodes of appropriate polarity to block or by-pass current flowing into the relay circuit in the wrong direction.

Overvoltage would not occur with a properly designed supply circuit but could be caused by external interference (see Chapter 5).

In large power stations and substations the leads connecting the c.t's to the relay may be a half mile or more long and a lightning stroke to nearby equipment could easily create a voltage spike which would go through the interwinding capacitance of an auxiliary transformer in the relay and destroy transistors. This can be avoided by using a copper screen around the primary winding of the auxiliary transformer and not only grounding the screen but connecting it to a point on the case as near as possible to the auxiliary transformer.

If the d.c. supply is derived from the station battery, voltage spikes can easily be caused by switching contacts in inductive or capacitative circuits [16, 19]. Hence it is necessary to protect the transistor circuits by connecting a diode or a low-pass filter across the d.c. supply terminals of the relay to absorb and short-circuit such transient impulses before they enter the relay. These spikes have undoubtedly always been present but electromagnetic relays have been protected against them by the inductance of their coils and their 2 kV insulation level.

In applying protective capacitors and grounded screens for transformers it must be realized that, at the very high frequency represented by the steep wave fronts of voltage spikes, a very small inductance such as that caused by a bend in a wire, or the proximity of iron, may result in a quite high voltage drop which may operate a transistor inadvertently or even damage it. Such spikes were ignored in electromagnetic relays because they were too short to operate the relay and well within the insulation voltage level of the insulation.

2.6.8. Personnel

Because of the colour coding of components, colour-blind operators should not work on static circuits. Furthermore, colours like violet tend to fade and to be confused with another colour by a person only slightly colour-blind.

Operators with a tendency to perspire should not work on static relay circuits. Air-conditioning is desirable in hot countries and the hands should be kept dry by periodic wiping on a rag or holding them over an air jet.

2.6.9. Redundancy Technique

As stated in Section 2.6.2, if a unit circuit or a component has one chance in a thousand of failure, there is only one chance in a million of failure if two of the units or components are worked in parallel. This is of course impractical in the use of components like resistors unless a 2: 1 change in value, due to the open-circuit failure of one, is acceptable; but it is perfectly feasible in most cases where a capacitor, transistor or a blocking diode is used and it is often possible to duplicate the whole functional unit of the circuit where maximum reliability is required. Furthermore, it is possible in some cases to duplicate a whole relay and this of course has been done in the past with electromagnetic relays and their attendant c.t's and p.t's and trip coils.

An example of this technique is in bus protection in England, where the C.E.G.B. has used three sets of static relays each having two output circuits. The latter are connected in three parallel pairs, each pair being from two different relays, i.e. connected 1-2, 2-3, 3-1 in parallel. This means that tripping will not be prevented by the failure of one relay and that tripping will not result from wrong operation of one relay.

2.7. TESTING

This is sometimes more difficult with static relays because they usually have very low power inputs and outputs so that the insertion of a meter would upset their calibration. For this reason it is preferable to test each unit of the circuit in a master circuit which corresponds to the rest of the relay.

Even more difficult is the problem of insulation testing because of the vulnerability of transistors to high voltage. No specifications for insulation testing have yet been agreed but many manufacturers are using 1,500 volts (at system frequency) between the case and all the terminals; it is essential to make sure that *all* the terminals are connected together because the omission of one may impress a transient voltage on a transistor circuit which would damage the transistor. An initial d.c. test with a high impedance source can avoid the destruction of semi conductors if a breakdown occurs.

2.8. STATIC RELAY CASES

Static relay circuit components are so small that a number of relay units, formerly in separate cases, can all be contained in a single case. Plate 2 shows complete distance protection for all faults (switched mho) in a case only 10.5 by 17.4 inches.

Such a case is shown in fig. 2.13. This case conforms to internationally accepted standard rack dimensions and provides facilities for testing.

The cases can be stacked in a cabinet (fig. 2.13c) and the interconnection means between modules and between cases are simple and minimise transient interference. Interlocks are provided to prevent modules from being extracted before the main connector is removed, de-energising them.

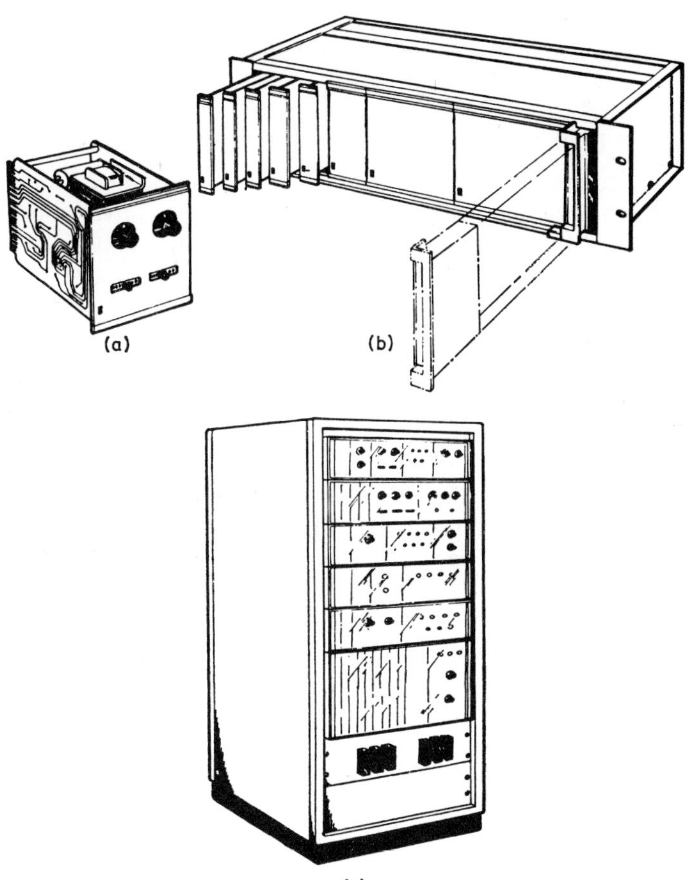

Fig. 2.13. Static relay mountings (E.E.Co)
(a) typical module
(b) modules mounted in case
(c) cases mounted in cabinet

3

Static Relay Tools

Simplified semiconductor theory. Types of semiconductor devices, Amplifier circuits using transistors. Level detectors, time delay circuits. Temperature and voltage compensation. Smoothing circuits and harmonic filters, Measurement circuits. Logic circuits

3.1. TRANSISTOR EQUIVALENTS OF RELAYS [33, 34]

The first semiconductor diode was the galena crystal used in early radio sets. The tungsten or platinum 'cat's whisker' was positive to germanium and negative to silicon. The point contact crystal was better for h.f. than the junction type but less consistent and more noisy for lower frequencies.

The transistor was discovered by Barden and Bratten in 1948 and circuitry was rapidly developed for it by W. Shockley. All three were in the Bell Telephone Laboratories [91].

To the average relay engineer without any electronic experience or training, a transistorized relay diagram is very difficult to understand at first but, if one considers that 95% of the transistors in protective relay circuits are used merely as auxiliary relays and level detectors, it is only necessary to picture the correct type of relay equivalent to be able to follow the relay circuit diagram. For this purpose a transistor can be regarded as a very sensitive high-speed relay which is polarized either magnetically or by series rectifiers.

Figure 3.1a shows a typical n-p-n transistor circuit and its relay analogy. If the input is such as to make terminal B positive relative to terminal E, the transistor will conduct and direct current will flow through the load via the collector and emitter. With an a.c. input the transistor will conduct only during the positive half-cycles and square impulses of d.c. will flow in the load. Both these conditions are easily followed in the relay analogy.

A similar analogy can be drawn up for the p-n-p transistor where the circuits are the same except that all polarities are reversed (see Fig. 3.1b). The arrow on the emitter electrode indicates the direction in which the junction is conductive.

In relay diagrams the positive supply busbar is usually at the top of the diagram and the negative busbar at the bottom. The contacts are usually

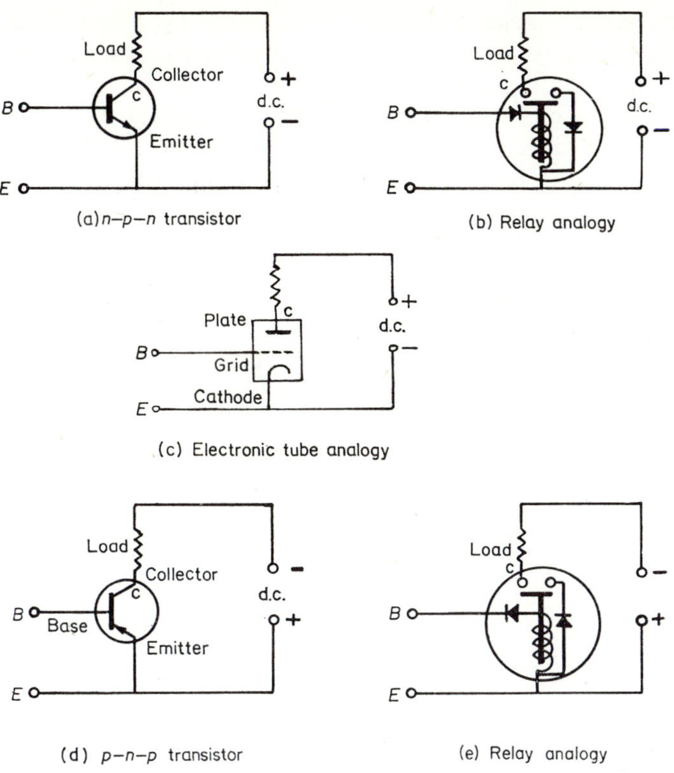

Fig. 3.1. Relay analogy for transistor operation

at the positive end and the coils at the negative end so as to minimize corrosion of the coil wire due to leakage currents. This favours the use of p-n-p transistors.

In transistor diagrams the input signal reference busbar is usually at the bottom of the diagram and can be either positive or negative according to whether a p-n-p or n-p-n transistor is considered. Since there are three electrodes for the input and output terminals, one electrode must be common to both; the common emitter connection shown in Figs. 3.1a and 3.1b is the one most used in relays because it gives the highest power gain.

Where a second transistor follows the first for amplifying or triggering purposes, it has to be connected so as to make use of the changes in potential in the circuit caused by the first transistor becoming conductive. For instance, in Fig. 3.2a the point C becomes less positive when transistor T_1 conducts; hence if T_1 is a p-n-p, T_2 must be an n-p-n in order to make T_2 conduct.

On the other hand, if it is desired to use two transistors of the same type (n-p-n), one of them (T_2) must be normally conducting as in the circuit shown in Fig. 3.2b. Here the output is normally short-circuited by transistor

T_2 and the output voltage is only the drop across T_2. When T_2 is cut off by the input signal through T_1 becoming conductive, the output voltage appears.

The snap action of electromagnetic relays is obtained in transistor circuits by positive feedback from the output to the input of T_1. Negative feedback is of no value in protective relays because they require a strong tripping signal directly the threshold of operation has been crossed.

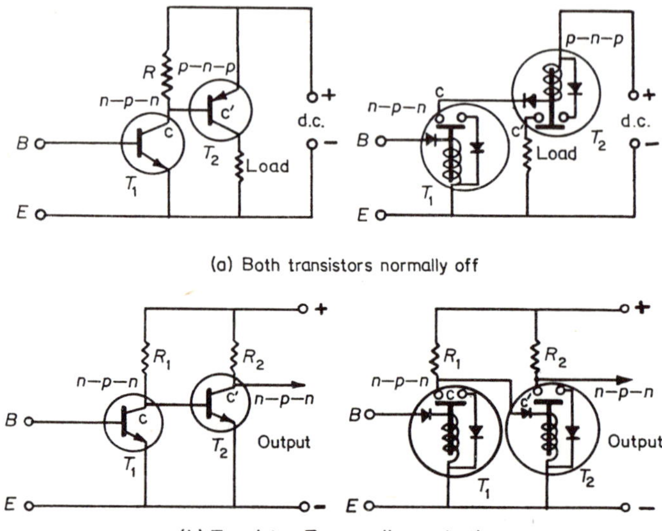

(a) Both transistors normally off

(b) Transistor T_2 normally conducting

Fig. 3.2. Addition of amplifying stage

Figure 3.3 shows how the positive feedback can be applied to the circuits of Figs. 3.2a and 3.2b through the resistor R_f. In Fig. 3.3a T_1 energizes T_2, as in Fig. 3.2a, and the point C' becomes more positive; this also makes B more positive through R_f so that T_1 is made fully conductive although the input may be only just above the threshold of operation.

Reference to the relay analogue of Fig. 3.3a shows exactly similar action. When relay T_1 operates, T_2 picks up and the load current makes C' more positive, which seals in T_1.

In Fig. 3.3b, which corresponds to Fig. 3.2b, T_2 is normally conducting while T_1 is off. When T_1 is made conducting by a positive signal at B it cuts off T_2 as before and the point D is then made more negative because it no longer has the IR drop in R_f due to the T_2 current. This again biases T_1 more in the conducting direction and again we have a mono-stable action.

Referring to the relay analogue of Fig. 3.3b, when relay T_1 picks up it short-circuits the coil of relay T_2 and the reduction of current in R_f makes the point D less positive, i.e. more negative, so that a larger voltage is applied ot T_1 making it more firmly in the operated position.

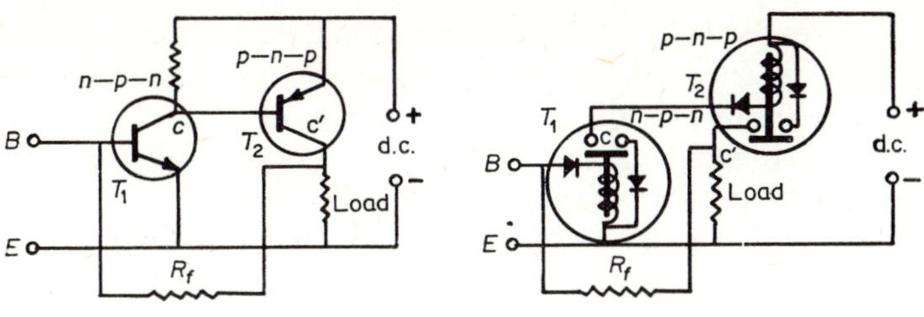

(a) Positive feedback added to fig. 3.2a

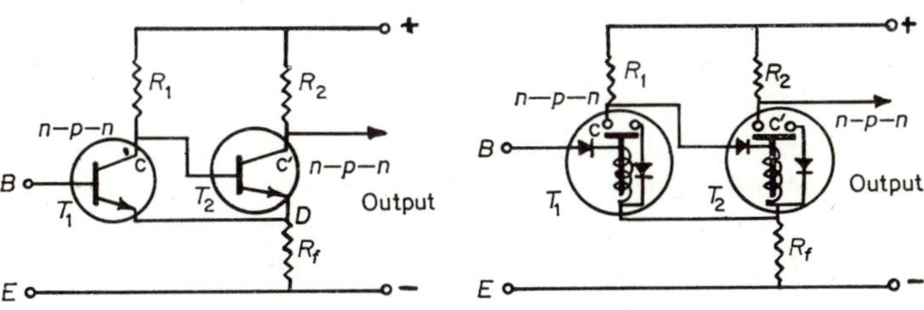

(b) Positive feedback added to fig. 3.2b

Fig. 3.3. Positive feedback for snap action

The other main use of transistors is as level detectors. If there is a reverse bias potential between the base and emitter, the input signal must exceed this in order to make the transistor conduct (see Overcurrent Relay in Fig. 1.5, Chapter 1). For instance, in order to use a transistor as an overcurrent relay, it is only necessary to turn the a.c. into a d.c. voltage and apply it to the base emitter circuit. When the a.c. exceeds a certain level, the rectified input voltage will exceed the bias voltage and it will conduct.

Figure 3.4 explains the operation of a thyristor (SCR). In the thyristor circuit (Fig. 3.4a), when the switch S is closed, putting a positive signal on

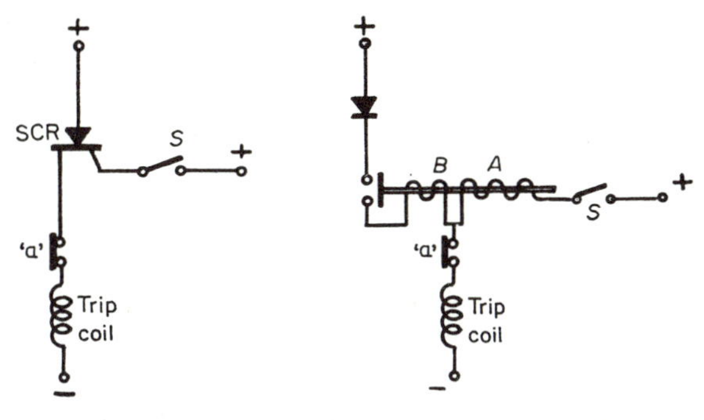

(a) Thyristor (b) Relay analogy

Fig. 3.4. Relay analogy for a thyristor
The thyristor is a bi-stable device

the gate, it makes the *SCR* conductive just as in the relay analogy (Fig. 3.4b) the closing of *S* picks up the relay and makes the diode circuit conductive. The relay is held in by the coil *B* so that opening *S* has no effect; this corresponds to the avalanche effect in the thyristor which keeps the trip coil energized until the breaker auxiliary switch *a* is opened.

The foregoing provides a very rough explanation of transistor and thyristor operation in protective relays. A more sophisticated explanation is given in Section 3.2.

3.2. SEMICONDUCTOR ELECTRONICS [91]

A semiconductor is a material which has a very high resistivity compared with that of most metals. In metals the resistivity increases linearly with the temperature Absolute; in the semiconductors with which we are concerned the resistivity is non-linear and is affected enormously (in decibels) by heat, light, magnetism and electric current. These effects are used for switching, amplification and measurement in electrical circuits.

There are two main types of semiconductor, the '*N*' type and the '*P*' type. In the *N*-type, electrons (negative current carriers) predominate; in the *P*-typye, holes (positive current carriers) predominate. The predominating carriers are called majority carriers and the others minority carriers. A hole has a positive charge equivalent to the negative charge of an electron so that, if it combines with an electron, there is no resultant charge and neither carrier is then available to carry current.

Semiconductors can be made of a number of metals, copper, selenium, germanium, silicon, etc., but the last two are the most common, with silicon tending to replace germanium for reasons to be explained later. These materials are processed to form either *P*-type or *N*-type semiconductors by controlling their impurity content. The basic semiconductor is refined to reduce the impurity level to less than 1 part in 10^{10} and a carefully controlled impurity (called a doping agent) is introduced to give the required *P*-type or *N*-type; arsenic, antimony, gallium and aluminium are typical doping agents.

To create an *N*-type semiconductor the doping agent should have a valency higher than that of the basic material; for example, arsenic has a nucleus and five valency electrons per atom compared with four for germanium. For the *P*-type the impurity should have a lower valency; for example, gallium has three valency electrons.

In the first example four of five valency electrons of the arsenic atom are strongly attracted to and fit into the diamond lattice of the surrounding germanium, but the fifth electron is only lightly held and is readily displaced, i.e. there can be movement of these extra electrons which means that current can flow. Since arsenic donates electrons it is called a donor impurity and the resultant semiconductor is of the *N*-type.

In the second example the three valency electrons of the gallium atom fit into the diamond lattice only with the help of an electron from a nearby germanium atom. The vacancy (hole) produced by this loan creates an unstable condition which permits electrons (current) to flow in the other direction and the material is *P*-type. Gallium is obviously not a donor impurity and is called an acceptor impurity.

The relation of electron flow to hole movement is best illustrated by the analogy of people moving up in a row of seats. If there is a vacancy at one end the person next to the vacancy moves into it, leaving a vacancy on the other side of him which is filled by the next person moving up, and so on. In this way it can be seen that the vacancy (hole) moves in the other direction from the people (electrons).

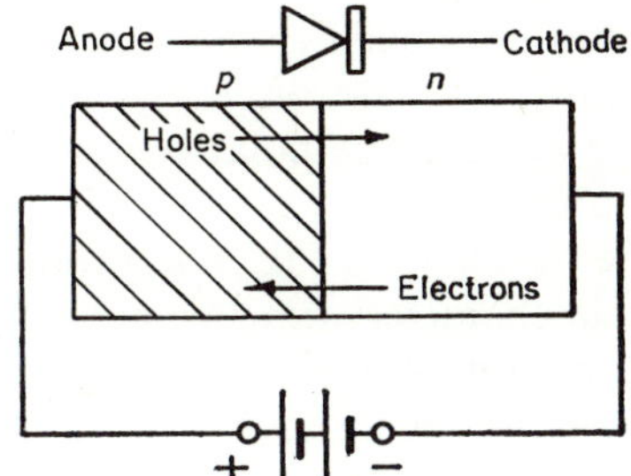

Fig. 3.5. *p-n* junction

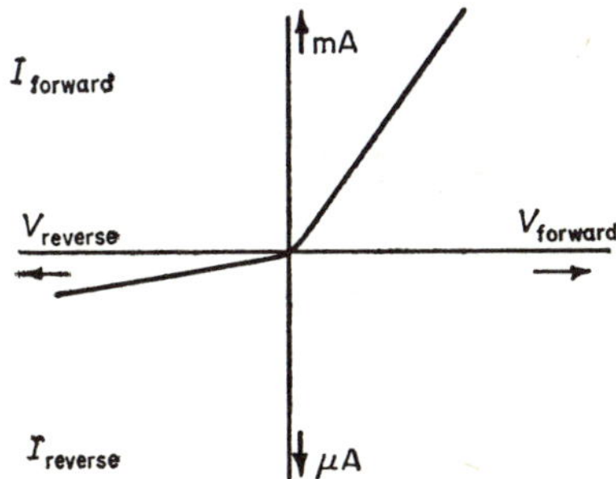

Fig. 3.6. Rectification effect of *p-n* junction

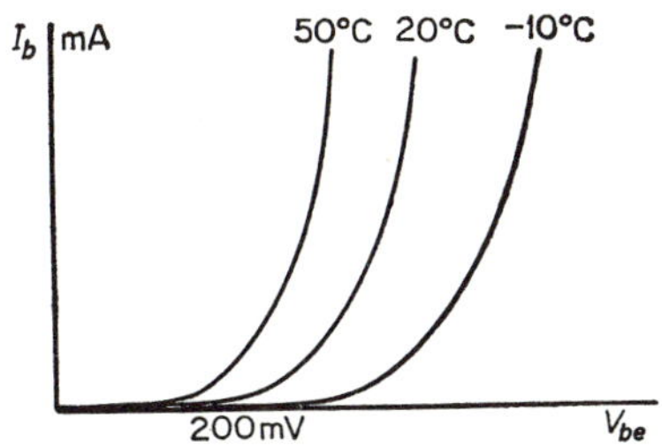

Fig. 3.7. Effect of voltage and temperature on *p-n* junction biased in reverse direction

3.2.1. Diodes [48]

If a piece of *P*-type metal is fused to a piece of *N*-type metal so that the crystal lattice is continuous, the junction has a rectifying characteristic (Fig. 3.5); it also has a non-linear resistance as shown in Fig. 3.6. Its resistance falls rapidly with increasing temperature (Fig. 3.7) or the incidence of light (Fig. 3.8).

If a source of e.m.f. is connected to the *p-n* junction, in the polarity shown in Fig. 3.5, then the 'spare' holes in the *P*-type region are drawn easily to the negative pole and the electrons are drawn from the *N*-type to the battery positive. If the battery potential is reversed, then the holes are repelled from the positive pole and likewise the electrons from the negative pole. Consequently, little current flows until such time as the electric stress is so high that a process similar to an electric discharge occurs. Hence, a current characteristic for forward and reverse polarity is as shown in Fig. 3.6 and has definite rectifying properties.

If a double junction be made, as in Fig. 3.9, a transistor is formed which has even more remarkable characteristics. An explanation of the theory of the operation of the transistor follows and a reference will be made to thyristoɪs which have three *p-n* junctions.

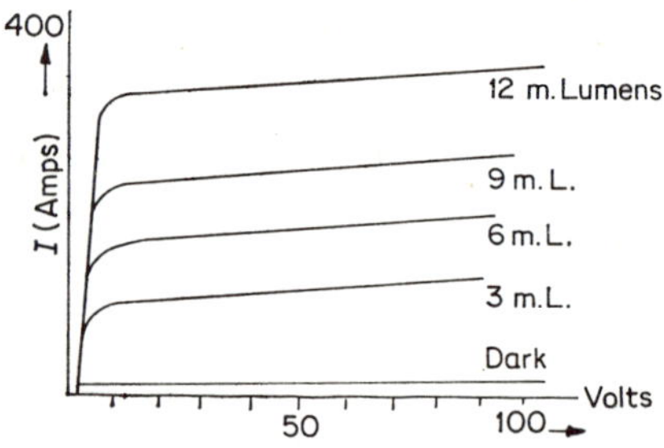

Fig. 3.8a. Typical junction photo-diode characteristics

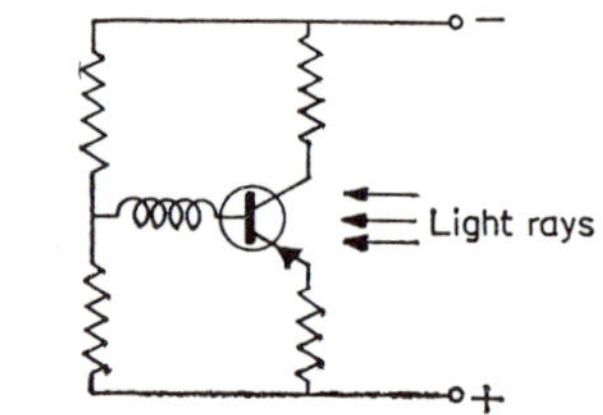

Fig. 3.8b. Typical photo transistor circuit

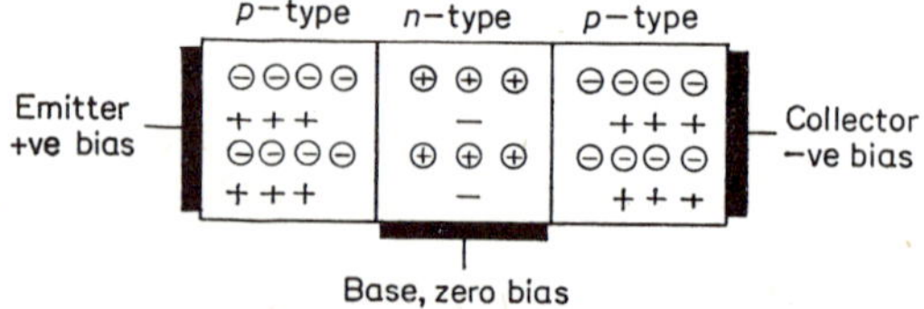

Fig. 3.9. *p-n-p* transistor

3.2.2. Transistors [33, 34]

A three-electrode junction transistor comprises three layers of semiconductor materials, the central thin layer of which is the *base*, and the two outer layers of the sandwich, as shown in Fig. 3.9, are the *emitter* and the *collector*. Each electrode may be either *P*-type or *N*-type semiconductor, as shown in Figs. 3.10 and 3.11.

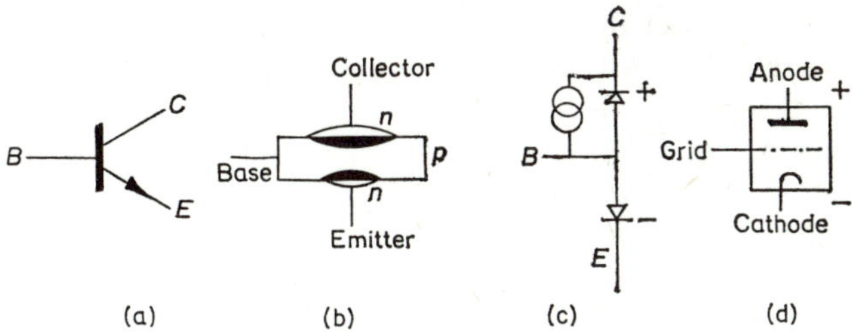

Fig. 3.10. *n-p-n* silicon transistor

(a) is schematic symbol
(b) is physical arrangement
(c) is electrical equivalent in terms of potential barriers
(d) is electronic tube equivalent

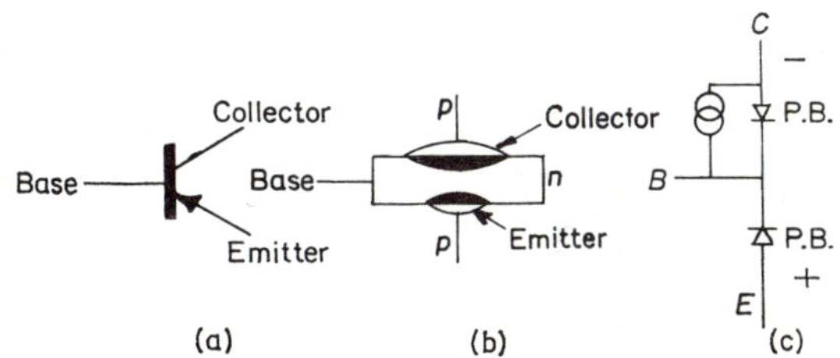

Fig. 3.11. *p-n-p* germanium transistor

(a), (b) and (c) are as in Fig. 3.10

For the practical user the operation of a transistor circuit can be likened to that of an electronic tube (valve). In Fig. 3.10 the emitter of an *n-p-n* transistor corresponds to the cathode of the valve, the base to the control grid and the collector to the anode (plate).

Unless the control grid has the requisite bias relative to the cathode, electrons will not pass through the grid to the anode; similarly in the *n-p-n* transistor the base must be positive relative to the emitter before electrons will flow through the base to the collector; the collector in turn must be still more positive to collect the electrons. Varying the base-emitter current will similarly vary the collector current.

As in the grid of the electronic valve, the electrons by-passed into the base are only of the order of 2% of those going on to the collector, i.e. the collector current is much greater than the base current and this results in current amplification (see Fig. 3.14a).

If we substitute holes for electrons and reverse all the voltages, we have the conditions for operation of the *p-n-p* transistor (see Figs. 3.11 and 3.12a). The direction of the arrow on the symbol for a diode, or on the emitter of a transistor, indicates the direction of the current to make the device conductive. The direction of the current is the same as the direction of the holes, i.e. the opposite direction from the electrons.

Referring to Figs. 3.10c and 3.11c, an alternative explanation of transistor operation is to consider the transistor as a pair of rectifiers joined back-to-back in a 3-wire circuit with a very thin common electrode, so that forward current in one rectifier affects leakage current in the other rectifier which is reverse-biased. This reverse current is the output current of the transistor.

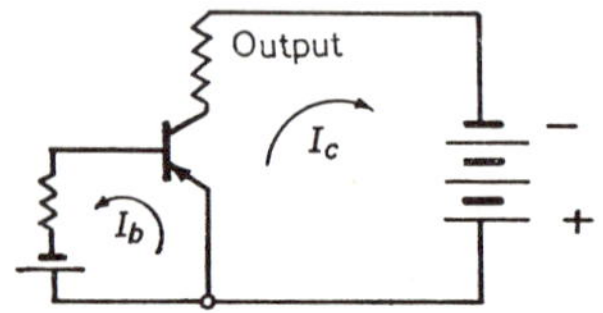

Fig. 3.12a. Polarities of *p-n-p* transistor circuit requires base to be negative relative to emitter in order to pass collector current

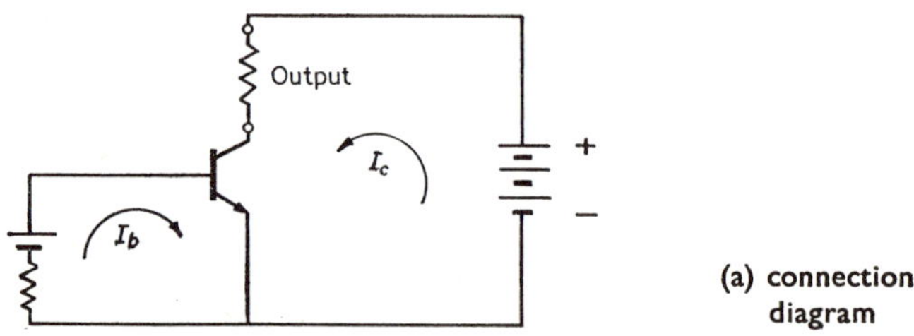

Fig. 3.12b. Polarities of *n-p-n* transistor (reversed from *p-n-p* transistor)

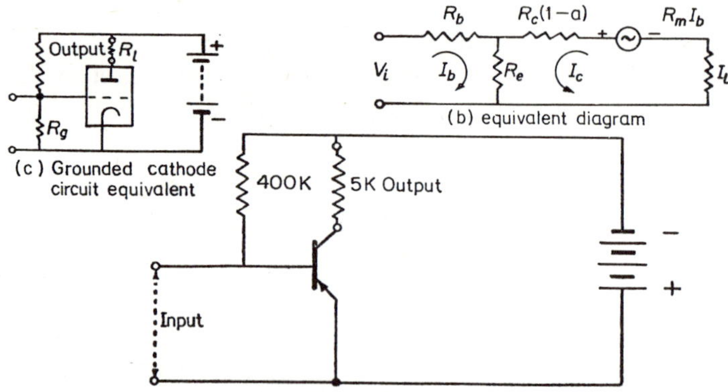

Fig. 3.13. Grounded emitter circuit: Alternative method of biasing *p-n-p* transistor to conduct I_c

Note: *n-p-n* transistor likewise except all polarities reversed

It is always within a few percent of the Forward current. The rectifier junction that carriers forward current is the emitter-base junction and the reverse-biased junction is the collector-base junction.

The collector current increases linearly with the emitter or base current but is negligibly affected by the collector voltage (Fig. 3.14b); it will be seen that the collector (output) current is in mA while the base (input) current is in μA, so that there is considerable current amplification. The curved line in

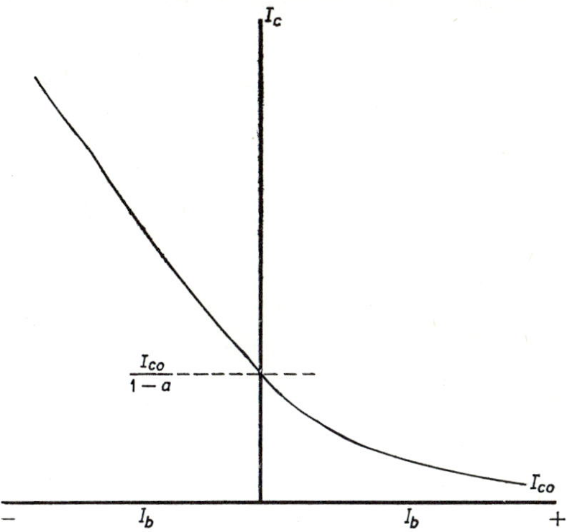

Fig. 3.14a. Collector current versus base current in the grounded emitter circuit

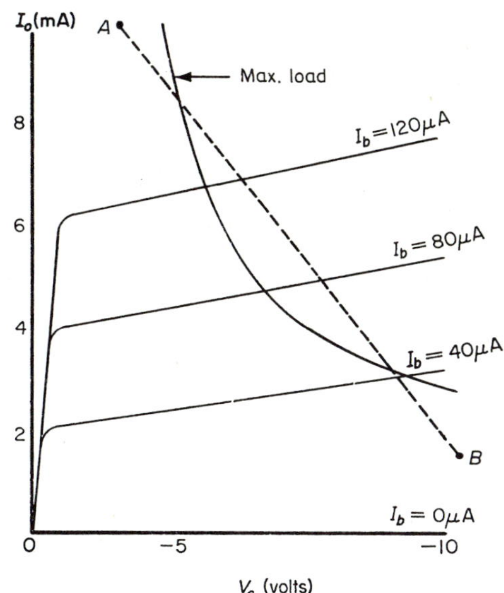

Fig. 3.14b. Collector current versus collector voltage in the grounded emitter circuit

Fig. 3.14b is the maximum power line, i.e. any continuous values to the right of this line would overload the transistor. The dotted line shows the load that can be switched from Condition A to Condition B; this does no harm because the switching is almost instantaneous, hence the I^2Rt value is negligible.

The voltage between the emitter and base varies between zero volts and the forward drop of a rectifier but, between collector and base, it varies from the supply voltage value (when the transistor is non-conductive) to a fraction of a volt as the base current is changed from zero to the saturation

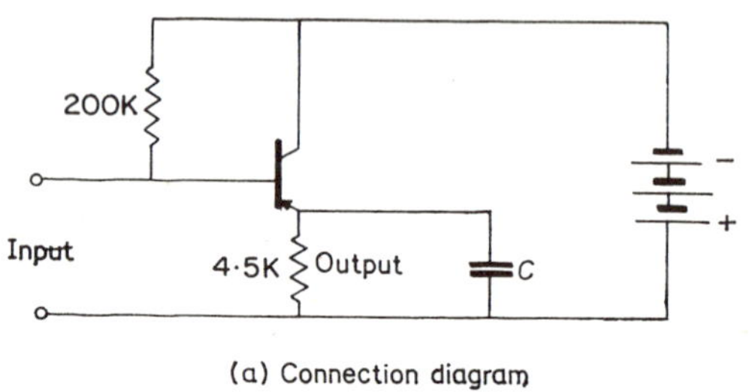

(a) Connection diagram

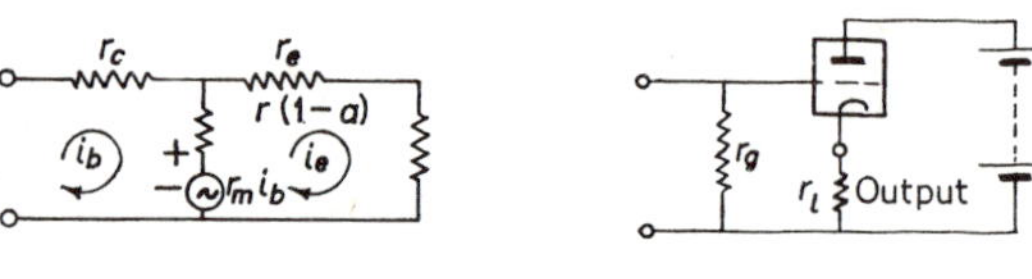

(b) Equivalent diagram (c) Cathode follower circuit

Fig. 3.15. Grounded collector circuit

Similar to Fig. 3.13 except load resistance in emitter circuit instead of collector. This is like a cathode follower circuit and gives high input and low output impedance

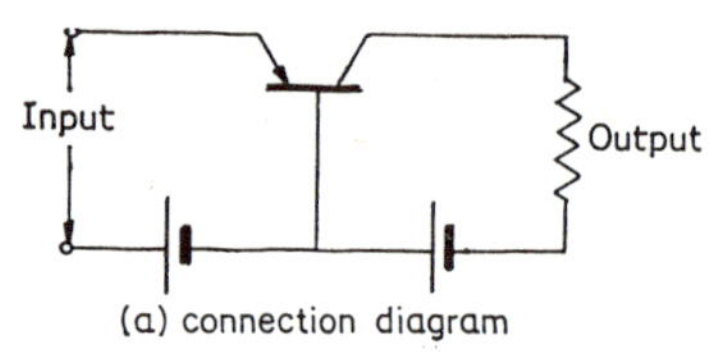

(a) connection diagram

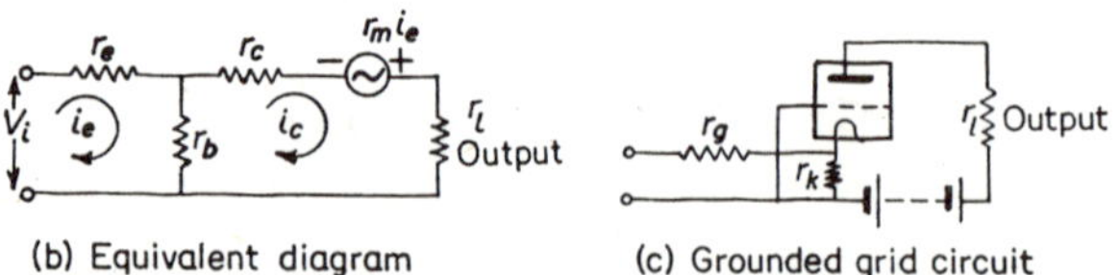

(b) Equivalent diagram (c) Grounded grid circuit

Fig. 3.16. Grounded base circuit

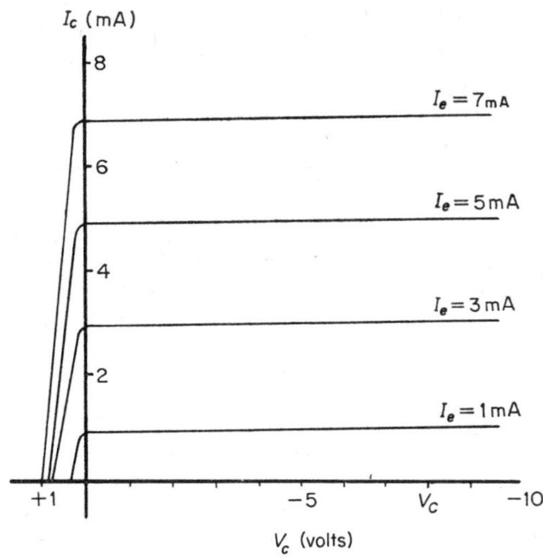

Fig. 3.17. Collector current versus collector voltage in the grounded base circuit

level. This voltage amplification, together with the current amplification, makes the transistor a high-gain power amplifier in a grounded emitter circuit (Fig. 3.13).

3.2.3. Thyristors [35, 46]

Whereas a rectifier (diode) has one *p-n* junction and a transistor (triode) has two *p-n* junctions, a thyristor (tetrode) has three *p-n* junctions (Fig. 3.18a). It acts like two transistors in tandem, one *p-n-p* and one *n-p-n*. With the polarities shown in Fig. 3.18a, injection of some holes into the base (P-gate) of the lower transistor will cause current to flow in its *N*-collector which enters the *N*-base of the upper transistor and causes a much larger electron current to flow in its *P*-collector (anode) and this can only flow as hole current back into the *P*-base of the lower transistor. When this return current reaches the same value as the current originally fed into the gate, (which is a matter of microseconds), the action will be self-sustaining even though the gate current be removed and the anode current will rise until limited by external impedance.

This action is similar to that of a thyratron, i.e. the thyristor can be triggered by a very brief signal and the current continues to flow until interrupted by a switch in the anode circuit (e.g. circuit-breaker auxiliary switch in a trip-coil circuit) or a reverse bias. Figure 3.4 shows a relay analogy which was explained at the end of Section 3.1.

Figure 3.18b shows the anode current/voltage characteristic for different gating currents. With no gate current the characteristic is like the reverse characteristic of a diode in either direction. In the forward direction there is one (middle) *p-n* junction; in the back direction there are two in series

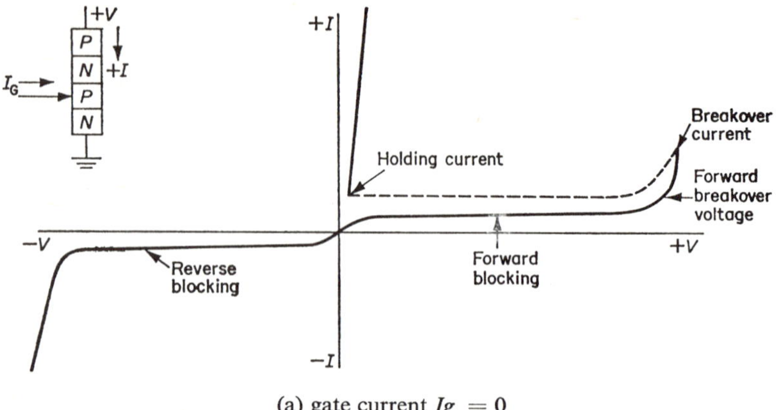

(a) gate current $Ig = 0$

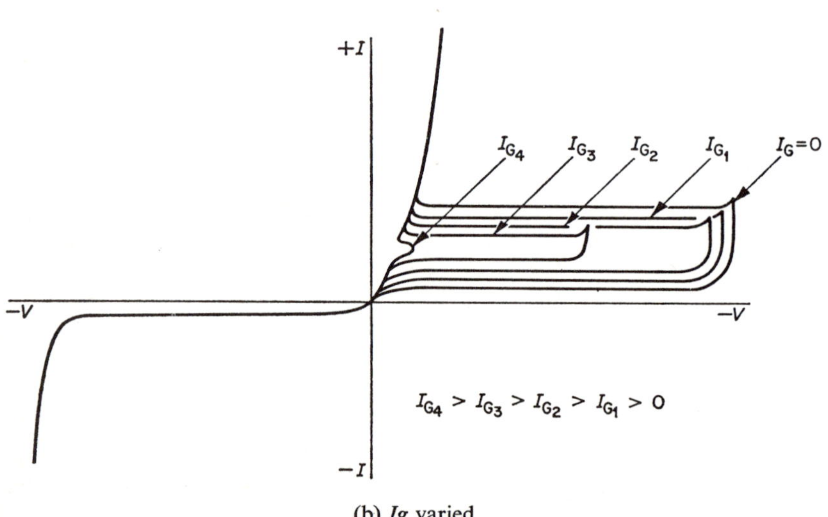

(b) *Ig* varied

Fig. 3.18. Static characteristic of a thyristor

(Fig. 3.18a). With increasing gate current the triggering voltage decreases until, at a very high gate current, the characteristic is the same as that of a p-n rectifier. To energize a trip coil, a positive bias can be applied to the gate to reduce the signal necessary to effect tripping.

Because of its high speed a thyristor is easily triggered by a small voltage signal (5 V, 0·1 mA). To prevent wrong operation it is customary to design the circuit so that the tripping signal will not trigger the thyristor unless it lasts for at least 2 ms. This prevents operation on interference spikes since they last only microseconds.

3.2.4. The Field Effect Transistor (Technetron) [47]

The closest semiconductor equivalent of the electronic vacuum tube is the technetron. Unlike the transistor, it is a high-impedance device with characteristics like those of a pentode tube. It consists of a small rod of *N* germanium, one end of which is the anode and the other the cathode. Near the middle is a deep annular groove around which is an Indium ring forming the control grid (Fig. 3.19a).

A few volts negative bias on the grid blocks current between anode and cathode by virtue of the field effect. Because of its high impedance (megohms) it is a voltage amplifier like a vacuum tube so that, when used in an *R-C* timing circuit, less capacitance is required to produce a given time delay. It is particularly useful for high frequency. because, unlike the transistor, mutual conductance increases with frequency. At 110 Mcs it is 22 db for a bandwidth of 1·7 Mcs; at 430 Mcs the gain is 9 db for a bandwidth of 30 Mcs.

Another similar field effect device uses a metal oxide planar silicon transistor (M.O.S.T.) and has an input impedance of the order of a million megohms (Fig. 3.19b). The drain, source and gate are the anode cathode and grid respectively.

3.2.5. Germanium versus Silicon Transistors

The first transistors were made of germanium but, when satisfactory silicon transistors became available, they offered the advantages of (*a*) being able to

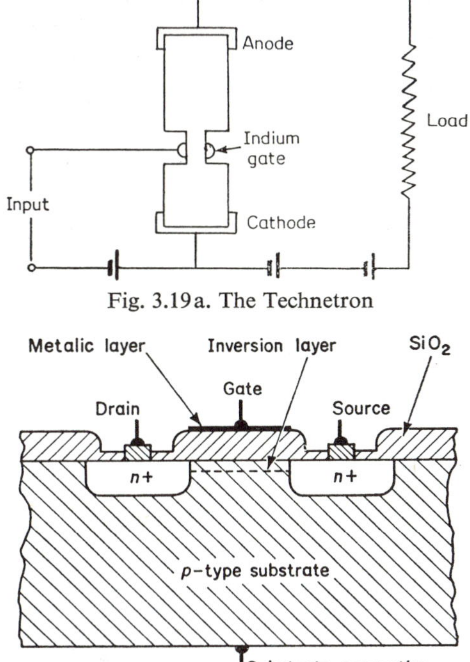

Fig. 3.19a. The Technetron

Fig. 3.19b. The metal oxide planar transistor

stand much higher ambient temperatures, (*b*) having less effect of temperature upon characteristics (*c*) having lower leakage currents (see Section 3.4.1), (*d*) being operable at high collector voltages and high power ratings. On the other hand, germanium transistors are still used for low V_{be} and V_{ce} saturation voltages, provided that the temperature conditions are not too arduous, and also for high power (*p-n-p*) where silicon transistors are sill too expensive.

The forward bias voltage across the base-emitter junction which is necessary for conduction is greater for silicon than for germanium. For applied voltages of less than 400 mV the junction may be considered to be reverse-biased. When fully conducting the base-emitter voltage V_{be} tends to be constant about 0·6 V for a silicon transistor. For a typical germanium transistor 0·2 V forward bias is required. In both germanium and silicon transistors V_{be} changes at a rate of about 2·5 mV per degree *C*, but the permissible range of operating temperatures is much greater for silicon than for germanium.

The grounded-base collector slope resistance, r_c, at ordinary operating voltages is lower for typical silicon transistors than for similar germanium types. At low values of collector-base voltage, however, r_c is appreciably greater in silicon than in germanium.

3.3. MODERN TRANSISTOR CONSTRUCTION

The alloy type of construction used for germanium transistors was not applicable to silicon transistors because of the difference of the thermal expansion coefficient of silicon and aluminium. The diffused method of construction used for silicon transistors is considered to be more efficient and is more controllable. The diffused construction of Fig. 3.21 led to the epitaxial construction of Fig. 3.22.

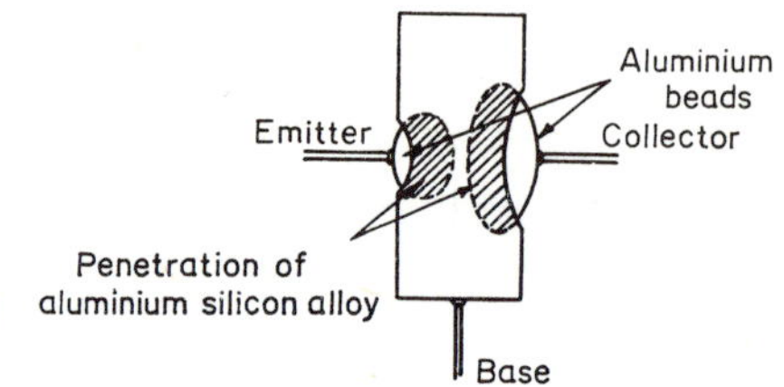

Fig. 3.20. Section through alloy type *p-n-p* transistor

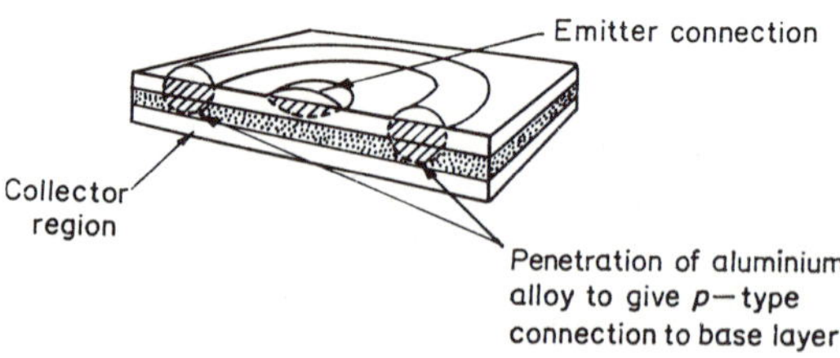

Fig. 3.21. Construction of diffusion type silicon transistor showing connections to emitter and base

Planar silicon transistors are the most sophisticated type no wavailable. They offer good reliability because their junctions are covered by a layer of silicon oxide which theoretically prevents molecular diffusing of impurities from outside; hence they are inherently leakproof. Furthermore, they are more robust, have extremely low leakage currents, low capacitance, higher power dissipation and a higher gain at very low currents.

Epitaxial means a layer construction with a thick layer of low resistance (collector) silicon and a thin layer of high resistance silicon into which the emitter and base are diffused (Fig. 3.22). This provides low leakage, low offset voltage and low saturation resistance.

Mesa is Spanish for 'table' and refers to the cutting away of the collector silicon layer (Fig. 3.23) leaving a flat top or table; this is to reduce collector-base capacitance and hence improve v.h.f. performance; it is also advantageous for low-input applications.

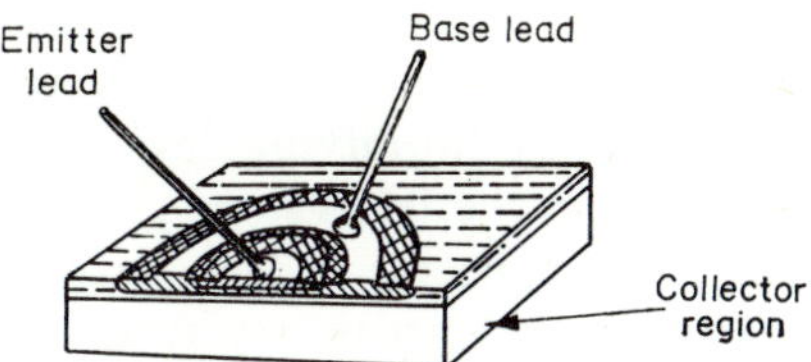

Fig. 3.22. A typical section through the planar epitaxial construction

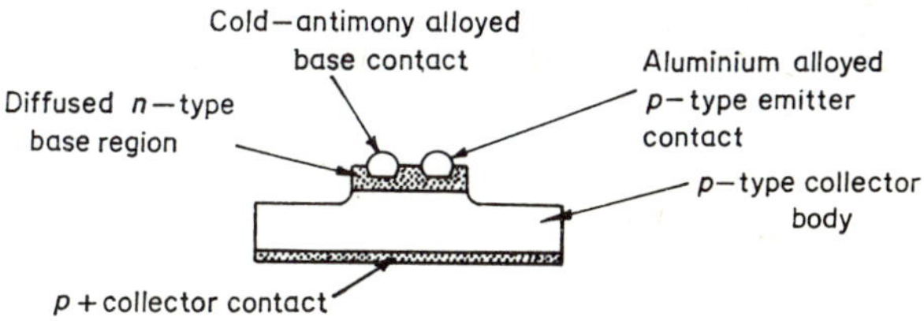

Fig. 3.23. Structure of diffused-base mesa transistor

Since the epitaxial and mesa transistors also have their junctions covered with silicon oxide they also have low leakage The manufacturers also claim that the characteristics of all three types are not only more controllable but more stable, so that the relay would be correspondingly more accurate and stable.

All these devices are made of silicon, which is safer than germanium in most practical applications because the junction will easily stand the maximum temperatures necessary for relay applications. They also have a higher voltage rating and better characteristics.

3.4. TRANSISTOR CHARACTERISTICS

For those who are content to regard a transistor as a polarized relay in order to understand static relay circuits the analogies in Section 3.1 and

3.2.2. will be sufficient. For those who are interested in a more detailed explanation of transistor electronics the following brief discussion may be useful.

3.4.1. Basic Operation [33, 34]

The most important characteristic of an electronic tube (valve) is the transfer characteristic which shows the variation of anode (output) current with control-grid (input) voltage and, since this is linear over a considerable range, it is expressed in mA per volt.

The corresponding transfer characteristic of the transistor shows the linear variation of collector (output) curient I_c with the current in the input electrode (base or emitter). Voltage is not involved because the resistance of semiconductors is non-linear and the ratio of the currents (and/or voltage) would be indeterminate.

Where there is a voltage input there must be sufficient impedance in the input circuit (source impedance) to swamp the varying input impedance of the transistor, since it is a current amplifying device.

As indicated by the arrow in the transistor symbol, input current flows into the emitter and out of the collector and base of a p-n-p transistor; the directions of these currents and the arrow are all reversed in an *n-p-n* transistor. Hence,

$$I_e = I_b + I_c \tag{3.1}$$

where the subscripts *e*, *b*, and *c* refer to emitter, base and collector respectively. At low current levels the direction of flow may change as the leakage current becomes a larger fraction of the total; for instance, when the emitter current I_e is zero, the collector leakage current I_{co} flows out of the collector into the base.

The collector leakage current I_{co} is never quite zero (Fig. 3.14a) even when the emitter is disconnected. It is highly temperature sensitive (Fig. 3.7) and is normally between 10^{-4} A and 10^{-9} A for modern silicon transistors and between 0·5 and 5 mA for germanium transistors.

Normally the area of the collector is about three times that of the emitter, to afford efficient collection of carriers, and the roles of the emitter and collector are not usually interchanged. There are, however, some special symmetrical transistors (unijunction type) whose emitters and collectors are interchangeable; these are used for switching purposes (Fig. 3.43).

3.5. AMPLIFIER CIRCUITS

The d.c. performance of a transistor can be described by its static characteristics when connected in three different ways, viz:

(*a*) grounded-emitter (Fig. 3.13)
(*b*) grounded-collector (Fig. 3.15)
(*c*) grounded-base (Fig. 3.16).

The most useful characteristic is the curve of collector current versus collector voltage. This is plotted for various values of I_b for the grounded-emitter circuit or I_e for the grounded-base circuit.

The fact that current amplification is linear, whereas the voltage amplification depends upon the load impedance, means that transistor amplifiers are linear only as current amplifiers, i.e. the input must be a current source or the source impedance must be high compared with the transistor input impedance if a voltage source is used.

TABLE 3.1

Comparison of Transistor Connections

Grounded Electrode	Emitter	Base	Collector
Input Electrode	Base	Emitter	Base
Current Gain	High	None	High
Voltage Gain	High	High	None
Power Gain	High	Medium	Low
Input Impedance	Medium	Low	High
Output Impedance	Medium	High	Low
Phase Shift	180°	0°	0°

3.5.1. Grounded-emitter Amplifier

Figure 3.24 shows a p-n-p amplifying circuit in which the base bias is obtained without a separate battery by means of a leakage resistance connected to the collector end of the battery; for a.c. the base must be biased in the forward direction well above the toe voltage. This circuit is called the grounded (or common) emitter circuit; it gives high current and power gains and is the circuit most commonly used for amplification, especially in circuits where the source impedance is high compared with the

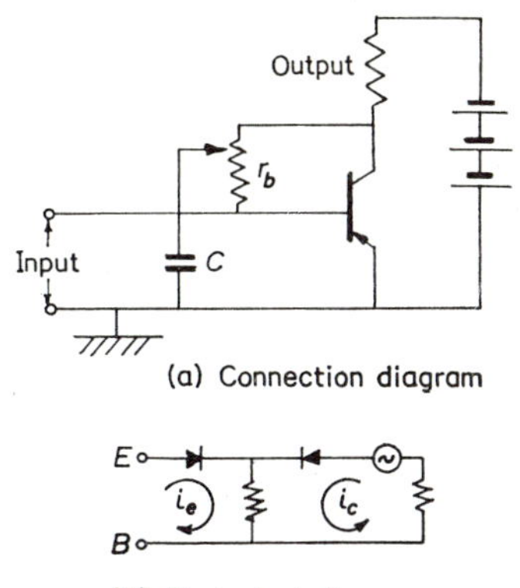

Fig. 3.24. Grounded emitter amplifier with compensation

input impedance Z_{be}, as in the case where the output of a comparator is to be amplified. It presents a low imput impedance Z_{be} and a fairly high output impedance; as can be seen from Fig. 3.13 it causes 180° phase shift.

The current gain ($\alpha' = i_c/i_b$) is not much affected by the load impedance (Fig. 3.14a) but the power gain increases with the load up to an optimum value where the load impedance matches the output impedance of the transistor; the voltage gain is high because the input voltage is only the base-emitter drop; this also contributes to the power gain.

The collector resistance, for a given value of i_b (Fig. 3.14b) is $r_c = (V_c/I_c)$. When $I_b = 0$, the leakage current $I_{co}/(1 - \alpha)$ flows in the collector and increases slightly with increasing voltage. The pronounced knee in each characteristic where the family of curves tends to converge occurs, typically, at collector voltages in the range 200–300 mV.

3.5.2. Grounded-collector Amplifier

Figure 3.15 shows the grounded (common) collector circuit. This is like the cathode follower electronic tube circuit; when the transistor becomes conductive the current through the output impedance reduces the voltage between the emitter and the base, providing a limiting action like negative feedback; this stabilizes the d.c. working point but does not affect the a.c. amplification if a small capacitor is connected across the output, as shown in Fig. 3.15 (p. 58). It has about the same current gain as the grounded-emitter circuit but the voltage gain is negligible because the drop across the output is also the input circuit, hence the power gain is small.

The input impedance is high and the output impedance is low; hence this circuit can be used as a matching stage to match the high output impedance of one grounded-emitter amplifier (Fig. 3.13 p. 56) stage to the low input impedance of another. It also causes 180° phase shift between input and output; hence a grounded-collector buffer stage between two grounded-collector stages causes no phase shift except that due to the capacitance of the transistors, which is apparent only at high frequencies.

3.5.3. Grounded-base Amplifier

Figure 3.16 shows a grounded (common) base amplifier. This has a very low input impedance and high output impedance. If gives no current amplification with d.c.; the amplification is small with a.c. and depends upon the position on the i_c/i_e transfer curve; hence it is no good for R-C amplifiers but can be used for power amplification with transformer coupling. It has no phase reversal and is temperature stable.

The static characteristic is as shown in Fig. 3.17 (p. 59). Beyond the knee-point on each curve shown, the curve has a very small slope. The collector resistance $r_c = V_c/I_c$ is generally greater than 1 megohm.

The difference between the emitter current and the collector current flows in the base and, though the external base connection is earthed, the flow of current in the internal base resistance produces an effective bias voltage. To reduce the collector current to zero it is therefore necessary to make the collector positive with respect to the base.

3.6. COMPENSATION

The output of a transistor is greatly affected by light, temperature and input bias. Light can be excluded by the container but the temperature effect must be compensated for and the bias voltage must be held constant. The methods of achieving this are as follows.

3.6.1. Temperature Compensation

Computing equipment and instrumentation is usually in an air-conditioned room, but protective relays may be in an unattended station or even mounted on the equipment protected, so they must be designed to operate correctly over a range of $-10°C$ to $+50°C$. Semiconductors are considerably affected by temperature, hence compensating means are necessary where the operation of the relay is dependent upon the transistor characteristics. Also the semiconductor can be destroyed if adequate arrangements to conduct heat away are not provided (heat sinks).

The collector current of a transistor should theoretically be zero when the transistor is not conductive but it has a large leakage current, just like the reverse current of a diode, which increases exponentially with temperature, doubling with about 8°C rise.

The fact that there is a current gain of the order of 50 in the circuits of Figs. 3.13 and 3.15 introduces considerable temperature error. If the leakage between base and collector is 100 μA this will cause a collector-emitter current of the order of 5 mA. At a higher temperature both these values will increase in proportion, causing an increasing error which can only be eliminated by compensation or a push-pull circuit.

The simplest compensation is done by connecting the biasing resistor through the load resistor in the collector circuit as shown in Fig. 3.24. This also provides some d.c. negative feedback (series degeneration) which gives d.c. stabilisation of the working point. This circuit is used when the source impedance is low compared with the input impedance Z_{be}. The capacitor is for preventing a.c. negative feedback, i.e. to preserve the a.c. amplification in spite of the d.c. negative feedback; it is usually connected to the mid-point of the biasing resistor.

A very stable method, which is common for photo-transistors and amplifiers, is shown in Fig. 3.25a; the compensation is provided by the emitter resistor R_e which is shunted by a capacitor to prevent negative feedback on a.c. For extreme temperature ranges R_2 should be shunted by a thermistor. Figure 3.25b shows a similar circuit with transformer coupling, which is more expensive but necessary where uniform output is required over a

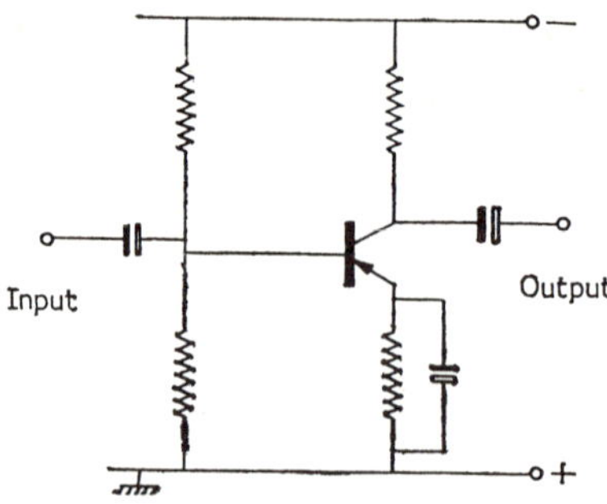

Fig. 3.25a. Alternative compensation of grounded-emitter amplifier

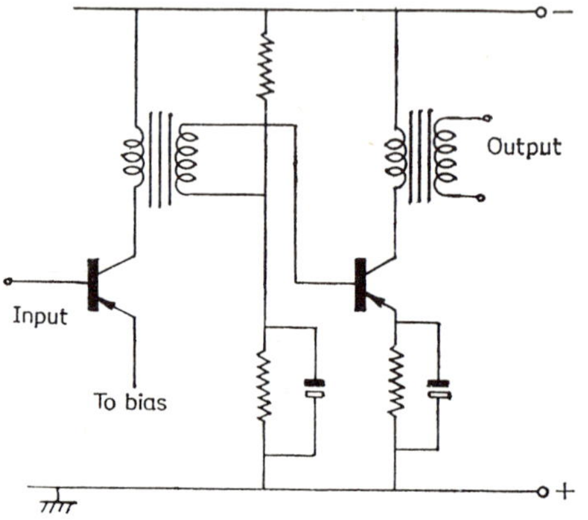

Fig. 3.25b. Transformer coupled version of Fig. 3.25a

wide band of frequency such as in carrier relaying (50–500 kc); it is more efficient because it does not have the power loss in the resistors.

The direct method of compensation is by a temperature sensitive diode, connected as shown in Fig. 3.26 for germanium transistors (or in series with the base-emitter resistor for silicon) so as to match the resistance of the collector and its parallel resistor and hence neutralise the base current over a wide temperature range.

This compensation is not necessary in the push-pull amplifier of Fig. 3.33. because the voltage drop due to leakage through one collector is balanced by a similar leakage through the other one, so that their bases and emitters maintain a constant relative potential.

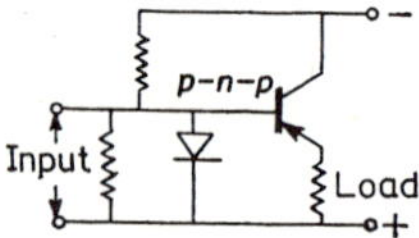

Fig. 3.26. Temperature compensation by temperature sensitive diode

3.6.2. Voltage Stabilization

When measurement is involved it is important to keep the bias of the input (base-emitter) circuit constant and a zener diode can be used for this purpose, as shown in Fig. 3.27. This also eliminates variation due to temperature because, no matter how much the collector leakage increases due to temperature, the base-emitter voltage is held constant by the zener diode.

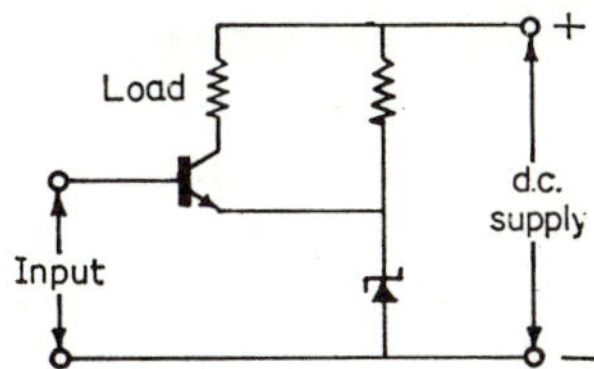

Fig. 3.27. Zener diode for stabilising base potential

In protective relays measurement should not be dependent upon transistor characteristics. The transistors should preferably be used only as switches and amplifiers.

3.6.3. Feedback

Positive feedback in a static relay is analogous to snap action in an electromagnetic relay. In both cases, when the relay operates, it tends to be held in the operated position until the operating quantity is reduced to a very low value. In thyristor circuits the collector circuit actually has to be opened before the thyristor will return to the non-conducting state.

Figure 3.29 illustrates positive feedback action. When transistor T_2 becomes conductive the drop across the load (relay) in the collector circuit increases the bias on T_1, causing more current to flow in the thermistor *Th*, increasing the bias on T_2 and thereby still further increasing its output.

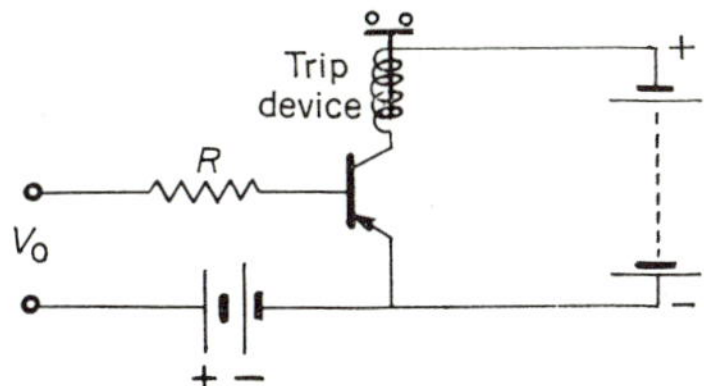

Fig. 3.28. Basic level detector

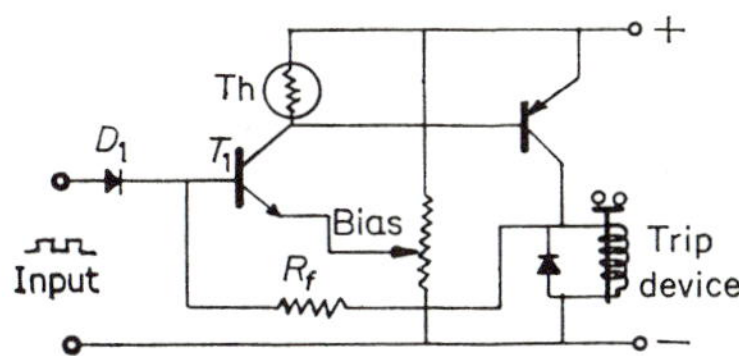

Fig. 3.29. Practical level detector with snap action

Figure 3.30 shows how a high ratio of reset/operate values of input can be retained in spite of positive feedback. The operation of this circuit is described in Section 3.8.

Negative feedback is used only for stabilization of operation. It is common in instrumentation and metering circuits but it is seldom used in protective relays except in fault locators and voltage or power factor regulators.

Figure 3.31a shows a simple method of obtaining negative feedback. When the collector-emitter current increases, the voltage drop across the feedback resistor R_f increases and reduces the base-emitter bias so that it takes a bigger input signal to keep the transistor conductive. Hence there is a tendency to keep the output at a fixed level.

Figure 3.31b shows the same idea applied to two transistors in cascade. Figure 3.31c shows a shunt method of applying bias to the first transistor T_1.

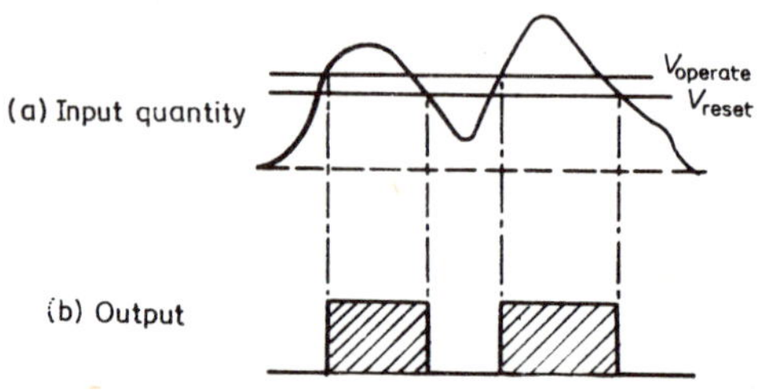

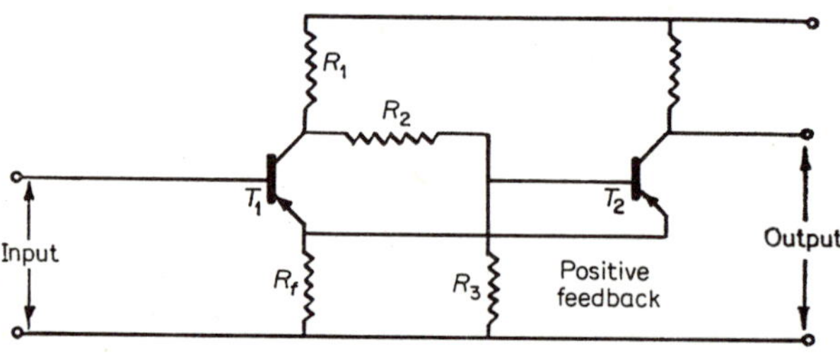

(c) Schmitt trigger circuit

T_2 normally conducting; T_1 not; vice versa when input signal exceeds $V_{operate}$ in diagram (a)

Fig. 3.30. Snap action retaining high drop-out pick-up ratio

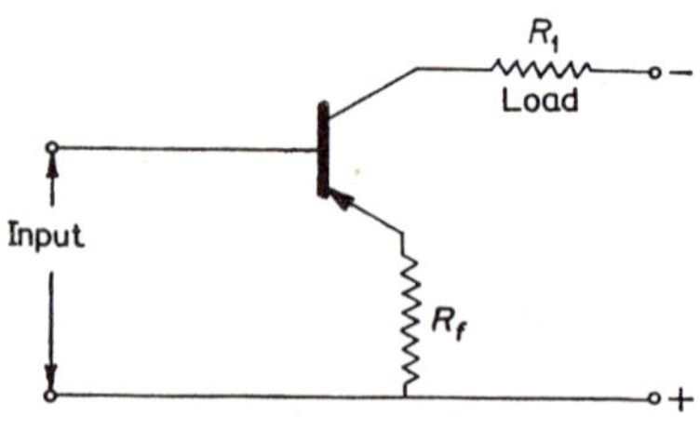

Fig. 3.31. a. Series

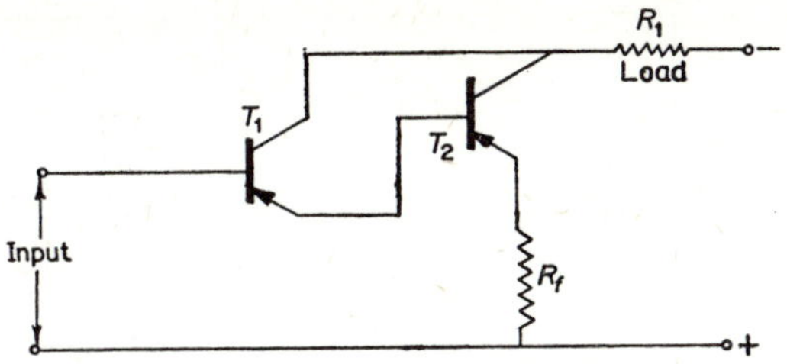

b. Multistage series

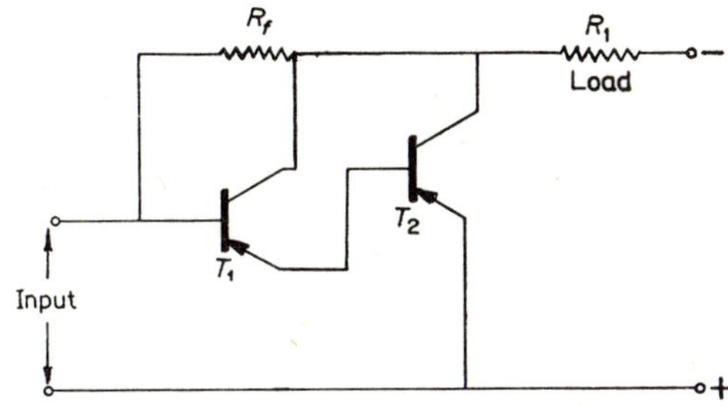

c. Multistage shunt

Fig. 3.31. Negative feedback

Note: A voltage feedback is used in vacuum tube circuits. A current feedback is used in transistor circuits

An increase in the I_c of T_2 increases the voltage drop across R_1 and, through R_f, decreases the bias potential of the base of T_1, limiting its output.

3.7. PUSH-PULL AMPLIFIERS

Mid-tap transformers facilitate push-pull amplifier circuits, as shown in Fig. 3.32, because they inherently provide reversed as well as forward input signals. Resistance amplifiers can also be used in push-pull (long-tailed pair) as shown in Fig. 3.33. When the potentiometers are balanced the push-pull connection prevents base-biasing current from getting into the phase comparator, as would be the case in a normal grounded-emitter circuit.

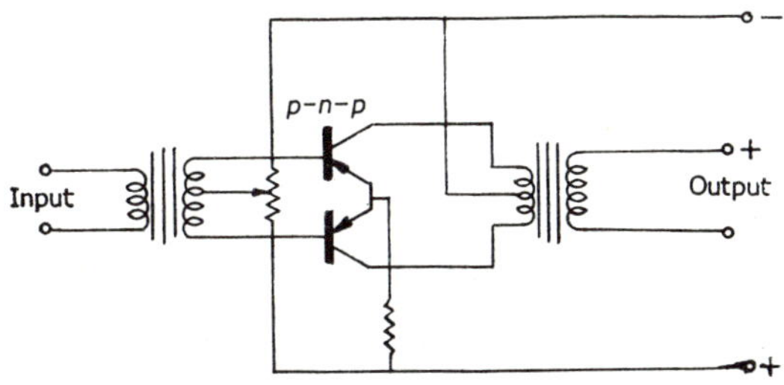

Fig. 3.32. Push-pull class B amplifier

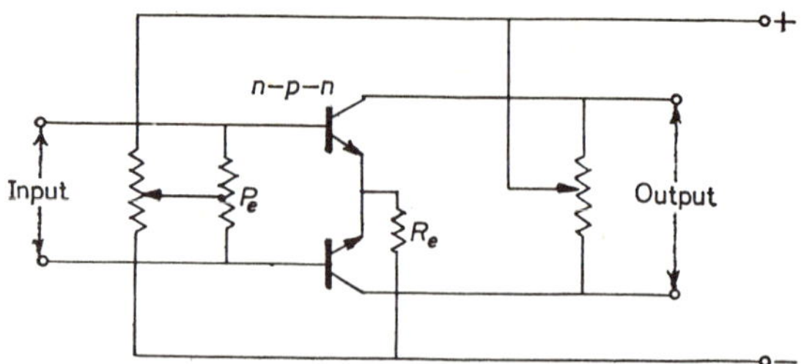

Fig. 3.33. Long-tailed pair (resistance coupled equivalent of Fig. 3.32) circuit

The transistors in push-pull have to be a matched pair because they work down to a few millivolts input and their emitter potentials are balanced by potentiometer P_e. The resistor R_e provides temperature stabilization because an increase in output current due to temperature produces a bigger drop in R_e and reduces the emitter bias and tends to reduce the output current to its former value. The second stage of common emitter amplification is used to provide more positive action of the relay.

3.7.1. Composite Transistors

Transistors can be ganged to increase amplification. In fact two transistors are sometimes supplied in the same can with three leads so that they act as one, as shown in Fig. 3.34.

Unlike transistors (p-n-p and n-p-n) can also be used in complementary fashion so as to save components and permit direct coupling (Fig. 3.35). However, this circuit has d.c. drift and temperature error unless suitable

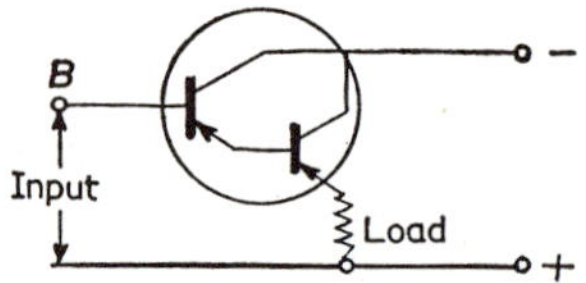

Fig. 3.34. Composite transistor (cascaded emitter follower)

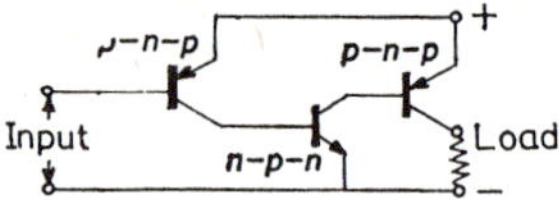

Fig. 3.35. Direct-coupled complementary transistors

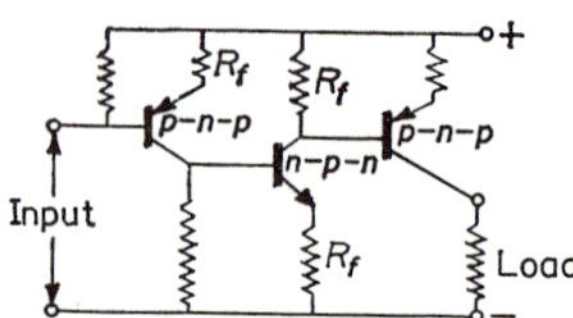

Fig. 3.36. Complementary transistors with negative feedback

resistors are inserted between the d.c. supply and the transistor electrodes (Fig. 3.36) at the expense of efficiency.

Complementary transistors can also be used in push-pull, as shown in Fig. 3.37.

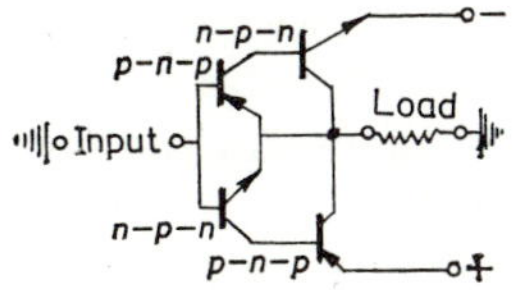

Fig. 3.37. Complementary transistors in push-pull

3.8. LEVEL DETECTORS

Figure 3.28 shows a simple level detector in which the input voltage has to exceed the opposing voltage of the bias battery before any output is produced. In practice the bias is usually obtained from a potentiometer, as shown in Fig. 3.29, rather than from a battery.

The circuit of Fig. 3.28 (p. 69) would be suitable for a fault detector but, where some precision is required, such as for measuring the voltage across the capacitor of an R-C timing circuit, Fig. 3.29 shows a precision level detector which has snap action and is suitable for operating a tripping relay. The thermistor Th has a negative coefficient of resistance and enables the timing unit to be accurate over a wide range of temperature. The diode D_1 is to protect the transistor T_1 against the bias voltage of the emitter when the capacitor is discharged since the transistor can in some cases stand a reverse base-emitter voltage of only 1 volt. The resistor R_f gives a positive feedback to provide snap action.

Figure 3.30c (p. 70) shows the Schmitt trigger circuit which retains a high reset value (high drop-out/pick-up ratio) although it has snap action. Transistor T_1 is normally non-conductive and T_2 conductive. The voltage at the base of transistor T_2 is determined by the supply voltage V_s and the potential divider $R_1R_2R_3$; when the input signal voltage attains a value which is sufficiently negative to begin to deflect current into T_1, the circuit regenerates and the emitter current is rapidly switched out of T_2 and into T_1. The base of T_2 is is now held at a voltage determined by the collector voltage of T_1 and, by its choice, the input voltage required to return the circuit to its original state (the resetting voltage) can be controlled.

3.9. TIME-DELAY CIRCUITS [30]

For very short delays (microseconds) a delay line is generally used (Fig. 3.38). For medium delays (milliseconds) a resonant circuit is common (Fig. 3.39) in which the delay is obtained by the building up to maximum oscillation.

For longer delays (seconds) R-C circuits can be used (Fig. 3.40). Because of the small magnitudes of the currents in transistor circuits, a delay of minutes can be obtained with a few microfarads. With tantalum capacitors of a few hundred microfarads, delays of several hours are practical. Such R-C circuits are used in time-current relays and industrial timers.

In Fig. 3.40 the transistor becomes conductive when $V_c > kV + V_{be}$ and it is obvious that capacitor or transistor leakage will affect the accuracy of timing. For this reason it is important for the circuit to be aged by heat-soaking before calibration.

Other conditions that can affect the accuracy of a timer are temperature and supply voltage variation. The first can either be compensated for by a thermistor or by arranging the circuit so that the transistor does not operate until the timing capacitor is charged; in Fig. 3.41 the circuit is arranged so that a diode prevents any current from reaching the transistor detector until the capacitor is charged to a greater voltage than the emitter of the transistor. This ciicuit is also immune to supply voltage variation because the charging current at any moment is proportional to the supply voltage.

Figure 3.41 has several other refinements [45]. A second diode is added to provide for instantaneous resetting of the timer when the voltage is removed, while a second transistor provides snap action to give positive tripping with a smaller input signal. This is the Schmitt trigger circuit and

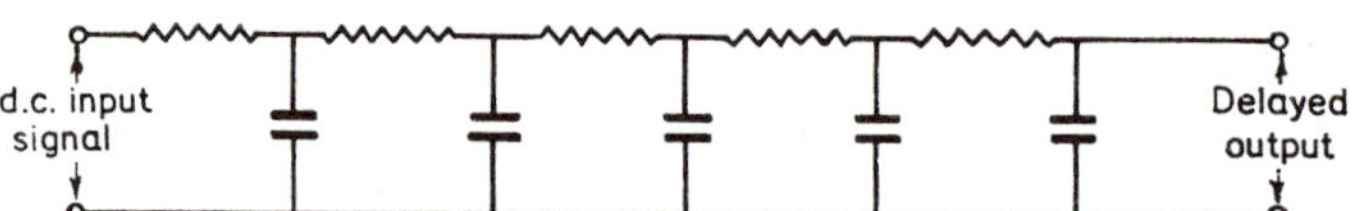

Fig. 3.38. Equivalent circuit of a delay line

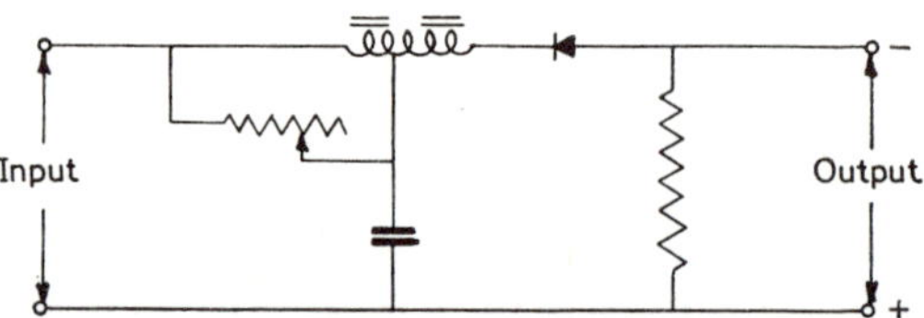

Fig. 3.39. Basic millisecond time-delay circuit

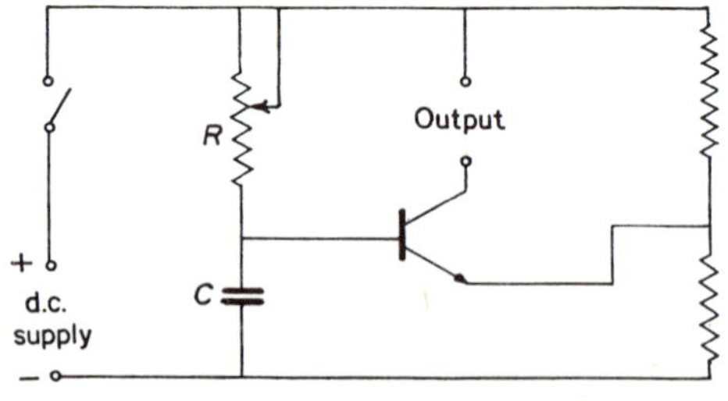

Fig. 3.40. Basic long-time-delay circuit

it will be noticed that initially T_1 is conductive and T_2 is not; when the timing capacitor C is charged sufficiently to extinguish T_1 it causes T_2 to fire and operate the relay. Finally, the potentiometer P_e compensates for variation in capacitor C if the time control rheostat R has a printed time scale.

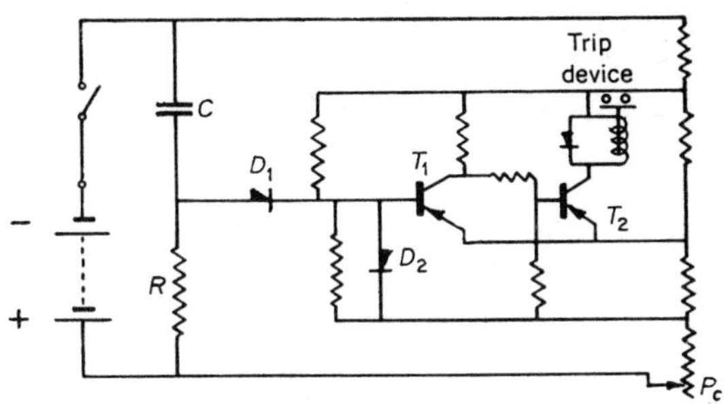

Fig. 3.41. Timer with instantaneous reset and capacitor compensation T_1 initially conducting

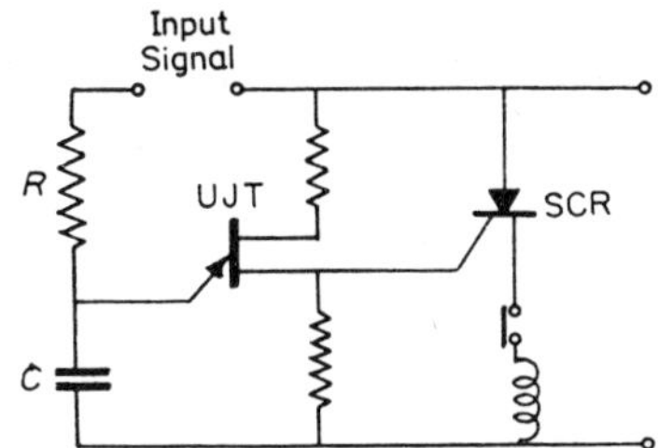

Fig. 3.42. Thyristor triggering circuit
R-C circuit provides 2 mS delay to override interference spikes

A thyristor can be used for tripping instead of a relay if only one circuit is to be controlled. The triggering voltage of a thyristor varies with temperature so, to avoid error, it may be necessary to let the capacitor voltage trigger a unijunction transistor (UJT) which, in turn, fires the thyristor (Fig. 3.42).

Figure 3.42 also shows how a unijunction transistor can be used for delaying the operation of a tripping thyristor. The tripping impulse from the relay circuit short-circuits the input terminals and the capacitor C is charged through R. As soon as V_c exceeds V_p, the peak point voltage of the unijunction transistor, it becomes conductive and the capacitor discharges through it, providing the gate current for the thyristor and firing it. The thyristor now energizes the trip coil until the breaker switch 'a' interrupts its circuit.

3.10. SQUARE-WAVE GENERATORS

A square wave is produced when diodes are connected across an a.c. source, because the tops of the sine waves are diverted through the diodes. Another square-wave generator using an a.c. source is a transistor circuit, such as Fig. 3.13, which produces rectangular output blocks during the period when the baseemitter bias is slightly positive with an n-p-n transistor, or slightly negative with a p-n-p transistor.

Figure 3.43 shows two circuits for producing a continuous square wave from d.c. Figure 3.43a is a relaxation oscillator based upon the trip impulse delaying circuit of Fig. 3.42. The capacitor C charges at a rate controlled by the value of R. When V_c exceeds the peak point voltage of the unijunction transistor it becomes conductive and the capacitor discharges through it and the primary of the transformer. The capacitor then starts to charge again and the process is repeated, giving a series of impulses whose frequency is controlled by R. This output can be fed into a multi-vibrator circuit for making square waves. The output of fig. 3.43a can be used directly for providing a.c. to supply a magnetic amplifier; such an a.c. supply would not fail during system faults. Magnetic amplifiers are cheaper and more robust than transistor amplifiers but have not yet been used in protective relaying.

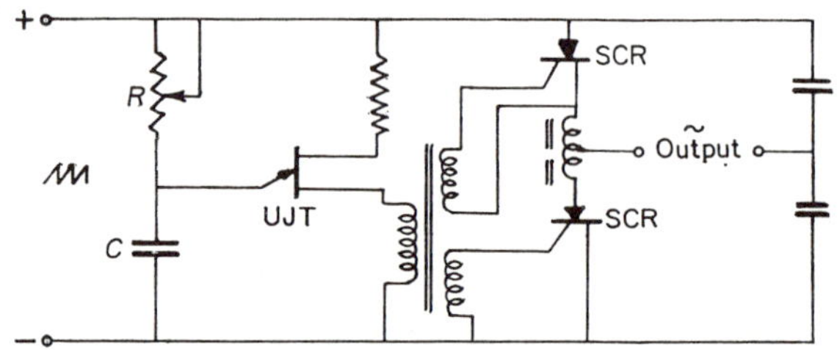

Fig. 3.43a. Sine-wave inverter using UST relaxation oscillator

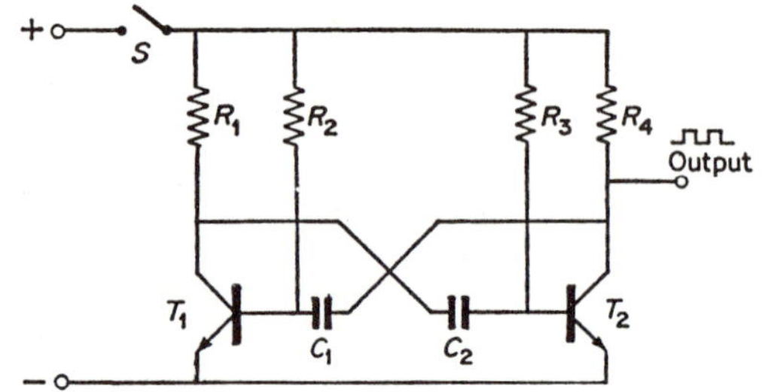

Fig. 3.43b. Astable multivibrator $R_2 = R_3$; $R_1 = R_4$; $R_2 = 10R_1$; $C_1 = C_2$

Fig. 3.43b is a continuous square wave generator using transistors. When S is closed C_1 and C_2 will charge and, if T_1 conducts, its collector goes more negative, making the base of T_2 more negative, cutting it off. C_2 then discharges through R_3 until the base of T_2 becomes sufficiently positive to make it conduct. Then T_2 collector goes negative, shutting off T_1 and so on. The circuit is astable but it can be synchronised by recurrent signals such as from the output of fig. 3.43a.

3.11. SMOOTHING CIRCUITS

The d.c. output from a rectifier bridge consists of a series of unidirectional half-waves of current or voltage. It is generally desirable to smooth this output but the degree to which it is necessary to eliminate the a.c. component depends upon the application.

The smoothing filter in Fig. 3.44a eliminates virtually all a.c. but tends to be expensive. The cost can be reduced by replacing the reactor by a resistor and possibly eliminating one of the capacitors, as in Fig. 3.44b.

An alternative method is to use a transformer to cancel out the a.c. component, as in Fig. 3.44c. This method is not suitable where the input is limited because the primary winding provides a low resistance shunt path for the d.c., diverting it from the output.

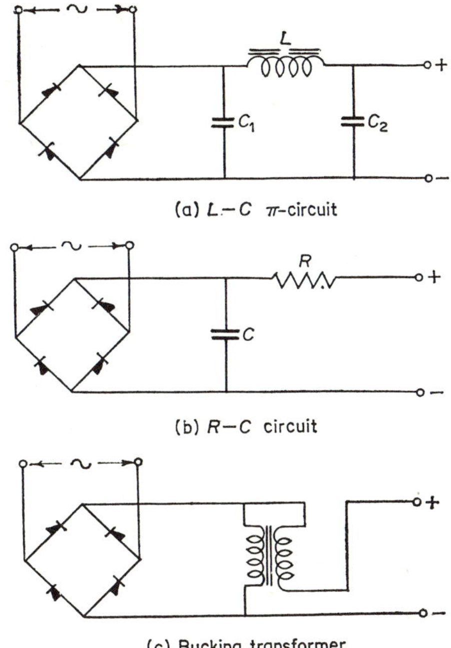

Fig. 3.44. Smoothing circuits for d.c. supply

3.11.1. Frequency Filters

In static relays filters are used for blocking certain frequencies and passing others. Table 3.2 shows the basic circuits. These can be elaborated in various ways; for instance, the high-pass circuit can be made more effective by providing another series capacitor on the left side of the choke, or another shunt choke on the right side of the capacitor.

A large number of duals appear when the circuit is made more elaborate, since a shunt capacitor has the same effect as a series choke and vice versa. These are given in radio engineering handbooks.

TABLE 3.2.
Basic Filter Circuits

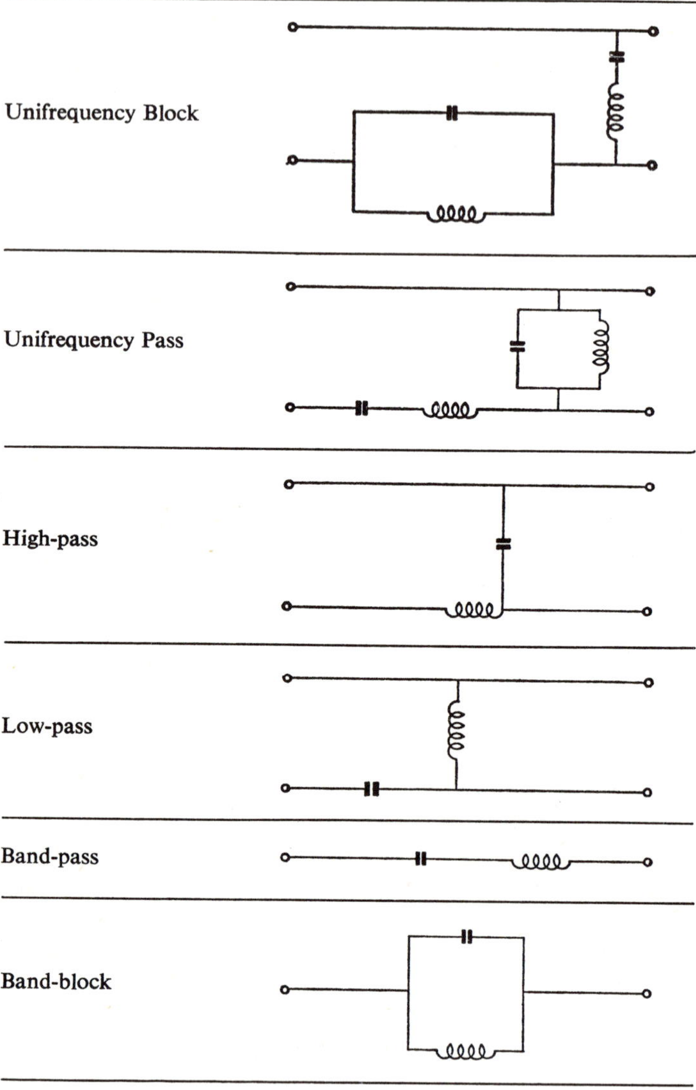

3.12. VOLTAGE REGULATORS

Zener diodes are most commonly used for holding a voltage constant over a wide range of applied voltage (see Fig. 3.45a), such as where the d.c. supply for a static relay is obtained from the c.t. circuit (see Chapter 5). Alternatively the rectifier bridge may be composed of zener diodes.

Figure 3.45b shows how a zener diode may be employed in a grounded-base amplifier circuit to provide a constant voltage from an unregulated d.c. supply. It shows how a lower voltage supply can also be regulated (by T_2) in spite of a varying load.

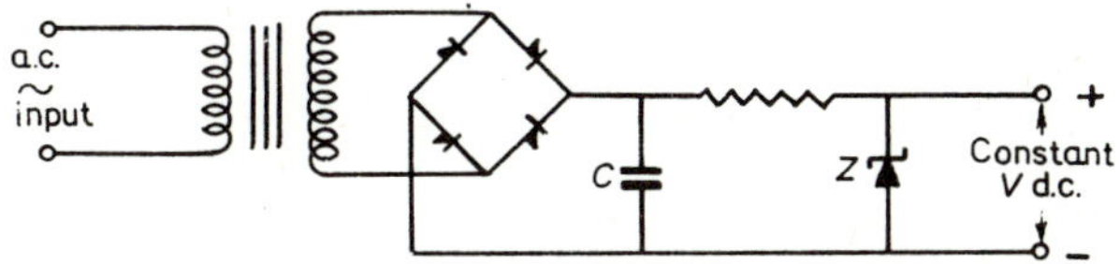

Fig. 3.45a. Zener diode for regulating d.c. supply voltage

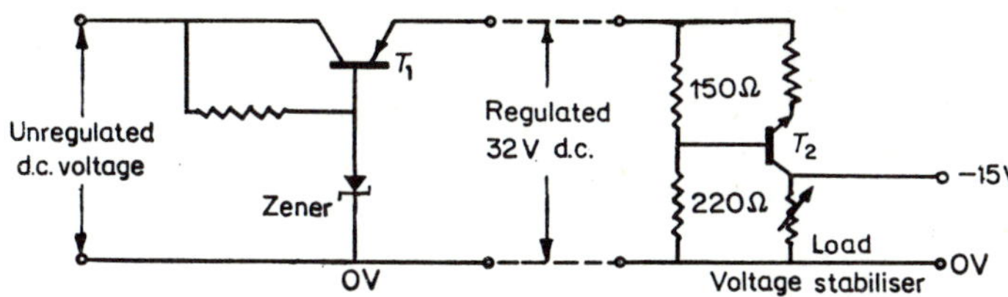

Fig. 3.45b. Transistor voltage regulator

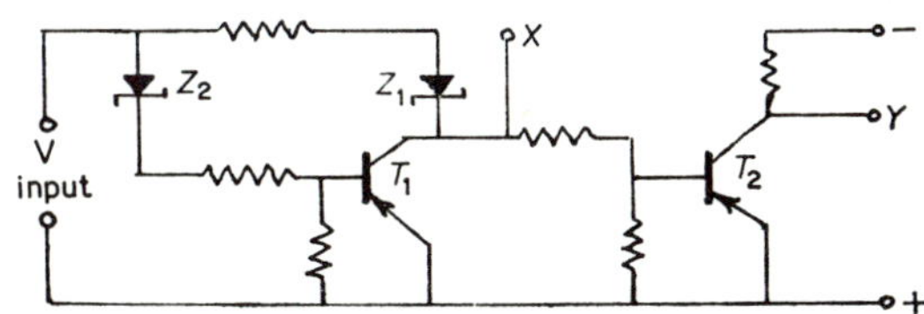

Fig. 3.45c. Under- and overvoltage relay

If $V > Vz_2$ — Z_1, Z_2 and T_1 conduct
If $Vz_2 > V > Vz_1$ — Z_1 and T_2 conduct
If $V < Vz_1$ — all are off

Figure 3.45c shows how an under- and overvoltage relay can be made with two zener diodes and two transistors. The zener diodes start to conduct backwards at two voltages V_{z1} and V_{z2} which are close together in value. If $V_{z2} > V_{z1}$ the following operation occurs.

Input	Z_1	Z_2	T_1	T_2	X	Y
$V \leq V$	off	off	off	off	—	—
$V_1 \leq V \leq V_2$	on	off	off	on	—	+
$V \geq V_2$	on	on	on	off	+	—

Terminal X becomes positive on overvoltage and terminal Y on undervoltage. The input voltage can be d.c. or rectified a.c.

3.13. TRANSDUCERS [31, 58]

In order to make a static relay respond to temperature, frequency, etc., a transducer is necessary to turn the measured quantity into a current or a voltage which can be fed into a level detector or a comparator. The following are examples of these transducers.

3.13.1. Temperature

The most direct method is to use the output of a thermocouple with a 1-stage magnetic amplifier or transistor amplifier if necessary.

Where an RTD (resistance temperature detector) is used the most effective means is the d.c. resistance bridge, Fig. 3.46. The resistors forming the other arms of the bridge should have the same resistance as the TRD which is usually 10 ohms. In Fig. 3.46 the thermistor is provided for temperature compensation of the transformer windings and rectifier bridge; alternatively a zener diode can be connected across the d.c. output of the bridge.

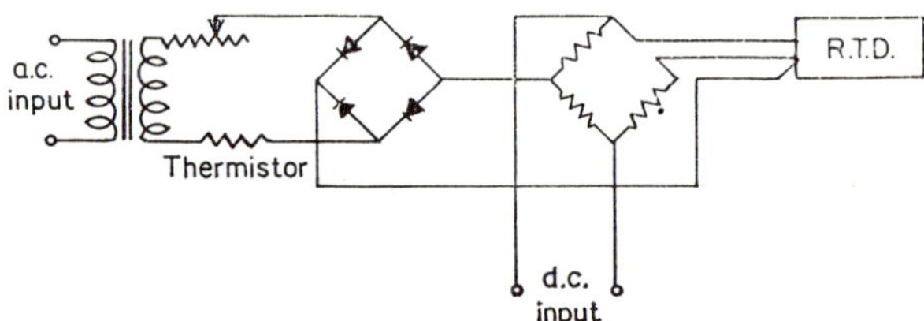

Fig. 3.46. Temperature transducer

3.13.2. Frequency

In order to obtain an output dependent only on frequency and to minimize the effect of voltage and temperature variation, a differential resonant circuit is used with a stabilizing voltage source (two zener diodes as in Fig. 3.47 a).

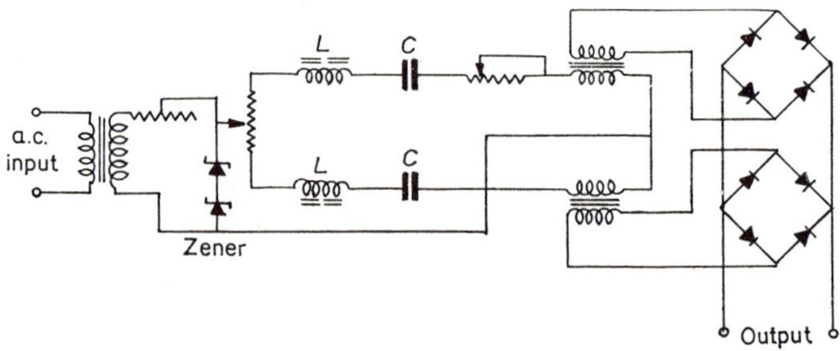

Fig. 3.47 a. Frequency transducer

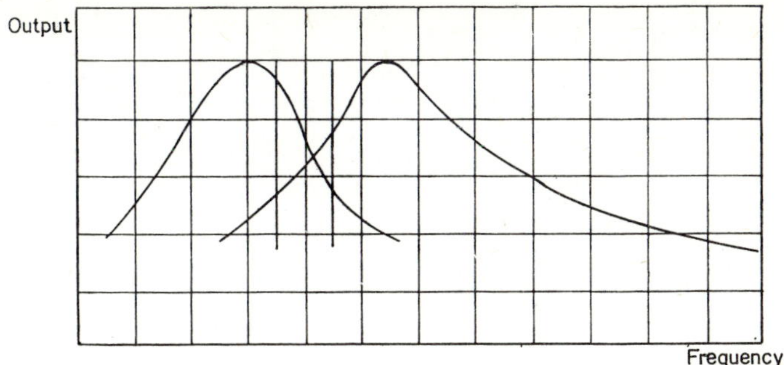

Fig. 3.47b. Resonance curve of frequency transducer

3.13.3. Power and Power Factor

Figure 3.48 shows a circuit for producing a d.c. voltage proportional to a.c. power which can be used for instrumentation or as a relay. Ignoring *S*, the diodes would pass $(I + V)$ and $(I - V)$ in opposite directions in the potentiometer *P*. *S* is a non-linear resistance which produces a current proportional to the square of the voltage across it.

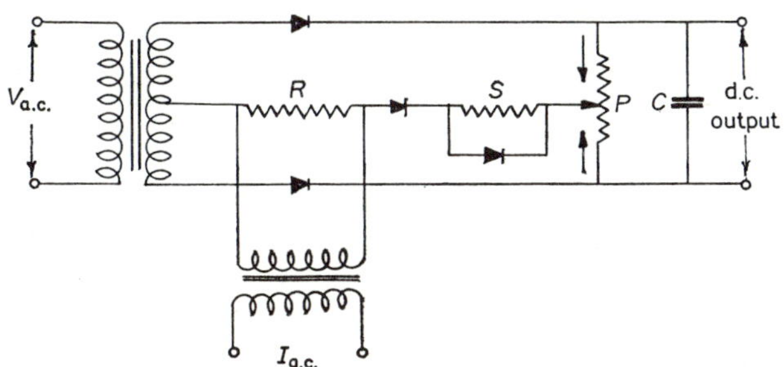

Fig. 3.48. Power transducer

Hence the voltage across the capacitor

$$V_c \propto (I + V)^2 - (I - V)^2 \propto 4VI \cos(\phi - \theta)$$

where ϕ is the angle between *I* and *V* and θ depends on the circuit constants. In Fig. 3.48 $\theta = \phi$, so the output is watts. θ can be adjusted to other values by suitable phase shifting means in the current or voltage circuits.

For measuring power factor or power factor angle a limiting resistor (parallel reversed diodes) is connected across the input current, as shown in Fig. 3.49a, which is in fact a more elegant circuit. In order to cover $\pm 90°$ range, the current is shifted 90° in a transactor in order to give zero d.c output at unity power factor. Figure 3.49b shows that the output is then proportional to sine ϕ where ϕ is the angle between *V* and *I*.

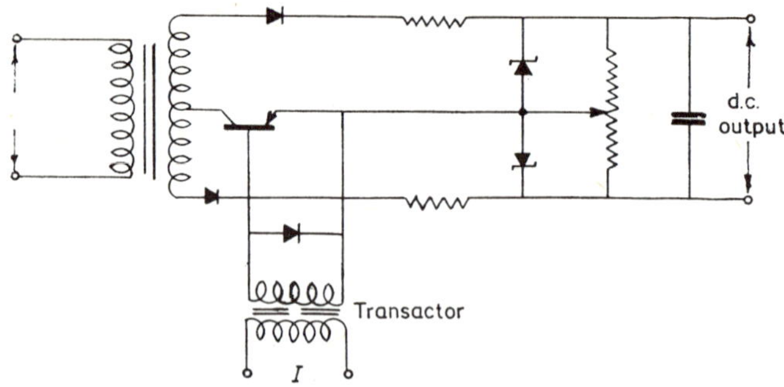

Fig. 3.49a. Power factor transducer

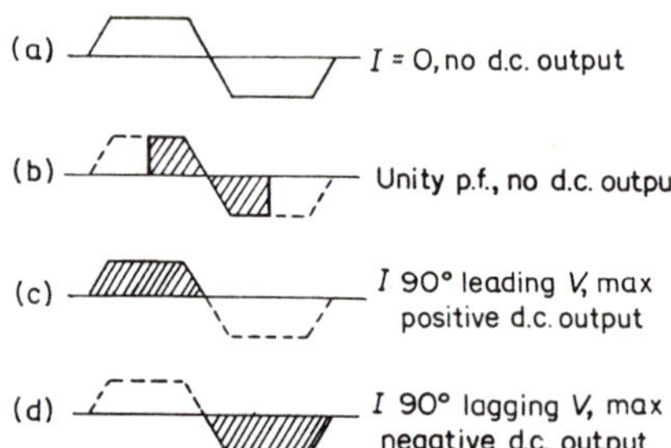

Fig. 3.49b. Output of power factor transducer

3.14. D.C. VOLTAGE DOUBLERS

Sometimes it is required to have a d.c. voltage supply of more than one value. Figure 3.50 shows how four steps of voltage can be obtained without a transformer, the maximum being four times the a.c. peak.

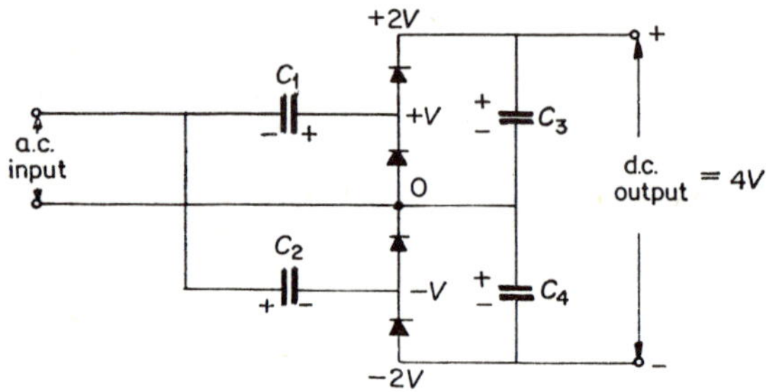

Fig. 3.50. a.c.—d.c. voltage quadrupler

3.15. SYMMETRICAL COMPONENT FILTERS [52]

It is extraordinary that, so far, protective relays have always been predominantly single-phase devices in spite of the facts that (*a*) they protect a three-phase system, (*b*) they cost much more and take up more space than polyphase relays.

A few relays are in existence which operate on a polyphase basis and there is no doubt that they will become more numerous as the advantages of polyphase measurement become recognized. This subject will be dealt with in more detail in Chapter 12.

One way of making polyphase measurement is to make relays which are responsive to one or both of the symmetrical components and this subject will be discussed in Sections 9.6.3 and 13.4.1.

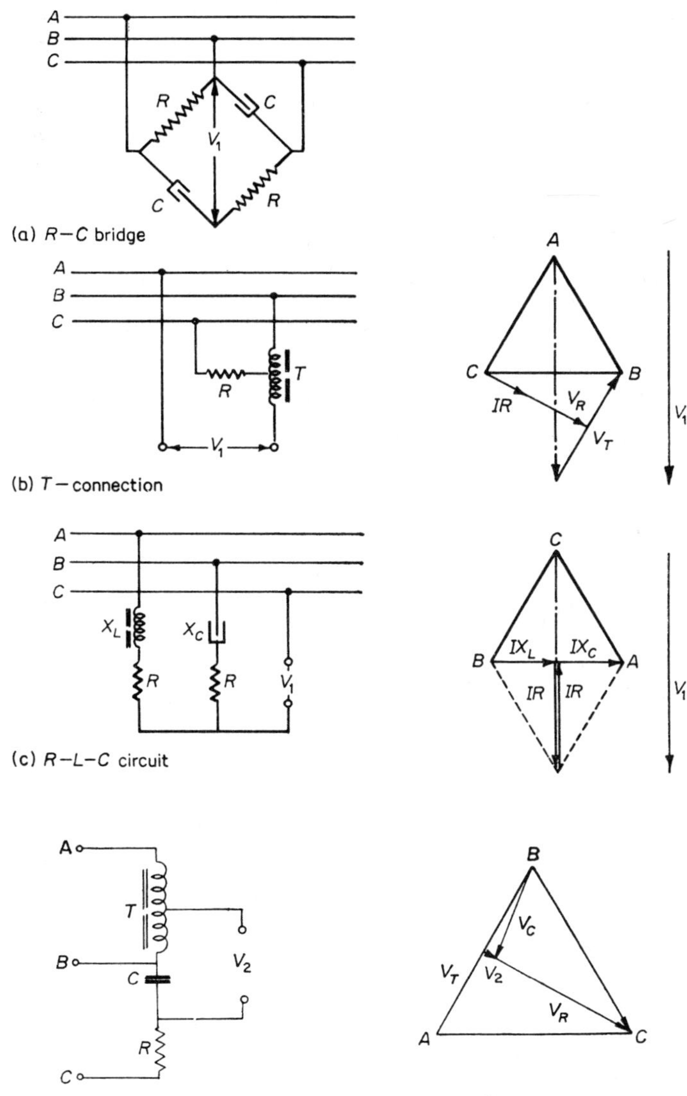

(a) $R-C$ bridge

(b) $T-$connection

(c) $R-L-C$ circuit

(d) Negative sequence voltage filter

Fig. 3.51. Positive sequence potential filters

In order to extract a symmetrical component for use as an input to a static relay, a filter is required. Typical negative sequence filters are shown in Fig. 3.51 and 3.52. If two of the input leads are interchanged, their phase sequence is reversed and the filter now has a positive sequence output.

Where the load is very small the mid-tapped choke in Fig. 3.51 d can be replaced by a resistor. This does not apply to the transactors (transformer-reactors) of Fig. 3.52.

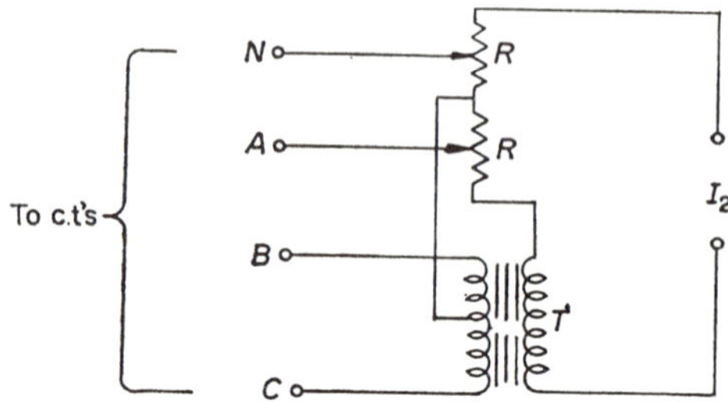

Fig. 3.52a. Negative sequence current filter output galvanically connected to c.t. citcuit

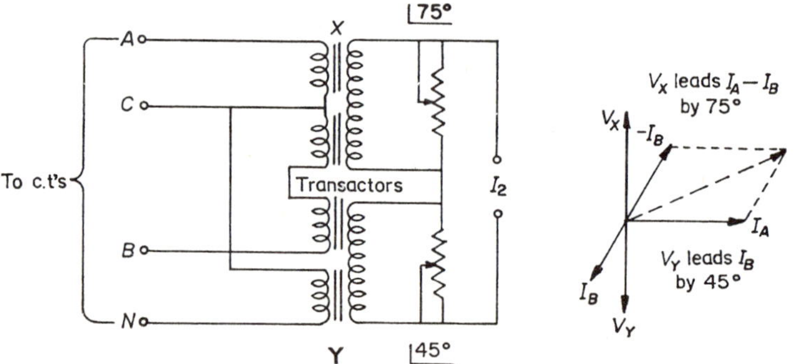

Fig. 3.52b. Negative sequence current filter output insulated from c.t. circuit

3.16. LOGIC CIRCUITS

In complex protective relays there may be a number of relaying functions which may be related by what is called 'logic'. Table 3.3 shows simple equivalent circuits for the AND, OR, NAND and NOR functions. Each circuit assumes that all the bases have a small negative bias to start with and the circuit is operated by positive inputs at A and/or B. Positive inputs to the AND or OR circuits produce a positive output at *C*. Positive inputs to the NAND or NOR circuits produce a negative output at *C*.

In a distance relay there may be a starting unit, a measuring unit and one or more monitoring units. Both the starting unit and the measuring unit must operate to cause tripping so that these are connected by AND logic (see Fig. 3.53) through the series transistors T_1 and T_2. There may also be a directional unit which prevents the starting unit from operating unless the

TABLE 3.3. *Logic units*

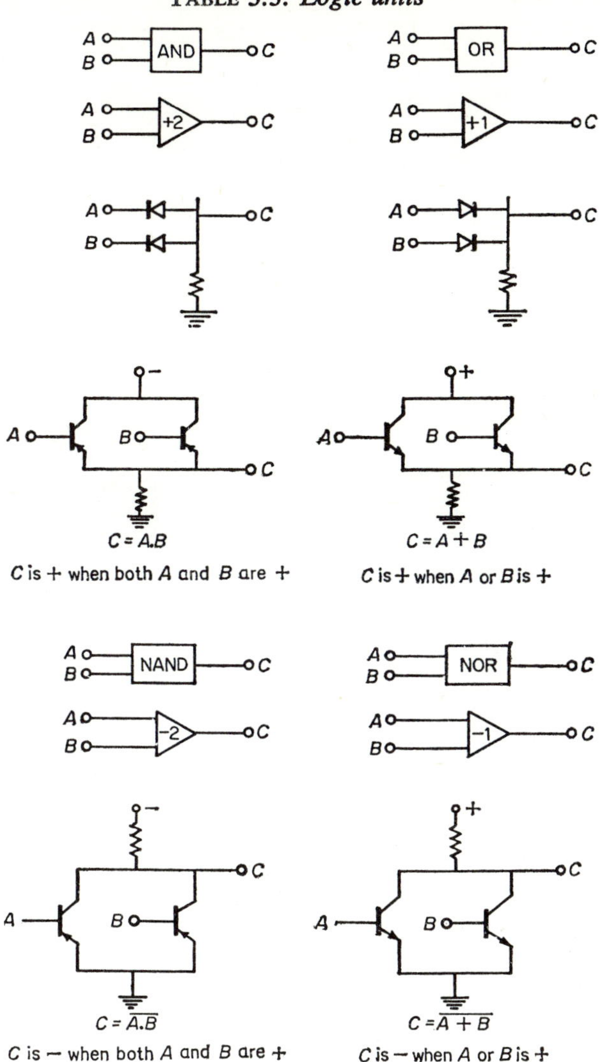

direction of the fault current is into the protected line section. Figure 3.53 shows how the NOT function is incorporated in the circuit.

It will be noticed that the circuit will operate equally well with all polarities reversed and n-p-n transistors instead of the p-n-p transistors shown.

3.17. SERIES AND PARALLEL RESISTANCE AND REACTANCE

In Chapter 6 it is shown in the Appendix, Section 6.7, how the equivalent value of two series of parallel impedances can be found very quickly by a simple graphical construction.

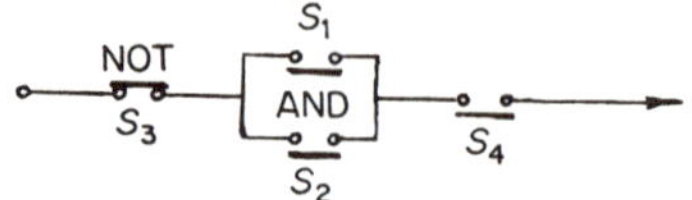

(a) Relay analogy of four inputs

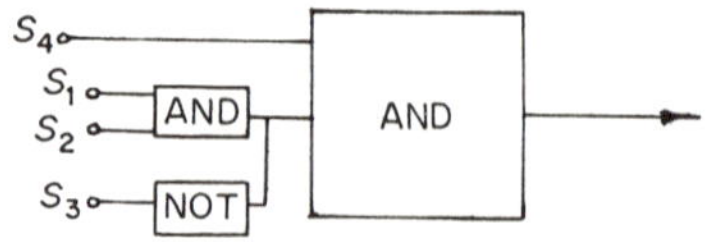

(b) Block diagram with four inputs

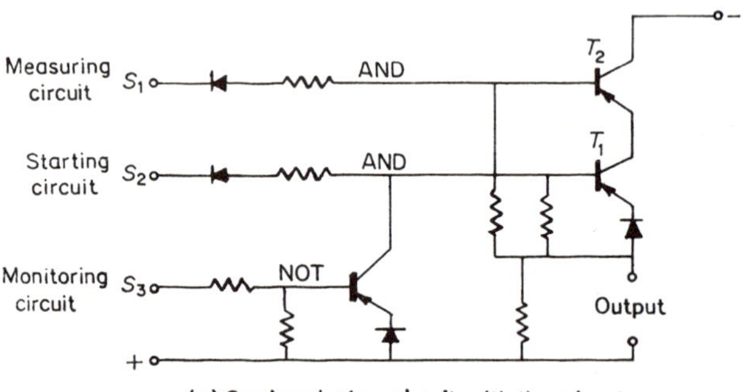

(c) Semiconductor circuit with three inputs

Fig. 3.53. Typical logic circuit

In the design of a relay, special values of resistance or capacitance are sometimes required which lie between the available standard values. The standard values follow a system of preferred numbers which increase approximately in geometrical progression in steps of about 20%, viz. 10, 12, 15, 18, 22, 27, 33, 39, 47, 56, 68, 82, 100.

To obtain intermediate values two resistors can be connected in series and parallel. If the smaller resistor is r and the larger is kr, the obtainable values in ascending order of magnitude are:

$$\frac{r\,kr}{r + kr}\text{(parallel)},\ r\text{ (smaller)},\ kr\text{ (larger) and } r + kr\text{ (series)}$$

i.e. $$\frac{k}{1 + k}r \quad , \quad r \quad , \quad kr \quad \text{and} \quad r(1 + k)$$

For instance, if the two resistors are 68 Ω and 82 Ω, they will give 37 Ω in parallel and 150 Ω in series, both of which are also preferred numbers.

The four resistance values can be arranged to increase in geometrical progression if $k = k(1 + k)$, i.e. if $k^2 - k - 1 = 0$, whence $k = (1 \pm \sqrt{5})/2 = 1{\cdot}618$. It is interesting to note that this is the famous ratio originated by Fibernocci and used by artists for obtaining ideal proportions.

4
Comparators

Amplitude comparators. Phase comparators. Direct comparison. Integrated comparison. Phase–splitting methods. Vector product devices. Effect of offset waves upon comparators

In Chapter 1 it was explained that, in order to detect a fault or abnormal condition of the power system, electrical quantities or groups of electrical quantities were compared in magnitude or phase angle and the relay operated in response to an abnormal relation of these quantities. The quantities to be compared are fed into a comparator as two or more inputs; in complex relays each input is the vectorial sum or difference of two currents or voltages of the protected circuit, which may be shifted in phase or changed in magnitude before being combined.

4.1. TYPES OF COMPARATORS

The amplitude comparator compares the magnitudes of two inputs by rectifying them and opposing them. If the inputs are A and B, the output of the comparator is $|A| - |B|$ and this is positive if $|A| > |B|$, i.e. if the ratio $|A/B| > 1$. Theoretically, the comparison should be purely scalar, i.e. the phase relation of the inputs should have no effect on the output, but this is only so if at least one of the inputs is completely smoothed as well as rectified.

The phase comparator achieves a similar operation with phase angle; its output is positive if arg A − arg B is positive, i.e. if $\psi = \arg A/B < \pm\lambda$, where λ is an angle determining the shape of the characteristic; $\lambda = 90°$ for a circular characteristic. This can also be expressed as $\lambda > \psi > -\lambda$. Both types of comparator can be arranged either for direct (instantaneous) comparison or to integrate their output over each half-cycle. The two methods will be discussed in the ensuing sections.

In this chapter only two-input comparators will be discussed. Multi-input comparators will be discussed in Chapter 12.

The actual circuitry of electromagnetic tripping relays is generally much simpler than that of static output devices and they have certain advantages (discussed in Chapter 5, Section 5.1); for the sake of simplicity the output devices in the circuit diagrams of this chapter will sometimes be shown as attracted armature relays. It is obvious that an attracted armature relay can be replaced by a triggering circuit, such as a thyristor (SCR), and that an amplifying stage (not shown) may be necessary to operate such a relay in practice.

4.2. OPERATING TIMES OF COMPARATORS

In both phase and amplitude comparators correct measurement in less than half a cycle presents a problem; in the amplitude comparator this is because the ratio of the magnitudes of the two inputs varies throughout the cycle (see Fig. 4.1) unless they are in phase; in the phase comparator the phase angle difference which is being measured is equivalent to time, hence a period of time is required which is at least as much as the phase difference to be measured (see Fig. 4.22).

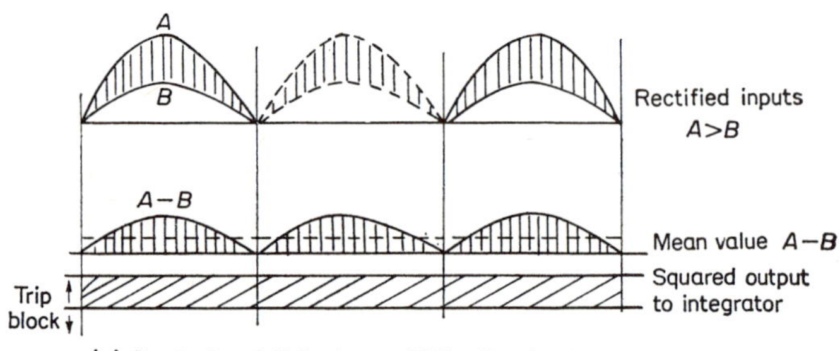

(a) Inputs A and B in phase: unidirectional output

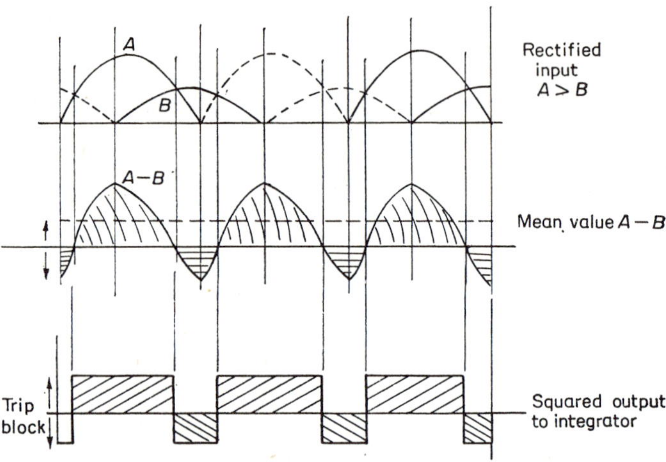

(b) Input B lags input A: alternating output

Fig. 4.1. Effect of phase angle upon direct amplitude comparison

A further delay of up to half a cycle is required if the fault occurs just after the moment in the cycle which is chosen for measurement. This will depend upon the inputs chosen. Faults usually occur 0° to 45° before voltage maximum and this will favour the phase comparator mho relays and delay phase comparator reactance relays for faults near the threshold of operation. Other types of relays can be examined by comparing their measuring period with the voltage wave.

The output of either comparator is a series of positive and negative impulses of double frequency if both half-waves are compared (Figs. 4.1 and 4.22). Their α-plane and β-plane characteristics are two sectors of a circle which can be made more or less than semicircles, as will be explained in Section 4.3.4.

To make a valid comparison of the two inputs at least one of them must be modified or else the alternating output of the comparator must be integrated, as shown in Fig. 4.2. These two methods will be discussed in Sections 4.3.1. and 4.3.2.

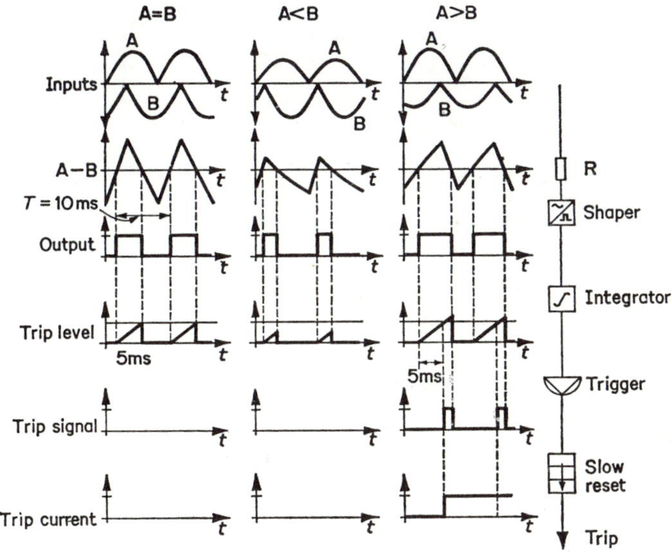

Fig. 4.2. Operation of Integrating amplitude comparator (B.B.Cie.)

4.3. AMPLITUDE COMPARATORS

Figure 4.3a shows how two currents can be compared in magnitude only, using rectifiers and, in Fig. 4.3b, two voltages are compared. The current comparator is usually the more practical because the rectifiers provide a limiting action so that (*a*) the relay can be made very sensitive, (*b*) the voltage across the rectifier bridge remains substantially constant (see Fig. 4.4f) and hence the rectifiers and the sensitive relay are protected at high currents. In the voltage comparator the limiting action is the wrong way, i.e. the increase of resistance at low voltage makes the relay less sensitive at low input and the rectifiers are not protected at high currents.

Figure 4.3c is a compromise using a moving coil relay as the comparator as well as the output device. It has linear characteristics and the double coil enables an auxiliary p.t. to be omitted in some cases; but it is not as efficient as the Fig. 4.3a arrangement because the volt-amperes consumption of the relay coils are added but their pulls (i.e. outputs) are subtracted.

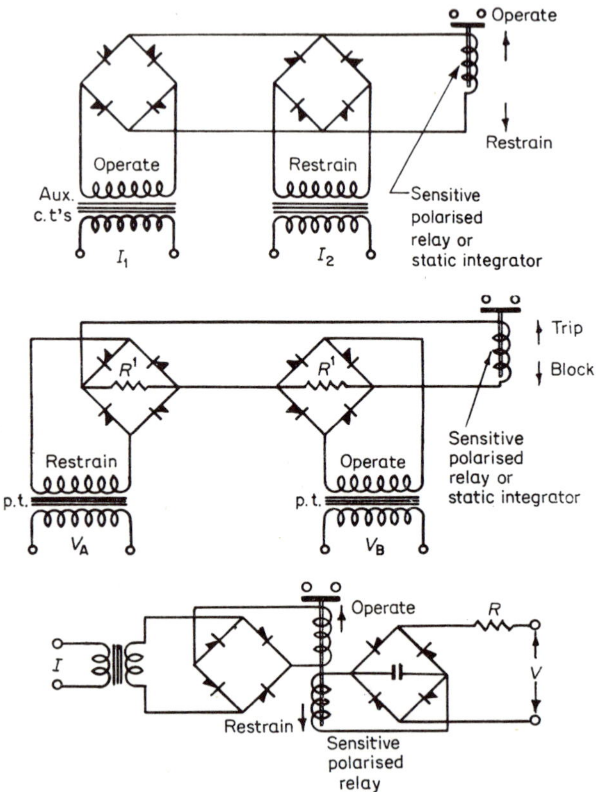

Fig. 4.3. Amplitude comparators

(a) circulating current
(b) opposed voltage
(c) current versus voltage

The operation of the circulating current bridge is as follows. Normally the restraining current preponderates and current flows in the winding of the polarized relay in the blocking direction. Small values of i_r will cause a current to flow in the output relay in the blocking direction, as in Fig. 4.4. b; the voltage drop, $-V$, across the relay serves as a bias in the forward direction of bridge 1. If i_r is increased further, the voltage drop across the relay will rise to a value $-V_t$, the threshold or toe voltage of bridge 1, and it will

TABLE 4.1

Comparators

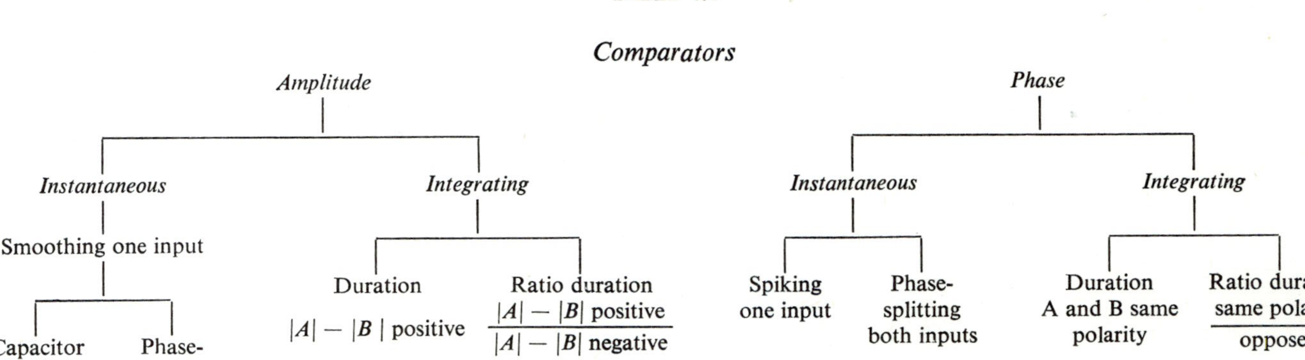

* In the integrating method either the inputs or the output can be squared without affecting phase comparison.

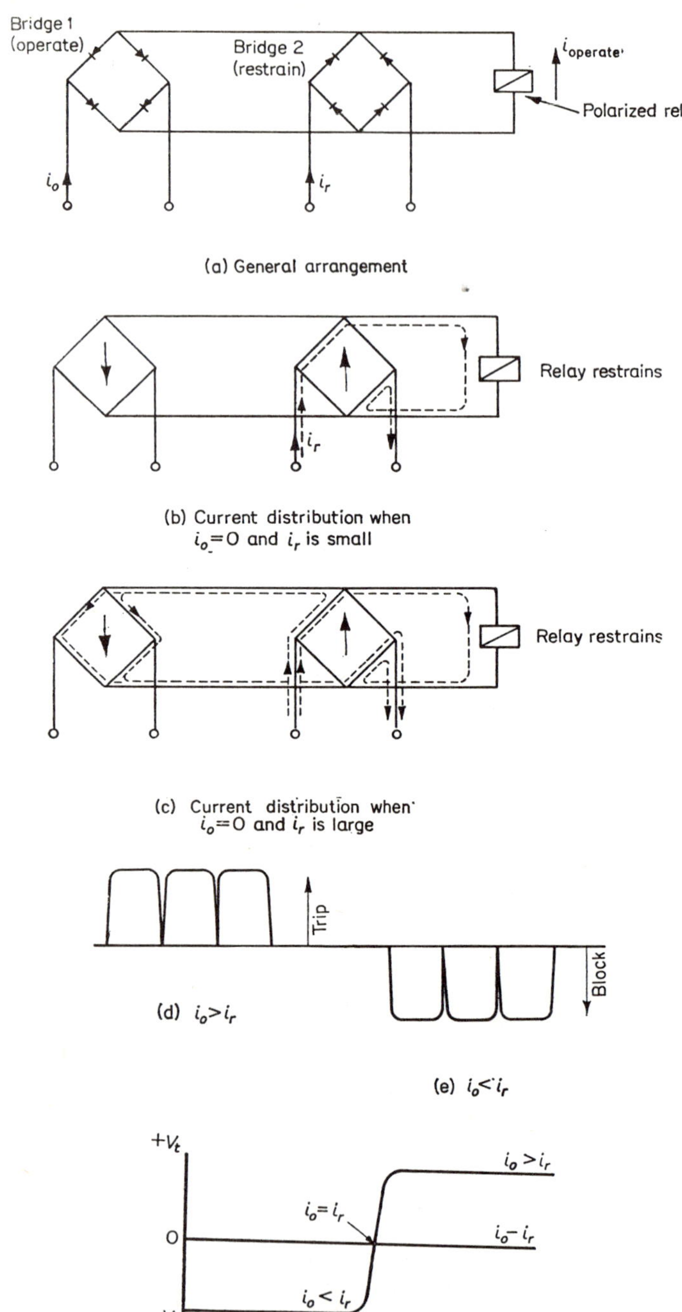

Fig.4.4. Operation of circulating current amplitude comparator

conduct (Fig. 4.4c). The current through the relay consists of fairly flat-topped half-waves corresponding to the case of $i_0 < i_r$, as in Fig. 4.4e.

The reverse is true if i_0 flows alone; the voltage drop across the relay will now be V and this will bias the restraint rectifier in its forward direction. When the voltage drop across the relay attains a value V_t, corresponding to the threshold voltage of two rectifiers in series, the surplus current from bridge 1 is spilled through bridge 2. This corresponds to the case of $i_0 > i_r$ in Fig. 4.4d.

When both bridges are energized simultaneously the relay is responsive to small differences between i_0 and i_r without requiring a very sensitive output relay. The composite characteristic for the relay is shown ideally in Fig. 4.4f. [130]

From the foregoing it can be seen that the current in the relay is a function of the difference between i_0, and i_r. Owing to the non-linear resistance of the rectifiers, the current through the relay is limited to a fixed maximum value (Fig. 4.4f) and the rest of the surplus flows through a rectifier bridge with the smaller current. The voltage across the comparator cannot exceed twice the forward drop (toe voltage) in one of the rectifiers, which is about 0·6 for S_i. The maximum current that can flow in the relay is the saturating voltage of the rectifier V_s divided by the relay coil resistance.

The linearity of the output characteristic can be improved by the use of different semiconductors in the two bridges, such as germanium in the operating bridge and silicon in the restraining bridge.

4.3.1. Direct Comparison

Because of the varying ratios of the instantaneous values of the sinusoidal a.c. inputs during the cycle, direct (instantaneous) comparison is possible only if the restraining (blocking) input is rectified and almost completely smoothed, providing a level of restraint which must be exceeded by a peak of the operating input in order to cause tripping. The smoothing can be done by a capacitor, but this delays the operation somewhat.

A faster method is phase-splitting before rectification, as shown in Fig. 4.5b, where the input is split into six components 60° apart, so that it is smoothed within 5%. Theoretically this would provide circular characteristics even with an instantaneous output relay or level detector such as the trigger circuit [76] shown in Fig. 3.30. The operating time here is determined by the time constant of the slowest arm of the phase-splitting circuit and by the speed of the output device. CR phase-splitting (Fig. 4.5a) is preferable to LR because a 60° CR circuit has a lower time-constant than a 60° LR circuit. $X = 3R$ for a 60° shift.

The time constant with LR is $L/R = X/R\omega = \sqrt{3}/\omega = 5.5$ mS. The time constant with CR is $CR = R/\omega X_c = 1/\omega\sqrt{3} = 1.84$ mS.

This method obviously applies to scalar measurement such as impedance or current differential relaying. To measure a component such as reactance

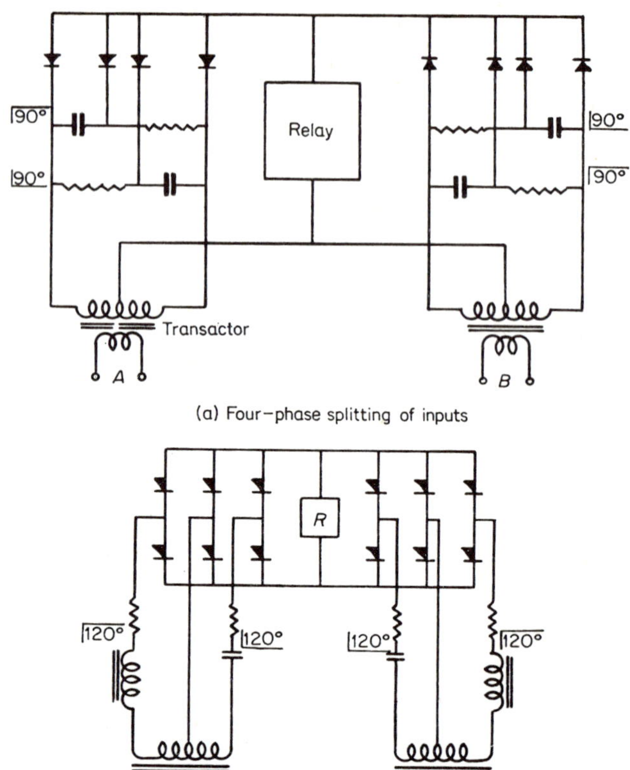

Fig. 4.5. Phase-splitting of inputs to amplitude comparator
(a) two-phase full-wave
(b) three-phase full-wave

or to obtain the offset current characteristic used in a.c. pilot wire relaying, it is necessary not only to rectify one input but to compare it with the value of the other input at a particular moment in the cycle (Fig. 4.6a).

For instance, in a reactance relay, the comparator is made operative only during a pulse at the moment of current zero. In Fig. 4.6b if A is potential and B is current, the potential is $V \sin \phi$ at the moment of current zero and, if this instantaneous value of potential is compared with the rectified current, KI, the comparator operates when $V \sin \phi < KI$, i.e. when $X < K$.

In the U.S.A. a mho characteristic has been obtained [4] by comparing the rectified potential $K'V$ with the instantaneous value of the current $I \cos (\phi - \theta)$ at the moment of voltage maximum; the relay operates when $I \cos (\phi - \theta) > K'V$, i.e. when $Y \cos (\phi - \theta) > K'$. This is illustrated in Fig. 4.6c.

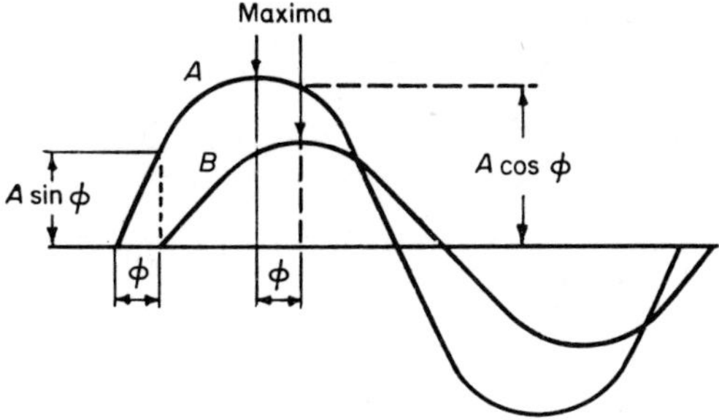

(a) Value of A at B max is $A \cos \varnothing$
Value of A at B zero is $A \sin \varnothing$

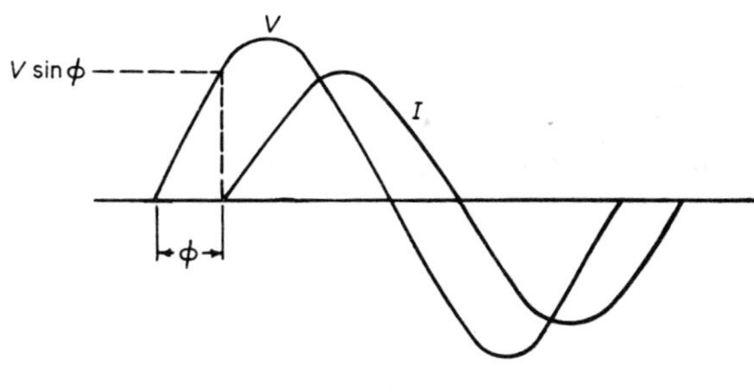

(b) Reactance relay

$$X = \frac{V \sin \phi}{I}$$

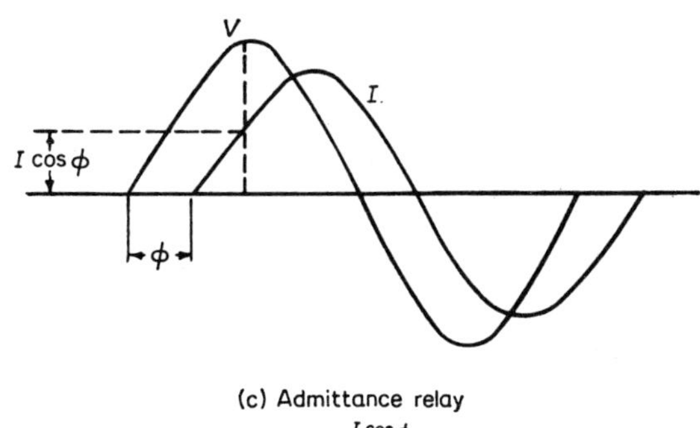

(c) Admittance relay

$$Y = \frac{I \cos \phi}{V}$$

Fig. 4.6. Instantaneous measurement of impedance complex quantities

4.3.2. Integrating Comparator

Integration means measuring ψ, the duration of the positive (tripping) impulses, or comparing it with the duration of the negative (blocking) impulses in the output of the comparator. When the duration is $\frac{1}{4}$ cycle, $\psi = 90°$ and the tripping and blocking periods are equal so that the output relay responds to the mean value of the output and a circular characteristic results.

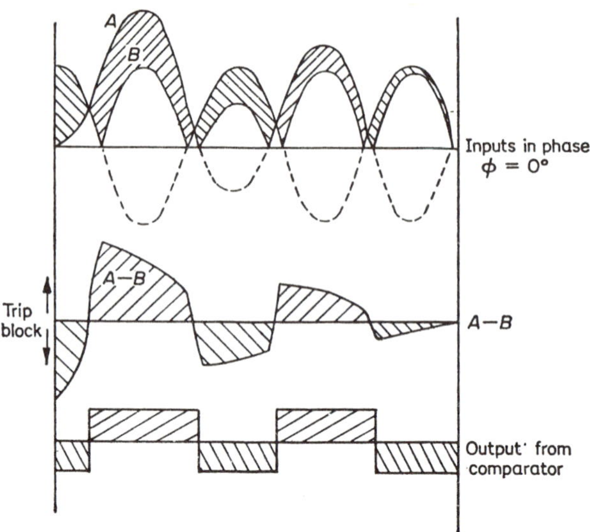

Fig. 4.7a. Inputs in phase and equal. Effect of 100% offset input wave upon output of amplitude comparator

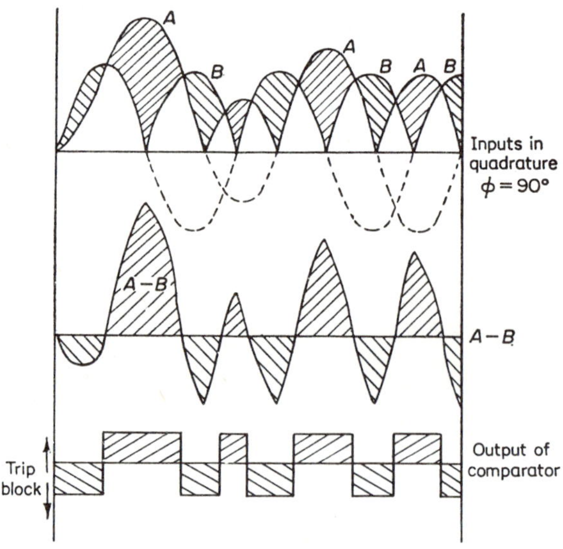

Fig. 4.7b. Inputs equal but in quadrature. Effect of 100% offset input wave. upon output of amplitude comparator

The output of the comparator (Fig. 4.9a) is of the form of rectangular blocks of constant voltage. In one type of integrator (Fig. 4.8b) this voltage charges a capacitor through a resistor and the capacitor voltage increases in an exponential manner because the charging current is proportional to the difference between the capacitor voltage and the impressed voltage.

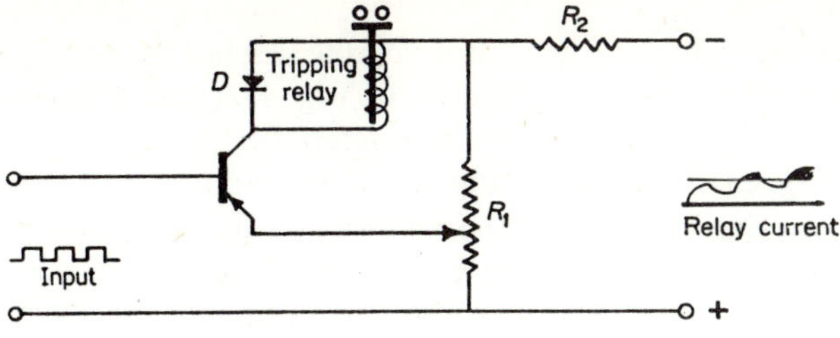

(a) Using inductance of relay coil

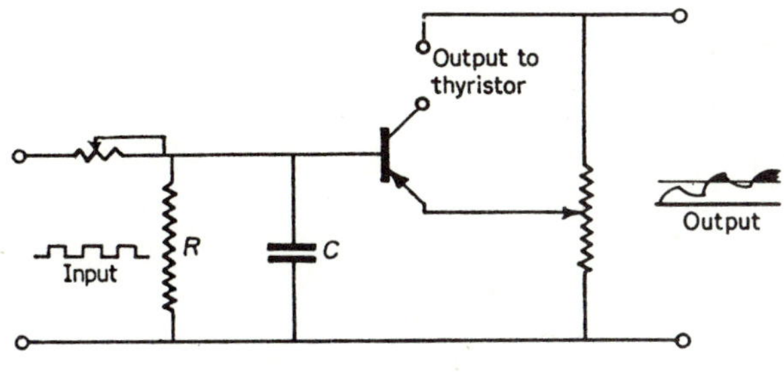

(b) Using capacitor

Fig. 4.8. Integrating circuits for outputs of comparators

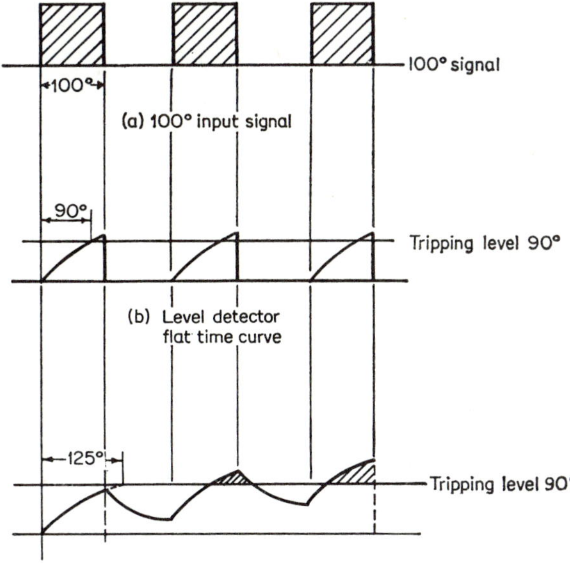

(c) Time integrator

Time curve rounded near cut-off prevents transient overreach with one input offset

Fig. 4.9. Output devices for comparators

Tripping occurs when the capacitor voltage reaches the setting value of the level detector and triggers a thyristor. In Fig. 4.9b the tripping level is set at 90° in order to obtain a circular characteristic [76].

In another type of integrator (Fig. 4.8a) the inductance of the coil of the tripping relay is used to control the build-up of the coil current in an exponential manner and the relay trips when the current reaches its pick-up value. This method is slower than the capacitor method by the operating time of the relay. It would be the same time if a separate inductance were provided and a thyristor used for tripping.

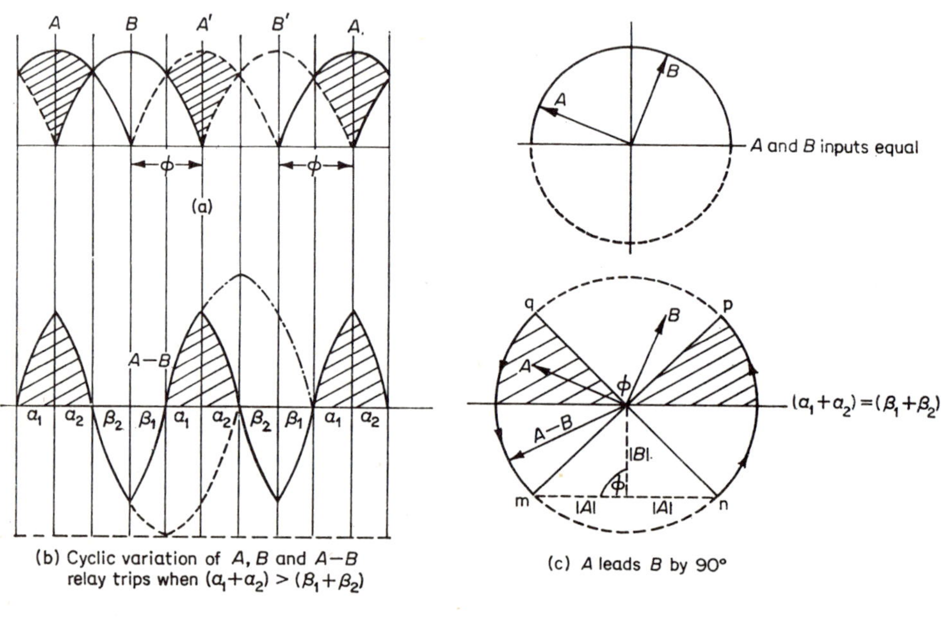

(b) Cyclic variation of A, B and $A-B$ relay trips when $(\alpha_1+\alpha_2) > (\beta_1+\beta_2)$

(c) A leads B by 90°

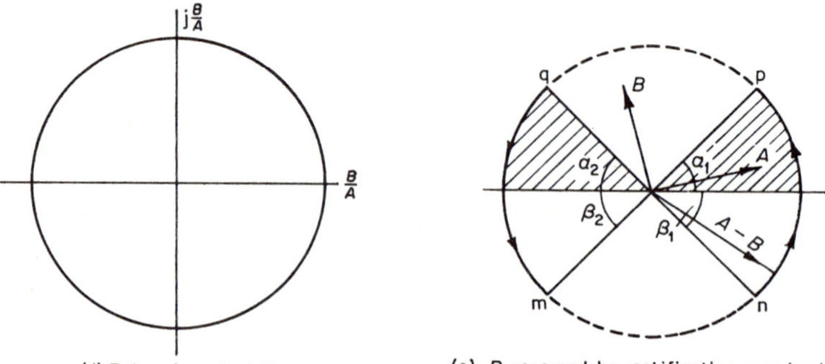

(d) Relay characteristic

(e) B, reversed by rectification, now leads A by 90°

Fig. 4.10. Amplitude comparator when $|A| = |B|$ and $\varnothing = 90°$

It is clear from Fig. 4.9b that the capacitor integrator trips in a time between $\frac{1}{4}$ and $\frac{1}{2}$ cycle depending upon the moment of inception of the fault, but it is susceptible to overreaching if one input is offset by a d.c. component as shown in Figs. 4.7 (amplitude comparator) and 4.31a (phase comparator) because the output of the comparator is spuriously lengthened by the offset.

To prevent this overreaching, an English company has delayed the discharging of the capacitor so that the charging may accumulate over more than one $\frac{1}{4}$-cycle block, as shown in Fig. 4.9c. In this system the peak values of the capacitor voltage approach the pick-up level assymptotically for 90° comparator output blocks and the operating time is one cycle near this threshold condition. On the other hand, the operating time is between $\frac{1}{2}$ and $\frac{3}{4}$ cycle for a fault condition well within the characteristic or on its axis.

This subject will be dealt with more fully in Section 4.4.6, which considers the effect of offset waves upon the output of comparators.

4.3.3. Wave-shape of Output

The peaked wave form of the output of the amplitude comparator is due to the sudden change of direction of the rectified inputs at the end of each half cycle.

From Fig. 4.10 it will be seen that the vector A–B rotates from q to m (Fig. 4.10c) and the output wave follows the sinusoidal locus $\alpha_2\beta_2$ (Fig. 4.10b) and then jumps to n continuing to p on another sinusoidal locus $\beta_1\alpha_1$.

In Fig. 4.10 the two inputs A and B are equal but 90° out-of-phase. In Fig. 4.11 they are again equal but only 45° out-of-phase, so that the wave form is not symmetrical. Figures 4.12 and 4.13 are similar but $A > B$ in Fig. 4.13, giving a mean value of output in the tripping direction and making the tripping duration $(\alpha_1 + \alpha_2)$ exceed the blocking duration $(\beta_1 + \beta_2)$.

4.3.4. Non-circular Characteristics

If the integrator is set to trip on positive impulses of less than 90° duration the characteristic will be apple-shaped, i.e. it will consist of two equal circular arcs of more than a semicircle (Fig. 4.14). If the setting is more than 90° the characteristic will be lemon-shaped, i.e. it will consist of two equal circular arcs of less than a semicircle.

These characteristics on the α-plane correspond to a bent straight line on the β-plane. This subject will be dealt with in more detail in Chapter 10 on Distance Relays.

For a circular characteristic the charging and discharging rates of the integrator are made equal for a full-wave comparator which operates in each half cycle. The discharging rate has to be made one-third of the charging rate if the comparison is made only during one half-cycle, e.g. the phase comparator of Fig. 4.24.

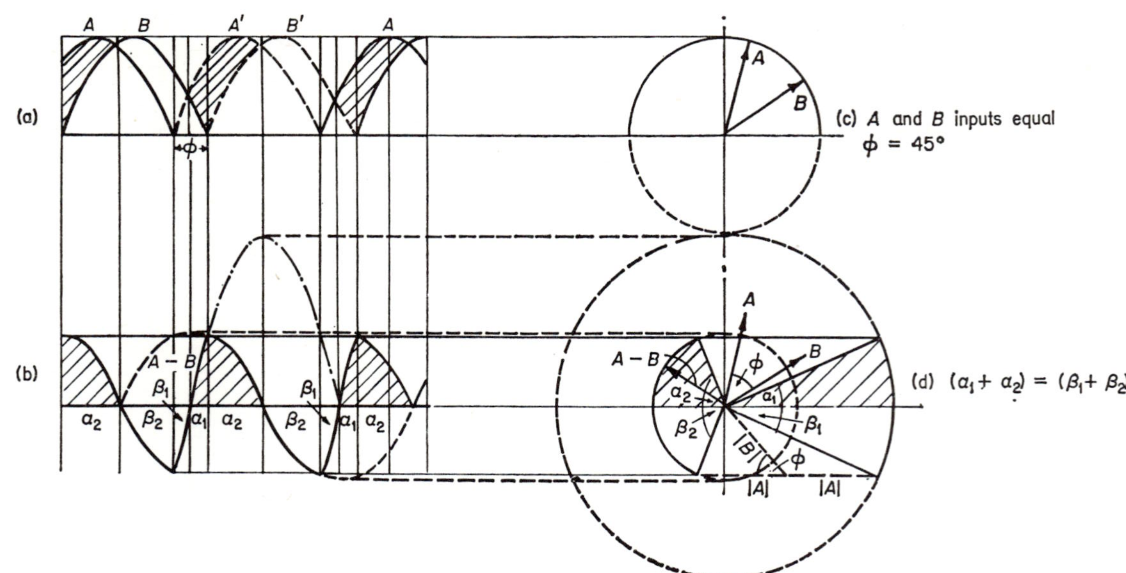

Fig. 4.11. Amplitude comparator when $|A| = |B|$ and $\varnothing = 45°$

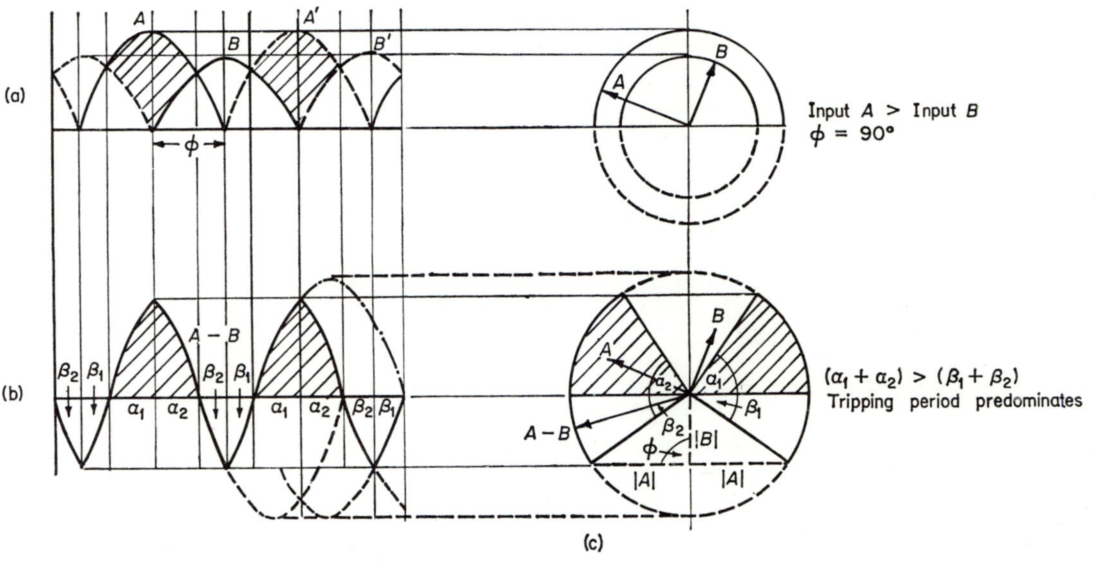

Fig. 4.12. Amplitude comparator when $|A| > |B|$ and $\varnothing = 90°$

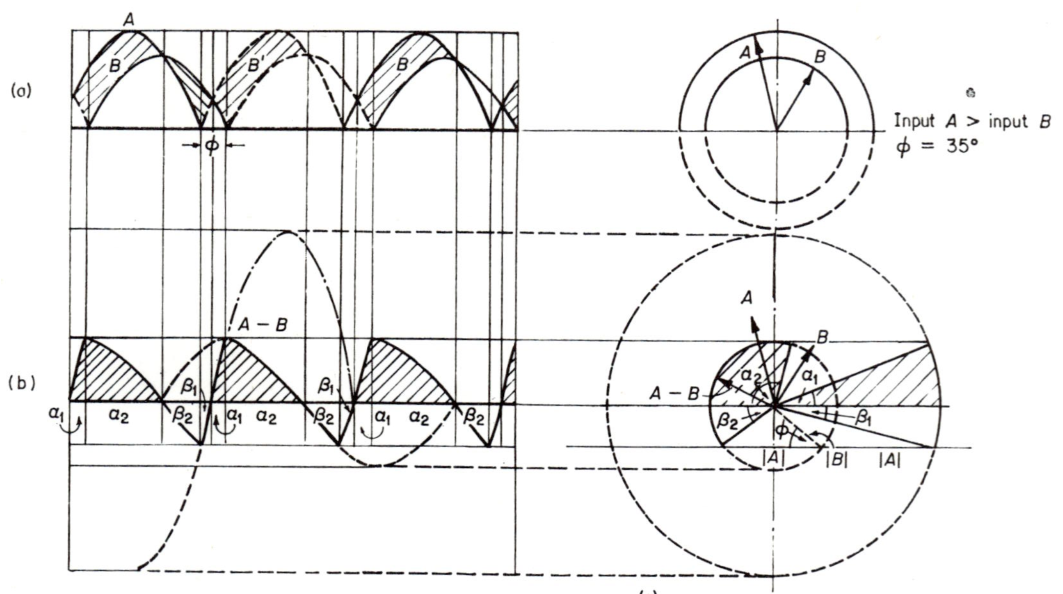

Amplitude comparator

Fig. 4.13. Amplitude comparator when $|A| > |B|$ and $\varnothing = 45°$

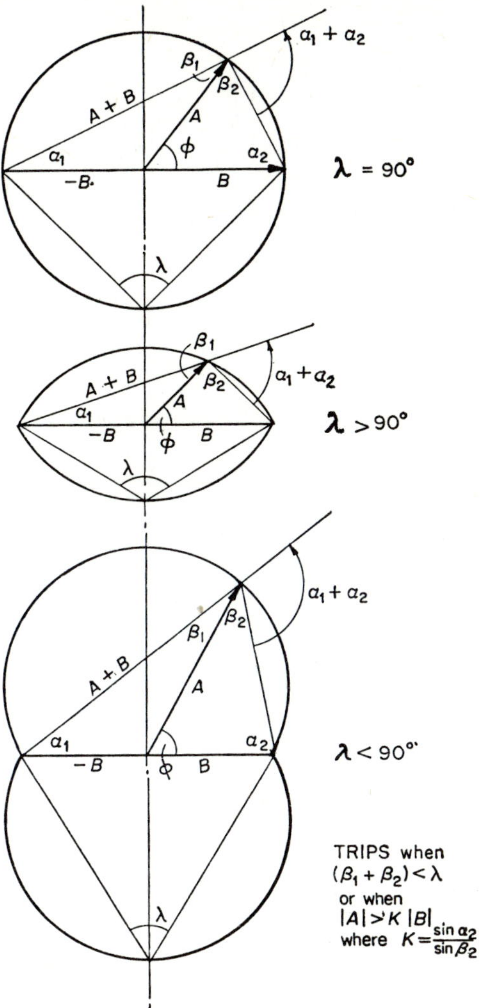

Fig. 4.14.

4.3.5. "Squared" Inputs

The inputs may be converted to rectangular wave form before comparison, as shown in Fig. 4.15, and the difference of the two squared waves will again be a series of blocks of which the duration of the positive blocks can be measured, 90° duration indicating equality of the inputs as before.

This method has been extended in Europe to turn the amplitude comparator into a phase comparator by supplying the amplitude comparator with the sum and difference of the two squared input quantities, as shown in Fig. 4.15. When $|A + B| = |A - B|$ it indicates that A and B are 90°

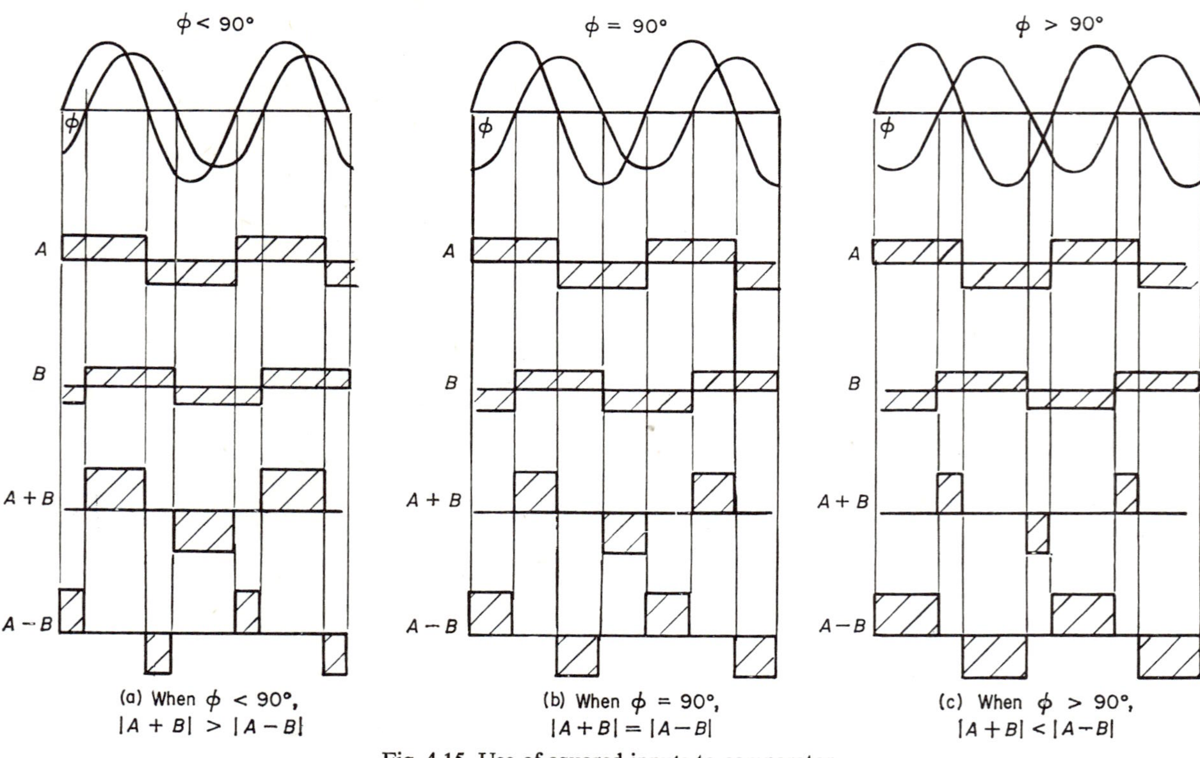

Fig. 4.15. Use of squared inputs to comparator

apart. There appears to be no particular advantage to this method and it requires somewhat more semiconductors than a phase comparator [49].

In Chapter 12 it will be shown that the use of a third rectifier input enables elliptical or hyperbolic characteristics to be obtained [13, 74].

There appear to be no practical amplitude comparator circuits using transistors directly for comparison. They could be used in circuits similar to Fig. 4.3, replacing the rectifiers, but the resulting circuit would be complicated and would be subject to inaccuracy due to variations in the transistor characteristics.

4.4. PHASE COMPARATORS

There are two main types of static phase comparator; (1) those whose output is a d.c. voltage proportional to the vector product of the two a.c. input quantities, (2) those which give an output whose polarity depends upon the phase relation of the inputs. The latter are sometimes called 'coincidence' type and can be direct-acting (instantaneous) or integrating.

4.4.1. Vector Product Devices

These devices produce an output proportional to the vectorial product of the a.c. input quantities.

(a) *Hall effect.* The principle of the Hall generator is illustrated in Fig. 4.16a. If a magnetic flux $|\Phi| \sin \omega t$, proportional to one a.c. electrical quantity, is arranged to pass through the surface of a flat crystal of N-type germanium (or any metal with a good Hall effect) and a current $|I| \sin (\omega t + \alpha)$, proportional to another a.c. electrical quantity, is passed through the crystal from the middle of one edge to the middle of the opposite edge, then a d.c. emf E_h will appear between the midpoints of the remaining pair of edges.

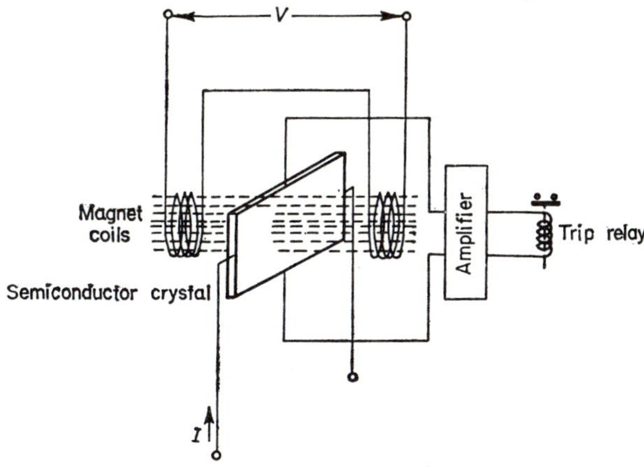

Fig. 4.16a. Hall Effect phase comparator

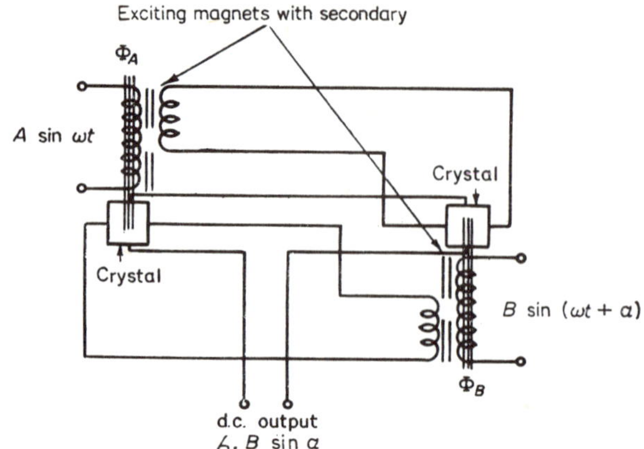

Fig. 4.16b. Hall generators cross-connected to give sine product output

If H is the field strength, I the current in amperes, ω the thickness of the germanium and R is the Hall coefficient, then the Hall effect voltage is

$$E_h = \frac{RIH}{\omega} = K\Phi I \tag{4.1}$$

$$= K' [|\Phi| \sin \omega t] [|I| \sin (\omega t + \alpha)]$$

$$= K' [|\Phi| |I| \cos \phi - |\Phi| |I| \cos (2\omega t + \alpha)]. \tag{4.2}$$

In Eq. (4.1) and (4.2) if the field flux density is expressed in maxwells/sq cm, H in oersteds, I in amperes and ω in cm, then the Hall coefficient can be expressed in volt/cm^3/amp. maxwell and has a value of about $8 \cdot 10^{-5}$ for germanium. In current applications to instruments, H is usually about 450 oersteds and ω is about 0.1 cm.

It will be noted that the first term is a d.c. voltage proportional to the vectorial product of Φ and I and the second term is an a.c. voltage of double frequency proportional to the scalar product of Φ and I. If the flux Φ is produced by a voltage V across the magnet coil, the first term can be arranged to be a measure of watts or a vector component of VA and the second term a measure of scalar VA. Either can be suppressed by suitable circuitry

The double frequency a.c. term can be eliminated by cross-connecting two Hall generators, as in Fig. 4.16b, so that one unit has flux A and current B and the other has flux B and current A. If their outputs are then opposed the double frequency components are cancelled and the output is the sine product of A and B. The proof is as follows:

$$i_A \propto B \cos (\omega t + \alpha)$$

because the secondary voltage from the exciting magnet is in quadrature with the primary current.

$$\phi_A = A \sin \omega t.$$

Similarly $i_B \propto A \cos(\omega t + a)$ and $\phi_B = B \sin \omega t$

$$e_H = e_A - e_B \propto A\frac{dB}{dt} - B\frac{dA}{dt}$$

$$e_A \propto AB \sin \omega t \cos(\omega t + \alpha)$$

$$\propto \frac{AB}{2} \sin(2\omega t + \alpha) - \sin \alpha \quad (4.3)$$

$$e_B \propto AB \sin(\omega t + \alpha) \cos \omega t$$

$$\propto \frac{AB}{2} \sin(2\omega t + \alpha) + \sin \alpha \quad (4.4)$$

therefore $$e_H \propto AB \sin \alpha. \quad (4.5)$$

This is a sine product (algebraic) instead of the cosine (cross) product which a single Hall generator produces Eq. (4.2). In Chapter 8 it will be seen that this is useful in the design of polyphase directional relays. It should be noted that the operation of the cross-connected Hall generator is analogous to that of the induction cup in which the torque is derived from the interaction of the current induced in the cup by the magnetic flux from one pole pair with the flux from the other pole pair [6]. It follows that the static equivalent of an 8-pole polyphase directional relay can be made from three such cross-connected generators, i.e. six Hall generator interconnected in sequence (Fig. 8.12).

Another interesting fact is that the Hall generator can be made to measure the reciprocal of a current and two interconnected Hall generators can thus be made to measure the quotient of two inputs [58]. The reciprocal is obtained by measuring the current which is fed back through the crystal when the voltage output of the crystal us subtracted from a constant direct reference voltage and the difference voltage controls an amplifier whose output current goes through the crystal; this current is then the reciprocal of the current in the magnet coil.

The Hall effect has been used in Russia for protective relays but not elsewhere because of cost, temperature error and low output. However, it has been found that the temperature effect is only very slight with indium arsenide and the output is not only higher than with germanium but has been further increased by replacing the crystal by a film of the semiconductor a few microns thick, sandwiched between two glass plates 0·00001 in. thick, so that the total magnetic gap is about 0·00002 in. and a very high flux density can be achieved with quite a low power input. The output is about 2 volts/amp/kilogauss.

(b) *Magneto-resistivity*. The Hall effect is associated with the effect of a magnetic field upon the resistance of the material chosen (magneto-resistivity) and with its electron mobility. Indium arsenide (InAs) has the highest value of electron mobility; indium antimonide (InSb) has about half this value and germanium about 5%.

The magneto-resistivity of a thin disc of semiconductor can be used directly as a phase comparator as an alternative to passing polarizing current through it. The disc diameter should be a maximum and its thickness a minimum in order to have the strongest possible magnetic field through it; this is of course limited by the brittleness of the material, but the Russians have made practical discs with a conducting metal rim as one terminal and a wire soldered to the centre of the disc as the other.

If one a.c. input voltage produces the magnetic field through the disc and the other passes current radially through it, this current will be a maximum when the two voltages are in phase and zero when they are in quadrature (±90°), so that the action is similar to that of the phase comparator rectifier bridge.

Figure 4.17 shows circuits developed in Russia [7] for mho and reactance relays which are similar in operation to the phase comparator rectifier bridge described under Section 4.4.3 . It is understood that this principle is used in Russian relays actually in production.

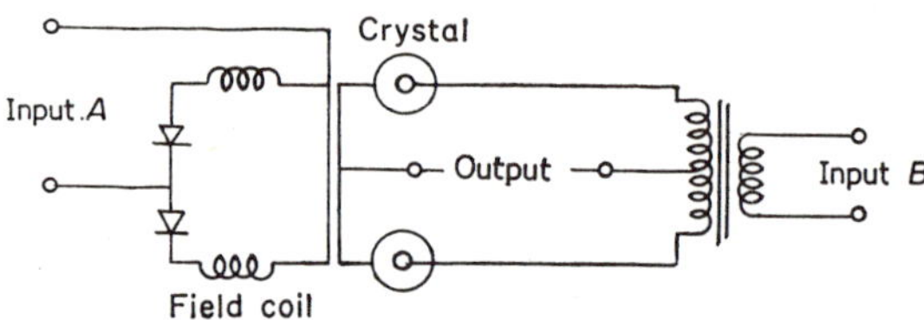

Fig. 4.17. Magneto-resitivity phase comparator

Advantages olaimed for the magneto-resistance wattmeter over the Hall Effect type include:

(*a*) Much simpler construction of semiconductor element.
(*b*) No need for d.c. as well as a.c. balancing between terminals.
(*c*) Less trouble due to rectifying action because of large areas of contact.
(*d*) Absence of thermo-electric e.m.f. due to local heating effect.
(*e*) Much higher gain in output.
(*f*) Possible elimination of induction error.
(*g*) Lowest cost because only two terminals and less need for dimensional symmetry.

4.4.2. Coincidence Circuits

A coincidence type of phase comparator operates if the phase angle ϕ between two inputs is less than a predetermined angle, usually 90°.

From Fig. 4.22a it will be seen that, if the two inputs are less than 90° apart, their sine waves will be of the same polarity for more than 90° of the cycle, i.e $\frac{1}{4}$ cycle or $\pi/2$ radians. Hence $\psi = 90° - \phi$.

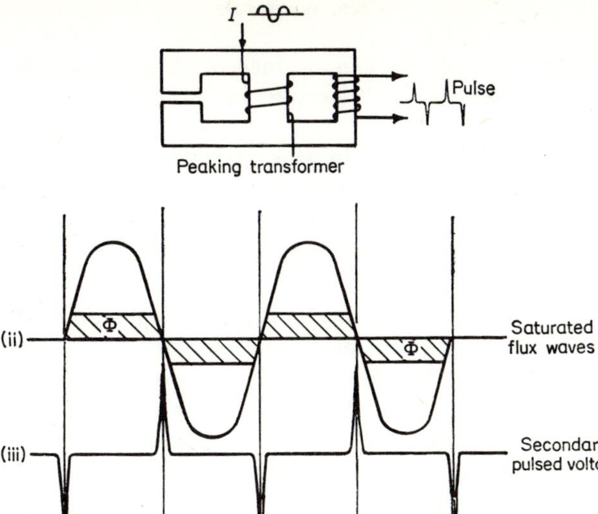

(a) Magnetic pulsing device

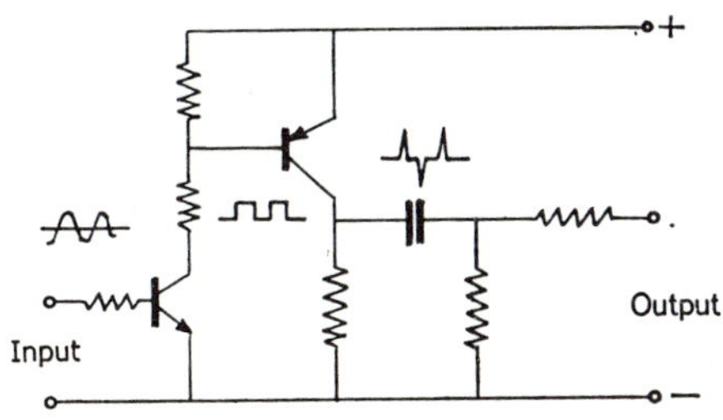

(b) Capacitative pulsing circuit

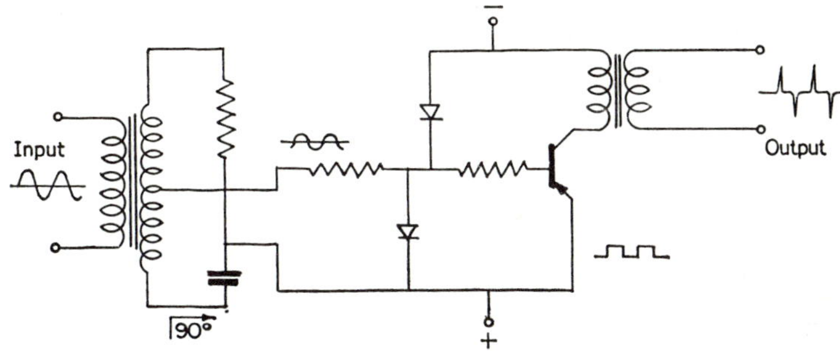

(c) Diode pulsing circuit with voltage step-up and 90° shift

Fig. 4.18. Pulsing devices

(a) transformer saturation
(b) capacitor
(c) transistor saturation

This period of coincidence can be measured in a number of different ways; the simplest methods are described in the following:

(a) *Direct phase comparison.* In this method one input is "squared" and the other turned into a spike (Fig. 4.18) at the instant of its peak value; the spike and block signals are then fed through an AND gate, as shown in Fig. 4.19c. If the signals are coincident at any time, the two inputs must be within $\pm 90°$.

The 'squaring' of one input means first limiting it and then amplifying it to form a rectangular wave or block in phase with the original sine wave. The spike is produced either by a peaking transformer (Fig. 4.18a), by a capacitor charge/discharge circuit (Fig. 4.18b) or by squaring the wave with diodes and feeding it through a transformer (Fig. 4.18c).

Since all these circuits produce pulses as the wave crosses the zero line, the input wave must be shifted 90° before pulsing in order to make it occur at the peak of the wave. This shift makes it a sine output device. In other words, the output of a phase comparator is inherently $\cos(\phi - \theta)$ where θ is 0 but, in the spike and block comparator, $\theta = 90°$ and the output is $\cos(\phi - \pi/2) = \sin\phi$, i.e. the relay operates when $180° > \phi > 0°$.

A *pulse-stretching* circuit, such as is shown in Fig. 4.19f, makes the output signal last long enough to operate a tripping device.

This method enables an instantaneous output relay to be used and permits tripping in less than half a cycle. Like all instantaneous measurements it is affected by harmonics and by spurious spikes caused by external interference. However, this becomes a calculated risk if the circuit is adequately shielded against interference.

This circuit is the dual of the direct amplitude comparison circuit, i.e. turning one input into a spike for phase comparison is the dual of smoothing one input completely for amplitude comparison.

The spike and block method was first used in an electronic distance relay with which the author was associated [26]. A transistorized version was first described by Loving in the U.S.A. [8] and by Adamson and Wedepohl in the U.K. [9]. Fig. 4.19c shows the block diagram of the circuit. Figure 4.19d shows the transistor comparator used in the Loving circuit and Fig. 4.19e the Adamson/Wedepohl circuit.

Figure 4.19f shows how the input single duration can be increased from the original spike width of about 5° to a period long enough to trip a relay.

These circuits operate only during one half of the cycle. By duplicating the circuit for the other half-wave polarity, the operating time can be reduced to less than half a cycle.

Some trouble has been experienced with this method due to d.c. offset in the input signals which changed the duration of the block and the moment of the spike, affecting the characteristic. Another source of transient error is the 90° phase shifting circuit required for the spike input.

In a more recent transistorised mho distance relay of this type developed

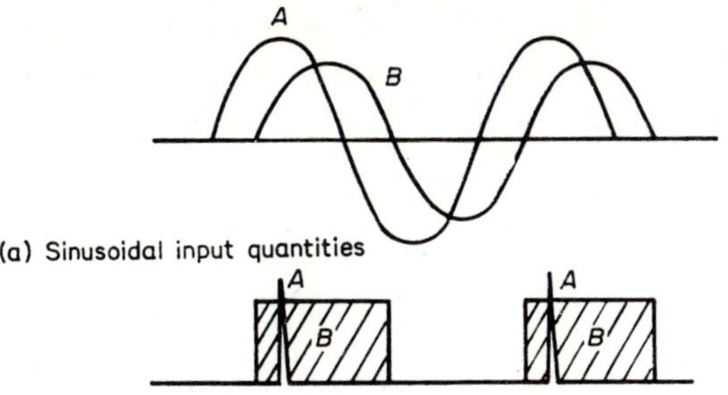

(a) Sinusoidal input quantities

(b) Corresponding spike and block

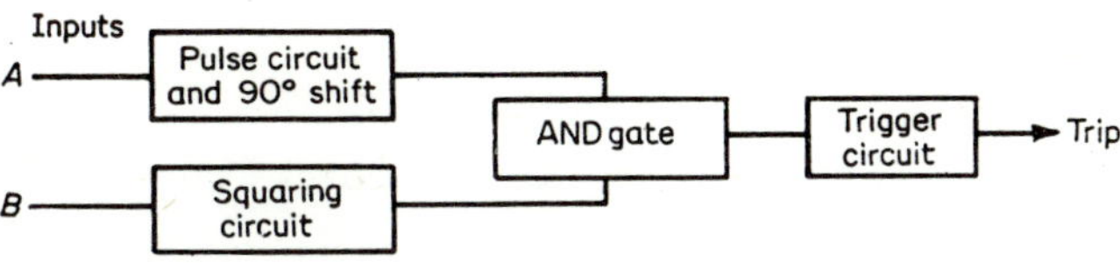

(c) Block diagram of circuit

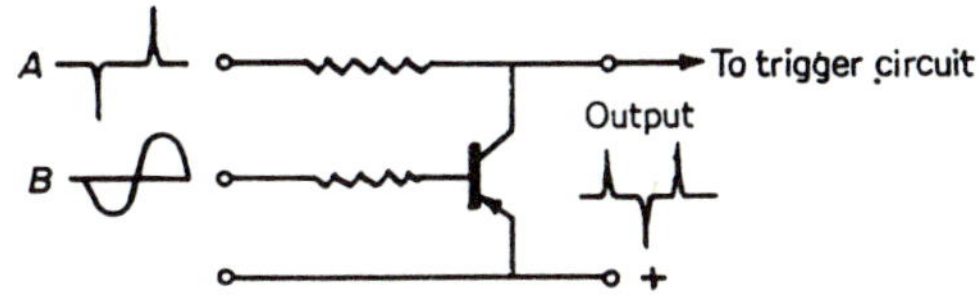

(d) Loving comparator

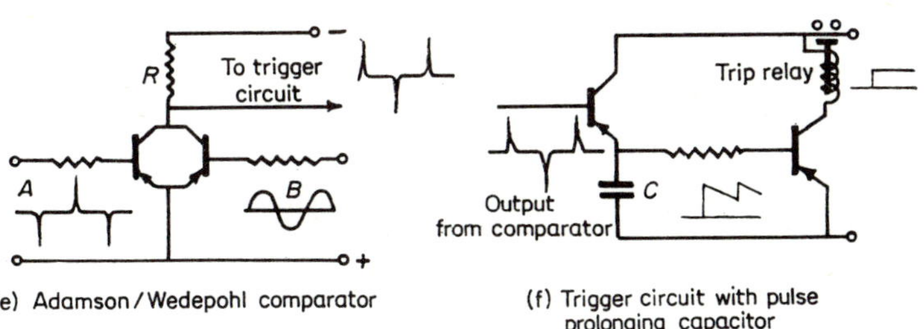

(e) Adamson/Wedepohl comparator

(f) Trigger circuit with pulse prolonging capacitor

Fig. 4.19. Direct phase comparators
Trips when the rectified pulse from one input is coincident with the rectified block from the other input

in the U.S.A. [32], the manufacturers claim that the overreach due to d.c. offsets can be limited to less than 10% by making the phase angle of the replica impedance 80° and adjusting the phase angle of the mho characteristic to equal that of the protected line angle ϕ, thus making the spike occur at θ from the voltage zero, where $\theta = \phi + 10°$.

Reactance relays have traditionally used phase comparators for measuring the angle between $IZ_r - V$ and IX and tripping when this angle ($90° - \alpha$ in Fig. 4.20) is less than 90°. As in the mho relay, if one of the inputs is a pulse, it has to be shifted 90°. This presents no problem in the reactance relay since it means only replacing IX by IR, to provide the necessary 90° shift; this eliminates transient overreach and permits faster tripping which occurs when $\alpha > 0°$.

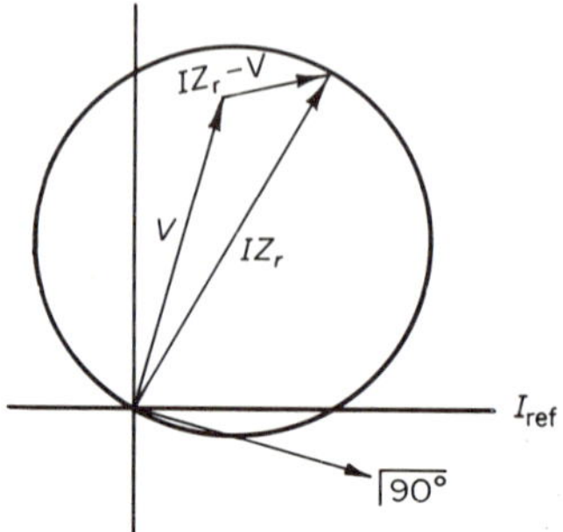

(a) Instantaneous Mho relay Trips if $(IZ_r - V)$ leads $V\,\overline{|90°}$

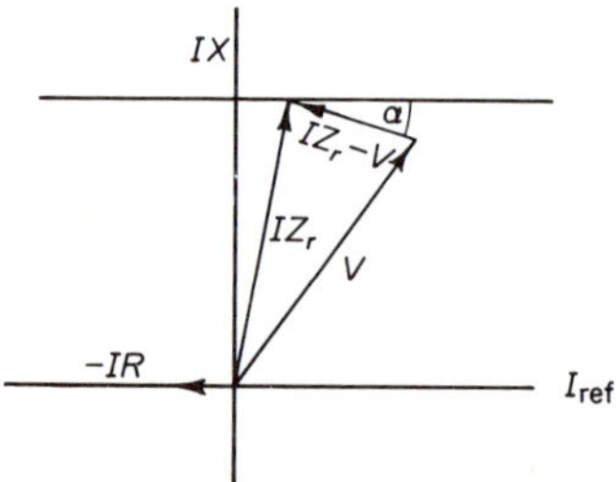

(b) Instantaneous reactance relay
Trips if $IZ_r - V$ lags $-IR$

Fig. 4.20. Application of direct phase comparator to distance relaying

(b) *Phase-splitting*. Another method of direct phase comparison is to split both inputs into two components shifted ±45° from the original wave, as shown in Fig. 4.21. The four inputs now produced are fed into an AND gate and tripping results if all are simultaneously positive at any time in the cycle. Figure 4.21a shows that, when input A leads input B, the coincidence limit will be determined by $A\,\underline{|45°}$ and $B\,\overline{|45°}$. When B leads A the limits will be $A\,\overline{|45°}$ and $B\,\underline{|45°}$.

It will be seen that the coincidence of the four inputs will occur when $90° > \phi > -90°$ where ϕ is the angle by which A leads B.

This method is slightly slower than the spike and block method because of the time-constants of the 45° phase-shifting circuits but, unlike the spike and block method, it will not be affected by interference spikes. On the other hand it can have source transient error due to the different time constants of the LR and CR phase shifting circuits. Two comparators will be needed for operation in less than a cycle, one for each polarity.

An alternative method of 90° phase-splitting is to divide the input into two equal parts, one of which is lagged or led 90° and the other is not shifted at all. This method is shown in Fig. 4.21 b in a simplified diagram; it is somewhat more subject to transient error than the ±45° method.

(c) *Integrating phase comparison.* In this method the time overlap of the two sinusoidal inputs is measured for each cycle (Fig. 4.22 a) by integrating the output of an AND gate through which they are fed. This is slower than the direct comparison of spike and block signals but it has the advantage of immunity to spurious pulses from outside sources since they are too short to affect it.

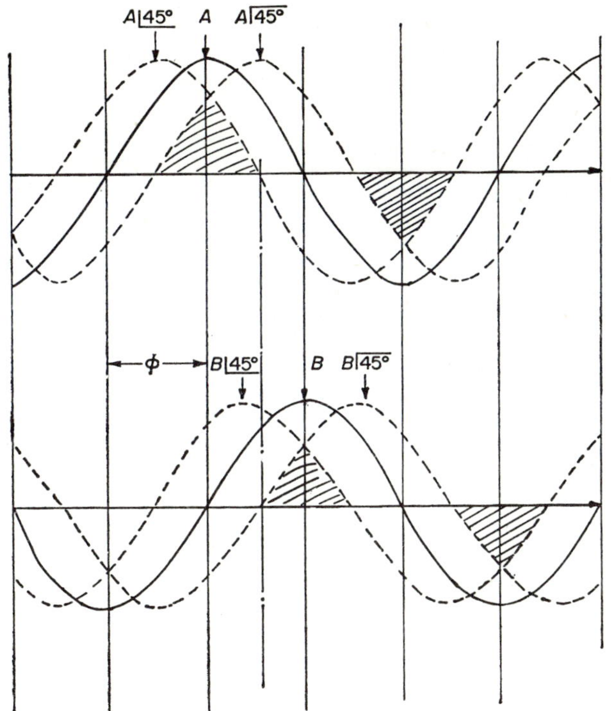

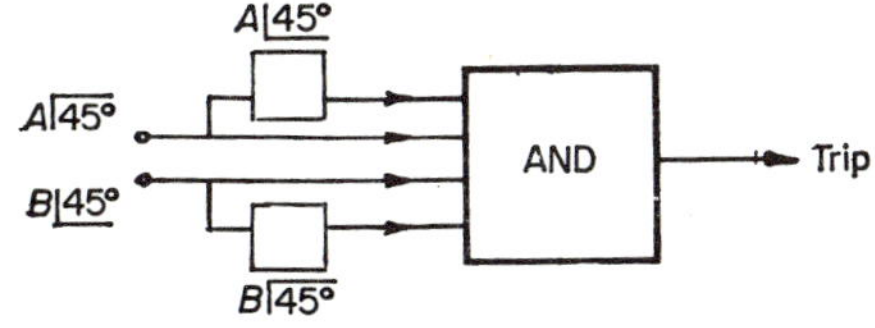

4.21 a. Phase-comparator with phase-split inputs.
Coincidence determined by $A\underline{|45^\circ}$ and $B\overline{|45^\circ}$ when A, leads B; or by $A\overline{|45^\circ}$ and $B\underline{|45^\circ}$ when B leads A

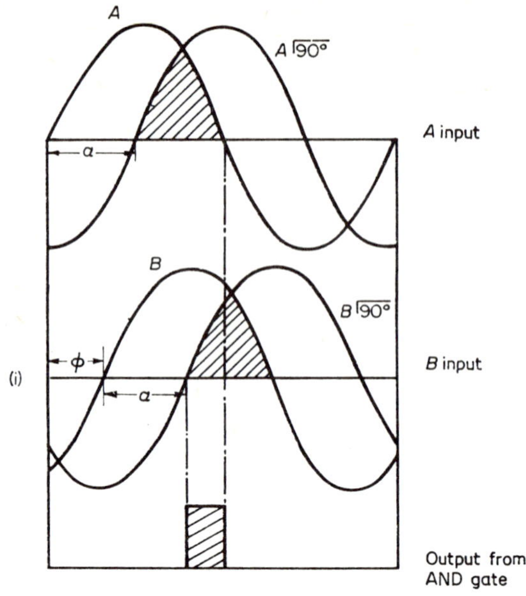

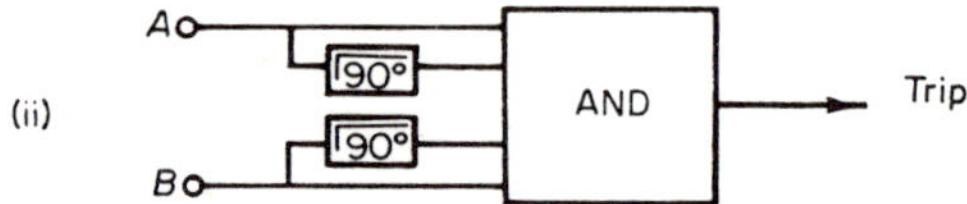

Fig. 4.21 b. Instantaneous phase comparison with phase-split inputs
Coincidence determined by A and $B\,\overline{|90^\circ}$ when A leads B; or by $A\,\overline{|90^\circ}$ and B when B leads A

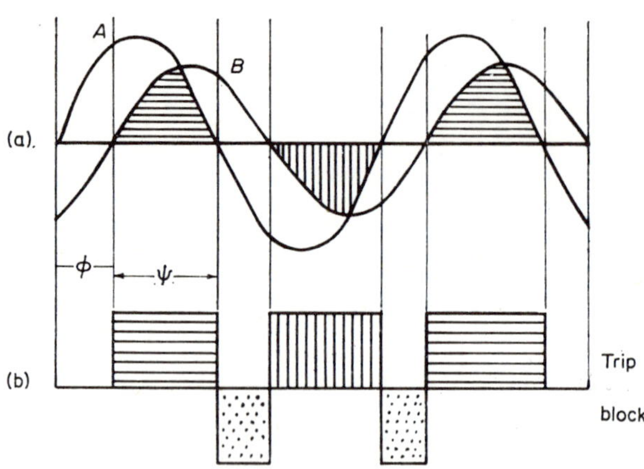

Fig. 4.22. Phase comparator measures duration of coincidence of sine wave inputs.

Trips when overlap $\psi > 90^\circ$
Balances when $\psi = 90^\circ$
Blocks when $\psi < 90^\circ$

4.4.3. Types of Integrating Phase Comparators

The early integrating phase comparators proposed for protective relaying used transistors. Rectifier bridge phase comparators have been used more recently and offer advantages of lower cost and greater flexibility.

With both direct and integrating types of phase comparators an AND gate is required. Different types of AND gates have been used by the designers of direct and integrating comparators, but it will be recognized that any type of AND gate can be used with either comparator.

a) Early relays using transistor-type AND gate. The integrating phase comparator was first described by Bergseth in 1957 in the U.S.A [10], followed by Adamson and Wedepohl in the U. K. [11]. The block diagram is shown in Fig. 4.23a, the Bergseth circuit in Fig. 4.23b and the Wedepohl circuit in Fig. 4.23c.

In the Bergseth circuit the period during which both signals were negative and the transistor was giving an output was measured by the integrating effect of an inductance. The transistor output was rectangular (squared) and of fixed amplitude so that the current through the inductance rose exponentially to a value dependent upon the duration of the signal; it

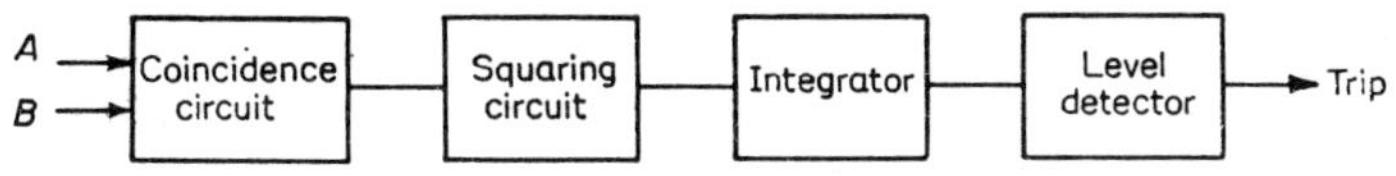

(a) Block diagram of direct phase comparator

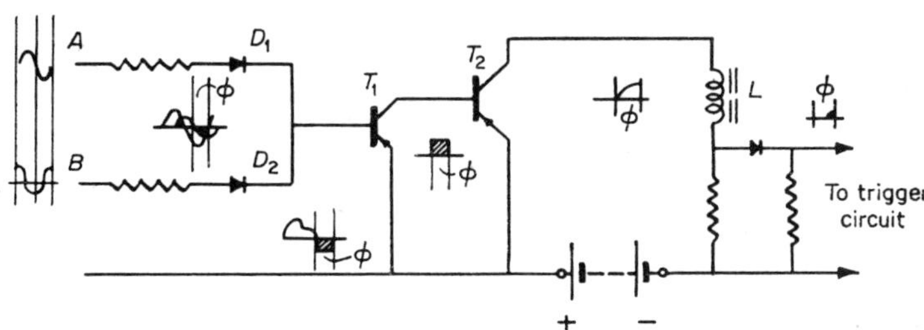

(b) Bergseth circuit (simplified) with inductance integrator

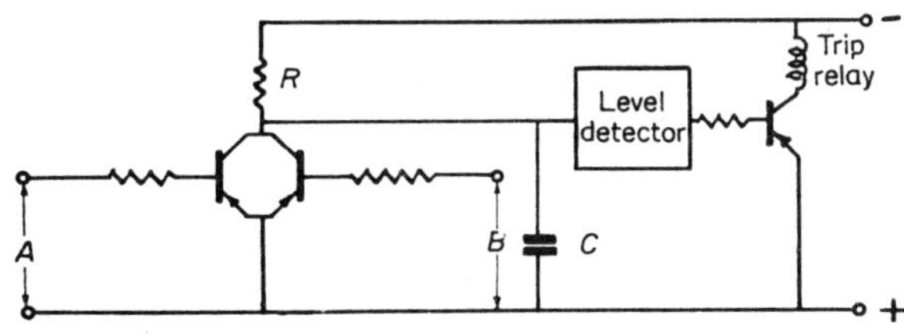

(c) Wedepohl circuit (simplified) with $R-C$ integrator

Fig. 4.23. Integrating phase comparator circuits

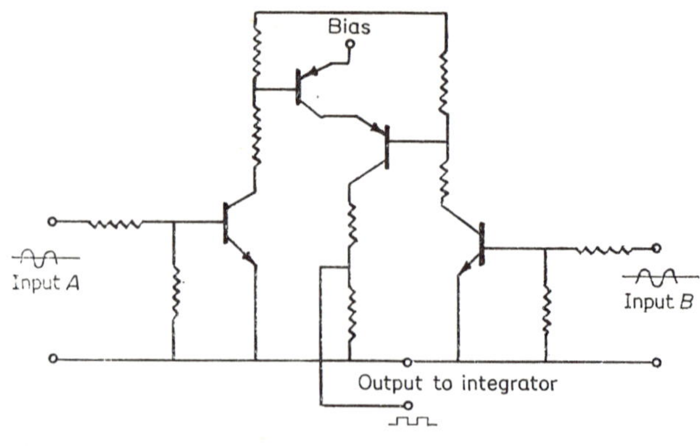

(d) Series transistor circuit

Fig. 4.23. Integrating phase comparator circuits [15]

operated a trigger circuit causing tripping when the current level reached a value corresponding to a coincidence period of 90°.

The Wedepohl circuit used a parallel (back-to-back) connection of transistors in the phase comparator instead of two rectifiers. The potential of the joint collector leads became sufficiently negative to operate the level detector only when both transistors were cut off, i.e. when both inputs were positive. During this condition the capacitor C in Fig. 4.23c is charged exponentially and then discharged while either input was negative; the level detector caused tripping when both input signals were positive for a duration of 90°. The $R-C$ integrating circuit was equivalent to Loving's inductance integrator.

The first relays were half-wave devices with variable operating times and a tendency to overreach on offset current waves. The use of the replica impedance and full-wave circuits in later models gave improved performance.

b) Rectifier bridge-type 'AND' gate. A polarized diode bridge can be used instead of a transistor AND gate with some advantage in economy and simplicity. The operation is explained in detail in the Appendix (Section 4.5).

Figure 4.24a shows a simple form of phase comparator using two rectifiers and giving one tripping signal per cycle during the period when both input quantities have the polarity necessary to pass current through the rectifier, i.e. the polarized relay trips only when both input quantities are positive. Figure 4.24b is an alternative arrangement.

Figure 4.25a shows how the circuit of Fig. 4.24a can be speeded up by adding two more rectifiers to give two tripping signals per cycle, i.e. when the inputs are both positive or both negative. Figure 4.25b is another way of drawing the same circuit.

Figure 4.25c shows how the circuit of Fig. 4.24b can be arranged to give two tripping signals per cycle. It is more economical than the circuit of Fig. 4.25a

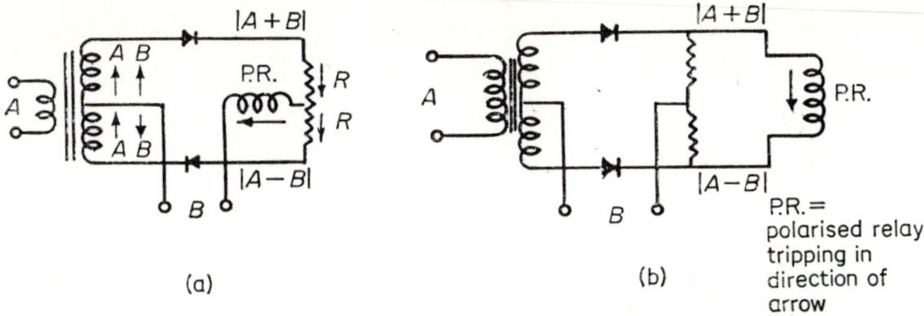

Fig. 4.24. Half-wave phase comparators (1 signal per cycle)
(a) and (b) alternative connections

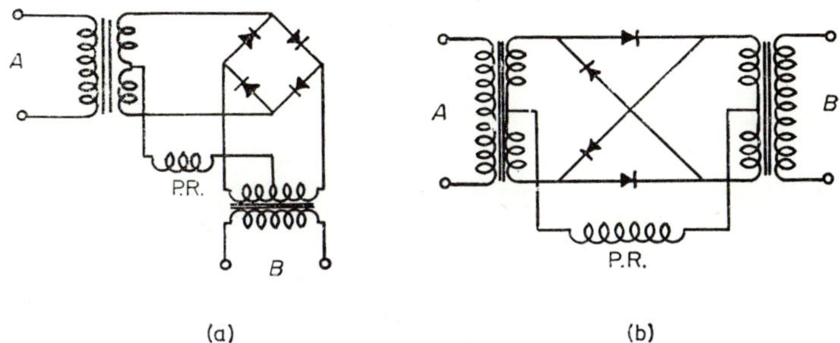

Fig. 4.25. Full-wave phase comparator (2 signals per cycle)
(a) and (b) alternative ways of drawing the same circuit

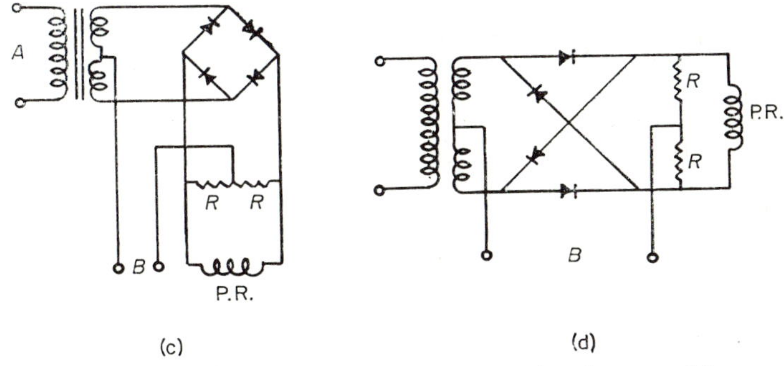

Fig. 4.25. Full-wave phase comparator (2 signals per cycle)
(c) and (d) same as in (a) and (b) except positions of P.R. and input B interchanged

because it can dispense with the *B* input transformer. Figure 4.25d is another way of drawing Fig. 4.25c, which corresponds to Fig. 4.25a.

It is interesting to note that in these circuits the phase comparison is actually done by amplitude comparison in the relay itself because the relay gets $|A + B| - |A - B|$ which is only positive if A and B are of the same polarity.

The duality between the rectifier bridge phase comparator and the rectifier bridge amplitude comparator can be seen by comparing Fig. 4.25d with

Fig. 4.26 which is an amplitude comparator in which the polarised relay is supplied with $|A + B| - |A - B|$ by using double primaries on the auxiliary transformers instand of double secondaries.

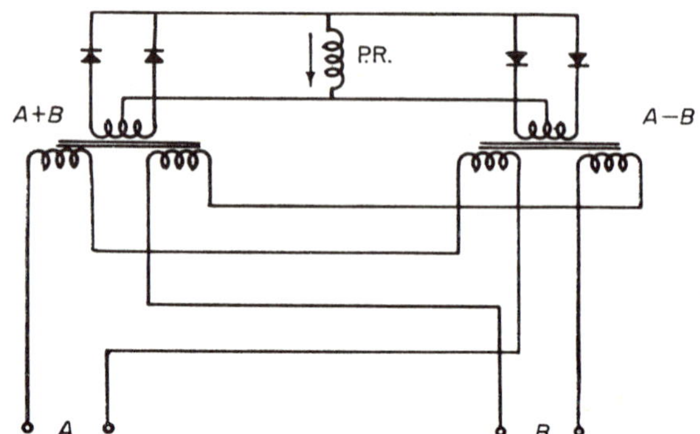

Fig. 4.26. Amplitude comparator rectifier bridge used as a phase comparator
P.R. is a polarised relay tripping in direction of arrow

The complete full-wave rectifier bridge comparator is shown in Fig. 4.29a. Its operation is explained in detail in the Appendix, Section 4.5. The limiter L is used where one of the quantities covers a very wide range of values, such as the voltage input to a mho relay. While the two input currents i_1 and i_2 have the same polarity, the output is in the tripping direction with signs as shown. The capacitor C charges during that portion of the cycle when the two inputs have the same polarity and discharges while they have opposite polarity.

The output is fed into a level detector and, if the charging period is long enough for the voltage across C to reach the setting of the level detector, tripping occurs. For a circular characteristic the setting is such that the charging period must exceed $\psi = 90°$.

This rectifier bridge phase comparator is in all respects the dual of the circulating current rectifier bridge amplitude comparator in Fig. 4.4. It has the same limiting action because the output is always paralleled by two diodes in series in each direction. This means that the wave shape of the output from this comparator is rectangular and of constant amplitude, except at extremely low input currents when it tends to be more like the shaded areas in Fig. 4.29b.

4.4.4. Time-Bias Type of Integrator

Another way of measuring the duration of coincidence is shown in Fig. 4.27. The output from the AND circuit is fed into another AND circuit (*a*) directly and (*b*) delayed by a time corresponding to the comparator angle δ, viz. 10 ms for $\delta = 90°$. This means that the output from the first AND gate has to persist for a period δ in order to operate the second AND gate and cause tripping.

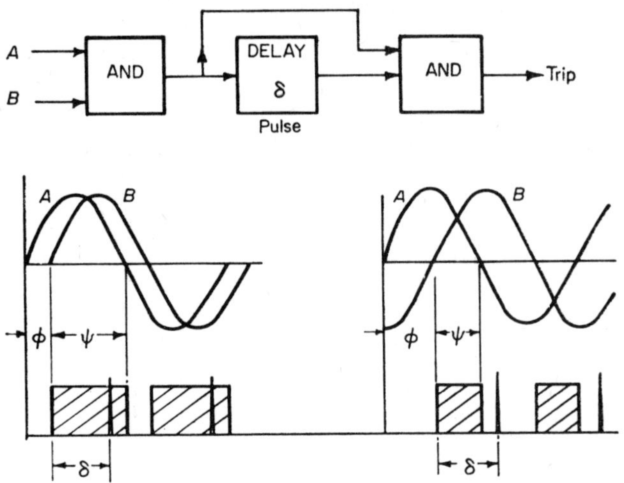

Fig. 4.27. Time bias type of integrator
ψ is the period of coincidence of inputs A and B
δ is the delay of the pulse forming circuit

This method is more convenient for multi-input comparators (see Chapter 12) but is subject to false tripping by an interference signal, whereas the inductive or capacitative types of integrator are not. It can be used for obtaining asymmetrical characteristics [75] by providing a different delay according to which input is leading the other.

4.4.5. Integrating Circuits

These circuits have already been discussed for amplitude comparators in Section 4.3.2. The rectangular output from the phase comparator is fed into an integrating circuit which operates a level detector and tripping device when the duration of the signal in the tripping direction is long enough to operate the level detector (Fig. 4.9).

4.4.6. Effect of Offset Waves on Phase Comparators

If one of the inputs is offset the effect on the phase comparator is to cause overreaching, as in the amplitude comparator. This can be prevented, in either comparator, by integrating over a full cycle.

Figure 4.31a shows an extreme case where the fault occurs at the moment when one input should be at its peak value so that it starts fully offset. It will be seen that, in a phase comparator, the offset causes the first period of coincidence to be lengthened and the next one to be shortened by the same amount so that, over 1 cycle, the average period of coincidence is correct.

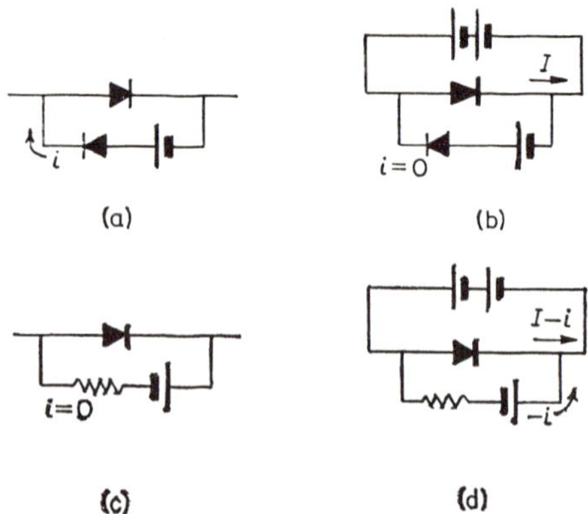

Fig. 4.28. Principles of diode switching

Consequently, if the integration could be carried out over a full cycle, the offset would cause no error. On the other hand, a relay set to trip at the end of the first 90° period of coincidence (i.e. set for $\psi = 90°$, to produce a circular characteristic) could trip on an external fault because the first period of coincidence could have been artificially lengthened by the offset to more than 90°, whereas the same fault with neither input offset would not cause tripping.

It will be noticed from Fig. 4.31a that the artificially long first tripping block of output from the comparator is followed by a similarly lengthened blocking period (shown below the zero line). Then follows a shortened tripping and a shortened blocking period. The integrating comparator takes advantage of this to avoid overreach.

This is done by comparing the tripping and blocking period in the integrator. The integrating circuit can be inductive or capacitative. In the inductive type the tripping relay coil is generally used as the inductance; during the tripping periods (shown above the zero line) the output voltage of the comparator is impressed on the tripping relay coil; during the blocking period the relay coil current continues to flow because of the inductence of the coil and a diode is connected across the coil to provide a path for this current (Fig. 4.8a).

In this way the output current increases during the tripping period and decreases during the blocking period and the time constants of the circuits are arranged so that the rate of increase is faster and the current gradually increases in a ratchet manner if the tripping period exceeds 90°. Figure 4.32a shows that the envelope of the peaks of the coil current are assymptotic to the tripping level when $\psi = 90°$ or the set value of λ.

Faster operation can be obtained if a thyristor is used instead of a tripping relay and the time integration is done by an $R-C$ circuit. The capacitor is charged during the tripping period, as explained in Section 4.3.2, and discharged during the blocking period through a parallel resistor, (Fig. 4.8b).

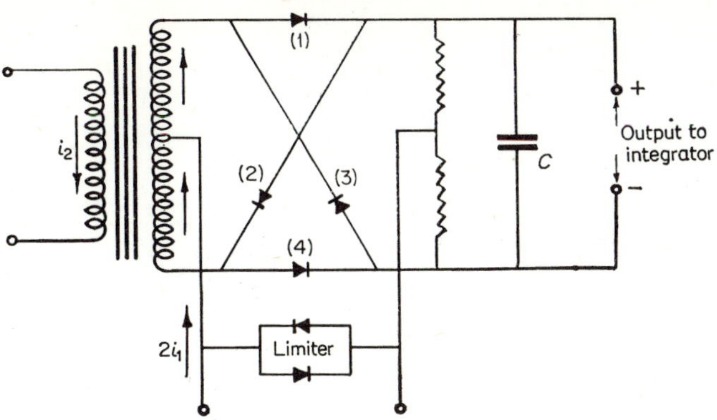

(a) Bridge rectifier circuit

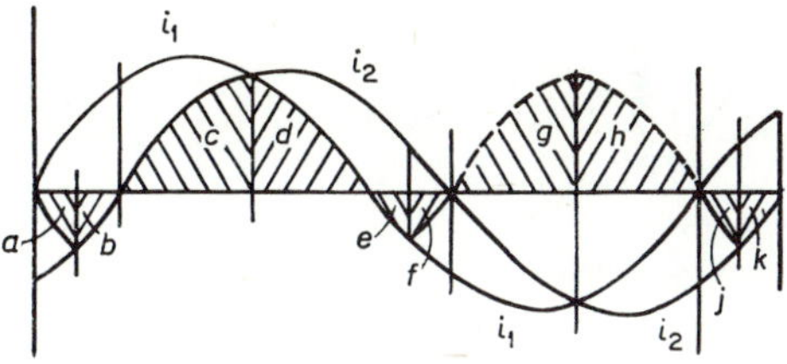

(b) Coincidence areas on input waves

Fig. 4.29. Rectifier-bridge phase-comparator

In this section the phase comparator has been considered but the same method can be applied to an amplitude comparator because, in rectifying and limiting the inputs, the offset wave appears as alternate long and short rectangular blocks (Fig. 4.31 b) and again it is necessary to integrate over a full cycle to obtain the correct average value.

The only difference is that, in the phase comparator, the first tripping and the first blocking signals are long and the next ones short whereas, in the amplitude comparator, a long tripping signal is followed by a short blocking signal and then a short tripping signal and a long blocking signal. In either case a full cycle is required to eliminate the effect of the offset for a fault near the relay setting. For faults well within the relay setting the time is faster because the tripping signals are longer (Fig. 4.33); in fact, faults on the diameter of the characteristic drawn from the origin (i.e. at the characteristic angle) cause a continuous signal and trip in just over half a cycle.

4.5. APPENDIX

4.5.1. Operation of rectifier phase comparator

The circuit arrangement of the full-wave rectifier phase comparator is shown in Fig. 4.29a. Its operation can be readily understood from the dia-

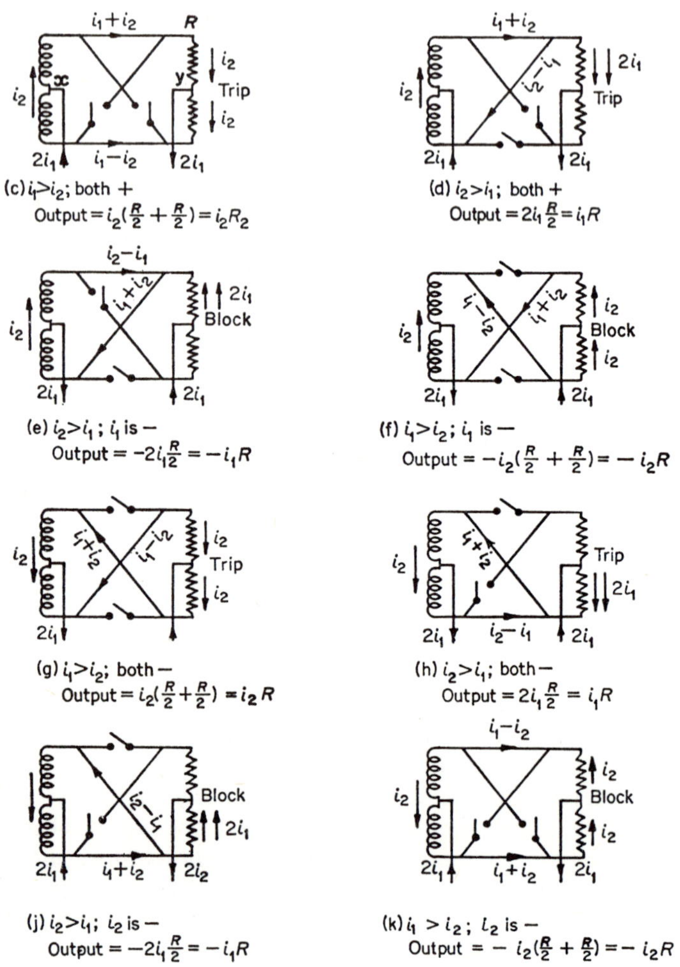

Fig. 4.30. Principle of phase comparator bridge

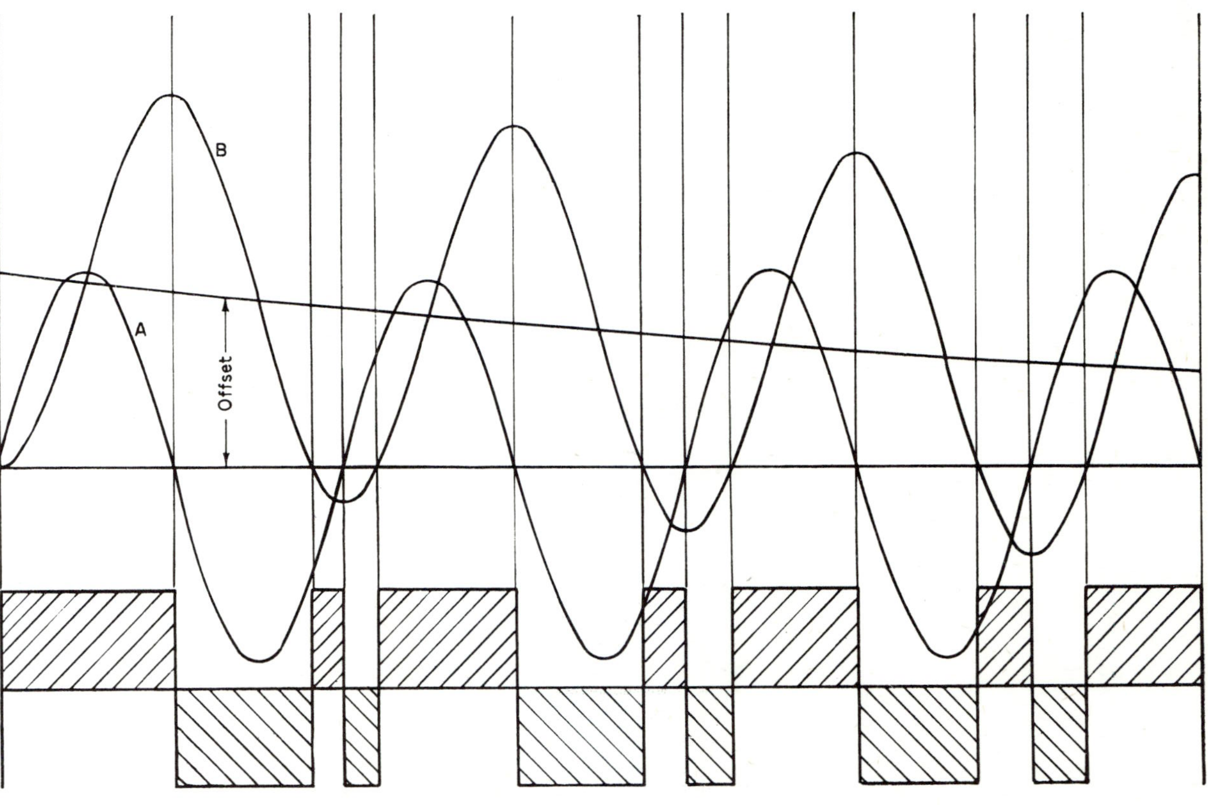

Fig. 4.31 a. Effect of one fully offset input upon phase comparator

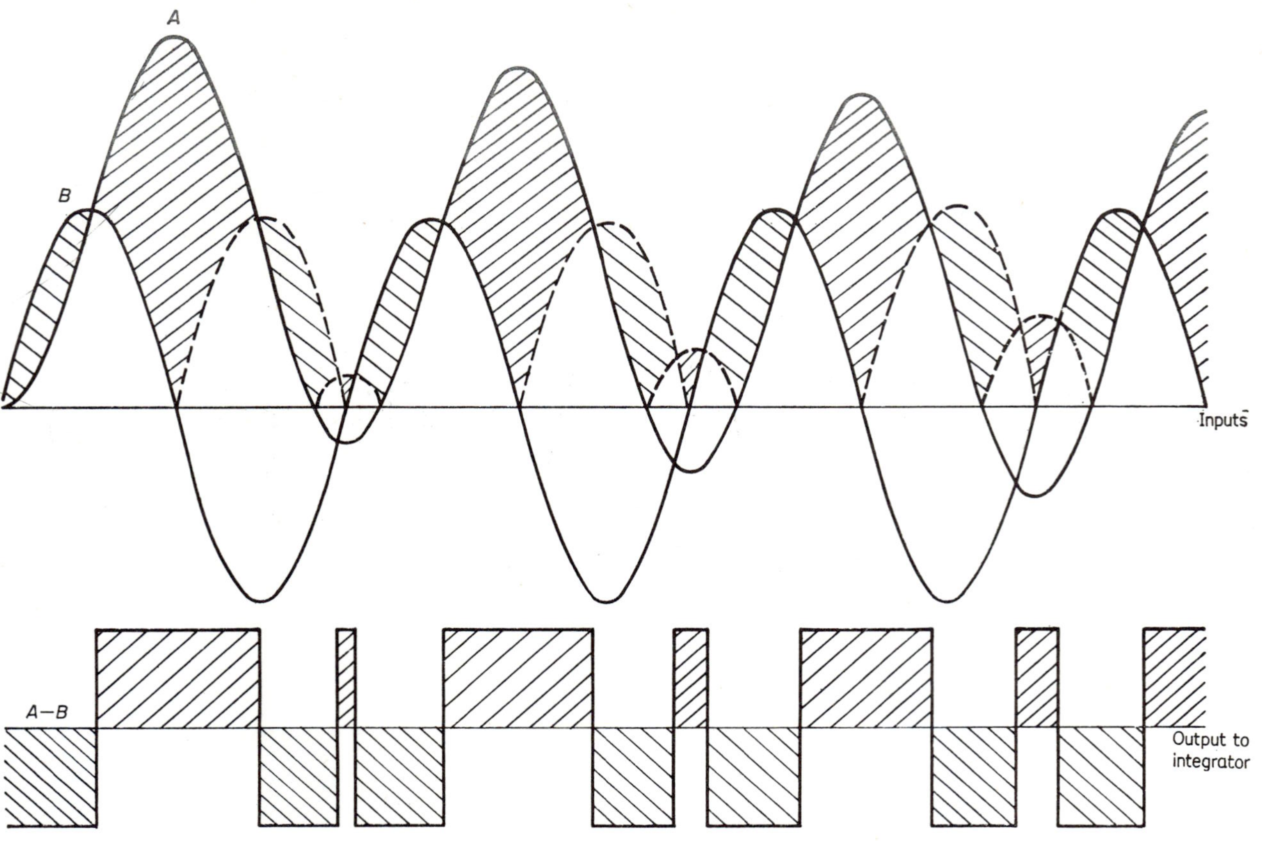

Fig. 4.31 b. Effect of one fully offset input upon an amplitude comparator

grams shown in Fig. 4.29b and Fig. 4.30, and is based on the following concepts:

(i) A large forward current flowing through a diode produces a potential difference across the diode which is sufficient to bias off diodes in neighbouring circuits which are carrying smaller currents (Fig. 4.28a).

(ii) A diode which is carrying a large forward current can effectively conduct a smaller current in the reverse direction by subtraction from the larger current (Fig. 4.28b).

The diodes may therefore be regarded as switches that open or close according to the path of whichever current is instantaneously larger, i.e.

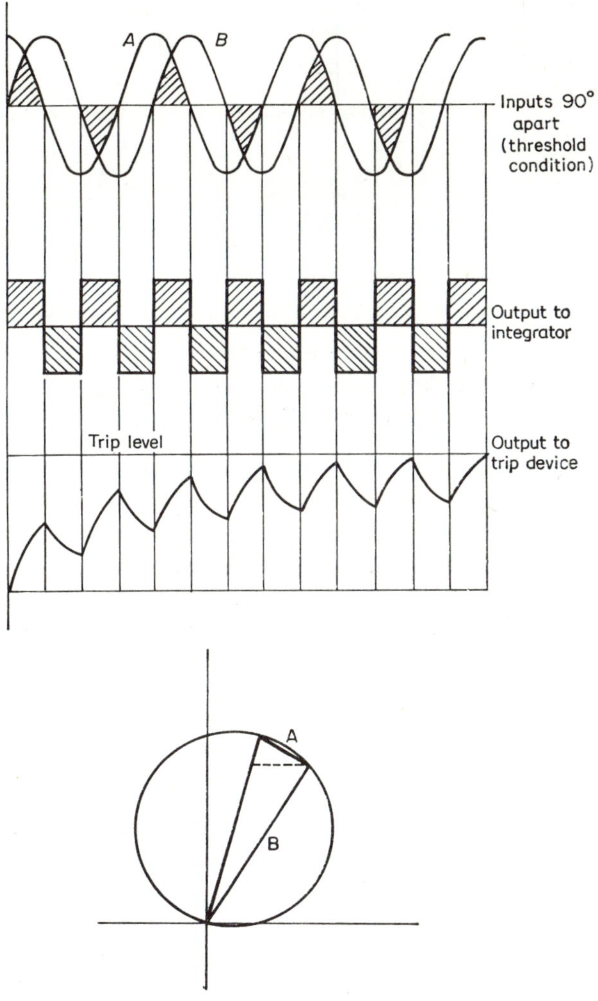

Fig. 4.32a. Effect of non-offset wave at setting of phase comparator

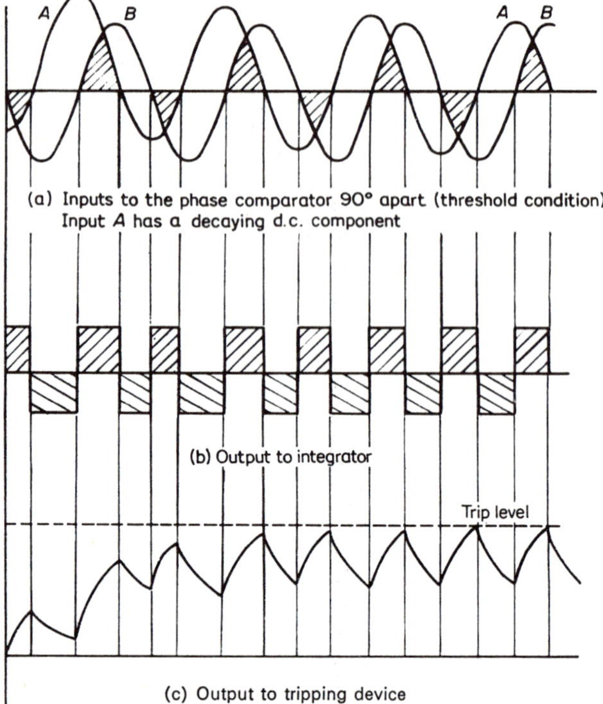

Fig. 4.32b. Effect of fully offset wave at setting of phase comparator

the larger current acts as a gating current that prepares the circuit path of the smaller current.

The operation of the phase comparator rectifier bridge can be understood if followed through a full cycle of both input quantities, as in Figs. 4.29 and 4.30. In Fig. 4.30 the various periods of like and unlike polarity are marked *c, d, e, f, g, h, j, k* and the conductivity state of the individual rectifiers of the bridge is given in Table 4.2 using corresponding letters.

Period (*c*): i_1 is greater than i_2 and both are positive. i_1 flows from X to Y through each of the two diodes 1 and 4; the nett output voltage due to i_1 across the output resistor R is zero because each current flows in opposite directions through the two halves of R. The potential difference due to the larger current i_1 in diode 1 appears across diode 2, in opposition to the e.m.f. driving the smaller current i_2; current i_2 is therefore prevented from flowing from X through 2 and flows from X through R (see (*i*) on previous page).
Diode 4 is carrying current i_1 in a forward direction and therefore the smaller current i_2 will pass through it in the reverse direction by subtraction (see (*ii*) on previous page).

The output voltage is $+\,i_2(^1/_2R + {}^1/_2R) = +\,i_2R$ hence tripping..

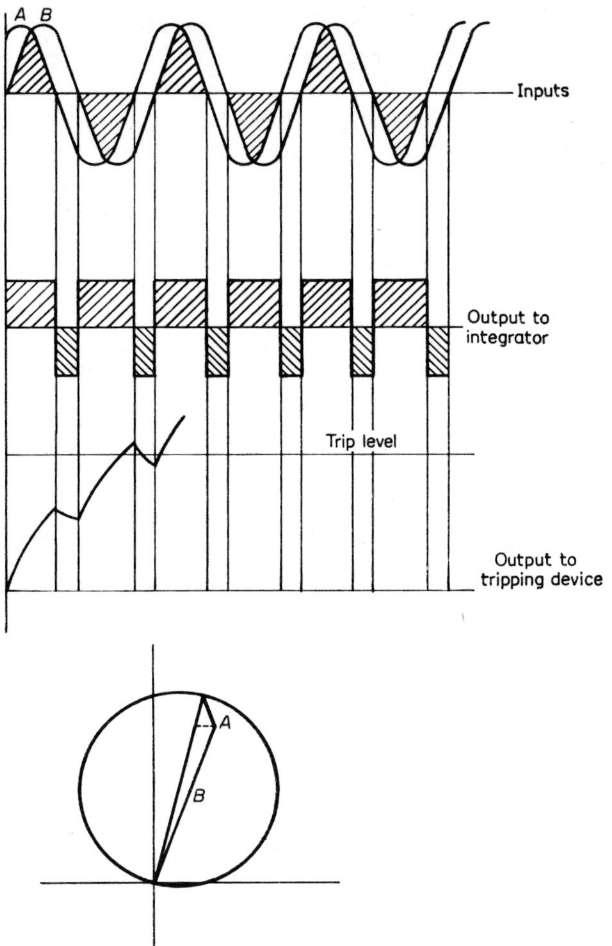

Fig. 4.33. Operation of phase comparator for fault well within setting

Period (*d*): i_2 is greater than i_1 and both are positive. i_2 flows through diodes 1 and 2, producing no output voltage across R. The potential difference due to i_2 in diode 2 appears across diode 4 in opposition to the e.m.f. driving i_1; current i_1 is therefore prevented from flowing through diode 4 to Y and flows in the reverse direction through diode 2 which is carrying forward current i_2. $2i_1$ therefore flows in the top half of R and the output voltage is $+2i_1\,(\frac{1}{2}R) = +i_1R$.

Period (*e*): The conditions are the same as for Period (*d*) except that i_1 reverses its polarity and flows in the opposite direction through the load resistor. The output voltage is $-2i_1\,({}^1/_2\,R) = -i_1R$, hence blocking.

Period (*f*): i_1 is greater than i_2; i_1 is negative; i_1 flows through each of the diodes 2 and 3; the nett output voltage due to i_1 is zero as in

Period (c). The potential difference due to i_1 in diode 2 appears across diode 1 in opposition to the e.m.f. driving i_2; current i_1 is therefore prevented from flowing through diode 1 and flows in the reverse direction through diode 3 which is carrying forward current i_1. Current i_2 then flows through R because diode 4 opposes it. The output voltage is $-i_2\ (^1/_2\ R + {}^1/_2\ R) = -i_2\ R$, hence blocking.

Period (g)*:* The conditions are the same as for Period (f) except that i_2 reverses its polarity and flows in the opposite direction through the load resistor. The output voltage is $i_2\ (^1/_2\ R + {}^1/_2\ R) = i_2\ R$, hence tripping.

Period (h)*:* i_2 is greater than i_1 and both are negative. The conditions are similar to those of Period (e) except that conduction is through the opposite pair of diodes and the output is developed across the opposite half of the load resistor. The output voltage is $2\ i_1\ (^1/_2\ R) = i_1\ R$, hence tripping.

Period (j)*:* The conditions are the same as for Period (h) except that i_2 reverses its polarity. The output voltage is $-2i_1\ (^1/_2\ R) = -i_1\ R$, hence blocking.

Period (k)*:* The conditions are the same as for Period (c) except that i_2 reverses its polarity. The output voltage is $-i_2\ (^1/_2 R + {}^1/_2 R) = -i_2 R$, hence blocking.

TABLE 4.2.
Operation of Phase Comparator Diode Bridge

Fig. 4.30	i_2	i_1	Larger	Paths	Output
c	+	+	i_1	1,4	i_2R
d	+	+	i_2	1,2	i_1R
e	+	—	i_2	1,2	$-i_1$R
f	+	—	i_1	2,3	$-i_2$R
g	—	—	i_1	2,3	i_2R
h	—	—	i_2	3,4	i_1R
j	—	+	i_2	3,4	$-i_1$R
k	—	+	i_1	1,4	$-i_2$R

Bigger quantity unlocks its path and closes the parallel path
Like signs + smaller + iR to trip
Unlike signs — smaller — iR to block

The output of the comparator given by the shaded area in Fig. 4.22 averaged over a complete cycle is proportional to the phase displacement between i_1 and i_2. The average is zero when the currents are 90° out-of-phase, negative when the displacement is less than 90° and positive when more than 90°.

To summarize, the output current of the bridge, supplied to the relay, is equal to the smaller of the two current inputs. The path of the current through the bridge is established by the larger of the two currents and depends upon their relative instantaneous polarity. If $i_1 > i_2$ the current will flow in the top and bottom rectifiers if i_1 is positive (Fig. 4.30c) and in the diagonal rectifiers (Fig. 4.30f) if i_1 is negative. If $i_2 > i_1$ the current flows in rectifiers 1 and 2 if i_2 is positive (Fig. 4.30d) and rectifiers 3 and 4 if i_2 is negative (Fig. 4.30j). If i_1 and i_2 have the same polarity the current in the polarised relay R flows in the tripping direction; if they have opposite polarity it will be in the blocking direction.

4.5.2. Non-circular characteristics

Fig 4.14 is based upon inputs A and B supplied to an amplitude comparator. It will be seen that, for a phase comparator, the inputs must be (A + B) and (A − B) to give the same characteristics. A circular characteristic is obtained when the angle between the inputs (A + B) and (A − B) exceeds 90°. This critical value is shown as in fig. 4.14; if $\lambda < 90°$ the characteristic becomes apple-shaped; if $\lambda > 90°$ the characteristic is lemon-shaped.

5

Output Devices; D.C. Supply; Transient Overvoltages

Tripping circuits and devices. d.c. supplies for transistors. Causes and effects of transient overvoltages–Methods of diverting or suppressing them

5.1. TRIPPING DEVICES

Owing to the fact that the tripping function of a relay is associated with other requirements such as remote indication, tripping another breaker, sealing-in, etc., the output device should have two or more electrically separate contacts. In an electromagnetic output relay it is an easy matter to provide the extra contacts for negligible extra cost, but static output devices as yet have only one equivalent of a contact so that a separate device has to be provided for each contact function; this is obviously expensive.

The thyristor or silicon controlled rectifier (SCR) is an excellent means of tripping a circuit-breaker from a low power signal of short duration; it not only operates in microseconds but, when once triggered, stays conductive like a thyratron until its anode-cathode circuit is broken by a switch. Against this the following advantages can be claimed for electromagnetic relays.

(*a*) Cheaper for multi-circuit control, e.g. generator protection.

(*b*) Higher insulation between circuits.

(*c*) Negligibly affected by voltage spikes.

(*d*) Position of relays is visible.

(*e*) Very fast (2 to 3 ms) relays are now available.

(*f*) One relay will handle a wide range of trip-coil ratings.

When a thyristor is used for tripping it should be chosen to block the d.c. trip-circuit voltage normally with a safety factor of at least 50% and it should have a thermal capacity sufficient to stand the trip-coil current for the normal tripping time with a margin sufficient to allow for a sticking trip latch. A reasonable rating is to stand a duty cycle of two 0·2 second periods 0·2 second apart.

Charts are available for determining the appropriate thyristor rating (continuous) from the short-time (0·1 second) trip-coil current rating [35]; a 10 amp SCR rating should be sufficient for any trip-coil rating up to 30 amperes, especially as the trip-coil current rises exponentially from zero; the de-energized L/R ratio of a typical breaker trip coil is about 0·05, falling to about 0·014 with the armature in the tripped position.

A good compromise between hinged-armature relays and Thyristors is the mercury-wetted reed relay, using its short-time rating as is done with thyristor tripping. Dry reed relays can also be used with a hinged-armature relay for sealing around its contacts; owing to their low resistance a 1 Ω resistor should be connected in series with the reed contacts in order to divert the trip current through the contacts of the seal-in relay.

Typical requirements for tripping devices are:

(i) A minimum operating voltage of not more than 60% of the nominal battery voltage.

(ii) Non-operation at currents below 100 milli-amps (to prevent tripping on wiring leakage currents).

(iii) Non-operation when a capacitance of 10 microfarads charged to 150 volts is discharged through the relay coil (to prevent tripping on discharge of wiring capacitance).

(iv) Operating time not to exceed 10 milliseconds at the nominal battery voltage of 110 volts.

5.1.1. Thyristor Triggering Circuits [14, 35, 46]

Figure 5.1a shows how a SCR can be fired by a small amplitude impulse [14] after amplification through a transistor. The charge on the capacitor C is discharged through the resistor R_3 into the SCR gate when the transistor is made conductive by a negative impulse through R_1. The leading edge of the input impulse supplies a positive impulse to the SCR gate, firing it.

Figure 5.1b shows a circuit in which the tripping impulse is given through a pulse (peaking) transformer which also serves to insulate the transistor circuit from the station battery. The SCR can be used equally well for trip coils connected to the positive bus. The capacitor C discharges through the pulsing transformer when the transistor T conducts, thus triggering the SCR. The trip current continues to flow through the SCR until the breaker auxiliary switch opens the trip circuit.

When an SCR is fired by a pulse [32], the current starts to flow through a small area of the junction and spreads rapidly over the whole surface. For this reason the initial rate of build-up should not be too fast, otherwise the current density in the initial conducting area may heat it up sufficiently to damage the junction. When used for controlling a circuit-breaker trip coil the inductance of the latter will protect the SCR. In less inductive circuits it is advisable to limit the initial current by a reactor.

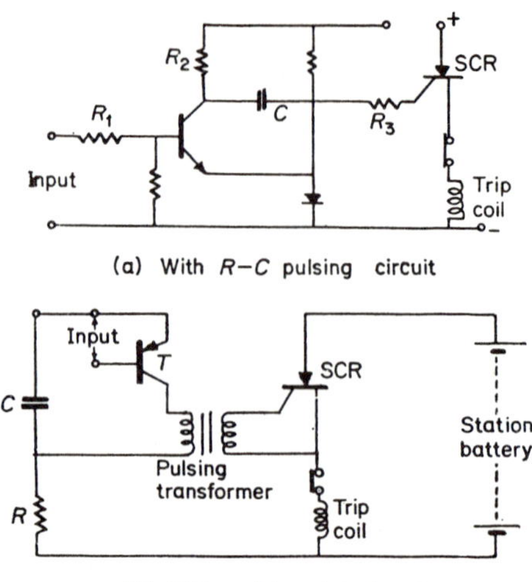

(a) With $R-C$ pulsing circuit

(b) With pulsing transformer

Fig. 5.1. Thyristor tripping circuits

Another precaution that is necessary in protective relays is to design the circuit so that the tripping pulse must continue for at least 2 ms in order to fire the SCR; in Fig. 5.1a the delay is provided by the time to charge up capacitor C through resistors R_1 and R_2. This eliminates wrong tripping due to interference voltage spikes because their duration is limited to microseconds.

Since the SCR operates in less than 5 μs its wiring must be shielded from inductive and electromagnetic interference as well as from voltage spikes from the d.c. source [16, 19]. A fuse can be used for protecting the SCR from excessive current due to some circuit component failure or wrong connection. Other precautions are a diode across the trip coil and a capacitor across the two '*P*' electrodes of the SCR (*p-n-p*) to minimize the 'water-hammer' voltage from the trip coil when its current is interrupted [18, 19].

5.2. D.C. SUPPLIES FOR TRANSISTOR CIRCUITS

In the early stages of the introduction of transistorized relays it was necessary to include the d.c. supply with each relay but, as they become more common, special d.c. supplies will be provided and eventually protection, control and telemetering will not only share a common d.c. supply but will tend to be brought together in one group of equipment in standardized cubicles.

The d.c. voltage required for transistors depends upon their design. Germanium transistors require 6 to 25 volts; silicon transistors require 6 to 50 volts. Alternative supplies are (*a*) a potentiometer across the station battery, (*b*) a d.c. converter from the station battery, (*c*) an auxiliary battery or (*d*) rectification of secondary potential current or voltage.

5.2.1. Station Battery

Figure 5.2a shows a typical circuit for reducing the station battery voltage by the potentiometer method. The capacitor *C* absorbs voltage spikes which may be caused by the opening of another (inductive) circuit supplied from the station battery. The zener diodes hold constant the voltage supplied to the relay circuit.

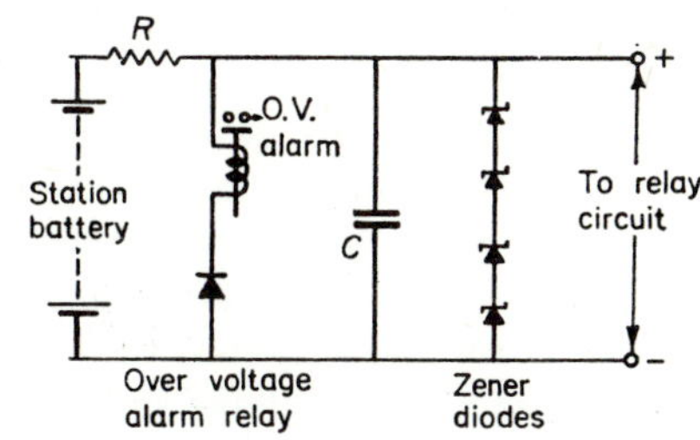

(a) Potentiometer method

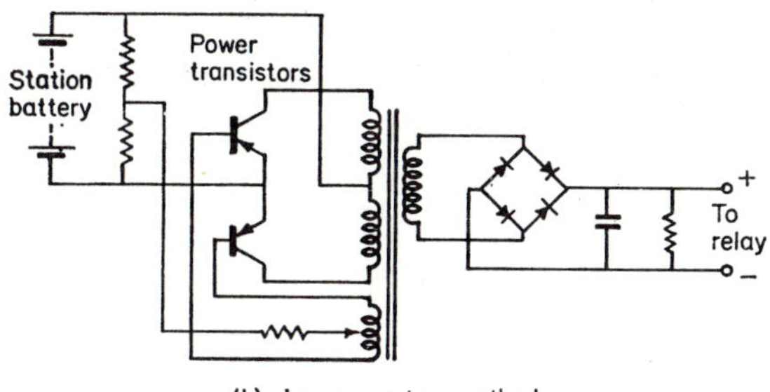

(b) d.c. converter method

Fig. 5.2. Relay d.c. supply from station battery

In England the generating stations of the C.E.G.B. use a 125 volt (55 cell) station battery for circuit-breaker control and a 30 volt 5 A.H battery for transistorized control and protection circuits; in substations the battery for circuit-breaker control may be 30 or 125 volts, but at present protective relays are expected to be self-powered. The normal battery drain is limited to 40 mA for 30 volt batteries and 4 A for 110 volt batteries; the maximum current during operation is limited to 2 A for 30 volt batteries and 10 A for 110 volt batteries.

The use of a potentiometer involves a constant battery drain. Since this may necessitate a larger station battery when most of the relays are static, the potentiometer method is not likely to be as popular as the d.c. transformer method, which imposes less drain on the battery, or the use of a separate battery for the relays. Where fault currents are always large, a fast overcurrent relay can be used to switch in the d.c. supply to the static relays only when there is a fault, thus eliminating the battery drain.

5.2.2. d.c. Converter

Figure 5.2b shows an alternative method using an a.c./d.c. converter which converts the station battery voltage to a.c. so that it can be stepped down by a transformer; it is then rectified by diodes, smoothed by the capacitor and regulated by zener diode. This circuit provides a constant relay voltage within $\pm 2\%$ which is free from voltage spikes.

5.2.3. Auxiliary Battery

The advent of small hermetically-sealed nickel-cadmium storage batteries made it practical to mount the d.c. supply battery inside the relay.

Ni-Cd batteries last about twenty years and can stand freezing without damage. They lose 35% of their momentary capacity at 0°F (−18°C). Overcharging does them no harm and they may be left idle for long periods. Fig. 5.3b shows how a stabilised lower voltage supply with good regulation can be obtained from a existing d.c. supply. S can be a zener diode

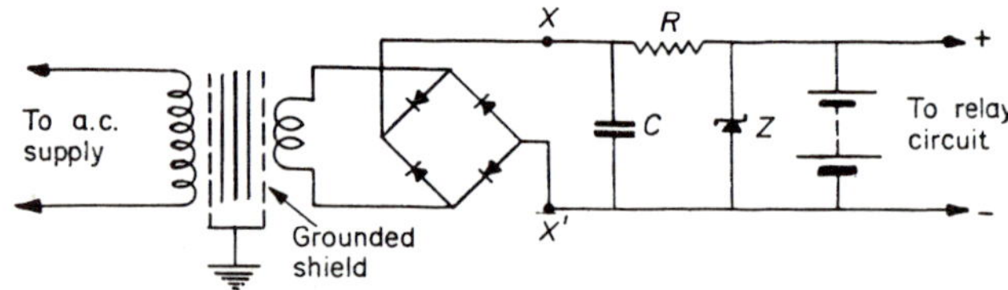

Fig. 5.3a. Internal battery with trickle charger

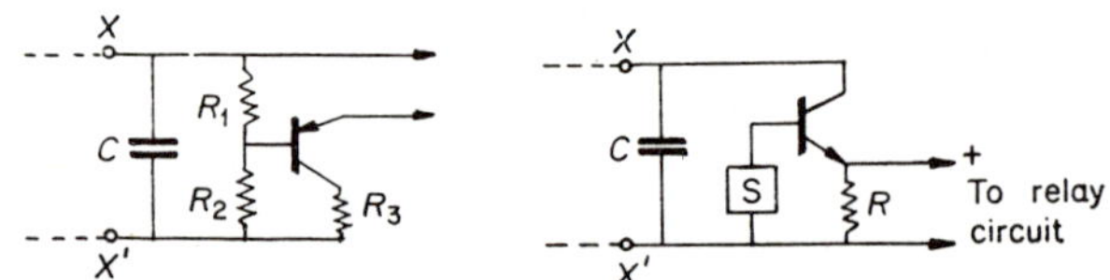

Fig. 5.3b. Methods of producing a regulated d.c. supply at reduced voltage

5.2.4. Rectified a.c.

Since the secondary current is a minimum during normal conditions and the voltage is a minimum during fault conditions, it is necessary to employ both quantities in the case of distance or directional relays. But, in the case of synchronizing relays, the a.c. potential may be used alone and, in the case of overcurrent relays, the fault current can be depended upon.

Figure 5.4 shows the combined current-voltage circuit. The c.t's are in the same phase of the two input lines, one of which is always energized. The p.t. is also used in the same phase but the station lighting circuit can equally well be used. The rectifiers must be equipped with a smoothing R-C or L-C circuit and zener diode, as shown in Fig. 5.3a.

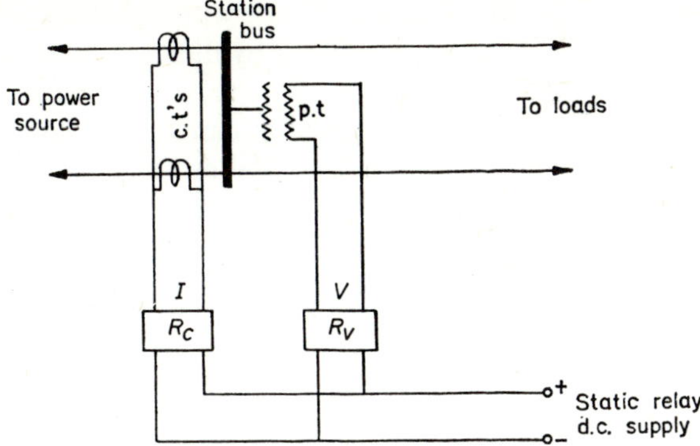

Fig. 5.4. Reliable d.c. supply from rectified a.c. a.c. inputs from c.t's and p.t's in the same phase

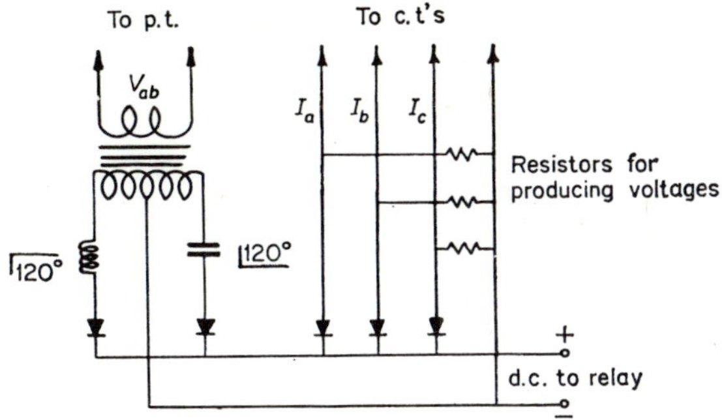

Fig. 5.5. Improved circuit for relay d.c. supply from p.t. and c.t's

The advantage of using secondary current and voltage from the same phase is that at least one of the quantities is always high. Figure 5.5 shows a somewhat more sophisticated circuit with better smoothing; all three currents are used and the voltage from one phase is split into three components 120° apart. The three resistors can be transactors which are more efficient.

5.2.5. The Fuel Cell

This device produces d.c. from chemical energy as long as the fuel supply is maintained. It is at present only in the development stage. Very small cells have been used in wristwatches and hearing aids.

5.3. SOURCES OF TRANSIENT OVERVOLTAGES [105]

Voltage spikes (Fig. 5.6) are transient overvoltages of very short duration (microseconds) whose appearance in a transistor relay circuit may cause wrong operation and, if large enough, damage to its semiconductor components.

Voltage spikes are caused by the sudden starting or stopping of currents in the relay circuit, or in a nearby circuit. For instance, the sudden interruption of the current in an inductive a.c. circuit will create a large back e.m.f. which may appear across a transistor either by direct connection or through capacity coupling. Similarly, the sudden increase in current caused by a short-circuit or by switching-in a capacitor can induce high voltage in a neighbouring circuit by mutual induction. Spikes can also come from inter-

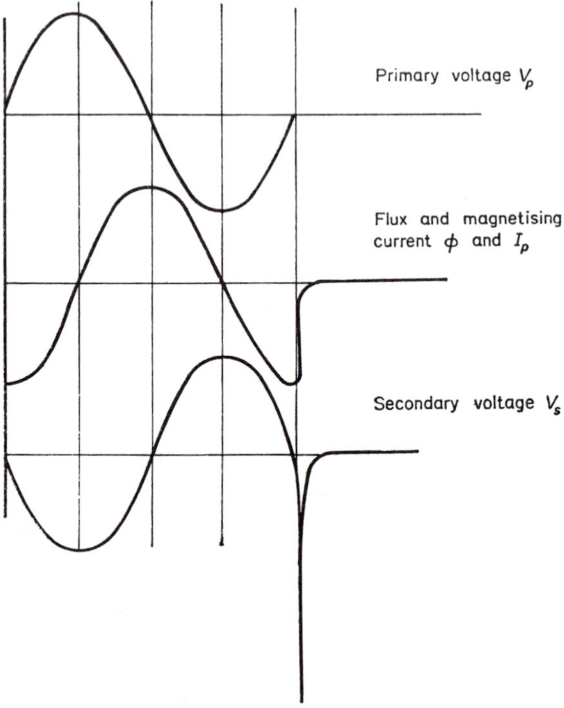

Fig. 5.6. Voltage transients due to interrupting inductive current at voltage zero

rupting a.c. inductive circuits at the moment of current maximum or closing a capacitative a.c. circuit at the moment of voltage maximum. Again the switching can be in a relay circuit or in a circuit coupled inductively or electrostatically with a relay circuit. Finally, an external transient voltage source such as lightning may enter the relay circuit through capacity coupling between wires and transformer windings.

Instead of a spike the transient overvoltage may be a burst of high-frequency overvoltage coming from an arc on the power system through the intercapacitance of instrument transformer windings.

These transients have caused negligible trouble in electromagnetic relays because (*a*) their level of insulation is too high for them to be damaged by voltage spikes, (*b*) the inertia of the armature prevents wrong operation due to such a short impulse.

The sources of transient overvoltages can be grouped under the following headings:

5.3.1. Direct effects
- (*a*) breaking of an inductive circuit
- (*b*) making of a capacitative circuit

5.3.2. Induced effects
- (*a*) inductive coupling
- (*b*) capacitative coupling

5.3.3. System faults
5.3.4. Lightning
5.3.5. System switching
5.3.6. Switching inside the relay

Interference from these sources is either electromagnetic or electrostatic. The former can be minimized by arranging the wiring in twisted pairs and, in some cases, by magnetic screening. The latter can be reduced by earthed copper screening (Section 5.5.3). Any interference signal that gets through can be suppressed (Section 5.5.2.) or diverted (Section 5.5.1).

5.3.1. Direct Effects

This refers to switching in circuits directly associated with the d.c. supply and a.c. inputs to the relay [16, 18, 19].

(a) *Breaking an inductive circuit.* Figs. 5.7, 5.8 and 5.9 show typical spike-producing circuits where the spike is produced by suddenly interrupting an inductive current. The inductance may be a solenoid, a choke, a transformer primary or even a wire-wound resistor. However, voltage spikes are not necessarily associated with solenoids. For instance, the blowing of a high-speed fuse due to a short-circuit in a circuit supplied by a battery may produce a considerable spike because, although the inductance of the circuit is only that of the wiring and the battery, the current may be high enough and the interruption sudden enough to make the value of $L\frac{dI}{dt}$ six times the battery voltage.

In Fig. 5.7 the breaker is manually tripped by the push-button. P.B. and, if the latter has a strong reset spring or if the finger of the operator slips off

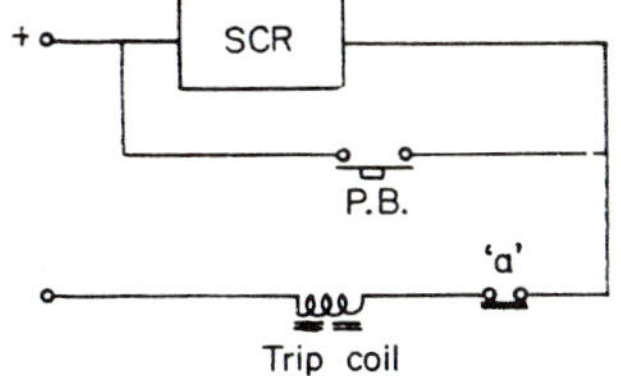

Fig. 5.7. SCR triggered by inductive kick from opening push-button suddenly

the button before the breaker auxiliary switch 'a' opens, a forward transient voltage will appear across the SCR which may trigger it, giving a wrong tripping (flag) indication; furthermore, the voltage spike may damage the SCR.

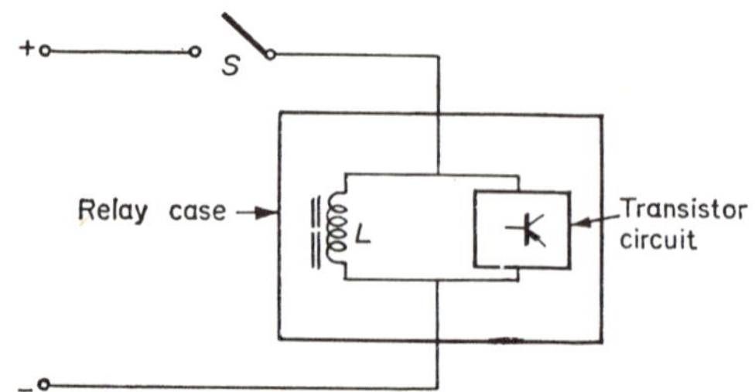

Fig. 5.8. Inductive back e.m.f. due to parallel inductance in relay circuit

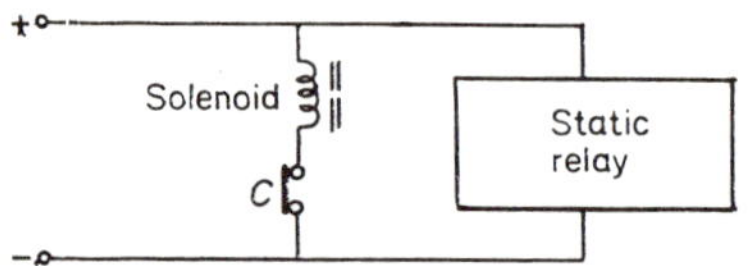

Fig. 5.9. Inductive back d.m.f. due to interrupting solenoid current

In Fig. 5.8, interrupting the d.c. supply by an external switch or contact *S* will cause an inductive coil inside the relay to maintain its current through any parallel path and hence impose a reverse voltage on the transistor circuit which could destroy the transistors.

In Fig. 5.9, when the contact *C* is opened, the current in the inductive coil *L* tends to continue to flow and this will create a high voltage across the contact which may cause a re-strike. Then the voltage across *C* will disappear and the whole sequence will be repeated. This can cause a series of voltage spikes to be superimposed upon the d.c. voltage supply to the relay.

(*b*) *Making a capacitative circuit.* During the first instant that a d.c. potential is impressed across a capacitor the current is limited only by any resistance that is between the capacitor and the potential source. Consequently it is necessary to see that there is sufficient resistance in the circuit to prevent the inrush current from being high enough to create a destructive potential across any semiconductor in the circuit.

High-speed tripping relays usually interrupt their own coil current by a fast auxiliary switch and the transient overvoltage generated across the contacts by the sudden interruption of the coil current is often suppressed by a capacitor connected across the contacts. Unfortunately, the decaying coil current charges the capacitor to a higher potential than the d.c. supply, thus reversing the polarity of the adjacent terminal of the coil. This reversal may be dangerous for any semiconductors in the circuit [18].

5.3.2. Induced Effects

(*a*) *Inductive coupling*. Wherever wires containing a.c. or d.c. input quantities to a static relay (which may be in mA) are in the same cable or in close proximity to wires carrying tens of amperes, such as circuit-breaker control wires, the sudden starting or stopping of the heavy control current can induce voltage spikes in the static relay circuit, which can cause wrong operation of the relay and damage to the transistors. A short-circuit in the breaker-control wiring, followed by the blowing of a fuse or the opening of a miniature circuit-breaker, can cause even higher induced voltages because of the much higher currents involved (Fig. 5.11). A spike can also be induced by the opening of a circuit-breaker near the relay, such as where relays are mounted on metal-clad switchgear cubicles.

Opening or closing a switch in the primary of an instrument transformer can also produce destructive voltage transients on the secondary. The primary of the transformer produces the magnetic flux in the core and, if the magnetizing current is interrupted near a peak value, the flux will suddenly decrease, causing a voltage spike on the secondary (Fig. 5.6 and 5.10a).

The equivalent of switching can be caused by a saturating c.t. because secondary voltage is generated only during the unsaturated part of the current wave. At very high currents this may be reduced to a small fraction of a millisecond and constitute a spike. To guard against these spikes, and those caused by switching, a non-linear resistance shunt can be connected across each c.t. secondary, consisting of two diodes in series, back-to-back.

The size of these voltage spikes depends upon the speed of primary circuit switching. $V_s = -M(dI/dt)$. Ordinary switches and circuit-breakers usually draw an arc which reduces the rate of change of current. Vacuum switches, however, have negligible arcing and switch the current extremely abruptly, causing maximum transients; this problem was serious in the early vacuum breakers using tungsten contacts, but modern vacuum breakers use contacts

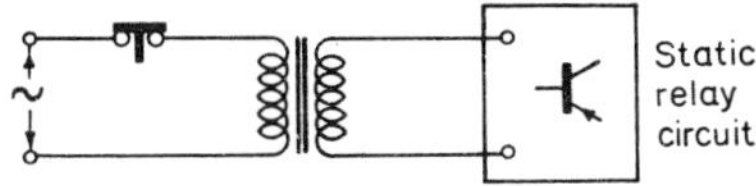

Fig. 5.10a. Transient due to interruption of transformer magnetising current

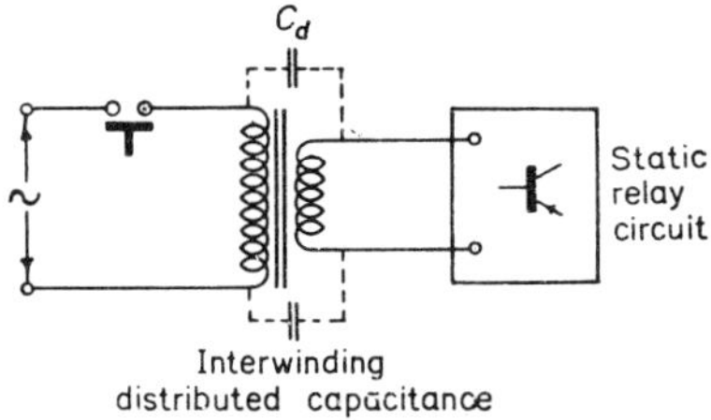

Fig. 5.10b. Transient upon energising transformer

alloyed to some metal such as caesium which gives up electrons under the high potential gradient which occurs as the contacts are parting. Air-blast breakers can also produce large secondary voltage spikes unless they are equipped with suitable resistors.

(*b*) *Capacitative coupling*. Voltage spikes and high frequency a.c. in primary circuits can reach the relay circuit via the capacitance between the two circuits. The most effective path is the interwinding capacitance of instrument transformers and auxiliary transformers. Although such a small capacitance presents a very high impedance to currents of system frequency, it would present a relatively low impedance to h.f. or to steep wave-front spikes generated in the primary circuit by faults or system switching.

5.3.3. Overvoltage due to Fault on Power System

Where the c.t's or p.t's are grounded in the switchyard while the case, circuitry and/or d.c. supply of a static relay are grounded at the relay panel at some distance from the switchyard, a fault near the station may cause a difference of potential of several kV between the two grounding points and this voltage can appear between the a.c. and d.c. circuits of the relay. Figure 5.12 shows such a case where a very high voltage can appear across an auxiliary transformer in the relay and between the relay circuit and the grounded case.

Another source of voltage transient is the sudden reduction in voltage caused by a line-to-ground fault. This sudden voltage change represents a very steep voltage surge which can go through the interwinding capacitance

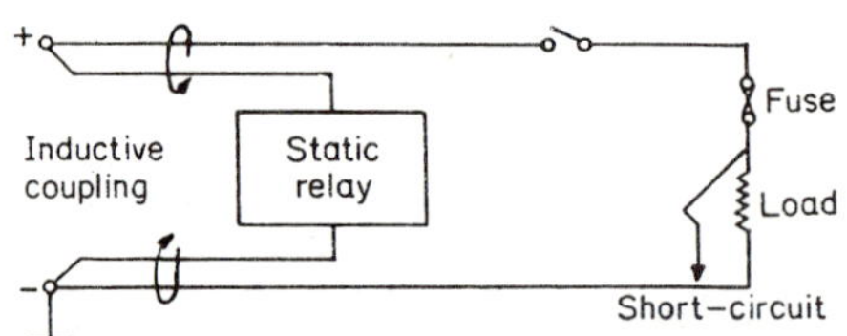

Fig. 5.11. Induced over voltage from a fault in another circuit

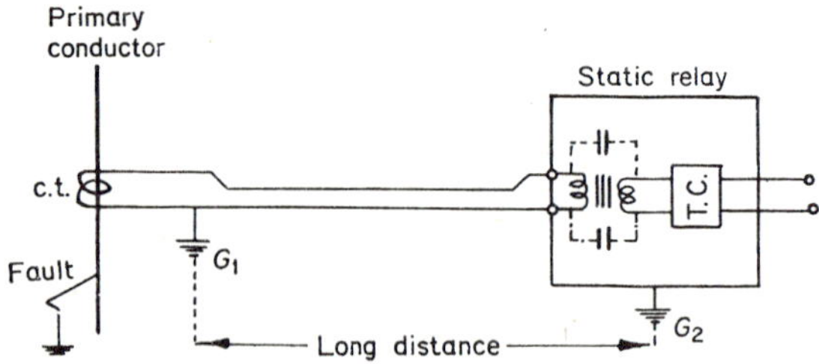

Fig. 5.12. Potential difference between grounding points G_1 and G_2 due to a fault near G_1

T.C. = transistorised circuit

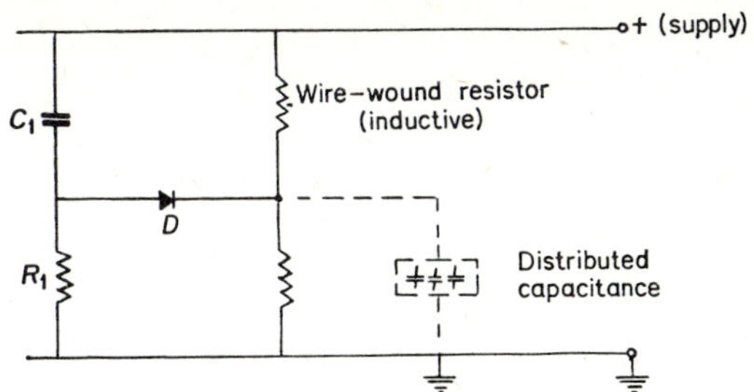

Fig. 5.13. Spike on d.c. supply voltage appears across a diode

of a nearby c.t. and damage the relays. The further the c.t. is from the fault, the more damped the surge will be.

5.3.4. Overvoltage due to Lightning

Lightning can also cause very high voltage gradients in the earth and the presence of a relay in the residual circuit of the c.t's can offer a very high impedance to the very high frequency lightning current, causing an overvoltage to appear between the grounding points of the c.t's and the relay case (Fig. 5.12).

Where pilot wires are involved, such as in pilot-wire relaying, the potential gradient in the earth due to a fault or lightning can appear between the grounded equipment at the two ends of the pilot wire, creating a similar condition to Fig. 5.12. In this case some protection can be provided by using insulating transformers between the pilot wires and the relay circuits as well as surge arresters at the pilot wire terminals, as shown in Figs. 5.22a and 5.22b.

5.3.5. Overvoltage due to System Switching

When an air-break switch is closed the circuit is made initially by a flashover between contacts at the first voltage peak, i.e. at the moment when the charge on the system capacitance is near its maximum. This initiates h.f. and a.f. oscillations, some of which are amplified by resonance and supported by subsequent restrikes and the harmonics in the arc current. The worst case is where a capacitance p.t. (C.V.T.) is discharged into other capacitances of an E.H.T. station, such as wall bushings, c.t's., etc. through a short length of bus bar by the closing of an isolator switch. The closing of a circuit-breaker is much less dangerous because it makes only one or two pulses whereas the isolator arcs over several times before making metallic contact.

The frequency of the oscillation is sufficiently high to pass easily through the interwinding capacitance of c.t's and p.t's and can appear as many kV at the terminals of a relay. In some cases bursts of h.f. potential as high as 20 kV have been measured at the relay at frequencies of 250 to 1500 kc s, due to amplification by oscillation between the capacitance potential divider and the leakage inductance of the auxiliary p.t.

These transient overvoltages (Fig. 5.18) may be reduced to safe values by grounding at the relay end the shielding of the secondary a.c. leads to the relay. A further improvement can be provided by an earthed copper foil screen between the windings of the c.t's and p.t's.

Harmonics in the c.t. secondary current can produce high voltages across the replica impedance of a distance relay, but these are usually suppressed by limiting devices in the measuring circuit.

Another possible source of h.f. transients in the relay circuit is by loops in the relay circuit acting as radio antennae if the relay is in the neighbourhood of the primary switching and if it is not adequately shielded and grounded.

5.3.6. Overvoltage from Components inside the Relay

A current or voltage spike has a steep leading edge which acts like a very high frequency; hence very small inductances, such as wire-wound resistors or event he circuit wiring, present a very high reactance (both self and mutual) to a current spike. Furthermore, a very small capacitance, such as that between parallel wires or between the winding of a transformer and its core, offers a very low impedance to voltage spikes.

A voltage spike, however, is dangerous only if it appears in the non-conducting direction of a diode or a transistor. In the other direction the spike would be short-circuited and do no harm; it might, however, trip a triggering circuit or an SCR unless a short delay is introduced.

In Fig. 5.13 a voltage spike in the d.c. supply would appear in full across the diode and could break it down. A similar condition could happen in a transistor circuit.

Smaller voltage spikes can be caused by reverse voltage recovery within a diode. When the voltage across the diode is reversed the current carriers on the junction face must be removed before the blocking can start. While this is being done a reverse current flows which suddenly stops when the process is complete. If the circuit is inductive, the sudden interruption of the current causes a back e.m.f. spike. This effect is worst in diffused junction diodes.

5.4. EFFECT OF SPIKES UPON SEMICONDUCTORS

Although the amplitude of a voltage spike may be no more than the peak value of a non-destructive sinusoidal wave, its sharp rate of rise can be destructive to semiconductors under certain circumstances. This is because:

(*a*) The initial inter-electrode capacitance is several times higher than normal and a sudden reversal of voltage to the blocking direction may destroy the junction due to a momentary inrush current.

(*b*) If the time taken for the voltage to reach its peak value is less than the time (nanoseconds) taken for the semiconductor to start conducting current, the full voltage will appear across the barrier. With a slower rate of rise the resistance of the semiconductor will fall to a small value before the voltage reaches a destructive value and very little of the spike voltage would appear across the semiconductor.

In addition to a high peak value and a steep wave-front, the spike has to be associated with a low source impedance, i.e. the power of the spike must exceed 10^{-10} watt-second, to damage the semiconductor (e.g. 100 microwatts for 1 microsecond).

Spikes up to 3·7 kV have been recorded at relay terminals but these are generally so brief that they have insufficient energy to damage transistors, especially if they are protected by a spike trap.

5.5. PROTECTION OF STATIC RELAYS AGAINST SPIKES

Any voltage or current transient can be absorbed by the appropriate connection of surge-diverting devices such as non-linear resistors, series inductances, parallel capacitances, etc., but proper arrangement of the relay circuit and supply leads can minimise the surge-diverting equipment needed.

Where possible, relay and switch contacts should have a non-linear resistor (VDR) such as back-to-back diodes across them and/or across any inductive coils in series with them. Relay cases should be grounded at a point as close as possible to the input leads.

5.5.1. Spike Diverters

Spike diverters are similar in action to surge diverters on power systems, i.e. they limit transient overvoltages by diverting surge current through a parallel VDR.

In an R-C surge diverter the resistor should be on the side of the capacitor away from the relay (Fig. 5.14a) because it reduces the surge voltage across the capacitor and because most capacitors have some inductive reactance.

The effectiveness of a capacitor to absorb a spike is reduced by long capacitor leads (Fig. 5.14a) because they have a certain amount of inductance. This can be minimized by using very short capacitor leads (Fig. 5.14b) or cancelled out by keeping them close together, preferably twisted, as in Fig. 5.14c; the mutual reactance between the twisted leads will be in parallel with the inductive reactance of the capacitance and hence will reduce it. A small parallel capacitor would do the same thing.

An alternative to Fig. 5.14a for connecting the R-C diverter, which is often used for rectified d.c. supplies, is shown in Fig. 5.15a; this connection, however, requires a larger capacitor than Fig. 5.14a. The capacitor required is given by $C = 600\ \mathrm{VA}/V^2f$ microfarads, where VA is the VA rating of the transformer, V is the secondary voltage rating and f is the circuit frequency. In Fig. 5.15b, C_1 is about 0·05 μf and C_2 is a large electrolytic capacitor.

The same precautions should be applied to a limiting (zener) diode which should preferably be connected in the position of the capacitor in Fig. 5.14a.

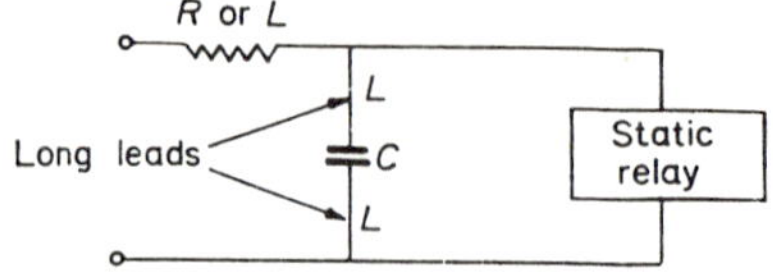

(a) Series inductance of capacitor leads

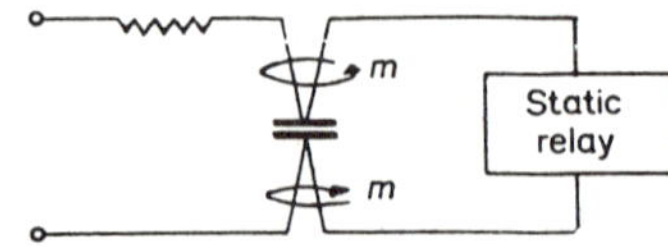

(b) Connection of capacitor leads to reduce their inductance

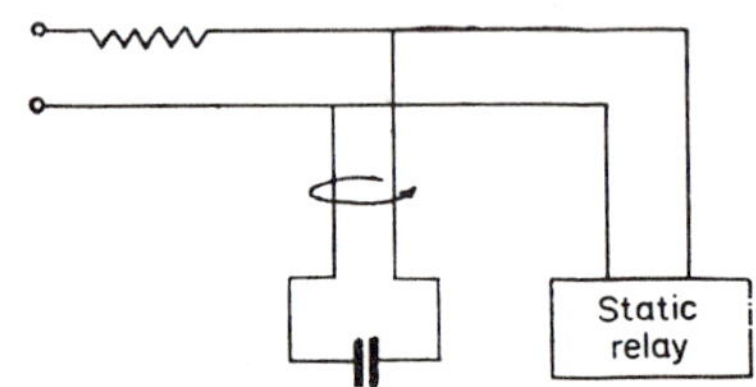

(c) Parallel inductance of leads offsets capacitor inductance

Fig. 5.14. Inductance of capacitor leads

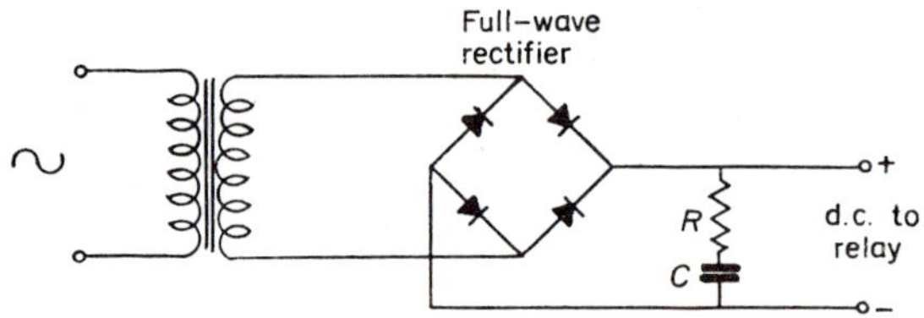

(a) Surge suppressor for full-wave rectifier

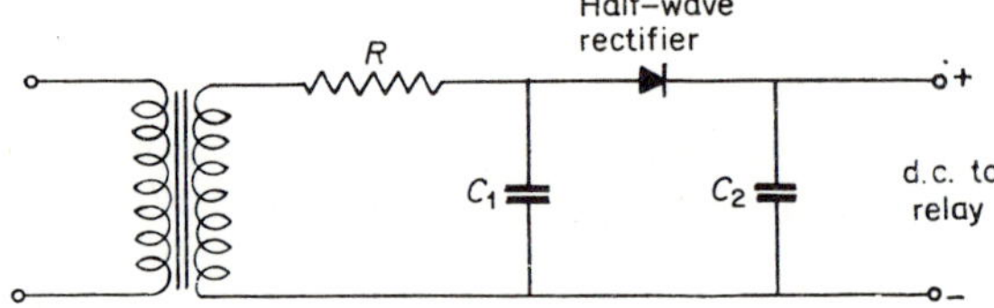

(b) Surge suppressor for half-wave rectifier

Fig. 5.15. Surge suppressors for d.c. supply

Selenium rectifiers can also be connected in this manner to act as voltage spike traps; the choice between them and capacitors depends upon the circuit voltage. Selenium diodes are used as spike traps by connecting them in series for 30 volts steps.

In the case where a diode is used to prevent a reversed d.c. supply from damaging a static relay, the diode should be connected on the other side of the capacitor, as shown in Fig. 5.17 which shows short capacitor leads.

Figure 5.16 shows how a single electrolytic capacitor C can be used to absorb spikes in a three-phase a.c. circuit, such as the potential supply to a distance relay. This circuit also absorbs voltage ripple, reducing it by about 10 to 1. The resistor R_1 prevents oscillation between C and primary circuit inductances; it should be about 10 times the resistance of the relay burden.

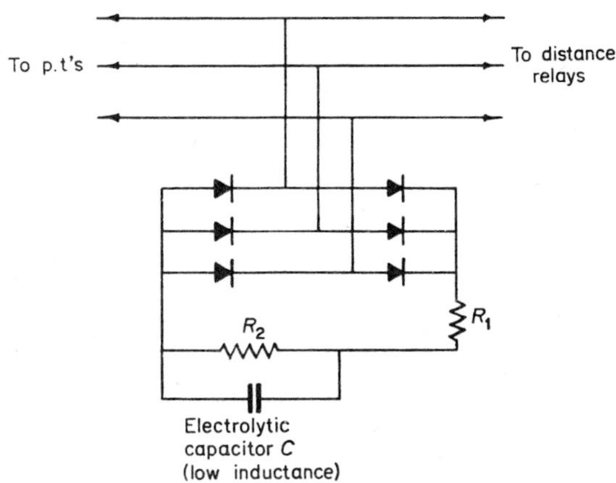

Fig. 5.16. Transient and ripple suppressor for three-phase voltage supply

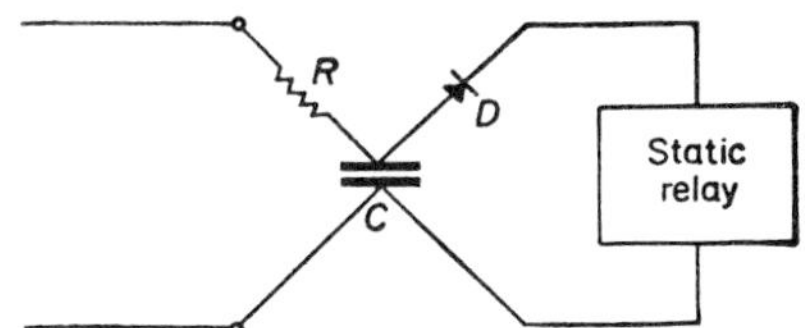

Fig. 5.17. Preferred positions for resistor and diode

Other forms of diverters are gaseous tubes (Fig. 5.22a) and spark-gaps (Fig. 5.22b) and non-linear resistors or diodes, (Fig. 5.22b). Modern gas-filled spark-gap arrestors are small and extremely effective [114]. They have a precise firing voltage and flash over in less than 50 nanoseconds, preventing the voltage surge from exceeding 40 volts.

Manufacturers in the United Kingdom, having provided suitable spike diverters or suppresors for the various inputs, then test the relay with several applications of h. v. spikes or bursts of h. f. as described in section 5.7.

5.5.2. Spike Suppressors

To suppress the voltage spike caused by sudden interruption of an inductive d.c. circuit, a capacitor can be connected across the inductive coil to provide a discharge path for the coil current. The value of the capacitance is given by $C = 0{\cdot}3(L/R_1)$, where L and R_1 are the inductance and resistance of the coil respectively.

A preferred alternative to the capacitor is a diode because it prevents any subsequent oscillation which may induce alternating voltage in the circuit of a nearby static relay. The maximum voltage across the switch or contact which interrupts the coil current is $-V\{(R_s + R_1)/R_s\}$, where R_s is the resistance in the suppressor circuit.

To prevent electrostatic interference from contact operation as well as to suppress the inductive surge voltage, an R-C suppressor across the contacts is effective. The values of R_s and C are given by solving the equations $R_s = 0{\cdot}2\ VC^{0\cdot 2}$ and $R_sC = L/R_1$. Here again diodes are preferable. In short, R-C circuits are effective as spike traps located at the relay but non-linear resistors are safer as suppressors at the source of the transient overvoltage.

5.5.3. Screening

In input leads and connections between relay units, where an interference signal could cause wrong operation or damage to a transistor, the connecting leads should be twisted and/or screened, as in Fig. 5.18, and the screen earthed and connected to the relay frame or case.

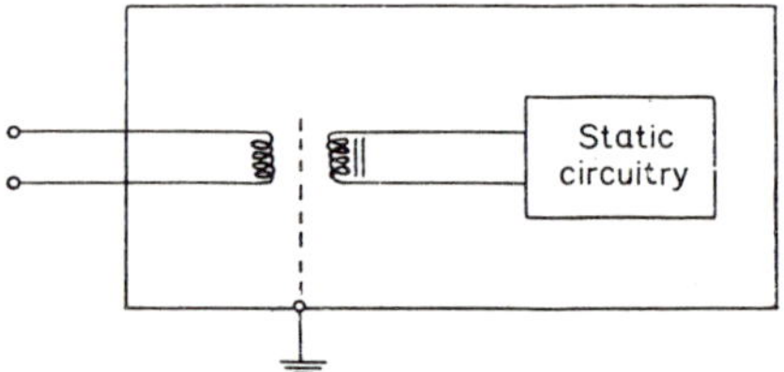

Fig. 5.18. Earthed copper screen between windings

Transformer coils should be earthed and, where possible, a split thin copper screen provided between the windings. The wire from the screen to the case should be as short as possible and preferably connected to the point on the case where it is earthed, as in Fig. 5.19. An alternative way of by-

Fig. 5.19. Screened wires for vulnerable interconnections

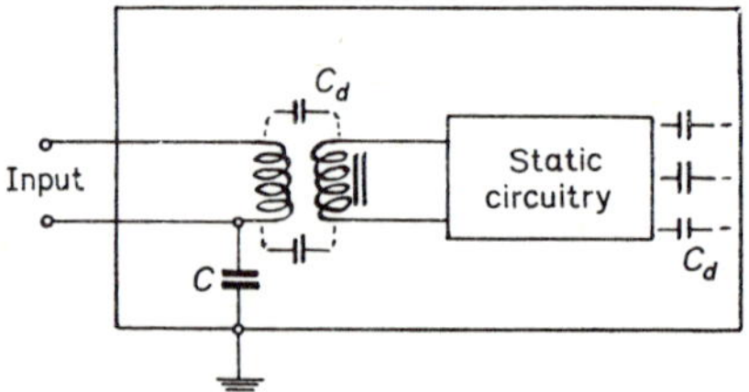

Fig. 5.20. Capacitor for by-passing spikes
Cd = distributed capacitance

passing voltage spikes from the primary is to connect a capacitor between the grounded side of the primary and the case of the relay, as shown in Fig. 5.20.

The ability of a transformer to pass a voltage spike from one winding to another via its interwinding capacitance can be reduced by connecting the grounded leads as far apart as possible on the windings, i.e. to the inside of the inside winding and to the outside of the outside winding, as in Fig. 5.21.

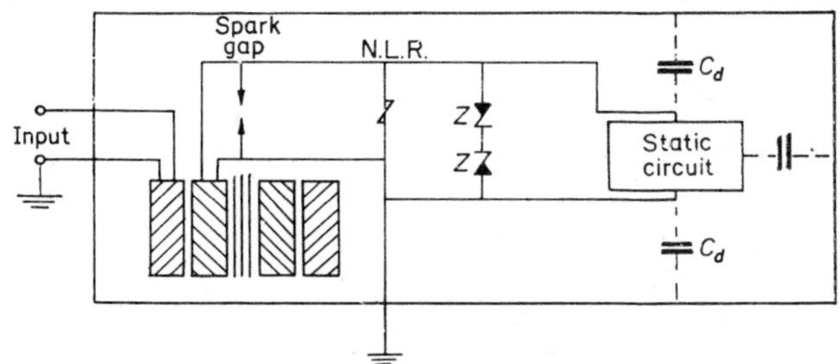

Fig. 5.21. Connections of windings of auxiliary transformer to minimise interwinding capacitance coupling for spikes
Cd = distributed capacitance

In the case of high voltage h.f. transients coming through a capacitance p.t., the most effective protection is (*a*) to provide a grounded copper screen between the windings, (*b*) to connect the relay to the p.t. by a shielded wire with the shielding grounded at the relay end.

5.6. TRANSIENT OVERVOLTAGE COUNTERS

In order to obtain statistics on the frequency and magnitude of voltage spikes appearing on a given circuit, transient counters have been developed which count the number of voltage spikes exceeding a certain magnitude. A number of these are used with different calibrations such as 1 kV, 1·5 kV, 2 kV, 3 kV, 5 kV, etc. Their records are helpful in the design of surge

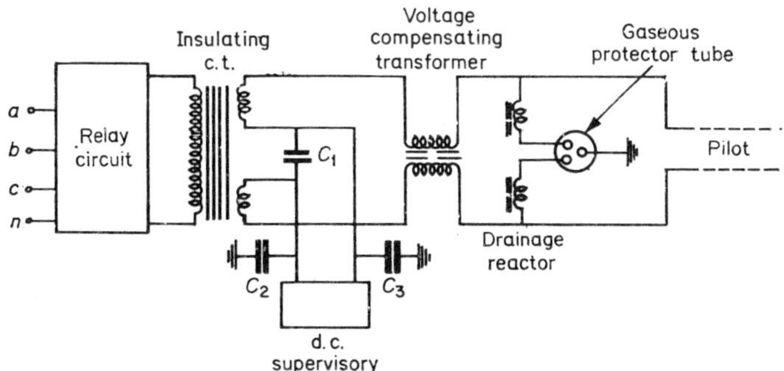

(a) Using gas-filled protector tube

Fig. 5.22a. Pilot wire protection against overvoltage on relays (W)

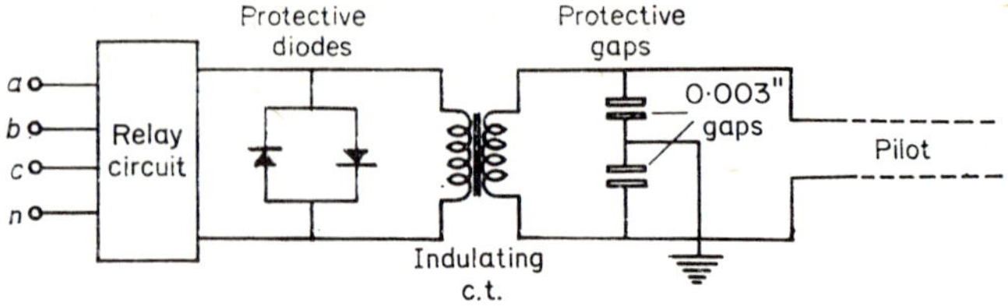

Fig. 5.22.b

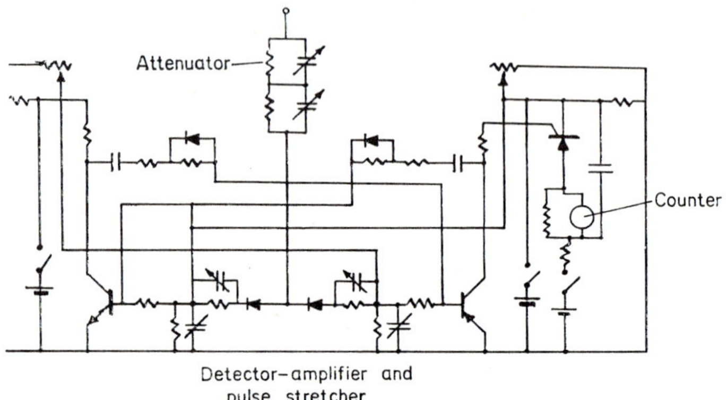

Fig. 5.23. Circuit of counter for voltage spikes down to 20 nano-seconds duration

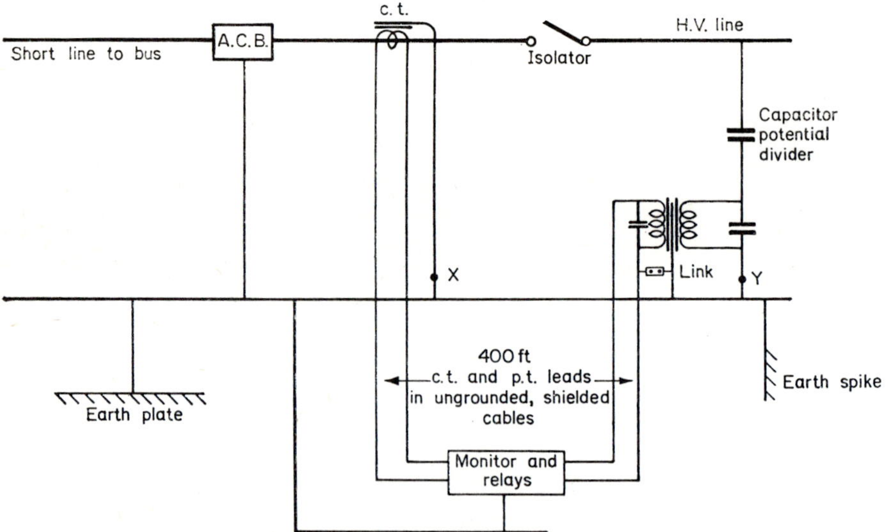

Fig. 5.24. Typical conditions for transient overvoltage on relays Monitor is a transient overvoltage counter recording signals over 1 μs duration

diverters for static relays and indicate to the power system engineer where steps should be taken to find the source of spikes and suppress them.

By adjusting the response time of the transient voltage counter some distinction can be made between destructive spikes and those which are too brief to have enough energy to damage semiconductors. The range of response time is from 1 microsecond down to 20 nanoseconds [71]. Where the actual counter is electromechanical, a pulse stretcher is necessary to make the signal last long enough to operate the mechanism. The complete equipment includes an attenuator, a pulse stretcher, a detector and a counter (Fig. 5.23).

5.7. TRANSIENT OVERVOLTAGE TESTING

The first method of testing the vulnerability of a static relay to transient overvoltages from external sources originated in the U.K. and consisted of applying to all the input leads several 5 kV, 1/50 μS spikes produced by a capacitor discharge of 0·5 coulomb through resistance.

An alternative method, originating in the U.S.A., is several applications of 2·5 Kv, 1·5 MHz damped waves initiated every 0·01 second through a 150 Ω source. This method is valuable for relays protecting E.H.V. systems where the operation of an isolating switch can create powerful h.f. oscillations, as described in section 5.3.5.

6
Power System Faults

Types, causes and frequency of faults on lines, cables, generators, motors, transformers and buses. Determination of relay currents and voltages during faults. Fault resistance

A system fault is a condition where the electric current follows an abnormal path due to the failure or the removal of the insulation which normally confines it to the conductors. Fault statistics also include faults on secondary circuits which result in the tripping of circuit-breakers.

Insulation is usually either air or high resistivity material which may also be used as a mechanical support. Air insulation can be accidentally short-circuited by birds, rodents, snakes, kite-strings, tree-limbs, etc., broken down by overvoltage due to lightning or weakened by ionization due to a fire. Organic insulation can deteriorate due to heat or ageing or can be broken down by overvoltage due to lightning, switching surges or faults at other locations.

Porcelain insulators can be bridged by moisture with dirt or salt, or can develop a crack due to mechanical forces. In such cases the initial lowering of insulation resistance causes a small current to be diverted which hastens the deterioration or ionization, causing this current further to increase in a progressive manner until a flashover occurs.

6.1. TYPES AND FREQUENCY OF FAULTS [50]

Faults can occur between (*a*) any one conductor to ground (single-phase ground fault), (*b*) any two conductors to ground (double-ground fault), (*c*) any two conductors (phase-to-phase fault) or (*d*) all three phases (three-phase fault). Most power system faults occur on overhead lines because they are more numerous and more exposed.

The incidence of particular types of faults depends partly upon the location; for instance, in extremely dry places like deserts, ground faults are rare, whereas elsewhere they are in the majority. The number of faults also varies with the weather; for instance, there are normally about 500 faults per year on the 30,000 MW system of the C.E.G.B. but in 1962/63 there were over

2,000 due to an unusually severe winter. Average figures for a typical large power system are as follows:

TABLE 6.1.
Annual faults

Equipment	Faults per Unit per year	% of Total
Overhead lines	2*	33
Cables	4†	9
Switchgear	0·005	10
Transformers	0·15	12
Generators	0·4	7
Secondary	1·5	29
TOTAL		100

In this Table 'Secondary' equipment means c.t.s and p.ts, relays, control circuits, etc.

* Per unit on lines and cables means per 100 circuit miles, but on other equipment it means per 100 MW.

† This figure is 7 if joints, sealing ends and trifurcated boxes are included

The percentage distribution of faults also varies with the voltage. The data in Table 6.1 is for a 132/275 kV system. At 33 kV the number of line faults may double and the number of station faults may be halved.

6.1.1. OVERHEAD LINES

There has been an average of one fault per 50 miles (80 km) of line per year on the C.E.G.B. lines over the last twelve years. In 1962/63 there was one fault per 14 miles due to unusual icing conditions. About half of the faults are caused by lightning, about 20% due to icing or moisture and the rest to various causes such as wind-borne vegetation. About 85% of line faults are single-phase to ground; 12% involve both circuits of a two-circuit line (275 kV).

In tropical countries the faults due to lightning are more numerous; in these countries large birds also cause faults by flying through the lines, excreting on insulators and by trailing nest-building materials such as vines over the lines. In the U.S.A. and the U.S.S.R. there is often a spate of winter faults due to icing.

Most faults on overhead lines are transient, so that service can be restored by isolating and then reclosing the faulty line section very rapidly, allowing only sufficient time for the air to be de-ionized after the fault arc has been extinguished so as to prevent re-striking, i.e. about 0·25 second.

6.1.2. Cables

About half of the faults are in the actual cables and half are at end junctions. Cable faults are invariably of a permanent nature so that automatic reclosing cannot be used. However, high-speed tripping is still valuable on transmission cables because it can prevent a system oscillation (power swing). On the other hand, it will not mitigate damage done to the cable; in fact too rapid tripping may make the fault difficult to locate.

6.1.3. Alternators

In Table 6.1, about half of the generator faults are usually in the exciter circuit or control equipment and about 40% in the auxiliary devices such as cooling equipment. Although the number of generator faults that occur are relatively few, their variety is greater than for any other equipment.

(*a*) *Stator faults:* The breakdown of conductor insulation may result in a fault between conductors or between a conductor and the iron core. The breakdown may be caused by overvoltage or by overheating which in turn can be caused by unbalanced currents. ventilation troubles, etc.; it may also be caused by damage done to insulation by conductor movement due to forces exerted by short-circuits or out-of-step conditions.

Because of the destructive effect of a ground fault (conductor to core) due to the high temperature of the arc, the fault current is usually limited by impedance in the neutral of the generator.

(*b*) *Rotor faults:* Rotor windings may be damaged by ground faults or open-circuits; structural parts of the rotor itself may be damaged due to overheating due to unbalanced stator circuits or blocked ventilation. Short-circuited rotor turns cause magnetic unbalance which may damage the machine bearings.

(*c*) *Overspeed:* Due to sudden loss of load.

(*d*) *Loss of field:* Due to failure of the pilot exciter, the main exciter or the field breaker.

(*e*) *Loss of synchronism:* Due to separation from the system by an inter-phase fault or wrong switching, or loss of field.

(*f*) *Bearing failure:* Due to failure of cooling or oil supply, or pedestal insulation.

(*g*) *Miscellaneous:* Voltage regulator failure, loss of boiler pressure or vacuum.

6.1.4. Three-phase Motors

These are subject to the same electrical faults as alternators. In addition they are subject to damage if started with one phase open and the load may be damaged if they are started with reversed phase rotation. They can also be damaged by overload, especially if the a.c. supply voltage is unbalanced.

6.1.5. Transformers

The distribution of transformer faults depends upon its type, rating, application and locality. Typical figures are as follows. Roughly 10% of transformer faults are inside the tank, about 65% are in the tap-changer and auxiliary control circuits and about 15% are bushing flashovers. Internal faults are very serious because there is always the risk of fire; these internal faults can be classified into two groups:

(*a*) *Electrical faults* which may cause immediate damage but are generally detectable by unbalance of current or voltage such as:

(i) Phase-to-earth faults or phase-to-phase faults on the H.V. and L.V. external terminals.
(ii) Phase-to-earth faults or phase-to-phase faults on H.V. and L.V. windings.
(iii) Short-circuit between several turns of one winding.
(iv) Earth fault on tertiary winding, or short-circuit between turns of a tertiary winding.

(*b*) *Incipient faults* (so-called) which are initially minor faults, causing slowly developing damage. They are not detectable at the winding terminals by unbalance; they include:

(i) A poor electrical connection of conductors or a core fault (due to a breakdown of the insulation of lamination, bolts or clamping rings) which causes limited arcing under the oil.
(ii) Coolant failure, which will cause a rise of temperature even below full load operation.
(iii) Related to (ii) is the possibility of low-oil content or clogged oil flow, which can readily cause local hot-spots on windings.
(iv) Regulator faults and bad load -sharing between transformers in parallel, which can cause overheating due to circulating currents.
(v) Turn-to-turn fault in a winding.

6.1.6. Bus Zone

The station bus is subject to the same faults as a line but, because of the interruption to service if it is disconnected, it is designed (e.g. earthed screening) so as to make faults on it extremely unlikely. A typical average incidence of bus faults is 1 in 15 years per installation; 25% of these are due to equipment insulation failure, 25% to flashovers caused by lightning, 30% to human errors and 10% to miscellaneous causes such as falling objects and circuitbreaker failures. 60% of the faults are to ground. Here again these figures vary with location, lightning faults being more numerous in tropical countries.

6.2. MOMENT OF INCEPTION

Over 95% of faults occur within 40° before voltage maximum, i.e. the peak value of the voltage wave. In cables, transformers and generators this is due to the fact that the deterioration of the insulation due to heat, age, etc., is a relatively slow process and a breakdown occurs when the phase-to-neutral peak voltage becomes too much for the insulation. On overhead lines a similar process goes on in the case of flashovers due to atmospheric pollution of insulators or to conductors swinging together, but faults due to lightning are more complex.

6.2.1. Lightning [51]

An overhead line is liable to be struck directly only by those lightning strokes which originate directly above it and over a width of about half its height above the ground. This number is usually small because the area under the line is small compared with the total area where lightning is possible. Assuming open ground in the vicinity, tall objects such as trees and buildings will reduce the number of direct strokes to the line. However, these indirect strokes can induce high voltages on the line which will flash over insulators if this voltage is above the impulse level of the line. High voltage can also be induced by flashovers between clouds (which are much more frequent than cloud to earth) caused by the release of the 'bound' charge which is induced on the line by a charged cloud above it. Figure 6.1

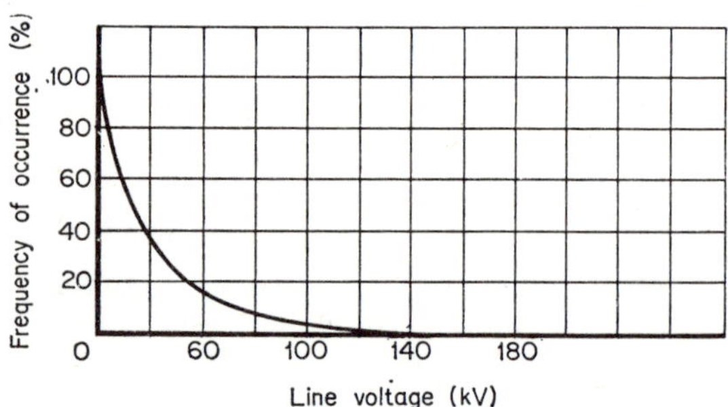

Fig. 6.1. Frequency of voltage surges exceeding 40 kV, induced by lightning
Note Line with earth wire below phase conductors and with insulator pins earthed at every third pole

shows the likelihood of the induced voltage attaining a particular value. The likelihood of an actual flashover occurring on a particular line depends upon its impulse level, for which typical values are given in the following Table 6.2.

The foregoing would appear to indicate that flashovers are less likely on high-voltage lines, but this is offset by the fact that the towers are higher and the lines more exposed to lightning. The most important factors for

TABLE 6.2.
Impulse levels

Rated kV	11	22	33	66	110	132	220	275	400
Impulse kV	80	150	200	300	450	550	900	1050	1500

preventing flashovers are overhead ground wires and low tower footing resistance.

In general there is a greater likelihood of flashovers on low-voltage (distribution) lines because of their low impulse voltage level. For the same reason there is more possibility of flashovers occurring away from the peak voltage; at transmission line voltage level (over 100 kV), all faults for which oscillograms are available occurred within 40° before voltage maximum.

The possibility of even a direct stroke causing a flashover near voltage zero is minimized by the fact that the lightning stroke lasts only one or two microseconds and, if the line voltage were near zero at that moment, there would be nothing to sustain the flow of power after the stroke. Although the stroke current may be up to 100,000 amps there is less than a coulomb in a stroke, so there would be no cloud of ionized air maintaining a low resistance path until the system voltage built up.

A flashover near voltage zero would cause a large d.c. offset in the fault current and, although offsets may appear in staged faults initiated by a test breaker, none could be seen in any of the automatic oscillograms of natural faults examined by the author.

The conclusion is that almost all faults occur within 40° before voltage peak and that the likelihood of overreach by instantaneous overcurrent or distance relays, due to faults occurring near voltage zero, is extremely small.

6.2.2. Discharge of System Capacitance

When a fault occurs at voltage maximum, the energy stored in the capacitance of the system is a maximum and is discharged through the fault. Because of the inductive reactances of the system the discharge is of an oscillatory nature and the current is also offset because it begins as d.c. with the polarity of the initial peak voltage, which is invariably negative.

This offset current can be mistaken for offset current due to the inductive reactance of the system that would occur with a fault initiated at voltage zero, but it can be recognized by the fact that the fault current started at or near a voltage maximum and the fact that its frequency is between 400 and 1800 c/s [98].

The frequency of the oscillatory component is

$$f = \frac{1}{2\pi}\sqrt{\left\{\frac{1}{LC} - \left(\frac{R}{2L}\right)^2\right\}} \tag{6.1}$$

where L, R and C are system parameters. The frequency decreases with the size of the power system (C) and the length of the faulted line (L), but does not change much with the kV rating since the inductance increases and the capacitance decreases as the logarithm of the conductor spacing.

The susceptance of a typical 50 c/s 132 kV system is 4·72 micro-mhos per mile. For a system with 1,000 miles of line the total capacitance in microfarads would be

$$C = \frac{4{\cdot}72 \times 1000}{100\pi} = 15\mu\text{F}.$$

The effective capacitance is much smaller because it is distributed and because the voltage is progressively less depleted as the source is approached. The capacitance per mile increases with the system kV but the number of lines decreases with the system kV, so that the effective value does not vary much.

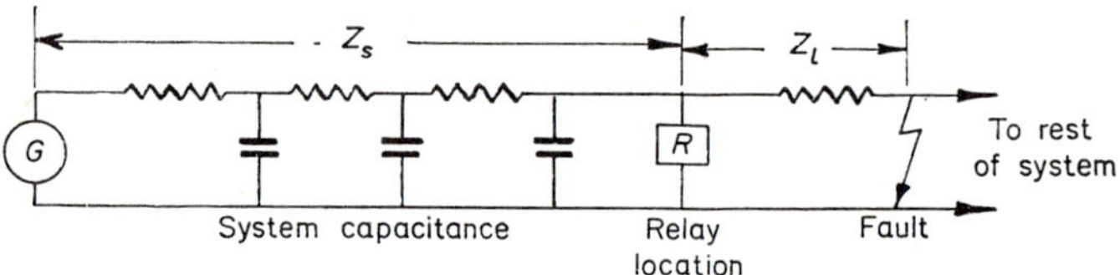

Fig. 6.2. Discharge path for system capacitance

The inductive reactance for a 20 mile 50 c/s 132 kV line is about 13 Ω or about 0·04 henry. The reactance of a power transformer might be 10% on a 100 MVA base which is about 17 Ω or 0·054 henry, making a total of 0·94 henry.

Assuming 2 ohms resistance in the circuit and $C = 1\ \mu$F, the natural frequency of the capacitance discharge oscillation from Eq. (6.1) would be

$$f = \frac{1}{2\pi}\sqrt{\left\{\frac{10^6}{0{\cdot}094} + \left(\frac{2}{0{\cdot}188}\right)^2\right\}} = 522 \text{ c/s}.$$

This is about the 9th harmonic of 60 c/s or the 10th harmonic of 50 c/s so that it is comparable with the harmonics associated with the non-linear resistance of arcs and of the earth, and can be seen on a fast oscillograph or detected by relays such as are described in Chapter 13, Sections 13.7 and 13.9.

The foregoing is important in the testing of high-speed relays because most modern distance and differential relays are designed to operate correctly with the offset current caused by initiating the fault current at voltage zero, whereas in practice the smaller offset current caused by starting the fault at voltage maximum is more important but, unfortunately, difficult to reproduce in a laboratory.

6.3. DETERMINATION OF RELAY QUANTITIES

During a fault, the current and voltage at any point on the system can be calculated by the standard method of symmetrical components. From the point of view of the protection engineer, however, only the currents and voltages at the relay are of interest.

The quickest method of calculation is as follows:

(*a*) Mark on a single-line diagram of the power system all the line and plant impedance values, calculated to a common MVA base. Also, mark all the grounding points and the position of the fault.

(*b*) Repeat for the positive, negative and zero sequence networks and calculate Z_1, Z_2 and Z_0, the positive, negative and zero sequence impedances of the system, viewed from the fault location.

(*c*) From the system impedances, Z_1, Z_2 and Z_0, and the corresponding impedances Z_1', Z_2' and Z_0' of the protected zone of the line or equipment, calculate the currents and voltages at the relay from the expressions in Table 6.3 in which R is the fault resistance and 'a' is the operator $\varepsilon^{-j\frac{2\pi}{3}}$ or $\left(-\frac{1}{2}+j\frac{\sqrt{3}}{2}\right)$ which rotates a vector 120° leading without changing its amplitude; similarly $a^2 = \varepsilon^{-j\frac{2\pi}{3}} = \left(-\frac{1}{2}-j\frac{\sqrt{3}}{2}\right)$ and rotates the vector 240° leading.

The method of calculation of the data in Table 6.3 is not given here because it appears in many references, notably Ref. [21]. However, in the Appendix of this chapter (Section 6.7) a simple method of calculating series and parallel impedances is described. The data are given for the sound phases as well as for the faulted phase (*s*) because it is necessary for the analysis of the behaviour of phase selectors and distance relays.

Figure 6.3a represents any power system simplified to the limit. G_A and G_B are generating sources whose e.m.f's are assumed equal to E. Z_s is the source impedance, Z_L the impedance of the protected line section, Z_f the impedance between the relay and the fault including the fault itself.

Figure 6.3b is a rearrangement of Fig. 6.3a to simplify calculation by reducing the variables to a minimum. The values V and I refer to the fault

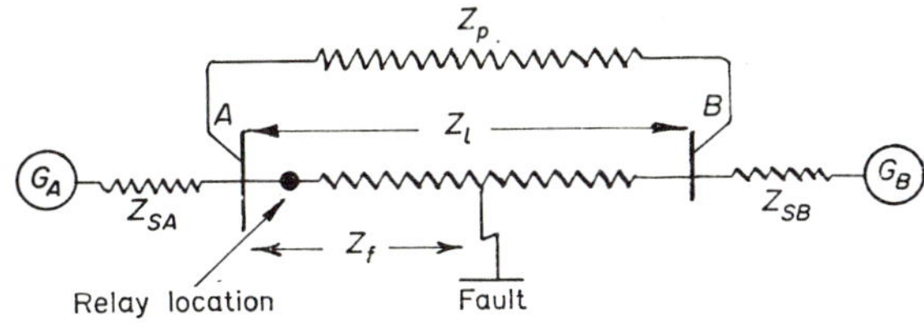

(a) Typical power system, simplified

Z_s = source impedance
Z_L = protected line impedance
Z_f = faulted impedance of line
Z_p = impedance of parallel line or network

Fig. 6.3.a Typical power system impedances (simplified for fault calculation)

TABLE 6.3

Fault	Three-phase	Phase b-Phase c	Phase a-Ground
KI_1	C	C	C
KI_2	0	$-C$	C
KI_0	0	0	C_0
KI_a	C	0	$2C + C_0$
KI_b	a^2C	$(a^2 - a)\,C$	$C_0 - C$
KI_c	aC	$(a - a^2)\,C$	$C_0 - C$
KV_1	$CZ' + R_F$	$Z_2 + CZ'_1 + R_F$	$Z_2 + Z_0 + CZ'_1 + 3R_F$
KV_2	0	$Z_2 - CZ'_2$	$CZ_2 - Z_2$
KV_0	0	0	$C_0Z'_0 - Z_0$
KV_a	$CZ'_1 + R_F$	$2\left(Z_2 + \frac{R_F}{2}\right)$	$2CZ'_1 + C_0Z'_0 + 3R_F$
KV_b	$a^2(CZ'_1 + R_F)$	$(a^2 - a)\,CZ'_1 + a^2R_F - Z_2$	$(a^2 - a)\,Z_2 - CZ'_1 + (a^2 - 1)\,Z_0 + C_0Z'_0 + 3a^2R_F$
KV_c	$a(CZ'_1 + R_F)$	$(a - a^2)\,CZ'_1 + aR_F - Z_2$	$(a - a^2)\,Z_2 - CZ'_1 + (a - 1)\,Z_0 + C_0Z'_0 + 3aR_F$
KV_{ab}	$(1 - a^2)\,(CZ'_1 + R_F)$	$(a - a^2)\,CZ'_1 + 3Z_2 + R_F\,(1 - a^2)$	$3CZ'_1 - (a^2 - a)\,Z_2 - (a^2 - 1)\,(Z_0 + 3R_F)$
KV_{bc}	$(a^2 - a)\,(CZ'_1 + R_F)$	$2(a^2 - a)\left(CZ'_1 + \frac{R_F}{2}\right)$	$(a^2 - a)\,(Z_0 + 2Z_2 + 3R_F)$
KV_{ca}	$(a - 1)\,(CZ'_1 + R_F)$	$(a - a^2)\,CZ'_1 - 3Z_2 + (a - 1)\,R_F$	$-3CZ'_1 + (a - a^2)\,Z_2 + (a - 1)\,(Z_0 + 3R_F)$
K	$\frac{1}{E}(Z_1 + R_F)$	$\frac{1}{E}(Z_1 + Z_2 + R_F)$	$\frac{1}{E}(Z_1 + Z_2 + Z_0 + R_F)$

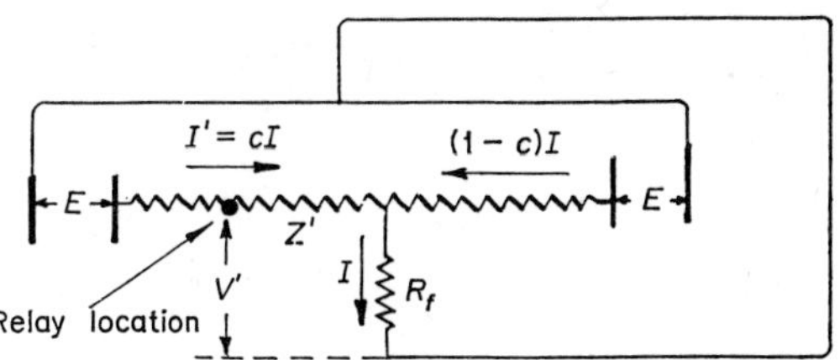

(b) System impedances arranged for calculation

E = generated e.m.f.
Z' = impedance relay to fault
I = fault current
I' = relay current = cI
V' = voltage at relay

Fig. 6.3.b

location. The primed values refer to their values at the relay. Similarly Z is the total system impedance viewed from the fault and Z' is the impedance of the line between the relay and the fault, so that $V' = I'Z'$.

6.4. DISTRIBUTION OF SEQUENCE COMPONENTS

Figure 6.4 shows how the phase sequence components of the system impedances can be arranged to simulate the various types of faults and this facilitates calculation of the phase sequence components of the currents and voltages at any point on the system, such as the location of a particular relay.

Figure 6.5 shows that only the positive sequence components are present in a balanced three-phase fault and confirms the network of Fig. 6.4a. The negative and zero sequence components provide the dis-symmetry of the other types of faults. In Figs. 6.5, 6.6, 6.7 and 6.8 a non-homogeneous system is considered, i.e. the typical case where Z_s is more lagging than Z_l or Z' since the X/R ratio of generators and transformers is higher than for lines, especially as Z' may include fault resistance.

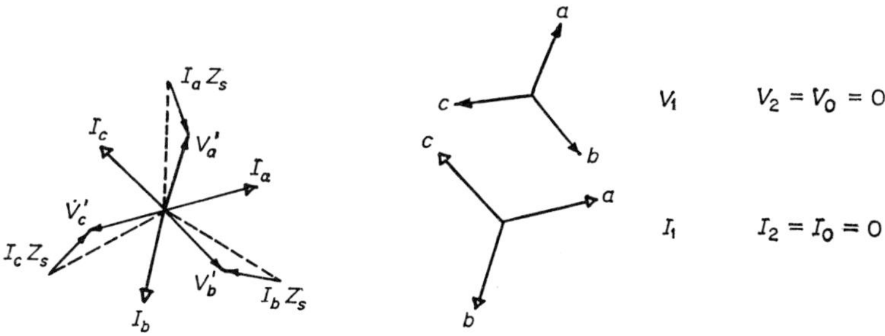

Fig. 6.5a. Phase sequence components in a balanced three-phase fault

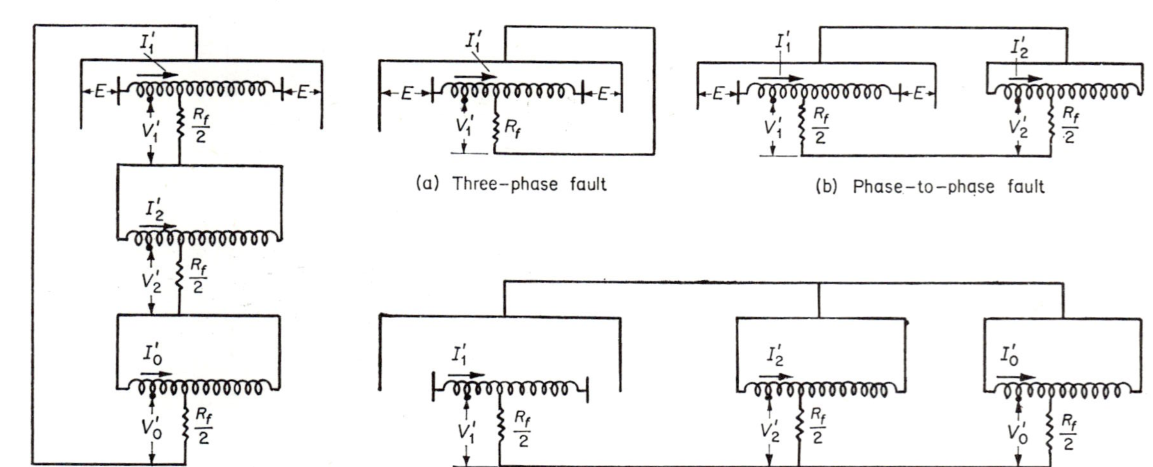

(a) Three-phase fault

(b) Phase-to-phase fault

(c) Single-phase ground fault

(d) Double-ground fault

Fig. 6.4. Arrangement of sequence networks equivalent to faults

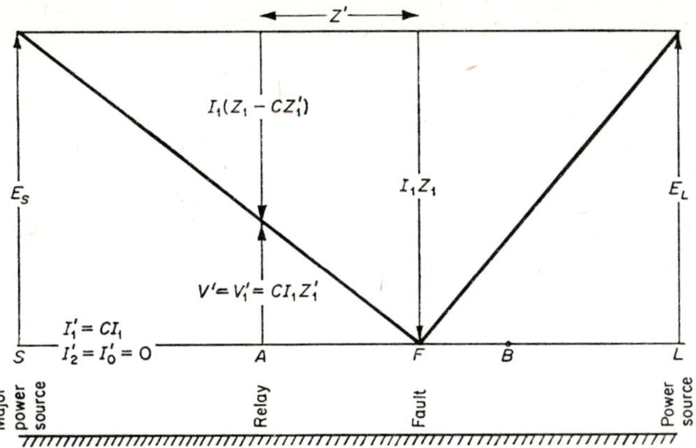

Fig. 6.5b. Distribution of phase-sequence currents and potentials during a balanced three-phase fault

Figure 6.6 shows how the positive and negative sequence components of current in a phase-to-phase fault cancel out in phase '*a*' and combine to produce the opposed currents I_b and I_c. The voltage components, on the other hand, add in phase '*a*' and are combined at 120° in phases '*b*' and '*c*'.

Figure 6.7 shows a single-phase-to-ground fault on a solidly grounded system. Since the three sequence networks are in series, the three sequence currents are equal at the fault. Hence the three sequence components of current add in phase '*a*', to produce the fault current, and cancel out in the other two phases where they are 120° apart. The negative and zero sequence voltage components oppose the positive sequence component in phase '*a*' and are combined at 120° in the other two phases. On a resistance or impedance grounded system there would be an additional factor of

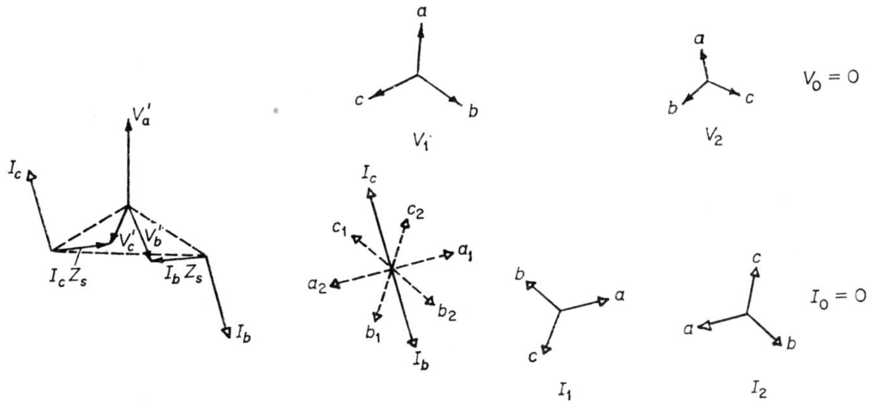

Fig. 6.6a. Phase sequence components in a phase-to-phase fault

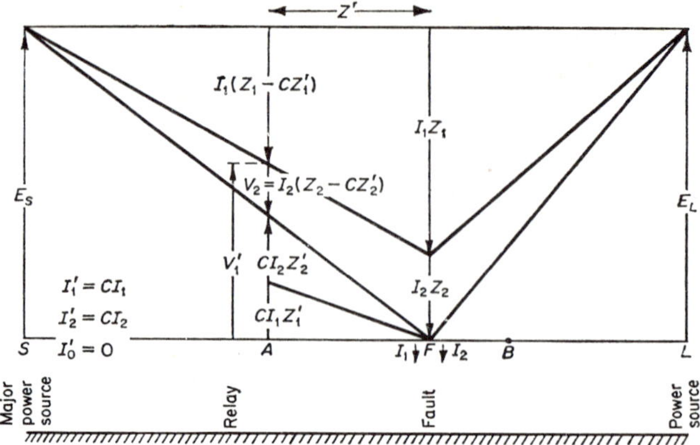

Fig. 6.6b. Distribution of phase-sequence currents and potentials during a BC fault

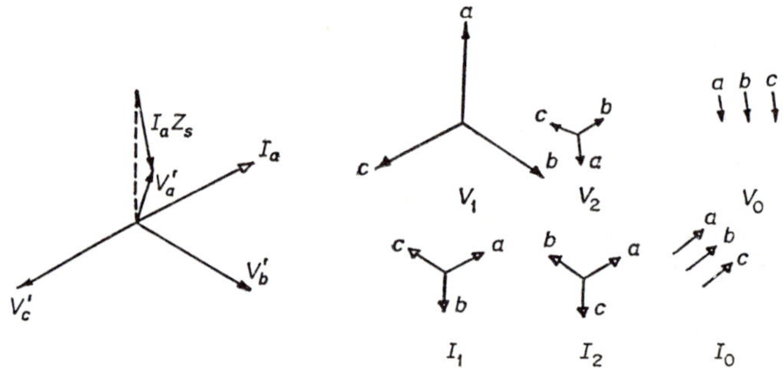

Fig. 6.7a. Phase sequence components in a single-phase ground fault

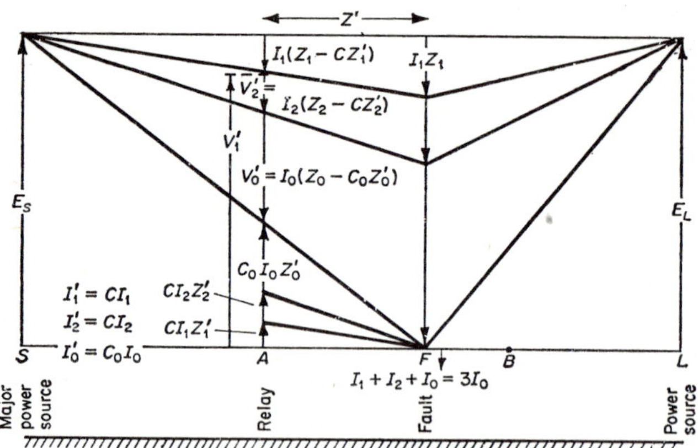

Fig. 6.7b. Distribution of phase-sequence currents and potentials in phase A during a phase A to ground fault

distortion due to the shifting of the neutral by the voltage drop in the neutral $3I_0Z_n$ where Z_n is the grounding impedance.

Figure 6.8a shows the more complicated condition of a double-ground fault. Here the negative and zero sequence networks are in parallel, so that the positive sequence current is equal to the sum of the negative and zero sequence currents at the fault, and the negative and zero sequence currents are equal. These facts simplify the distribution diagram 6.8b, although the fault itself is more complicated since it is a combination of two single-phase ground faults.

6.5. FAULT RESISTANCE

The impedance of the copper path through the power system is composed of the series inductive reactance and resistance of lines and equipment, and the shunt capacitance and leakance mostly of lines and cables; this impe-

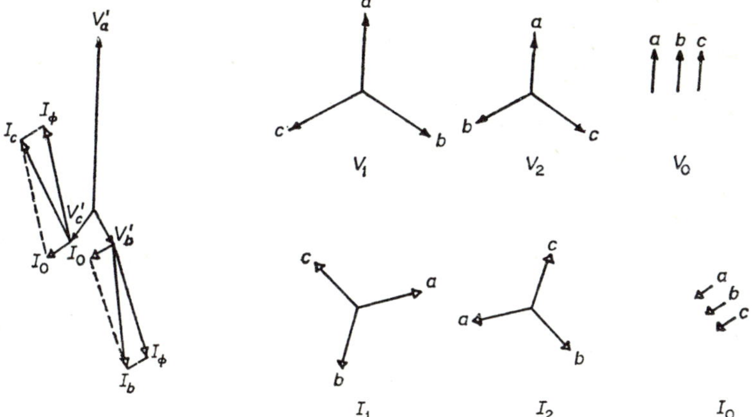

Fig. 6.8a. Sequence components in a double ground fault

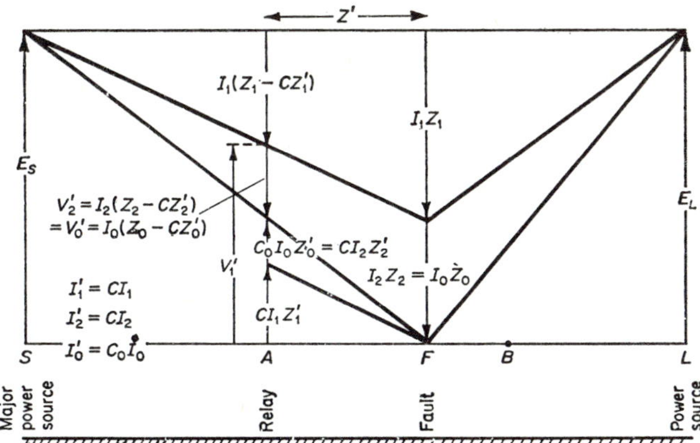

Fig. 6.8b. Distribution of phase-sequence currents and potentials in phase A for a double-ground fault on phases B and C

dance is linear and the fault currents which result from short-circuiting it are sinusoidal in wave form.

On the other hand, if the fault is not a metallic short-circuit but involves an arc or a path through the earth, non-linear impedances are introduced which tend to introduce harmonics into the current and/or voltage.

6.5.1. Arc Resistance

The resistance of a power arc is $8750(S + 3ut)/I^{1\cdot 4}$ ohms where S is the conductor spacing in feet (see Fig. 6.9), u the wind velocity in miles per hour and t the duration in seconds [22].

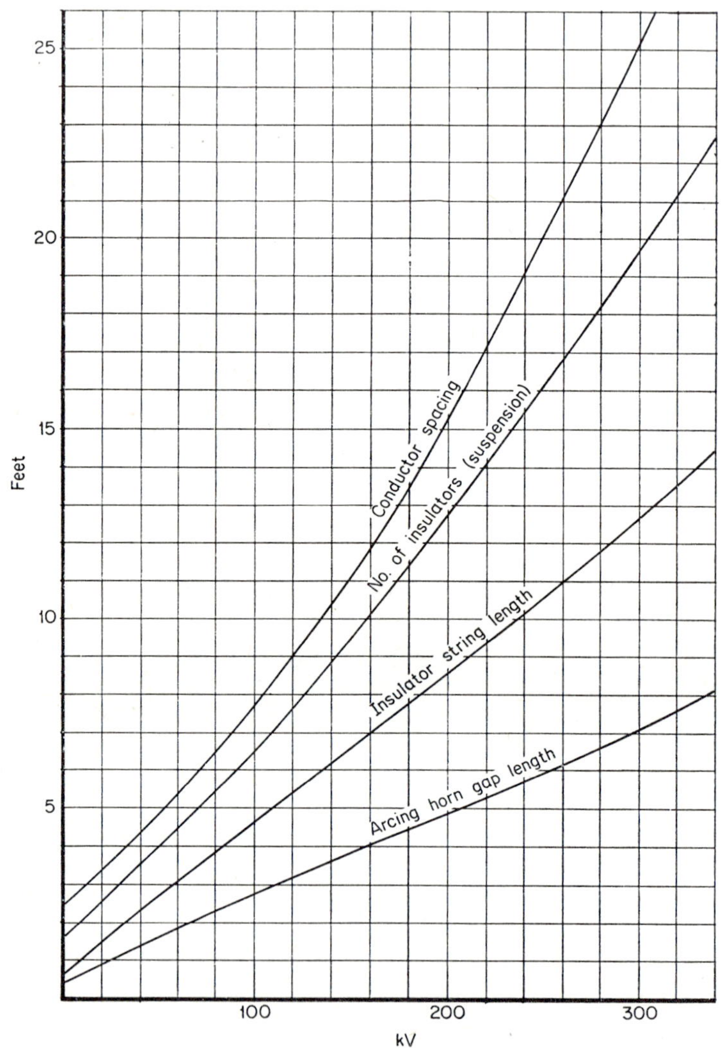

Fig. 6.9. Minimum arcing distances on overhead lines (based on average tower dimensions in U.K. and U.S.A.)

The relation between the current and voltage is

$$V_{\text{arc}} = \frac{8750(S + 3ut)}{I^{0.4}}$$

This non-linear relation means that, if an arcing fault occurs near the relay bus, the fault current may be sinusoidal but the voltage wave-form will have a rectangular tendency if the power source is remote; but with a nearby power source the voltage will be sinusoidal and the current peaked.

6.5.2. Ground Resistance

When a high resistance ground fault occurs, such as when a wood-pole line suffers a fault to earth or when a conductor breaks and falls to the ground, the dominant impedance in the fault circuit is the contact to earth and the path through it.

The resistance of the ground fault is somewhat non-linear because it consists partly of small arcs between conducting particles and partly of compounds of silicon, carbon, etc., which have a non-linear resistance of the form $R \propto 1/I_n$ where n varies from 0·5 to 3.

In a very high resistance ground fault the ground resistance will be the dominant impedance and, because of its non-linear nature, the current will be distorted and a Fourier analysis will show that it is rich in harmonics in the low audio range (see Chapter 13).

6.6. LIGHTNING [50, 51]

There seems to be no clear explanation for the creation of a lightning stroke but the following seems to be the most popular theory.

Clouds become charged to potentials of the order of 20 million volts due to movement of heated air upwards and tiny droplets of water moving downwards, which tends to make the top of the cloud positive and the bottom negative. This creates a voltage gradient between clouds, or between a cloud and earth, which can exceed the breakdown value of the air and cause a flashover.

Most flashes are between clouds but, when the flash is to earth, it starts with a low-current leader stroke to the positively charged air near the earth; the leader is about 2 mm in diameter with 50 V drop per cm length. There are usually several leaders a few milliseconds apart until earth is contacted, or another cloud, when a return stroke occurs along the same path; this may involve up to 100,000 amperes, rising at 40,000 amperes per microsecond, which can have very destructive effects.

Although the power stroke involves a colossal amount of current it lasts only for few microseconds and generally involves less than a coulomb.

After about 30 ms there may be another leader and another power stroke due to the oscillating nature of the surge impedance Z_0 of the lightning path which is of the order of $Z_0 = \sqrt{(L/C)} = 400\,\Omega$ and has a natural frequency of $1/\{2\pi\sqrt{(LC)}\} \simeq 3$ c/s. Each successive stroke contains about a quarter of the coulombs of the previous stroke.

If the stroke is to a transmission line tower, the tower will be raised briefly to a maximum potential of $E_t = I_g R_t$ where I_g is the peak current to ground; R_t is the tower footing resistance which can be anything from 5 ohms upwards. Assuming $I_g = 50$ kA and $R_t = 15$ ohms, we get $E_t = 50 \cdot 15 = 750$ kV. The impulse flashover voltage of a 132 kV insulator string is 650 kV, so that a flashover could occur on a transmission line of 132 kV or less with a direct stroke of this magnitude to the tower. A lower voltage line would flash over more readily and a higher voltage line only with the assistance of the conductor potential.

It should be borne in mind that lightning current lasts only a few microseconds and the flashover must be initiated in that time, otherwise the voltage charge on the tower will be gone. Hence a power arc is likely to follow only if the stroke occurs at a moment of sufficient phase-to-neutral system voltage to ionise a path for the power arc.

6.7. SERIES AND PARALLEL IMPEDANCE SUMMATION [103]

In the course of simplifying the system impedances for the purpose of fault calculation, it is usual to add them arithmetically, i.e. to assume that they all have the same phase angle. In some cases however, such as high resistance faults, or when considering the effect of system capacitance, this cannot be done. In such cases a quick graphical method of combining series and paralleling values of L, R and C is useful.

Figure 6.10a shows how an impedance Z can be split into its active and reactive components either [52] in series (R_s and X_s) or in parallel

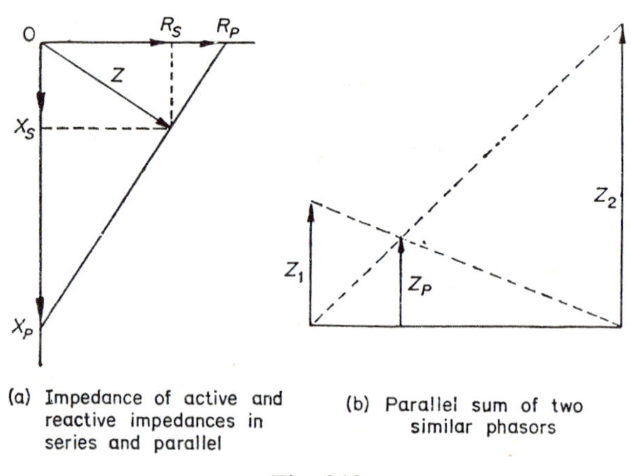

(a) Impedance of active and reactive impedances in series and parallel

(b) Parallel sum of two similar phasors

Fig. 6.10

R_p and X_p) so that

$$R_s + jX_s = Z$$

and

$$\frac{1}{R_p} + \frac{1}{jX_p} = Y$$

It will be seen that this construction also enables a series complex circuit to be changed into the equivalent parallel circuit and vice versa.

Two series impedances of the same angle can be added arithmetically and Fig. 6.10b shows how their equivalent parallel impedance can be quickly obtained.

The impedance of two series impedances of different phase angles is easily obtained by drawing their phasors head to tail, their series impedance being the line joining the beginning of the first vector to the end of the second one (Fig. 6.11).

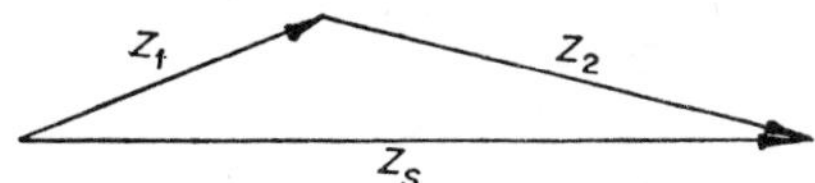

Fig. 6.11. Series sum of two dissimilar phasors

To combine in parallel two impedances Z_1 and Z_2 of different phase angle is slightly more complicated.

Their series impedance is $Z_s = Z_1 + Z_2$ (6.2)

Their parallel impedance is $Z_p = \dfrac{Z_1 Z_2}{Z_1 + Z_2}$ (6.3)

therefore $Z_p Z_s = Z_1 Z_2$ (6.4)

therefore $\dfrac{Z_p}{Z_1} = \dfrac{Z_2}{Z_s}$. (6.5)

Figure 6.12 shows how Z_s and Z_p can be obtained by a very simple construction based upon Eqs. (6.2) and (6.5). First draw Z_s by completing the phasor triangle. Then Z_p is found by erecting a similar triangle upon Z_2, i.e. by drawing lines at angles α and β fron the ends of Z_2. Where they meet is the end of phasor Z_p.

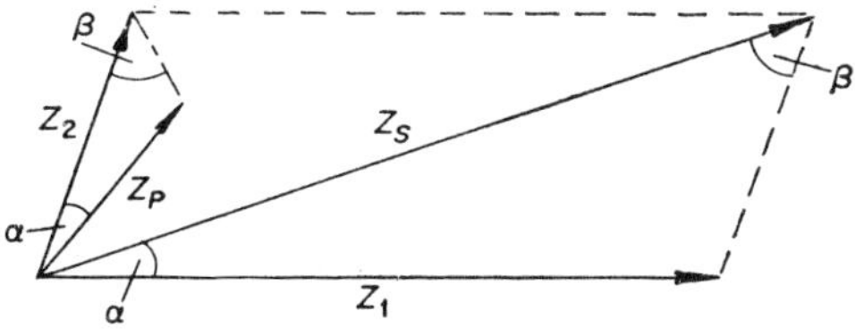

Fig. 6.12. Series and parallel sums of two dissimilar phasors

The foregoing operations can also be done mathematically if precision is required, viz:

$$Z_s = Z_1 + Z_2 = R_1 + R_2 + g(X_1 + X_2)$$

$$Z_p = \frac{Z_1 Z_2}{Z_1 + Z_2} = \frac{R_1 R_2 - X_1 X_2 + j(R_1 X_2 + R_2 X_1)}{R_1 + R_2 + j(X_1 + X_2)}$$

6.7.1. Conversion of a.c. Complex Times to Reciprocals

In Volume I a useful inversion chart is given in section 3.5 which converts terms of the form $\left|\frac{A}{B}\right|_p + j\left|\frac{A}{B}\right|_g$ to the reciprocal form $\left|\frac{B}{A}\right|_p + j\left|\frac{B}{A}\right|_g$; this is useful for converting impedance $|R + jX)$ to admittance $(G + jB)$ or vice versa.

6.8. WYE-DELTA TRANSFORMATION

In the process of simplifying the network for the purpose of calculating relay settings, it is often necessary to replace three delta-connected impedances by three wye-connected impedances, so that they can be added to impedances in series with them, and vice versa (Fig. 6.13).

The delta values D_a, D_b, D_c can be replaced by three equivalent wye values Y_a, Y_b, Y_c as follows:

$$Y_a = \frac{D_b D_c}{D_a + D_b + D_c}$$

$$Y_b = \frac{D_c D_a}{D_a + D_b + D_c}$$

$$Y_c = \frac{D_a D_b}{D_a + D_b + D_c}.$$

Similarly the wye values can be replaced by delta values as follows:

$$D_a = Y_b + Y_c + \frac{Y_b Y_c}{Y_a}$$

$$D_b = Y_c + Y_a + \frac{Y_c Y_a}{Y_b}$$

$$D_c = Y_a + Y_b + \frac{Y_a Y_b}{Y_c}.$$

Figure 6.13 shows how these transformations can be used to simplify a network in progressive steps to a single impedance.

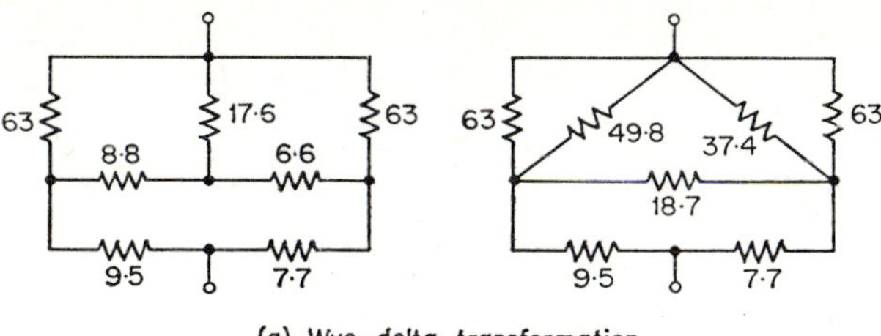

(a) Wye-delta transformation

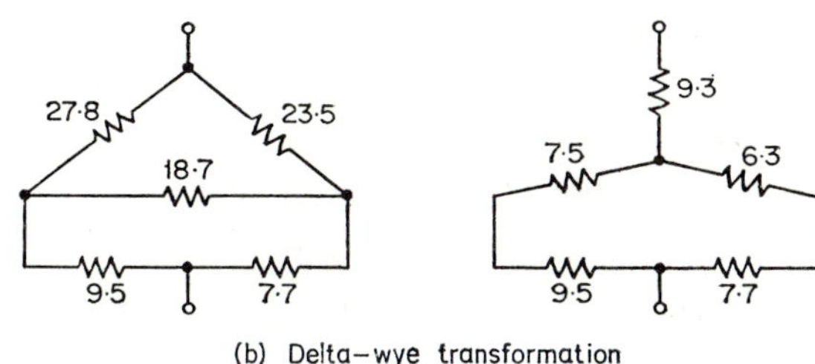

(b) Delta-wye transformation

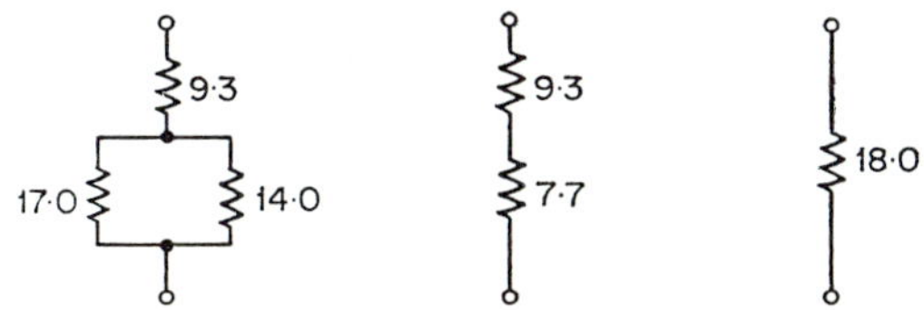

(c) Progressive simplification of system impedance

Fig. 6.13. Simplification of a network by wye-delta transformation

6.9. ISOLATION OF PHASE SEQUENCE COMPONENTS

When only the positive sequence component is present, the three phase vectors are equal in length and 120° apart (Fig. 6.14). The presence of the negative sequence component disturbs this symmetry but still results in a system which is mechanically balanced (Fig. 6.15) such as in a phase-to-phase fault (Fig. 6.15). It will be seen that I_2 has rotation *cba* instead of *abc*.

If the zero sequence component is present the vectors will not be mechanically balanced, i.e. the vectors will not sum to zero. If the three vectors are drawn in a delta (Fig. 6.16), the residual vector required to close the delta is a measure of the zero sequence component. It is in fact in phase with and three times the magnitude of the zero sequence component in any phase.

If there is no zero sequence component the delta will be closed and, if there is no negative sequence component, the delta will be equilateral (Fig. 6.17a). The difference between the apex of the delta and the apex of

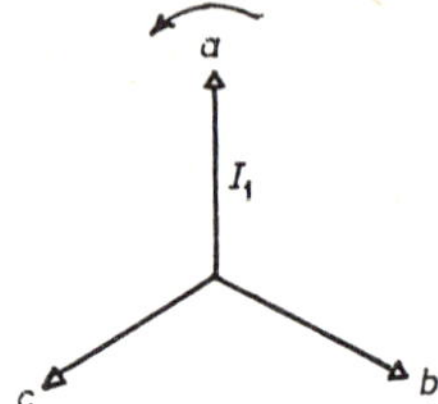

Fig. 6.14. Balanced three-phase currents

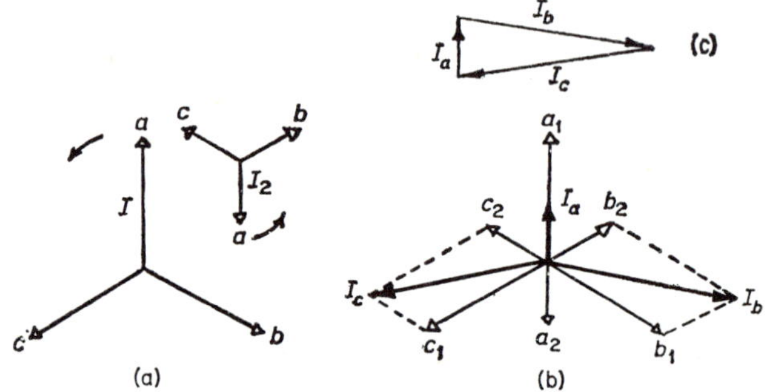

Fig. 6.15. Negative sequence component caused by short-circuit between phases

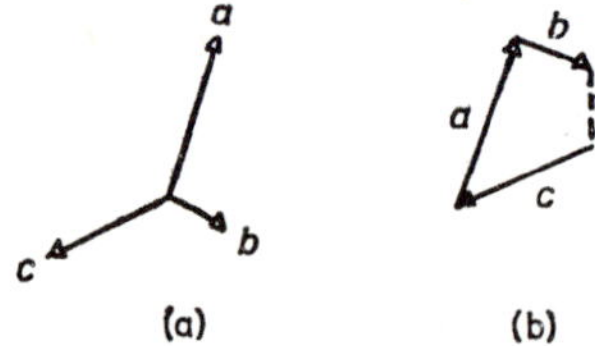

Fig. 6.16. Unbalance caused by zero sequence component

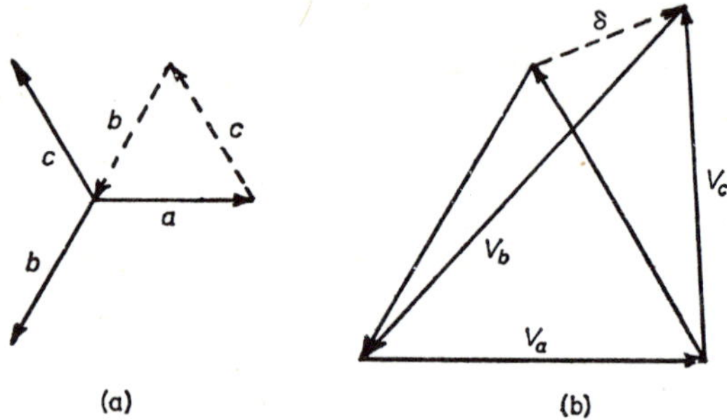

Fig. 6.17. Dissymmetry caused by negative sequence component

an equilateral delta (Fig. 6.17b) is a measure of the negative sequence component. In fact it is the negative sequence component of the delta sum of the two phase vectors (V_b and V_c) not forming the base of the delta. In other words it is $-j\sqrt{3}$ times the component of the base (V_a).

This can be demonstrated algebraically as follows. The negative sequence component of phase 'a' is V_{a2}

$$\begin{aligned} V_{a2} &= \tfrac{1}{3}(V_a + a^2 V_b + aV_c) \\ &= \tfrac{1}{3}\{V_a + a^2 V_b - a(V_a + V_b)\} \\ &= \tfrac{1}{3}\{(1 - a)\, V_a + (a^2 - a)\, V_b\} \\ &= \tfrac{1}{3}(\underline{|30^\circ}\, V_a + \underline{|90^\circ}\, V_b) \\ &= \frac{\underline{|60^\circ}\, V_a + V_b}{-j\sqrt{3}} \\ &= \frac{\delta}{-j\sqrt{3}} \end{aligned}$$

therefore $$\delta = -j\sqrt{(3)}\, V_{a2}$$

In Chapter 3, Fig. 3.51 shows typical networks for extracting the negative sequence component of current or voltage for the purpose of supplying a protective relay. A typical application is the relay for protecting three-phase induction motors against overheating due to an unbalanced voltage supply; this subject is discussed in Chapter 13.

7

Current and Potential Transformers

Error due to magnetizing current. Effects of magnetic saturation, d.c. offset and secondary burden upon c.t. accuracy. Linear and segmental couplers. Magnetic p.t's. Capacitor p.t's–Errors in p.t's.Loss of a.c. potential

7.1. GENERAL

Current and potential transformers are used (*a*) to reduce the power system currents and voltages to values low enough for safe measurement in protective relays, (*b*) to insulate the relay circuits from the primary circuit and (*c*) to permit the use of standardized current and voltage ratings for relays.

In all transformers there is a state of equilibrium between current, magnetic flux and voltage; part of the primary current produces magnetic flux which generates the voltage on the secondary and part of it balances the secondary ampere turns.

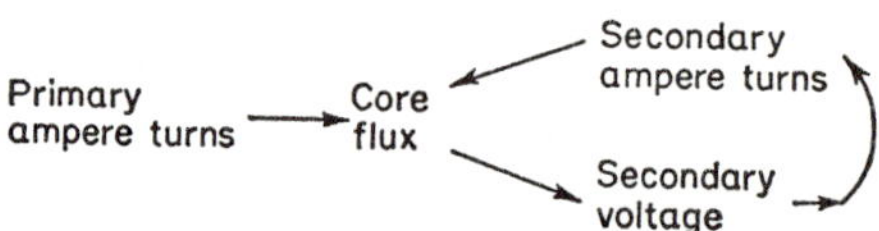

In current transformers the impedance of the secondary burden is very low so that the magnetizing AT are very small and the secondary AT are normally within about 1% of the primary AT.

In potential transformers the secondary impedance is very high so that the magnetizing AT are large compared with the secondary AT but the secondary voltage is within 1% of the primary voltage (corrected for turns ratio).

Current transformers are connected with their primaries in *series* with the protected circuit and, because the primary currents are so large, the primary winding has very few turns; usually the primary conductor merely passes through a ring core around which the secondary is wound as a uniform toroid. With this construction the secondary leakage reactance is negligible; the equivalent impedance diagram of the c.t. is shown in Fig. 7.1, while Fig. 7.2 shows the vector diagram of the currents and voltages.

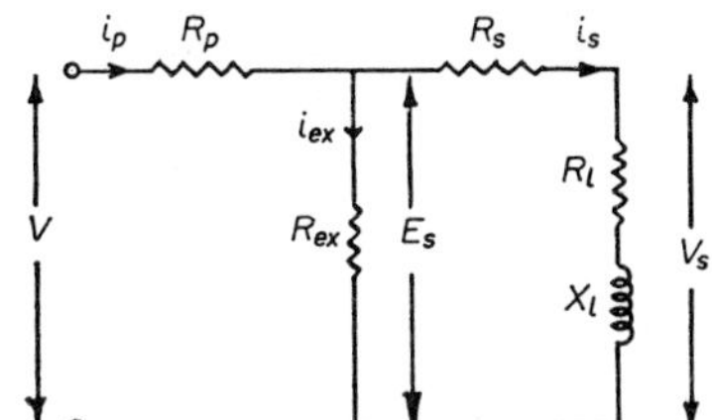

Fig. 7.1. Equivalent circuit for ideal 1 : 1 transformer

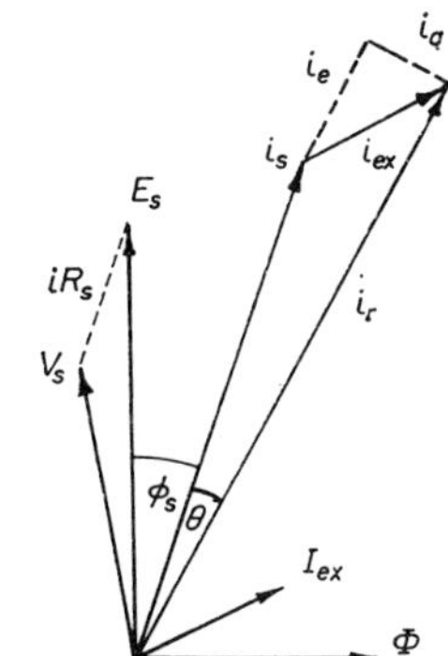

Fig. 7.2. Vector diagram of ideal 1 : 1 transformer
ie = ratio error component of i_{ex}
iq = quadrature error

The c.t. core may be of iron or air; iron-cored c.t's have a substantial power output but are subject to many errors, both static and transient, which will be discussed later. Air-cored c.t's also have errors and these too will be discussed later; they have a low power output which is generally inadequate for electromagnetic relays but suitable for static relays; they have linear characteristics with no transient errors and are called linear couplers [60]. Whereas in iron-cored c.t's the secondary current is theoretically propor-

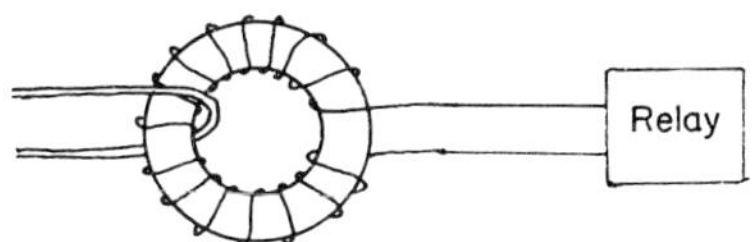

Fig. 7.3a. Toroidal c.t.

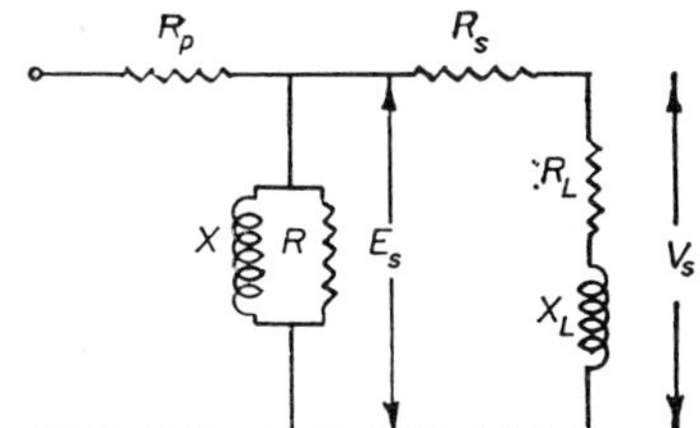

Fig. 7.3. Equivalent circuit for toroidal 1 : 1 c.t.

tional to the primary current, in air-cored c.t's the secondary voltage is proportional to the primary current.

Potential transformers are *shunt*-connected devices which are invariably iron cored; they are similar physically, and in electromagnetic characteristics, to a power transformer. At high voltage ratings they become so expensive that it is customary to use them with a capacitance voltage divider (Fig. 7.4) so that the wye voltage supplied to each p.t. is only about 10% of the system voltage. This arrangement is called a capacitor p.t.; its equivalent circuit is shown in Fig. 7.5.

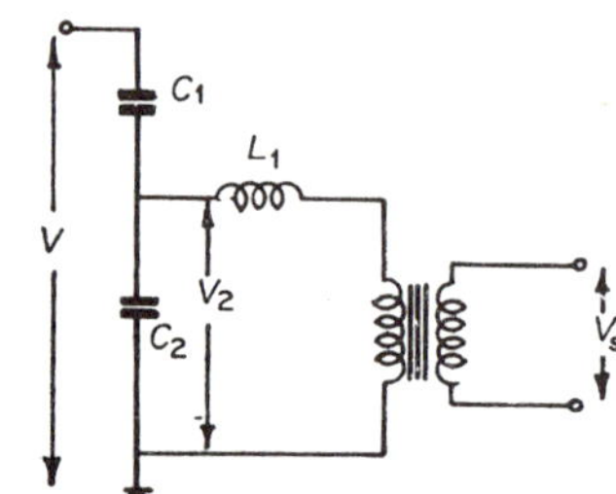

Fig. 7.4. Basic circuit of capacitor p.t.

C_1, C_2 = capacitance voltage divider
L_1 = tuning reactor
V = line potential
V_s = secondary potential

$$V_2 = \frac{VC_1}{C_1 + C_2}$$

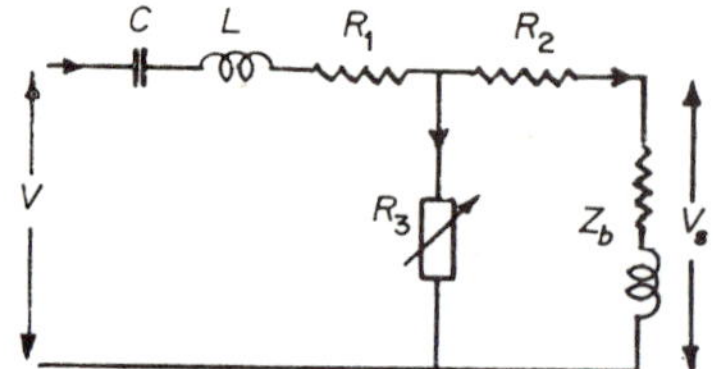

Fig. 7.5. Equivalent circuit of basic capacitance p.t.

$C = C_1 + C_2$
L = inductance of tuning reactor plus transformer leakage
R_1 and R_2 = circuit resistance
R_3 = dielectric and magnetic losses
Z_b = relay burden impedance

7.2. CURRENT TRANSFORMERS

Figure 7.6 shows the magnetic characteristic of the iron lamination materials commonly used for c.t. cores. It will be seen that steel with very low exciting current tends to saturate at lower flux densities. To overcome this some main c.t's and most of the auxiliary c.t's used inside static relays have composite cores made of laminations of two or more of these materials so as to produce a desired result, such as a more uniform permeability over a wide range of flux density, i.e. over a wide range or primary current.

It is necessary to keep the secondary of an iron-cored c.t always short-circuited through a relatively low impedance because, on open-circuit, the whole primary current acts as magnetizing current and secondary voltage can then be dangerously high; magnetic saturation of the iron does not limit the open-circuit voltage because it is proportional to the maximum rate of change of flux, which occurs as the flux passes through zero.

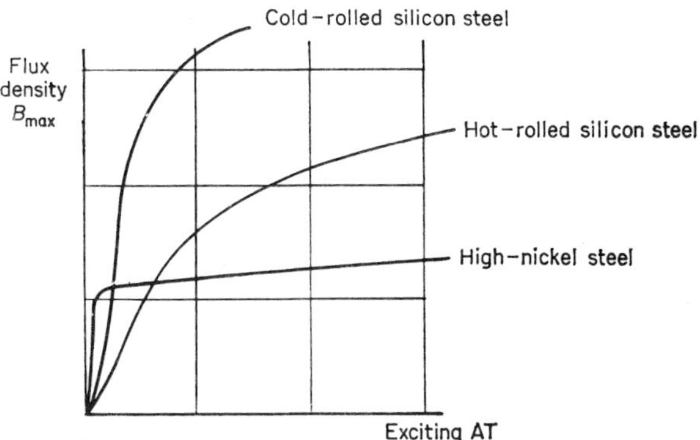

Fig. 7.6. Magnetization curves of c.t. cores

7.2.1. Error due to Magnetizing Current [61]

The transformation ratio of currents in a c.t. is theoretically the inverse ratio of the turns in the primary and the secondary coil turns, but the magnetizing current is supplied by the primary windings only so that $NI_s = I_p - I_m$ where I_p is the primary current, I_m the magnetizing current, I_s the secondary current and N the turn ratio.

I_m is a function of the core flux Φ which is proportional to E_s, the secondary e.m.f. it produces (Fig. 7.3). Since E_s is wholly dissipated in the impedance of the secondary circuit, $I_m \propto Z_s$; hence the c.t. error is a direct function of the total c.t. secondary burden. This error is usually small at currents and fluxes below the magnetic saturating value of the core (Fig. 7.7); hence the impedance of the relay and the secondary leads should be kept low enough so that, at maximum fault current, the secondary voltage does not cause the c.t. core to saturate; otherwise the ratio error would be excessive at high currents.

Saturation of the c.t. may cause no inconvenience and may even be desirable for its limiting action in the case of switchboard instruments or low-set current relays; but obviously it could seriously upset the performance of a time-current relay or any form of comparator relay, such as a differential relay or a distance relay, during heavy faults.

Saturation can be avoided either by increasing the cross-section of the iron cores of the c.t. or by reducing the burden. The first method is expensive

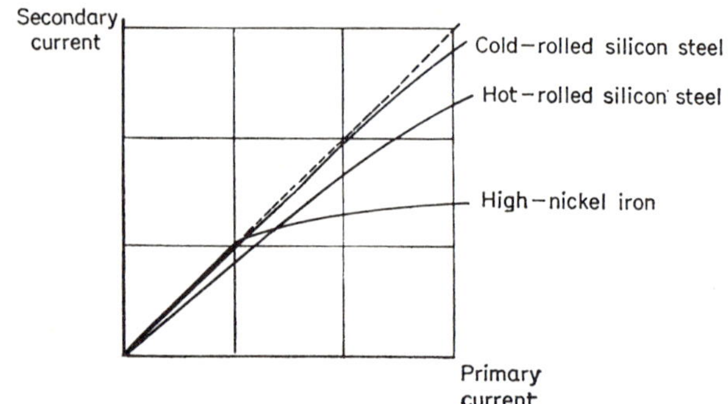

Fig. 7.7. Effect of iron on c.t. performance

and the second may be difficult. The burden on the c.t. is due to the resistance of the relay, the c.t. secondary and the leads. For a given performance, the relay burden cannot be reduced except by changing its design. On the other hand, the lead burden can be reduced by using a lower secondary current rating. For instance, in a large station with long runs from the switchyard to the relay panel, the lead resistance may be as high as 6 ohms.

With 5 A c.t's the normal burden imposed by the 6 ohms leads is $5^2 \cdot 6 = 150$ VA and, in order not to saturate with a fault current twenty times normal, the iron circuit would have to be large enough to avoid saturation below $20{\cdot}5{\cdot}6 = 600$ volts plus a margin for offset waves and remanence. This would require quite a large c.t.; on the other hand, with a 1 A secondary rating, the lead burden would be $1^2 \cdot 6 = 6$ VA. The resistance of the secondary of the 1 A c.t. would be about 5 ohms compared with 0·2 ohm for the 5 A c.t... Hence the total burden would be $1^2 \cdot (5 + 6) = 11$ VA, which is still relatively small, and the maximum voltage would be 20.11.1. = 220 volts. Actually, for mechanical reasons, the size of the c.t. wire is not scaled down in proportion to the current and the maximum voltage would be nearer 150 volts.

7.2.2. Effect of Saturation upon Wave-shape of Secondary Current

It will be shown that magnetic saturation of the iron core of a c.t. can cause appreciable errors in both phase and amplitude comparators, the former being greatest with a resistive secondary burden and the latter with an inductive burden.

At any instant the secondary e.m.f., E_s, necessary to push the current I_s through the total secondary impedance Z_s is:

$$E_s = I_s Z_s = I_s R_s + L_s\,(dI_s/dt) \tag{7.1}$$

The flux linkages necessary to produce this e.m.f. are:

$$N\Phi = -\int E_s\,dt = -R_s \int I_s\,dt - L_s I_s \tag{7.2}$$

Assuming an ideal c.t., the exciting current will be negligible until saturation occurs, when it will be equal to the total current. The secondary voltage will have its peak at the moment of maximum rate of change of flux which, in the absence of saturation, occurs as the flux passes through zero.

(*a*) *Resistive Burden.* If the c.t. secondary burden is wholly resistive, E_s will be in phase with I_s. With saturation due to remanence or high primary current and/or high secondary burden, the flux wave will be flat-topped as shown in Fig. 7.8a; this flat top will represent zero rate of change of flux and hence will produce no secondary voltage; consequently a part of the secondary current wave will be missing at the tail end [53].

In Fig. 7.8a the flux appears to be about 120° leading the secondary current and voltage instead of the 90° that might be expected. This is due to the fact that the sinusoidal part of the flux begins at the saturation level instead of its unsaturated peak level and hence is offset towards zero by the difference between the peak and saturation values. If the secondary load resistance is twice the value causing saturation, the last half of each half-wave of secondary current will be lost and this would obviously affect a phase comparator.

(*b*) *Inductive Burden.* If the burden is wholly inductive (Fig. 7.8b), which is of course a hypothetical case the flux wave Φ follows the primary current I_p (since $\Phi \propto L_s I_s / N$ from Eq. 7.2) until saturation is reached and then becomes flat-topped. The secondary voltage is a series of spikes in quadrature with

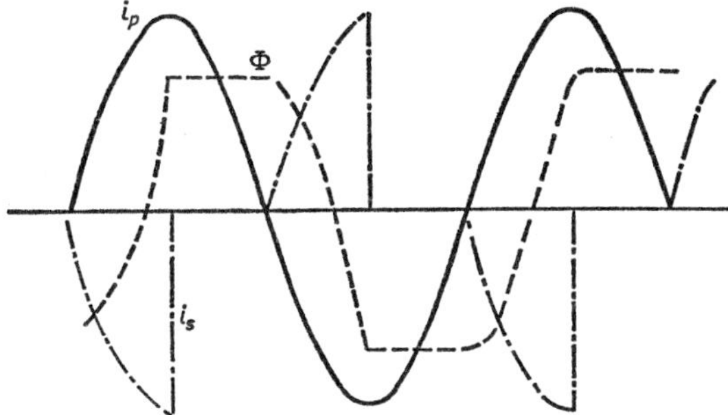

Fig. 7.8a. Effect of resistance load on saturated c.t.

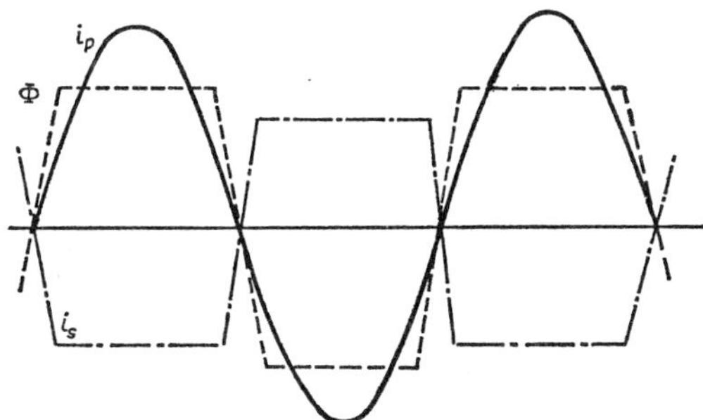

Fig. 7.8b. Effect of pure inductive load on saturated c.t.

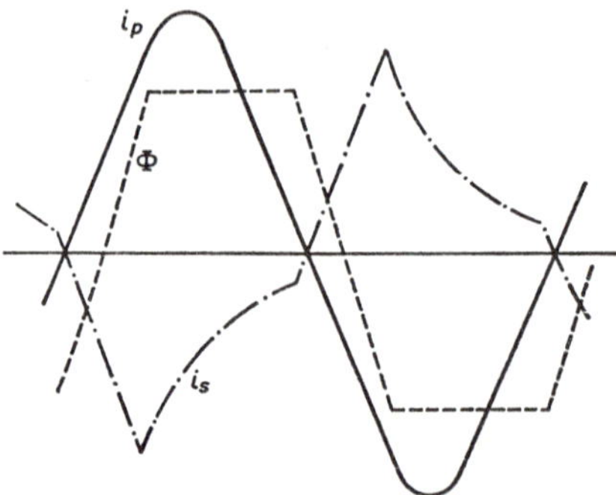

Fig. 7.8c. Wave-shapes of current and flux in saturated c.t. with 45° load

both the flux and the current and is in the form of separated rectangles since no e.m.f. can be produced during the saturated part of the flux wave.

The secondary current wave must also be flat-topped because it cannot change during the saturated period when the secondary e.m.f. is zero. Hence the secondary current is the same shape as the primary current but 180° out-of-phase with it. Its reduced height would affect an amplitude comparator.

(*c*) *Effect on comparators*. It is clear from the foregoing that magnetic saturation reduces the width of the secondary current waves for a resistive load and reduces the amplitude of the secondary current for an inductive load, while a 45° load produces an intermediate effect, as shown in Fig. 7.8c. This means that, if saturation is likely, the power factor of the burden should be adjusted according to the type of comparator used, otherwise the tripping characteristic of the relay will be reduced in area to a degree corresponding to the saturation ratio.

7.2.3. Effect of Remanence in Iron Core

The c.t. core may saturate prematurely at currents well below the normal saturation level due to the existence of remanent flux. Unfortunately, cold rolled silicon steel, which is favoured nowadays because of its high saturating level, has high remanence so that the recent occurrence of a heavy fault may leave a remanent flux high enough to cause saturation when a second fault occurs (Fig. 7.9).

Since circuit-breakers tend to interrupt the current as it goes through zero, the amount of residual flux left in the c.t. core depends upon the phase angle of its secondary burden [54].

With a purely inductive burden the voltage will be at a maximum at the moment of current zero and the flux will be zero; hence there will be no remanent flux. With a purely resistive burden the voltage will be zero at the moment of current zero and the flux will be at a maximum; hence there will be maximum remanent flux.

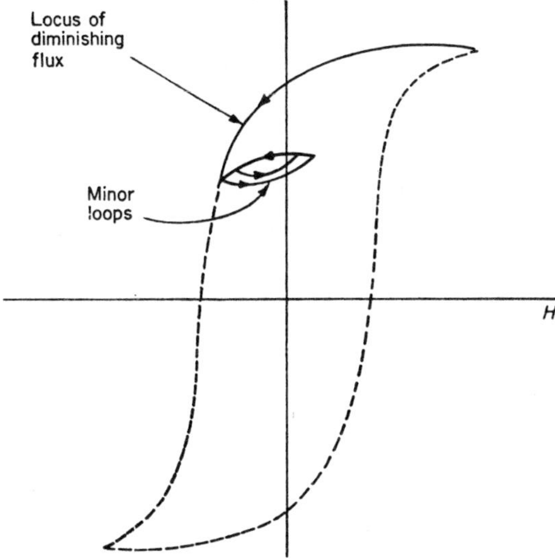

Fig. 7.9. Major and minor hysteresis loops

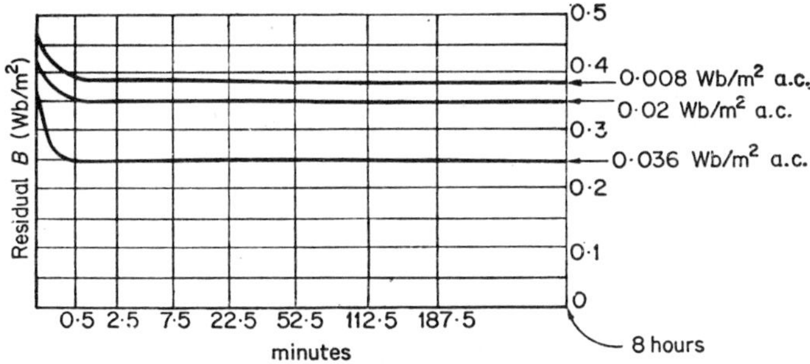

Fig. 7.10. Decay of residual flux density

Most electromagnetic relays impose a burden impedance with a phase angle of about 60° lag, so that the remanent flux is about 50% of the peak value. Most static relays impose a resistive burden and the remanent flux left in a c.t. after interrupting a fault current may be near the peak value. Hence the operation of the relay for a subsequent fault may be inaccurate.

Wright's investigations [54] show that the decay of the flux is negligible after the first 30 seconds and that high residual flux densities can remain indefinitely in spite of the demagnetizing effect of the a.c. flux due to load current after the breaker has been reclosed (Fig. 7.10).

7.2.4. Effect of d.c. Offset in Primary Current [61]

When a fault occurs the current is of the form:

$$I_{\max}\,[\sin(\omega t + \psi - \phi) + \sin(\psi - \phi)\,\varepsilon^{-(r/L)t}] \tag{7.3}$$

in which the first term is the symmetrical a.c. component and the second term is a d.c. component which starts at a maximum and decays exponentially. This of course refers to the primary current and hence does not include any c.t. characteristics because they affect only the c.t. secondary current [60]. In Eq. (7.3) ϕ is the phase angle of the primary circuit, i.e. $\tan^{-1}(L\omega/r)$, and ψ is the time in radians after voltage zero at which the fault occurs.

The formula indicates that the d.c. offset lasts longer if r/L is small, e.g. in a highly inductive circuit such as a power system above 100 kV. The maximum offset occurs when $\psi = 90° - \phi$, i.e. if the fault is initiated at the moment when V is going through zero. In Chapter 6 it was shown that it is almost impossible for a fault to occur near V_{zero} but there are automatic oscillograph records showing that flashovers due to direct lightning strokes have occurred at 45° after V_{zero}; this condition would give a 70% offset, since sin 45° = 0·7.

The voltage across the c.t. secondary is $I_s Z/N = d\Phi/dt \cdot 10^{-8}$ where I_s is the secondary current and Z the burden. Hence the total flux Φ is:

$$\Phi = KI_sZ\left(\int_0^{\pi} \cos \omega t + \int_0^{\infty} \varepsilon^{-(r/L)t}\right) 10^8$$

$$= KI_sZ\left(\frac{1}{\omega} + \frac{L}{R}\right) 10^8$$

$$= \frac{KI_sZ}{\omega}\left(1 + \frac{X}{R}\right) 10^8. \tag{7.4}$$

This shows that the flux associated with the d.c. component of the offset current wave is X/R times the symmetrical a.c. component for a 100% offset wave and the resultant flux is shown in Fig. 7.11.

The reproduction of the d.c. component in the secondary current is shown in Fig. 7.12. It will be noticed that the d.c. component decays more rapidly

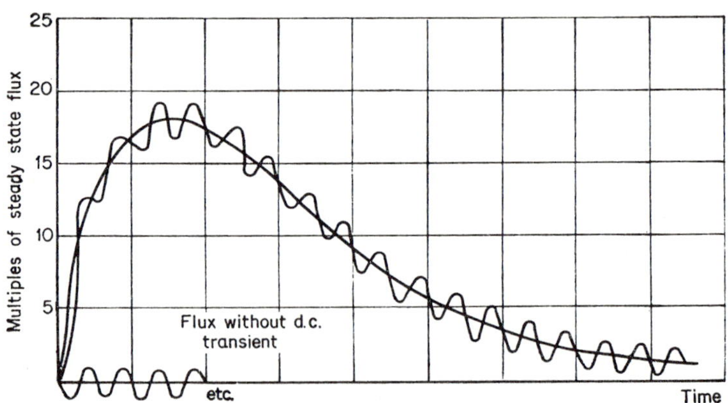

Fig. 7.11. Effect of transient d.c. component upon c.t. flux

in the secondary than in the primary; in fact it reverses in sign after about three cycles [60].

The error caused by the d.c. component can be deduced from Fig. 7.13 which shows the part of the primary current which is used for magnetizing the core and hence does not appear in the secondary current (Fig. 7.14).

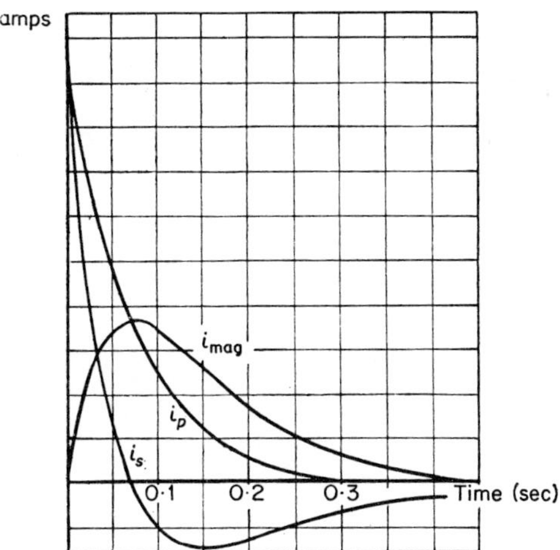

Fig. 7.12. Response of c.t. to d.c. transient

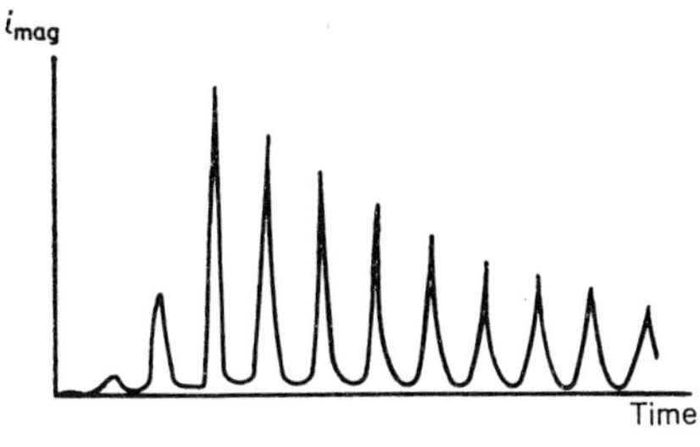

Fig. 7.13

The value of X/R increases with the system voltage because of the increased spacing of the conductors. At 275 kV X/R is at least 10, so that the d.c. flux is about 10 times the a.c. flux.

7.2.5. Accuracy

Current transformer accuracy is defined in terms of the departure from true ratio; it is called the ratio error in the U.K. and the ratio correction factor (RCF) in the U.S.A. It is expressed as a percentage, viz.

$$\% \text{ error} = \left(\frac{NI_s - I_p}{I_p}\right) 100 \tag{7.5}$$

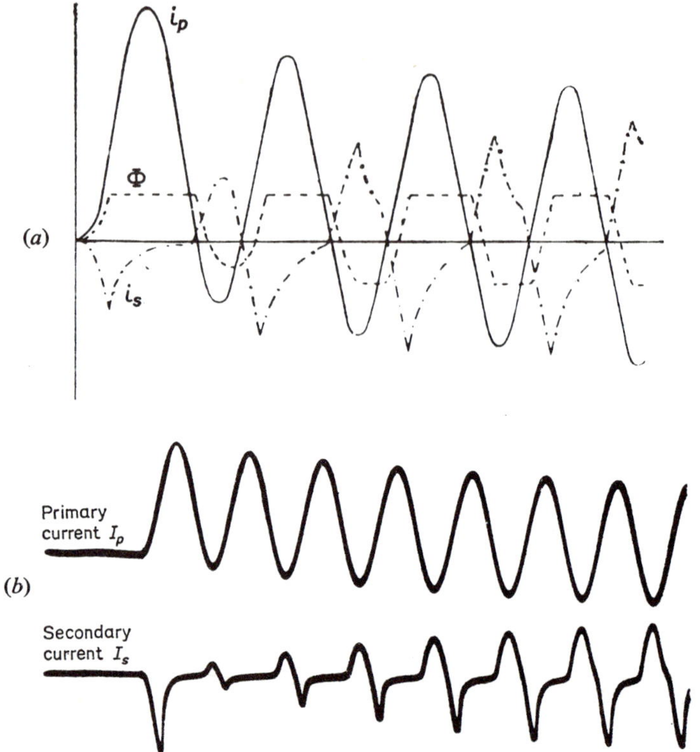

Fig. 7.14. Wave form of secondary current with offset primary current (a) theoretical (b) oscillogram

where N is the nominal ratio and the subscripts p and s refer to primary and secondary. The RCF is $(NI_s - I_p)/I_p$.

The accuracy varies with the secondary burden and is dependent upon the saturation level of the iron. This is expressed by a code whose form varies with the countries of origin.

For many relays an accuracy of ± 10 to 15% is tolerable; for instance, in a time-current relay with a definite minimum time at high currents it is not important if the c.t. output flattens off at high currents. On the other hand, with distance and differential relays a c.t. accuracy of ± 2 to 3% is desirable since an overall accuracy of about $\pm 5\%$ is usually required.

In the U.S.A. the code is of the form 2·5L 250 where 2·5 is the accuracy per cent and 250 is the saturation voltage. L means low leakage and refers to toroidal (bushing) type c.t's; the alternative, H, means high leakage and refers to c.t's with cylindrical windings ("through" or "wound" type).

The performance of a low reactance c.t. can be predicted from the turns ratio N and the exciting current I_{ex}, but this is not so with a high reactance c.t. because the voltage drop in the secondary is appreciable and not necessarily in phase with the load voltage. The standard U.S. accuracy categories are ±2·5 and ±10%; the standard saturation voltage values are 10, 20, 50, 100, 200, 400, 800. The accuracy figure is at twenty times rated current and the voltage figure is the secondary voltage at which the stated accuracy (error) is not exceeded.

In the U.K. the code is different but gives equivalent information. For instance, a 15 S 20 c.t. has a rated secondary burden of 15 VA at 0·7 lagging power factor; S means 3% accuracy from the rated current I_n up to the saturating value I_{sat}; 20 is the ratio I_{sat}/I_n. The U.S. code number corresponding to 15 S 20 would be 3 L 60 for a 5 A c.t. because the $V_{sat} = 15 \div 5 \times 20 = 60$.

The standard U.K. categories for c.t's are given in the following table.

TABLE 7. 1.

c.t. errors

Class	Ratio Error	I_{ex}/I_{sat}	θ
S	±3%	3%	2
T	±10%	10%	6
U	±15%	15%	9

The ratio error applies between rated current and rated saturation current. θ is the maximum phase angle error (leading).

7.2.6. Linear and Segmental Couplers

Linear couplers [60, 62] are toroidal c.t's with non-ferrous cores (Fig. 7.15), usually of air or plastic. The mutual inductance of the secondary is constant relative to the primary so that the secondary voltage is proportional to the primary current, i.e. $V_S = j_\omega M I_p$ which is generally about 5 V per 1,000 A.

The absence of iron eliminates ratio and phase angle error due to saturation at high currents and minimizes transient errors due to d.c. offset in the primary current. The power output is small compared with that of an iron-cored c.t. but it is adequate for any static relay and can be substantially increased by series tuning the secondary or by the compromise of a powdered iron core.

They were first introduced by the Westinghouse Co., U.S.A. in 1941 and were used for bus differential protection [66] where their linear characteristics were particularly valuable. No other applications have since been recorded but the advent of transistorized relays may stimulate their reconsideration.

Their cost is increased by their large number of secondary turns and the fact that the windings must be as symmetrical as possible in order to match them within 1·5% and to eliminate error if the primary conductor is non-central or not straight. In low voltage switchgear, where the distance between

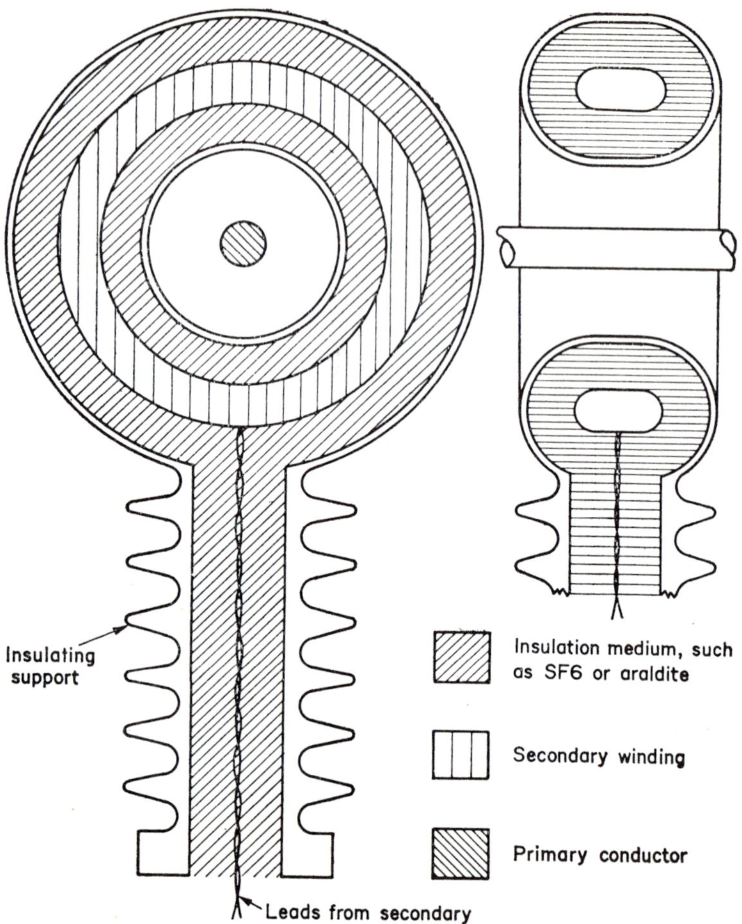

Fig. 7.15. E. H. T. linear coupler c.t.

phase conductors is small, there is some interference from the adjacent phases which could affect the accuracy of zero sequence and balanced current relays but should not be important in overcurrent and distance relays. They are also affected by the proximity of iron parts and by bends in the primary conductor. For these reasons they are not recommended for low-voltage switchgear.

Linear couplers are more susceptible than iron-cored c.t's to pick-up in their secondary leads because they have a very much lower effective impedance than iron-cored c.t's. This means that, with linear couplers, relays are somewhat more likely to operate incorrectly on faults in other circuits with adjacent conductors.

The fact that the secondary voltage is in quadrature with the primary current, and hence is nearly in phase with the line voltage during a fault, eliminates the large phase shift necessary with distance relays supplied from iron-cored c.t's and thereby simplifies their design (see Chapter 10). Also, the absence of core saturation by the d.c. component of fault current makes possible faster tripping and avoids the possibility of large error due to repeated transients of the same polarity.

The fact that $V_s = M(\mathrm{d}I/\mathrm{d}t)$ means that the long time-constant of d.c. offset current makes it an almost negligible component of the secondary current. In other words, a transient having a time constant greater than the periodic time of the supply system is attenuated in the secondary circuit of the linear coupler. On the other hand, where the time constant is less than the periodic time of the system frequency, the amplitudes of the corresponding transients are amplified relative to the fundamental. Harmonics in the primary current are likewise amplified in the secondary voltage in proportion to their frequency and this has to be considered in designing fast relays, so that they will not be adversely affected by transients.

Segmental couplers [59] or sensors are open-ended solenoids which are placed at right-angles to the primary conductor so as to correspond to a 5° or 10° segment of a linear coupler. By suitably locating them relative to other phases, interference from the latter can be minimized.

Their secondary output is clearly less than that of a linear coupler but this does not matter in static relays which have inherent amplification.Their cost is much less than that of either c.t's or linear couplers and they are easier to mount, since they are small enough to go inside insulators on distribution systems.

So far they have been used only for operating simple relays, such as localizing fault indicators, but they will probably supply more sophisticated relays in the future. Since it is possible to use them as voltage sensors also [59], they may in future be used in a computer system of protection, as mentioned in Chapter 15, 'The Future'.

Air-gap c.t's do not saturate and are negligibly affected by magnetic interference, but they have a limited output and a high magnetizing current. If, however, they are used as a voltage source and their power output is suitably increased by a solid-state amplifier, they are free from the limitations of both iron-cored c.t's and linear couplers and are suitable for high-speed relaying.

Their output is of the same form as the linear coupler and hence is proportional to frequency; this would not affect differential relays and could be compensated for in distance relays.

The increased cost of arranging air-gaps symmetrically in the core could be eliminated by the use of a powdered iron core.

7.2.7. High Voltage c.t's

Owing to the high cost of the insulation between their primary and secondary windings, c.t's for circuits of 400 kV and above are extremely expensive; furthermore their performance is somewhat limited by the large dimensional separation of the secondary winding from the primary conductor. These difficulties have been mitigated by the use of new forms of insulation, such as SF_6 (gas) and clophen (liquid) which reduce the size and cost of the c.t., but the ultimate solution seems to be in completely isolating the c.t. from grounded circuits.

The new technique is to use a low voltage toroidal c.t. encircling the E.H.V. conductor and to transmit a signal equivalent to its secondary current to the protective relay via a communication channel. The main problems associated with this technique are to provide accurate transient response and to avoid interference from outside sources of similar energy. Several countries have reported encouraging result from trial installations using a light beam through an optical rod coupling, a laser beam and an f.m. radio channel.

7.3. POTENTIAL TRANSFORMERS

Potential transformers are more standardized than c.t's in that there is only one value for secondary voltage, viz. 115 volts between phases or 66.4 volts line-to-ground in the U.K. and 120 V/69.4 V in the U.S.A. (Fig. 7.16). On the other hand, there are two types, magnetic and capacitor, which have somewhat different characteristics.

7.3.1. Magnetic p.t s

A magnetic p.t. differs from a power transformer physically in terms of cooling and conductor size, and 'performance-wise' in that its output is determined by accuracy requirements rather than by operating temperature limits.

The percent ratio error is:

$$\left(\frac{KV_s - V_p}{V_p}\right)100 \tag{7.6}$$

where K is the nominal voltage ratio. It is caused partly by the fixed voltage drop in the primary due to the exciting current (which is there whether or not any burden is imposed) and partly by the voltage drop in the secondary and the wiring due to the burden current.

In the U.K., magnetic p.t's are usually of three-phase construction with a 5-limb core. Each phase has two secondary windings, one for wye-connection to give three-phase voltage necessary for phase relaying and the other for connecting in series with the corresponding windings of the p.t's in the other phases so as to form a broken delta across which the zero sequence voltage will appear (Fig. 7.16a). In the U.S.A., single-phase p.t's are more common and, where zero sequence voltage is not required, only two p.t's are provided, connected in open-delta as in Fig. 7.16b.

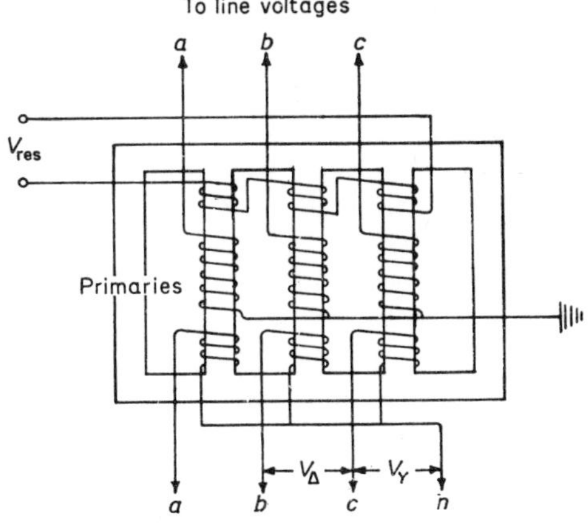

(a) 5–leg three–phase p.t.

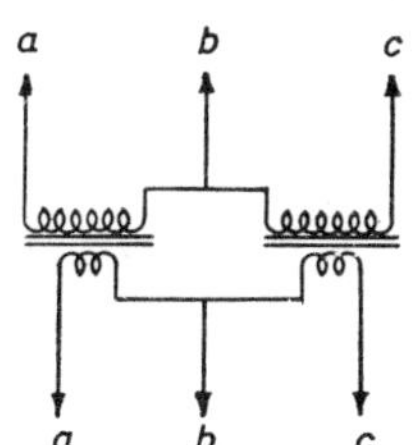

(b) Single–phase p.t's in open delta

Fig. 7.16. Connections of p.t's
(a) for wye potentials (b) for delta potentials

7.3.2. Capacitance p.ts [94]

Since its insulation requirements are the same as for a power transformer, the cost of a magnetic p.t. for circuits above 100 kV becomes prohibitive; hence it is common practice to step down the line voltage through a capacitance voltage divider before applying it to a p.t. The voltage rating of the p.t. primary can thus be reduced to about one tenth of the line voltage. Capacitor p.t's are usually preferred for indoor stations because they offer less fire risk.

The transient performance of the capacitance type of p.t. is not as good as that of the magnetic type. This is because, when a fault occurs, the secondary voltage does not immediately fall to the fault value but goes through a decaying oscillation at a frequency determined by the p.t. capacitance and the various inductances of the system, which is much lower than system frequency. Furthermore, the capacitance p.t. has a tendency to introduce harmonics in its secondary voltage due to oscillations between the p.t. leakage inductance, the divider capacitance and various reactances on the system which are triggered by transients or harmonics in the supply voltage.

For these reasons the performance of high-speed distance relays is less reliable with capacitor p.t's than with magnetic p.t's, to that a decision has to be made between p.t. economy and relay performance, depending upon the importance of the line and the speed of tripping required. Improved performance can be obtained with higher divider tap voltage and larger capacitor sizes, but the most effective solution is to reduce the burden on the device since it causes a phase shift, the compensation for which causes the oscillatory condition. Static relays have this low burden but the other loads such as metering can still be heavy; on the other hand, the latter are not affected by the short oscillations and could be separately supplied through an amplifier.

7.3.3. Other Methods

A cheaper form of capacitance p.t. is the bushing p.t. which uses high-dielectric sheaths inside the high-voltage bushing to form a capacitance divider, the tap-off being of the form of a sheet of metal foil between the insulating sheaths.

Bushing p.t's have similar electrical characteristics to those of capacitance p.t's but smaller burden capability. They are used mostly for synchronizing and voltage indication purposes.

Some progress is being made on still cheaper potential devices in which the voltage signal is picked up by a suitably placed aerial and boosted by a transistor amplifier so that a reasonable power output can be produced.

7.3.4. p.t. Errors

The accuracy classes specified in the U.K. are given in the following table. The rated burden (unity p.f.) is that at which the p.t. will have specified accuracy.

p.t. errors

TABLE 7. 2.

% V	% Rated u.p.f. va	Ratio error	θ
50 and 100	100 to 25	±2·5%	±2·5°
5 and 10	100 to 25	±5%	±5°
	Residual p.t's		
100	100 to 25	±10%	±10°

7.3.5. Accidental Loss of Potential

This is usually indicated by three alarm relays whose coils are connected across fuses in series with the potential supply. While the fuses are intact the relays are not energized, but the blowing of a fuse puts full voltage in the coil of its relay and the alarm is given. In Chapter 10, Section 10.7, it is explained how wrong tripping by distance relays is also prevented.

8
Overcurrent Directional Relays

Operating principles of static time-current relays–Timing and resetting circuits–Adjustment of characteristics–Template and computer setting–Directional relays–Voltage relays–Direct tripping devices–Static operation indicators

Current magnitude is widely used as a means of detecting faults on distribution systems but seldom on transmission systems. This is because the load current tends to be fixed at about 500 A irrespective of the system voltage (because about 0·6 in. dia. is an optimum size of conductor for mechanical reasons) whereas, for a given source impedance, the fault current must decrease as the voltage rating increases so that, on a high voltage transmission line, the minimum fault current may be less than maximum load current.

Fault current level detectors are called overcurrent relays and, in their electromagnetic form, are very simple reliable devices since they consist simply of an electromagnet and a contact-carrying armature. In their static form they are somewhat more complicated; Plates 3 and 4 show an induction disc time-current relay and its static counterpart.

The main advantages of static time-current relays over their electromagnetic counterparts are as follows:

(*a*) c.t. burden about one tenth, so that smaller instrument c.t's or lower ratio bushing c.t's can be used.
(*b*) Volume of single-phase relay halved; 3-phase relay reduced to about one third; the consequent reduction in panel space should reduce the overall cost of the installation and facilitate miniaturization of the control equipment.
(*c*) Instantaneous reset can easily be provided, which facilitates the application of automatic reclosing of breakers on radial or loop lines.
(*d*) Immune to vibration, dust and polluted atmospheres; hence they should not require maintenance.
(*e*) Great accuracy of time-current characteristics, making it possible to gang-test them and to set them by computer with smaller selective intervals.

The arguments generally brought up against some of these points, in the same order, are:

(*a*) Low burdens offer little advantage unless the burdens of other devices supplied from the same c.t's are also reduced, including that of the wiring from the c.t's.
(*b*) These advantages are offset by the increased complexity of static relays.
(*c*) Increased accuracy of characteristics offers little advantage because the magnitude of the fault current is not a precise function of the location of the fault relative to the relay.

For these reasons (although they are not always valid) it is unlikely that electromagnetic time-current relays will be wholly replaced by the static variety until there is general use of computer setting by customers and automatic testing by manufacturers.

8.1. OPERATING PRINCIPLES OF STATIC TIME-CURRENT RELAYS

Figure 8.1 is a block diagram of a static time-current relay. The current from the line c.t. is reduced about 1,000/1 by an auxiliary c.t. in the relay so as to be suitable for transistor operation. The auxiliary c.t. has taps on the primary for selecting the desired pick-up and current range and its

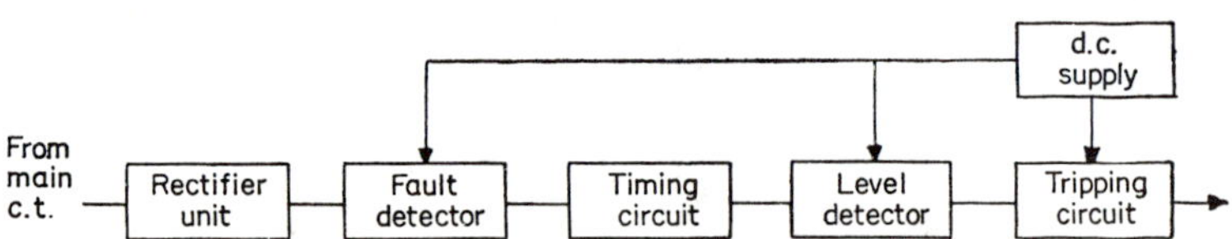

Fig. 8.1. Block diagram of time-current relay

rectified output is supplied to a fault detector and an *R–C* timing circuit. When the voltage on the timing capacitor has reached the value for triggering the level detector, tripping occurs.

For the strictly inverse characteristic, $t = K/I$, the timing capacitor would have to be charged directly by the rectified current but, since the standard inverse characteristic is asymptotic to the pick-up value of the current (Fig. 8.2) and has some tendency towards definite time at very high currents, the timing capacitor is generally charged by a voltage derived from the c.t. current, e.g. from the voltage across a non-linear resistor through which the rectified current flows.

This method also facilitates the shaping of the time-current characteristic to a desired curve by non-linear resistors and *R–C* networks. This shaping is made necessary by the fact that new static relays have to match the imperfect characteristics of existing induction type relays. In years to come it may be possible to use static relays with straight-line log t/log I charac-

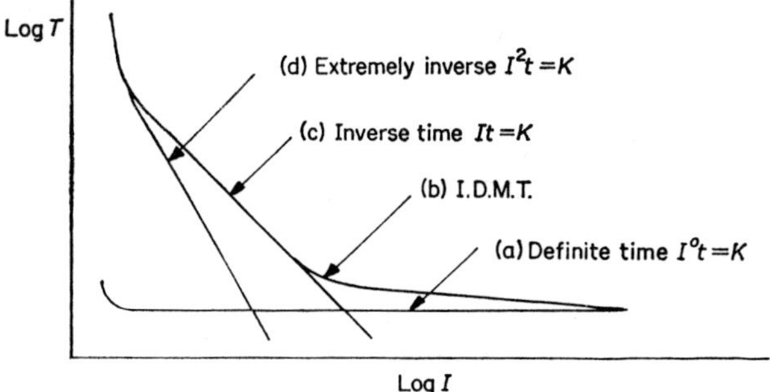

Fig. 8.2. Time current curves

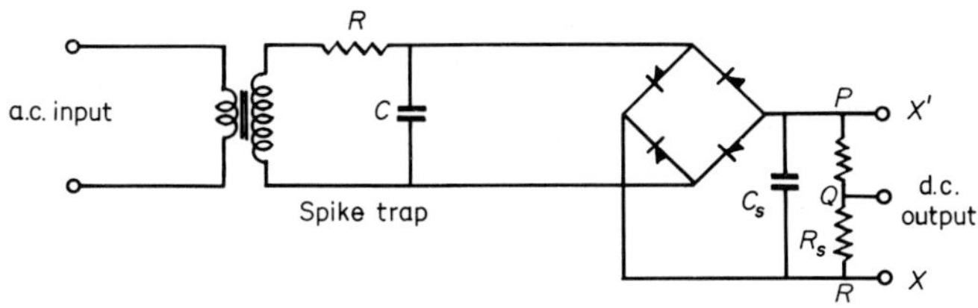

(a) a.c.—d.c. transformation

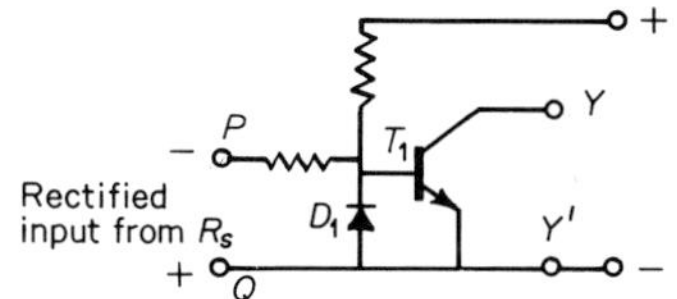

(b) Fault detector circuit

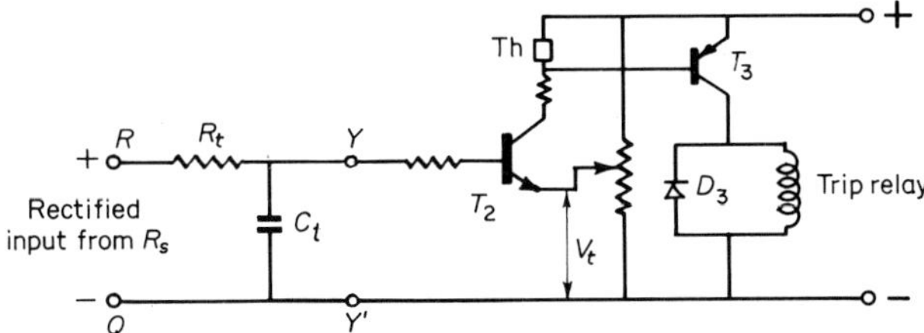

(c) Timing and trip circuits

Fig. 8.3. Static time-current relay

teristics which will make the application of time-current relays very much simpler and also easier to apply to computers.

8.2. OPERATION OF A TYPICAL STATIC TIME-CURRENT RELAY

The current from the main c.t. is first rectified (Fig. 8.3a) and partially smoothed by the capacitor C_s and then passed through the tapped resistor R_s so that the voltage across it is proportional to the c.t. secondary current. The spike filter RC protects the rectifier bridge against transient overvoltages in the incoming current signal.

8.2.1. Timing Circuit

The rectified voltage across R_s charges the capacitor C_t through the resistor R_t and, when the capacitor voltage exceeds the base-emitter voltage V_t o the transistor T_2 in Fig. 8.5c, T_2 becomes conductive, triggering T_3 and operating the tripping relay.

$V_c = E\{1 - \exp(-t/RC)\}$ where E is the voltage across R_s. The charging time $t = RC \log_e\{E/(E - V_t)\}$ where V_t is the value of V_c required to make T_2 conduct.

For a given setting of V_t it will be seen that, at high values of E, the time will tend to be constant but, at low values of E, the time will bear an increasingly inverse relation to E; in other words, since $E \propto I_s$, the auxiliary c.t. secondary current, the relay has an inverse-definite time characteristic. In order to meet the standard I.D.M.T. characteristic in terms of the main c.t. current, the definite time tendency is increased somewhat by arranging the auxiliary c.t. to saturate at high currents.

8.2.2. Resetting Circuit

In order that the relay shall have an instantaneous reset, the capacitor C_t must be discharged as quickly as possible. This is achieved by the fault detector (Fig. 8.3b) as follows.

The base of the transistor T_1 is normally kept sufficiently positive, relative to the emitter, to keep it conductive and hence short-circuiting the timing capacitor C_t at YY' in Fig. 8.3c. When a fault occurs, the overcurrent through the resistor R_s makes the base of T_1 negative and cuts it off, leaving C_t free to be charged. When the fault is cleared the current falls to zero and the negative bias on T_1 disappears, so that C_t is again short-circuited and discharged immediately.

8.2.3. Instantaneous Overcurrent Unit

The circuitry for this unit is similar to that of the time-current unit except that C_t is omitted and the voltage is applied directly to the transistor T_2. This voltage is obtained from tap PQ on the same resistor R_s.

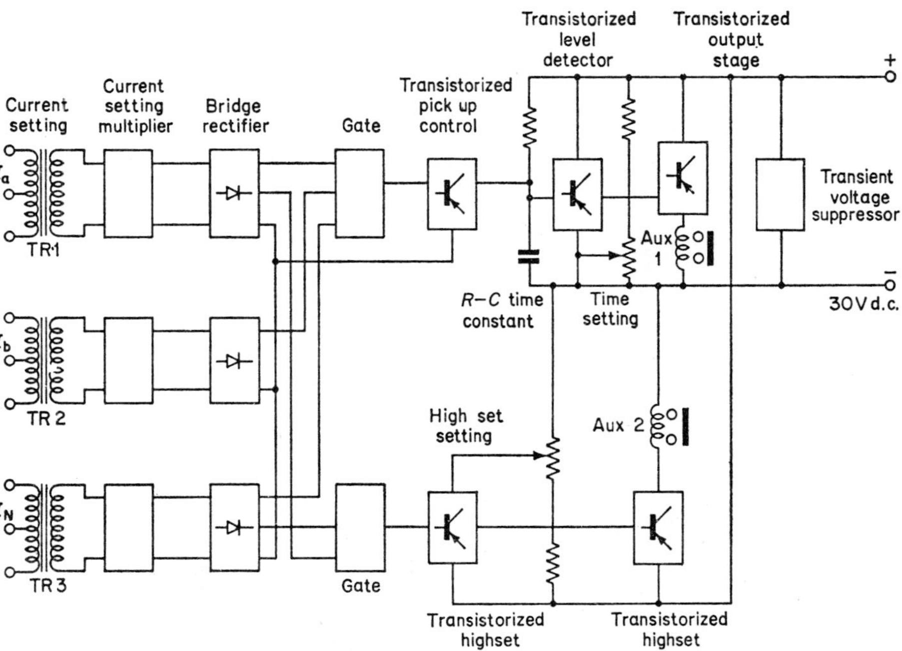

Fig. 8.4a. 2-phase and earth fault definite time-current relay with high-set unit

A weakness of very fast instantaneous units is their tendency to over-sensitivity on offset current waves. An analysis of this phenomenon is given in Chapter 11, Section 11.5.1. The instantaneous unit can be made insensitive to the d.c. offset component by making the auxiliary c.t. saturate just above the pick-up current value and connecting a capacitor and a resistor across the rectified input to the level detector. This prevents tripping until both halves of the current wave are above pick-up value, i.e. until the offset has gone. The short delay this entails is acceptable with time-current relaying.

8.2.4. Modifications

In Fig. 8.3 single transistors are used for the level detector and fault detector circuits. Figure 8.4 shows an alternative arrangement [55] which uses the Schmitt trigger circuit (described in Section 3.8) to provide more precise operation because of its positive feed-back feature. In another relay the rectified output of the c.t. is smoothed by a transformer which cancels out the a.c. component with minimum delay in the build-up of the d.c. component (Fig. 3.44c).

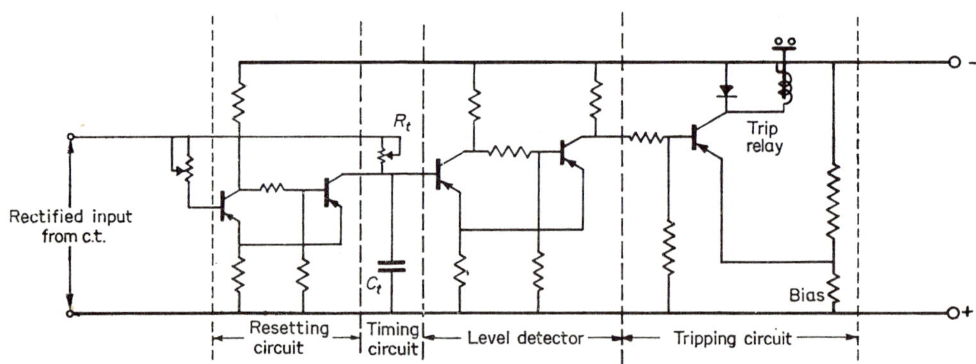

Fig. 8.4b. Partial time-current relay circuit using Schmitt level detectors

8.2.5. Economy of Components

In three-phase relays the timing circuit, level detector and tripping circuit can be common to all three phases, but they must be arranged so that the timing and tripping are controlled by the faulted phase, i.e. the phase with the highest current. This can be done by a parallel voltage arrangement or a series current arrangement of the phase rectifiers. Without such an arrangement the relay will have a different pick-up value for different types of faults.

In Fig. 8.5 the largest of the three currents passes through all three rectifier bridges into the relay circuit since all the rectifiers are open to it [93]. In a phase-to-phase or phase-to-ground fault the output will be a series of half-waves whose mean d.c. value will be $1/\sqrt{2}$ of the peak. In a three-phase fault there will be six half-waves 60° apart which will have a mean value of 96% of the peak. By providing a smoothing capacitor across the rectifier bridges this difference can be further reduced.

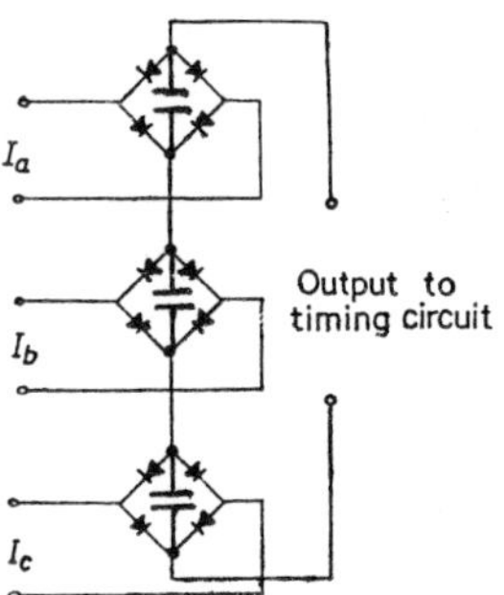

Fig. 8.5. Circuit for equalizing response to different types of faults

Alternatively the rectified currents can be turned into voltages in a parallel circuit. In practice this form of economy may not always be acceptable unless a separate ground relay is provided. For instance, in the U.S.A., it is generally required that, while the relay in one phase is being tested, those in the other two phases are left in service. Also, individual operation indicators are required for each phase in many applications.

8.3. TIME-CURRENT CHARACTERISTICS

Figure 8.2 shows commonly used time-current characteristics. Definite time-current relays ($I^0t = K$) are used where the level of generation is liable to vary considerably. Inverse time-current relays ($It = K$) offer better discrimination and permit lower time settings where the level of generation is reasonably constant and the fault current is controlled by the fault location. More inverse time characteristics are used where sharper discrimination is required and for protection against overheating ($I^2t = K$), or for matching the time characteristic of fuses ($I^3t = K$), or for the protection of power rectifiers ($I^8t = K$). This subject is dealt with in considerable detail in Volume 1, Chapter 4.

A general expression for the operating time of a time-current relay is

$$t = \frac{KM}{I^n - I_p^n} \tag{8.1}$$

where I is the multiple of tap current, I_p is the multiple of tap current at which pick-up occurs, K is a design constant of the relay and M is the time multiplier setting. If the relay picks up at tap value of current this simplifies the equation to

$$t = \frac{KM}{I^n - 1}. \tag{8.2}$$

Owing to the various non-linear design factors of induction disc relays (such as the B-H curve of the magnet iron), the time characteristics which have been accepted so far in the U.K. are more complex; their equations are approximately as follows:

$$\text{Standard Inverse (I.D.M.T.)} \quad t = \frac{0.14}{I^{0.02} - 1}$$

$$\text{Very Inverse} \quad t = 1 + \frac{16}{(I - 1)^{1.6}}$$

$$\text{Extremely Inverse} \quad t = 0.2 + \frac{19}{(I - 1)^{1.7}}$$

American I.D.M.T. curves (IAC 51 and CO-8 relays) are slightly less inverse at the low-current end. The U.K. very inverse and extremely inverse curves lie between the corresponding IAC and CO relay curves.

The curves now specified in the U.K. are based on the ideal form of Eq. (8.2) and the values of K and n have been chosen to give the following equations to which new (static) relays will conform:

$$\text{Standard Inverse (I.D.M.T.)} \quad t = \frac{0.14}{I^{0.02} - 1}$$

$$\text{Very Inverse} \quad t = \frac{13.5}{I - 1}$$

$$\text{Extremely Inverse} \quad t = \frac{80}{I^2 - 1}.$$

It is possible that future characteristics will be of the form $t = K/I^n$ which will give a straight line on a log t/log I graph. Since this curve will no longer be assymptotic to the pick-up value of current, the pick-up will be controlled by a separate device. Similarly, the I.D.M.T. curve will probably be eliminated and the definite time portion of the characteristic provided by a separate unit.

Initially there will be a tendency for power system engineers to demand static relays with time-current curves to match those of existing induction disc relays, but eventually the advantages of the simple curves of static relays will be appreciated. One great advantage is the saving of engineering time in calculating relay time settings and the possibility of doing the work semi-automatically by computer. Another advantage is automatic gang-testing of the relays, where a group of relays is tested against a master relay; the low burdens and simpler controls of the static relay will facilitate this.

8.3.1. Achievement of $I^n t = K$ Characteristic

With the exception of definite-time relays, electromagnetic relays are able to follow the simple time-current curve corresponding to Eq. (8.1) only up to a few times the c.t. rating because of magnetic saturation in their electromagnets. For this reason the inverse definite time curve (I.D.M.T.) has been commonly used. With static relays, however, the circuit components are linear so that it is actually easier to produce a curve following the simple law of Eq. (8.1) than an I.D.M.T. characteristic.

The circuit for producing an inverse or inverse-definite time-current characteristic is the straight R-C circuit shown in Fig. 8.3c for either current or voltage charging of the timing capacitor. The extremely inverse ($I^2t = K$) characteristic can be obtained by replacing the resistor R_t by an R-C network; Fig. 8.6a shows a typical circuit for current charging [56] and Fig. 8.6b for voltage charging.

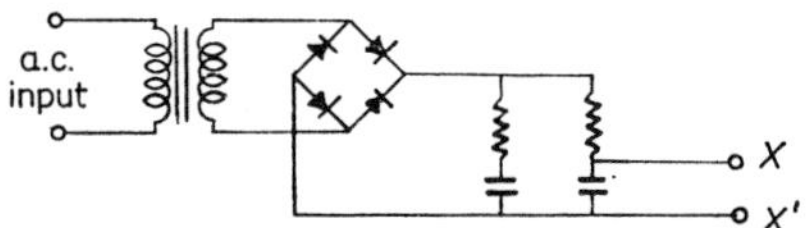

Fig. 8.6a. Current charging circuit for I^2t relay Terminals XX' go to the level detector and timing circuit

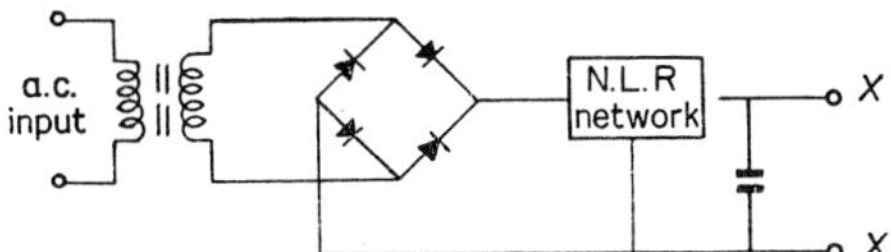

Fig. 8.6b. Voltage charging circuit for I^nt relay with n up to 3.5

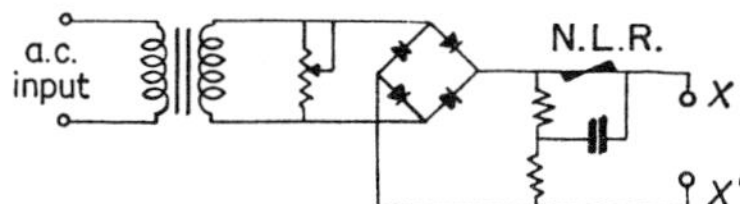

Fig. 8.6c. Voltage charging circuit for I^nt relay with n above 3.5

Figure 8.6b shows a circuit [108] for controlling the shape of the time-current characteristic for values of n up to 3·5 in the characteristic equation $I^nt = K$. Where higher values of n are required, such as for protecting power-rectifiers, the circuit of Fig. 8.6c can be used.

A circuit for producing any of the three most common standard time-current characteristics (inverse-definite, very inverse ($It = K$) and extremely inverse ($I^2t = K$) is an attractive idea because it would simplify stocking for customers. This is possible either by having three different plug-in R-C timing circuit modules or by a switching arrangement.

8.3.2. Choice of Characteristic

Loads are generally protected by fuses which have a very steep time-current characteristic ($I^3t = K$) and, if the relays near the load end of the system also have steep characteristics, the time setting of all the relays can be reduced without sacrificing selectivity.

Since the fault current $I_f = V/(Z_s + Z_L)$, where Z_s is the source impedance and Z_L the line impedance, it follows that variations in generation affect the fault current more at stations near the power source. Hence a steep time-current curve is advantageous only at points remote from the source.

Near the source a steep characteristic will cause high operating times during low generating conditions; this may not overheat equipment (I^2t) but an arc of any size has a temperature of about 3,000°C and this can burn through a line conductor if left on for several seconds.

The ideal arrangement then is to choose the slopes of the log time/log-current curves of the relays to match the fuses at the load end and progressively to diminish their slopes as the power source is approached.

The curves can be chosen by hand for relays at successive stations, using templates, or automatically by computer. With the present curves that are assymptotic to a pick-up value at one end and tend to definite time at the other end, this would require an analogue computer or a very large digital computer because of the complex equation of the curve.

If the log time/log current curve were a straight line, the pick-up value and the definite time being determined by separate relay units, the curve could be determined by two points and the job would be suitable for a digital computer.

To make this method practical the time-current relay would also require the choice of plug-in timing circuit modules, giving a choice of slope between $n = \frac{1}{2}$, 1, $1\frac{1}{2}$, 2 and 3. It would then be possible to use time-current relaying for applications where it is now impractical and to achieve tripping times of perhaps half those which are now tolerated.

8.4. SETTING BY TEMPLATE OR BY COMPUTER [27, 28]

The first step in this direction was the use of a transparent template of the time-current curve for choosing current taps and time scale multiplier settings which would give an adequate selective interval between the operating times of the relays at successive stations.

Because the time-multiplier scale is infinitely adjustable, an infinite number of time-current characteristics are possible for a given relay. These can be theoretically combined into one, however, by using logarithmic scales so that the curved edge of the template follows the time curve log t/log I. This is a legitimate procedure for modern static relays but it is conducive to error at low time multiplier settings (T.M.S.) in induction disc relays because of the effect of disc inertia, friction and imperfect spring compensation. Another advantage of static relays is that the overshoot is small and the time settings can therefore be reduced.

With a logarithmic curve a given movement along either scale represents a multiplying factor. For instance, the distance between the 0·1 and 0·2 second ordinates is the same as the distance between 1·5 and 3 seconds. The current scale should be marked preferably in multiples of pick-up; here again a movement along the current axis represents a multiplying factor. Values in amperes can be obtained by multiplying the multiple of pick-up the pick-up current.

In order to use a computer, the template curve has to be translated into an equation and, of course, the simpler the equation the easier it is for programming. The log t/log I curve will be assymptotic to the pick-up value on the current multiple scale; this is zero on the log multiple (M) scale. The curve can be of the form $\log t = f(\log M)$, where f is a mathematical function; f is a simple function in a simple relay, such as one in which the time t is inversely proportional to the multiple of pick-up current.

On the other hand, in relays conforming to the induction relay characteristics f is quite complex (see equations in Section 8.3) and would be better expressed as a mathematical series.

The flexibility and speed of the computer would make it practical to realize the full benefit of a relay with an adjustable time/current slope since it would make short work of the tedious job of trying various slopes at each relay location to nest the characteristics in for the minimum overall operating times. It could also vary the selective interval since the relay time error is not constant but a function of the time setting.

Another useful adjustment that would favour computer-setting and gang-testing of these relays is to have 14 taps instead of the usual 7. This would provide an 80 to 1 range of pick-up current settings instead of the usual 4 to 1 and would enable one relay to be used for any fault value and any c.t. ratio.

8.5. DIRECTIONAL OVERCURRENT RELAYS

In loop lines it is not possible for selectivity to be based on current magnitude alone because the current may be the same in two feeders on the same bus except for its direction (as shown in Fig. 8.7); furthermore, the direction may change with the direction of the fault.

For example, a fault at X in Fig. 8.7 would produce similar currents in the relays at D and E, except that the current in D is incoming while the current in E is outgoing. Similarly, a fault at Y causes similar current at G and H. Directional control of these relays would prevent the relays at D and G from disconnecting their sound lines. Reference 22 provides a great deal of data on the application of directional units.

The main problem of directional units is to preserve their discrimination at a very low voltage such as would occur with a fault very close to the bus. There are a number of ways of automatically amplifying the potential at low values which are used with distance relays and are described in Chapter 10, Section 10.4a; but for some strange reason, these are not used with the directional units of overcurrent relays.

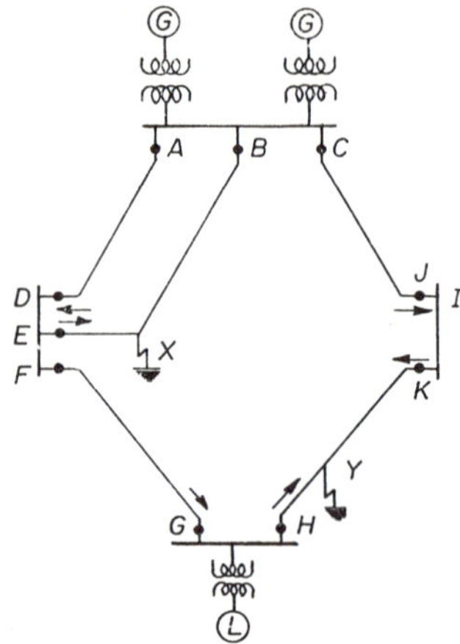

Fig. 8.7. Application of directional overcurrent relays

In static directional overcurrent relays the problem is less serious because static comparators are inherently very sensitive and it is not difficult to make a static directional unit reliable down to 1 % of system voltage, which is well below the minimum fault voltage.

8.5.1. Phase-Comparator Directional Units

The two kinds of static phase-comparator at present under consideration are the Hall generator, which gives a linear cosine product output, and the rectifier bridge phase-comparator which has a cosine output limited by the rectifier characteristics. Most of the work on Hall effect and magneto-resistivity comparators seems to have been done in Russia. European and American investigations appear to have been centred on rectifier bridge type comparators.

(*a*) *Hall Effect generator* [7] [58]. Figure 8.8a shows the basic connections of the simplest static directional unit, the Hall generator, which has already been discussed in Chapter 4, Section 4.4.1a. A tuned shunt is provided for eliminating the double frequency components of the output voltage. At present such relays are not available because of the high cost of the Hall generator units and because their limited output necessitates an amplifier.

The maximum output of this unit is produced when the magnet flux and the current input are in phase. Since the magnet coil is inductive, the maximum output would be when the current input lags the voltage across the field magnet by 60°, which would make it suitable for ground faults.

For phase faults, using the quadrature connection, it is necessary to connect a series capacitor in series with the magnet coil to lead the coil current and flux by 90° so that the maximum output occurs when I_a leads V_{bc} by 30°, as in Fig. 8.8b. Alternatively, the double connection of

Fig. 4.16b (Chapter 4) can be used. For ground faults, a series resistor in the potential circuit will reduce the lag of the current so that maximum output will occur with I_0 lagging V_0 by 45°.

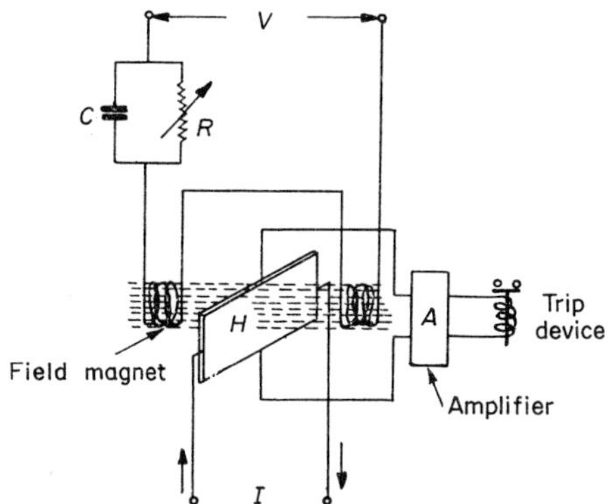

Fig. 8.8. Directional relay using Hall generator

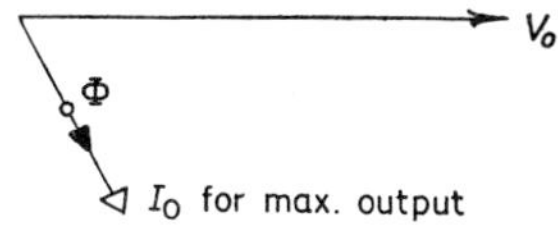

(i) Ground faults

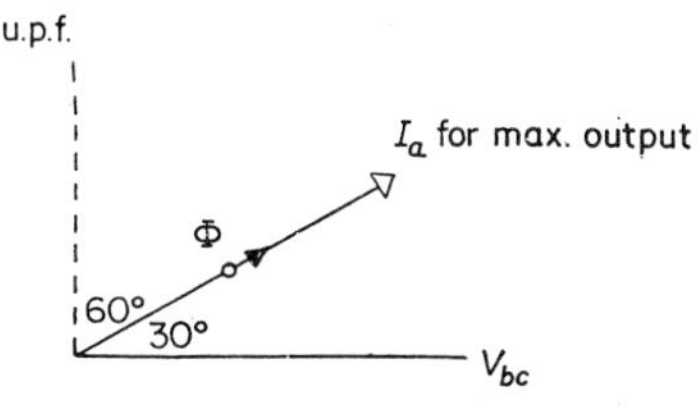

(ii) Phase faults

Fig. 8.8b. Vector diagrams of Hall generator magnet flux and crystal current in a directional relay

(*b*) *Rectifier bridge phase-comparator*. This comparator (Fig. 8.9) provides an economical directional unit, efficient in output-input VA ratio, which will operate a sensitive d.c. polarized relay directly without intermediate amplification. Its operation was explained in Chapter 4, Section 4.4.3b.

This unit inherently has a maximum output angle near unity so that, for phase faults using the quadrature connection, the current in the potential circuit must be shifted forward 30° by an R-C circuit, as shown in Fig. 8.9.

For ground faults it is necessary to lag the current in the potential circuit by 45° with an *R-L* unit. As before, the variable resistor provides a means of adjusting the phase angle for maximum output.

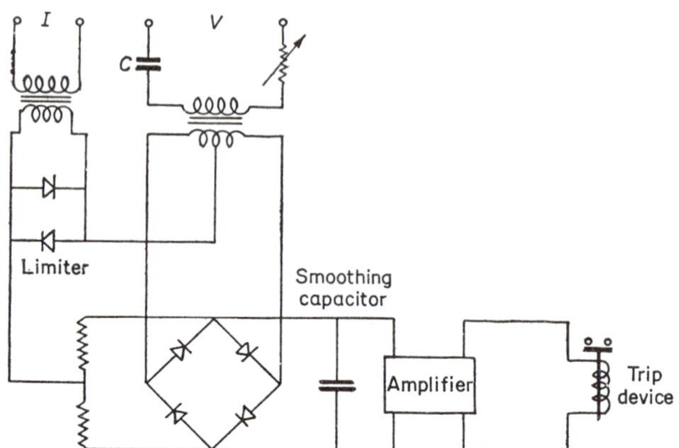

Fig. 8.9. Directional relay using rectifier bridge phase comparator

(*c*) *Instantaneous coincidence comparator.* This comparator was described in Chapter 4, Section 4.4.2a. Since it is operated by a pulse at the moment of current zero, its output must be prolonged, i.e. its reset has to be delayed for at least half a cycle. Any tendency to operate inadvertently due to an interference signal in the current circuit is of no importance in the directional control of a time-current relay since the operating time of the latter is much longer than half a cycle.

The fast operation of the coincidence comparator ensures the consistency of the operating time of a time-current unit and it is obviously advantageous for controlling an instantaneous overcurrent unit, if the chance of a spurious current signal occurring at the moment of a fault in the non-tripping direction is accepted.

(*d*) *Integrating coincidence comparator.* This comparator was described in Chapter 4, Sections 4.4.2c and 4.4.3. A typical arrangement used in a Swiss relay [88] is shown in Fig. 8.10.

8.5.2. Amplitude Comparator Directional Units

As explained in Chapter 1, an amplitude comparator can be adapted to perform the phase comparison inherent in a directional relay by supplying it with the sum and difference of the voltage and current. This is done in Fig. 8.11.

Here again the maximum output is with the currents in the two windings in phase. The phase shifting arrangements are therefore the same as for the phase comparator rectifier bridge. In all three of the foregoing directional units a transformer with opposed windings is advisable for eliminating the

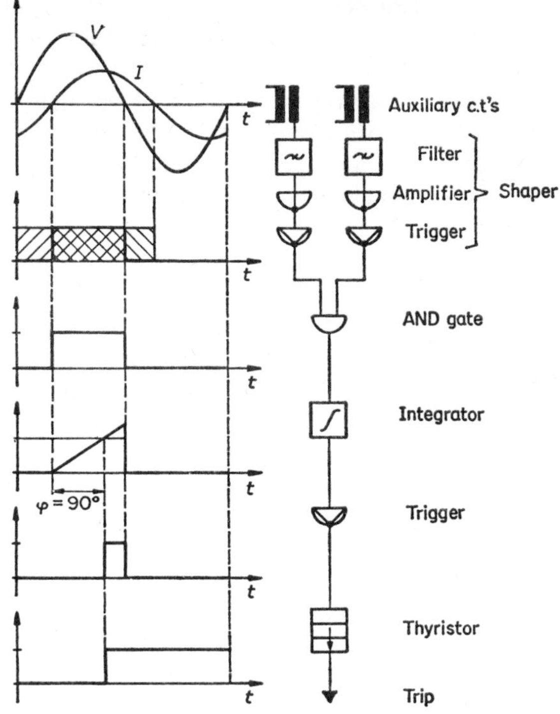

Fig. 8.10. Operation of integrating phase comparator (B. B. Cie.)

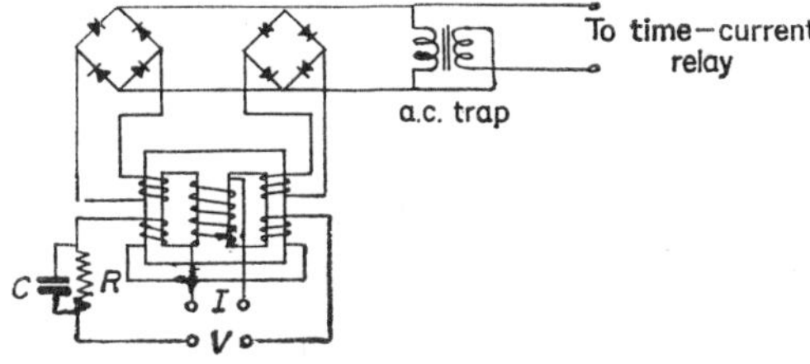

Fig. 8.11. Amplitude comparator directional relay for phase faults. For ground faults the R-C phase-shift circuit is replaced by an R-L circuit

a.c. from the output of the comparator instead of a smoothing capacitor; this is because faster action is required than for the overcurrent unit; on the other hand, this is a more expensive method.

8.6. POLYPHASE DIRECTIONAL RELAYS

In electromagnetic relays the eight-pole induction cup with current and potential from all three phases produces a three-phase torque proportional to $P_1 - P_2$, where P_1 and P_2 are the 60° lagging components of the positive and negative sequence VA [22]. This gives correct operation on all interphase faults; a separate zero sequence power relay is required, which is associated with a zero sequence overcurrent relay, for dealing with single-phase ground faults.

Figure 8.12 shows the direct static equivalent of the eight-pole induction cup relay using six Hall generators, but this gives no economy over the three pairs of cross-connected Hall units representing three separate single-phase directional units. Hence single-phase directional overcurrent relays are likely to be the standard with static relays.

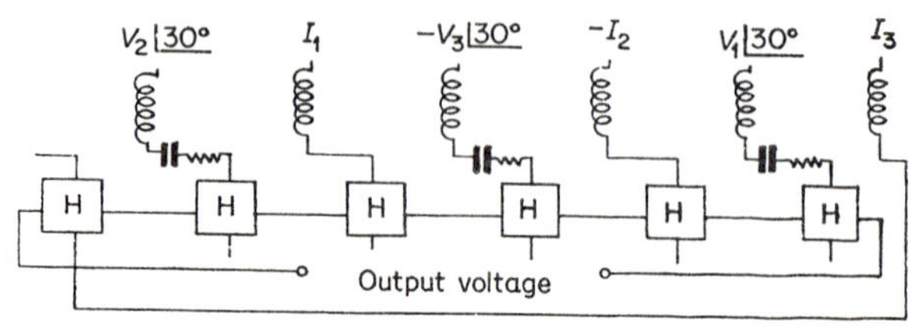

Fig. 8.12. Static equivalent of eight-pole induction cup polyphase directional relay

The advantage of the polyphase directional relay over single-phase directional relays under certain operating conditions is discussed in Ref. [22], but, with the restricted angle of operation that is possible with static directional relays (Chapter 4, Section 4.3.4), this advantage disappears.

An alternative is the negative sequence directional relay [104] which is more sensitive to unbalanced faults. It is of particular value for polarizing ground relays in cases where only two p.t's are available, connected in open-delta, or where a zero sequence directional relay would be unreliable because of strong mutual coupling with a parallel line.

8.7. VOLTAGE RELAYS

Overcurrent relays seldom require to have a low reset value because they are seldom set below twice normal current. Voltage relays on the other hand are usually set closer to the normal value and hence require to have more precise action and a smaller difference between their pick-up and drop-out values. To achieve this the Schmitt trigger circuit (see Chapter 3, Section 3.8) is useful. The basic circuit for an undervoltage relay is as shown for one phase on the left side of fig. 12.19.

8.8 DIRECT TRIPPING DEVICES

In low-voltage circuits, instead of a time-current relay on a switchboard panel operating a trip coil on the breaker, the time-current relay circuit is often mounted on the circuit-breaker and operates the trip coil directly.

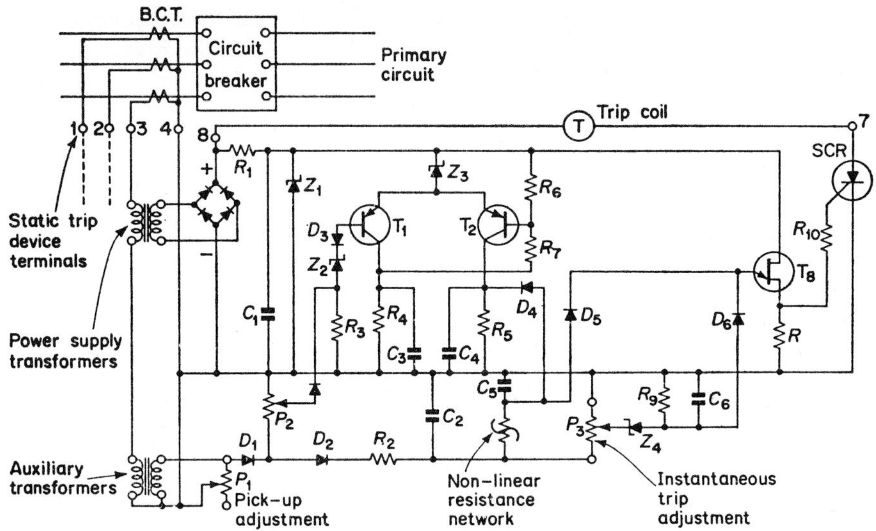

Fig. 8.13. Static direct trip device (I. T. E.)

Static relays are more suitable for this practice than electromagnetic devices because they are smaller and impervious to the mechanical shock of breaker operation. Figure 8.13 shows an American (I.T.E.) direct trip device circuit using an R-C inverse time circuit, a Schmitt level detector and a thyristor trip.

The current from the secondary of the bushing c.t. in each phase is fed into two auxiliary c.t's ,one for relaying purposes and the other for providing a d.c. power supply.

The relaying current is passed through a resistor P_1, the voltage drop across which is applied to a level detector of the Schmitt trigger type. P_1 is used to adjust the current to the setting of the level detector, i.e. to control the pick-up of the relay. P_2 is a factory adjustment to make the scale of P_1 agreee with the calibration.

The operation of the Schmitt trigger circuit has already been given in Chapter 3, Section 3.6.3. T_1 is normally conducting but, when the voltage across P_2 and P_3 is high enough to block the base current of T_1, T_2 conducts and the current through R_5 establishes a blocking potential at D_4 which permits C_5 to charge. Before this C_5 was short-circuited through D_4 and R_5.

When the charge on C_5 exceeds the peak point emitter potential of the unijunction transistor T_3 it fires, discharging C_5 through R_8 and producing a pulse which triggers the SCR and energises the trip coil. When the breaker opens the c.t. current ceases, T_2 stops conducting and C_5 is short-circuited through D_4 and R_5, resetting the circuit ready for another operation.

The non-linear resistor determines the shape of the time-current curve of the device. Three time-current curves are available from three different values of C_5.

The instantaneous trip setting is controlled by the potentiometer P_3. When the potential of the slider is high enough, the zener diode Z_4 conducts and the voltage across R_9 and C_6 is sufficient to operate T_3, causing tripping.

8.9. STATIC OPERATION INDICATORS

There are at present two types of static indicators, one using a reed relay and the other a d.c. voltage amplifier. Both can be used with low-voltage d.c. supplies down to 6 volts.

In the first type a reed relay controls a lamp or a luminescent indicator. The reed relay is not static but, being sealed, it requires no maintenance and is not affected by ambient conditions. The lamp may be of neon or argon type or a low-voltage (and hence robust) filament type. The luminescent indicator consists of a sheet of material which becomes luminous when a voltage is impressed across it. It can be used with a mask to give a message such as ϕ_A for Phase A trip or Z_1 for Zone 1 trip.

In the other type a neon lamp is energized through a d.c. voltage amplifier. When it receives the appropriate signal from the relay circuit it starts a transistor oscillator whose output is stepped up in voltage through a small transformer to operate a neon lamp. The whole equipment is only 1 in. ong and ½ in. diameter.

9
Differential Relays

Basic principles: Longitudinal, transverse and multiinput differential comparator circuits – Applications to generators, transformers, buses and lines – Effect of c.t. errors and primary current transients – Summation c.t. versus sequence filter

9.1. BASIC PRINCIPLES OF DIFFERENTIAL PROTECTION

By Kirchoff's law the vector sum of all the currents entering a circuit should be zero unless an additional current path is added (i.e. a fault) whose current is not included in the vector sum. Consequently, if the secondaries of the c.t's in all the connections to the protected circuit are paralleled with each other and a relay, no current should flow in the relay unless there is a fault in the protected circuit which provides an additional shunt path (see Fig. 9.1).

This method is called Unit Protection. It is not only simple but positive in action, since current should flow in the relay only if there is a fault

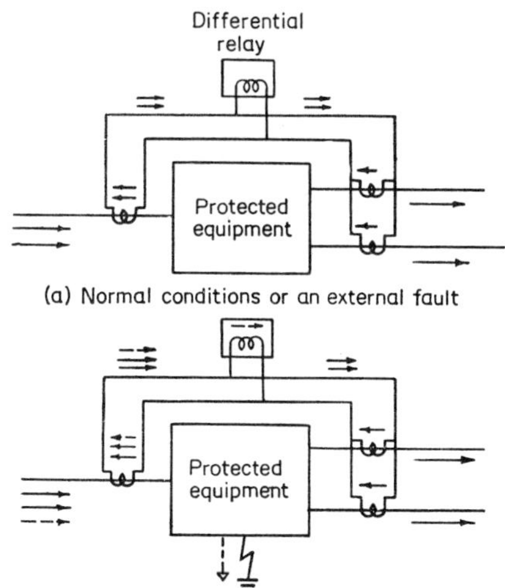

Fig. 9.1. Basic principle of longitudinal differential protection

within the protected section; this permits the use of a sensitive relay and provides a high discriminating margin, i.e. a high ratio between the stability on external faults and the sensitivity on internal faults.

9.2. APPLICATIONS OF DIFFERENTIAL PROTECTION

The various applications were considered in detail in Volume I. In this chapter the basic principles will be reviewed and the descriptions of relays limited to those of the static variety.

The simplest application is to generator protection (Fig. 9.1) where the currents at the two ends of each phase of the stator winding are compared through c.t's which are as nearly as possible identical. The secondary currents should balance during normal load and external faults, so that the relay receives no spill current unless there is a fault in the stator windings.

A similar arrangement is used for overall protection of a transformer except that in a delta-wye transformer the c.t. secondaries are connected in the opposite manner on the two sides of the transformer, e.g. wye-delta, to compensate for the delta-wye transposition (Fig. 9.3). In addition, means are required [63] for preventing the magnetizing inrush current from operating the relay when the transformer is first energized from one side.

Figure 9.2 shows a simple method of protecting the windings of a transformer against earth faults. This is called restricted earth protection; it avoids the problems of magnetizing inrush currents and each side of the transformer is protected separately. This was discussed in some detail in Chapter 7 and in Vol. I Section 4.5.2 d.

Where it is not possible to balance the c.t's closely the relay is generally provided with a restraint which increases with the magnitude of the fault current and hence with the divergence in the secondary currents so that it provides stability, i.e. it prevents wrong operation during a heavy fault external to the protected section. The through (scalar sum) current is used for this purpose using rectifiers.

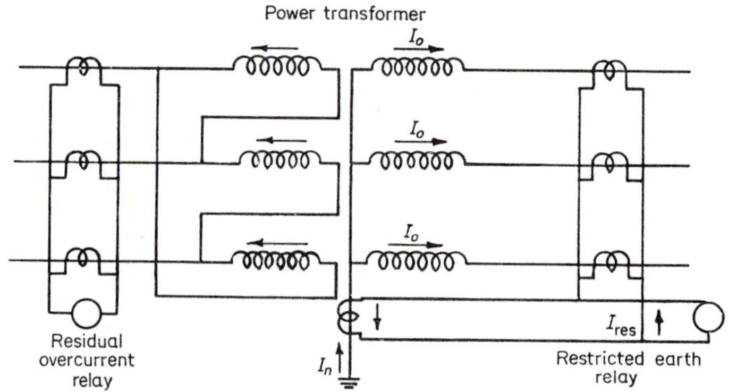

Fig. 9.2. Restricted earth protection of power transformer

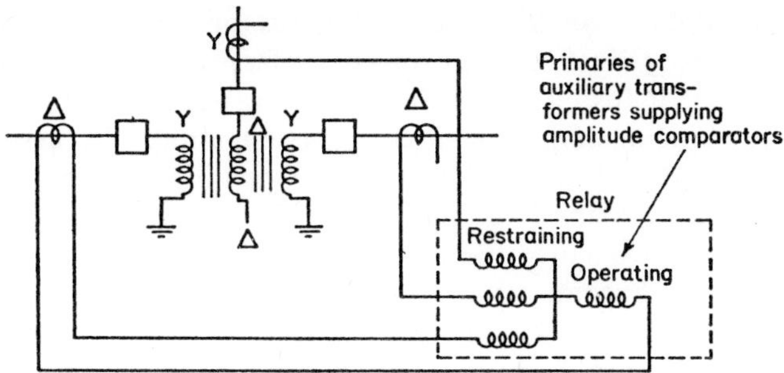

Fig. 9.3. Biased differential protection of a three-winding transformer

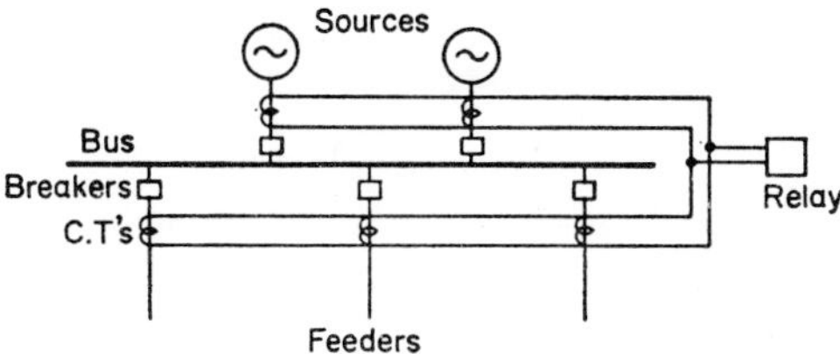

Fig. 9.4a. Unbiased differential protection of a bus. The relay is connected to trip all the breakers. Only one phase shown

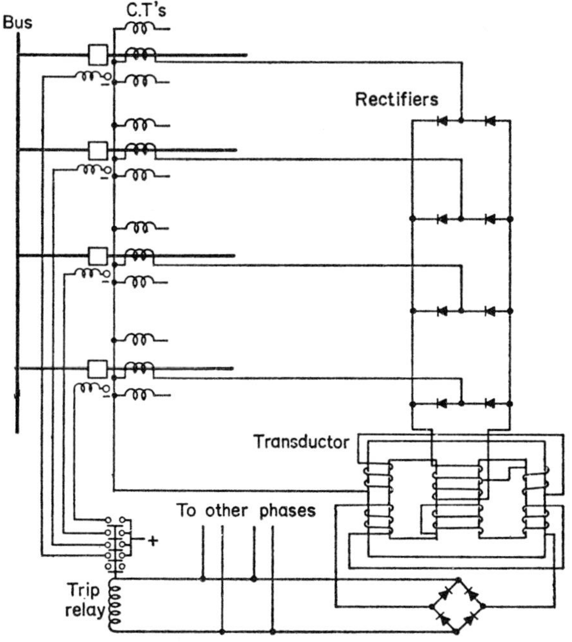

Fig. 9.4b. Biassed differential relay bus protection (Reyrolles)

Where many c.t's are involved, as in bus protection, the restraint is generally omitted (Fig. 9.4a) because stability on external faults can be achieved without it (using a high impedance relay) and because there is a considerable gain in simplicity, and hence reliability, by paralleling the c.t's at the bus instead of bringing them in separately to the relay. Where the ratios of the c.t.s are dissimilar an auxiliary matching c.t. is used with appropriate taps to balance the currents from the main c.t.s so that they sum to zero under normal conditions.

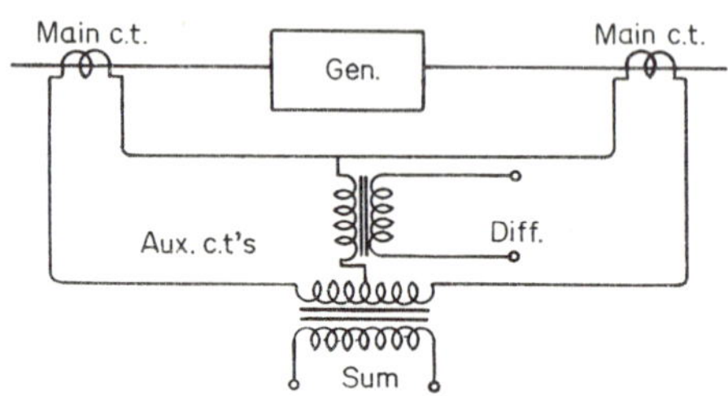

(a) Inputs to relay from one phase

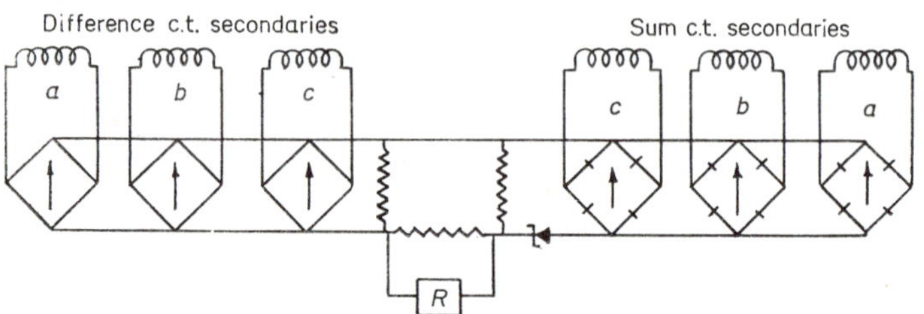

(b) Three-phase comparator circuit

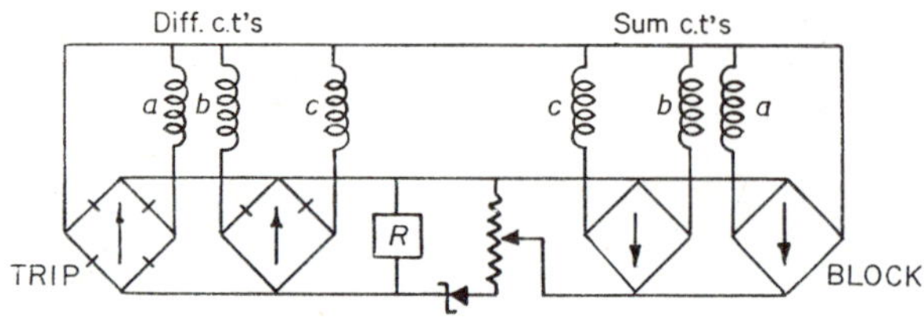

(c) Alternative comparator circuit

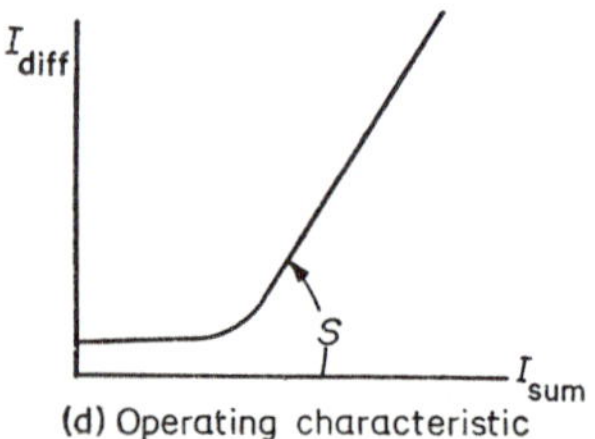

(d) Operating characteristic

Fig. 9.5. Static generator differential relays

9.2.1. Generator Differential Relays

The rectifier bridge amplitude comparator is the most convenient device for this application. Furthermore, it lends itself to polyphase operation so that a very compact relay results.

Figure 9.5b shows how the rectified sums and differences of the currents in the three phases are paralleled in an American relay [64] and fed into a sensing circuit (Fig. 9.6a) which controls a thyristor tripping circuit. The voltage outputs from the operating (difference) and restraining (sum) circuits are the maximum values from the three phases; hence the tripping signal is automatically derived from the faulted phase, while the restraint signal is based on the through current in the sound phases.

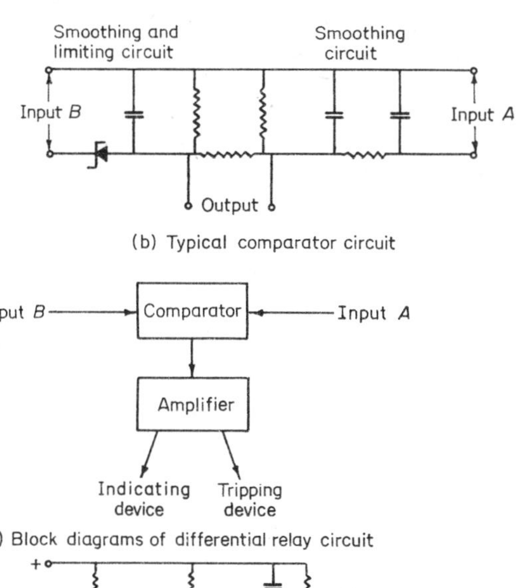

(b) Typical comparator circuit

(a) Block diagrams of differential relay circuit

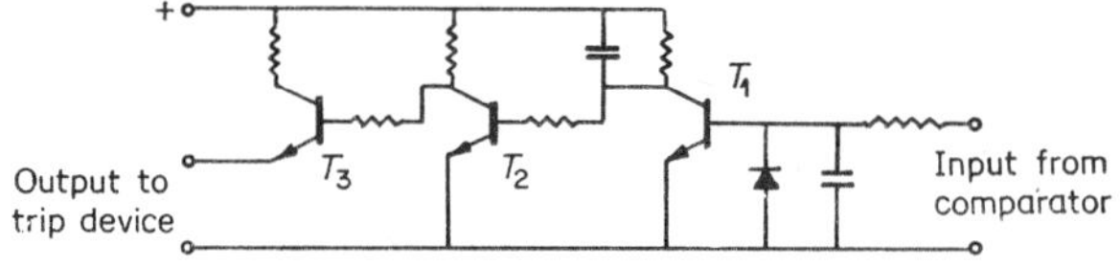

(c) Typical amplifier circuit (T_2 normally conducting)

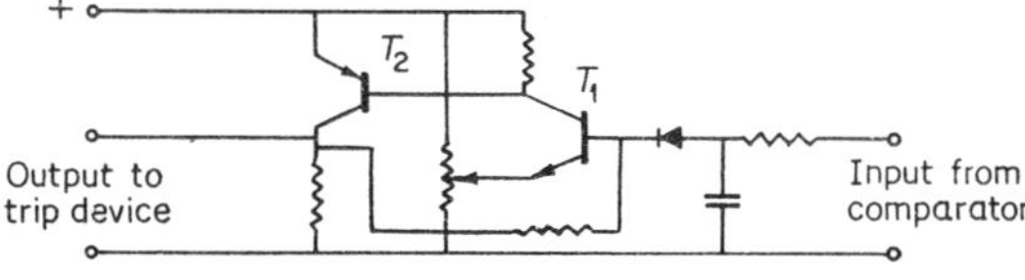

(d) Alternative amplifier circuit (no transistors normally conducting)

Fig. 9.6. General arrangement of static differential relay

Figure 9.5c shows how, in a European relay [70], the circuit is further simplified and the slope of the tripping characteristic is made adjustable. This relay uses a sensitive polarized relay (*R*) for tripping but the circuit is obviously adaptable to a static tripping circuit. Figure 9.6 shows details of a static relay to replace the polarized electromagnetic relay marked *R* in Fig. 9.5. Figure 9.6d shows an alternative tripping circuit which avoids having one of the transistors normally conducting.

9.2.2. Transformer Differential Relays

These relays were among the earliest using semiconductor comparators. Figure 9.7a shows the basic arrangement of the first relay having harmonic restraint [63]. More recent relays [65] have simplified the circuit by replacing the harmonic restraint with a harmonic blocking relay (Fig. 9.7b).

In harmonic restraint, the harmonic content of the differential current adds to the restraint provided by the through current. In harmonic blocking a separate relay prevents tripping if the 2nd harmonic exceeds 15% of the fundamental. In a transformer fault the 2nd harmonic of the fault current never exceeds 5% of the fundamental but, in the inrush current, it is never less than 23%.

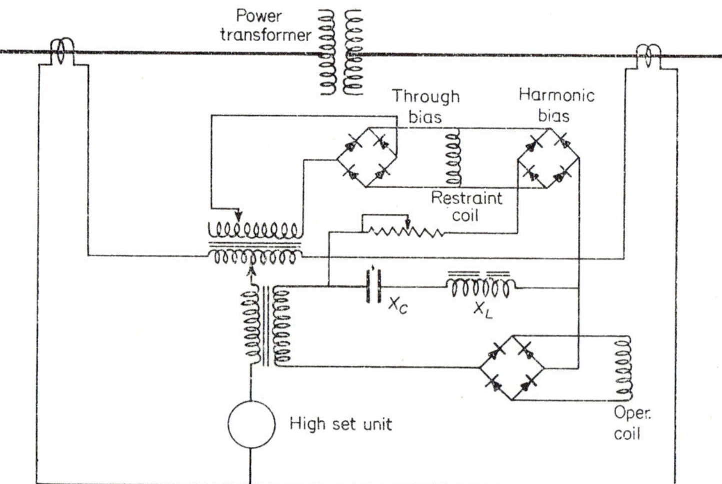

Fig. 9.7a. Transformer differential relay with harmonic restraint

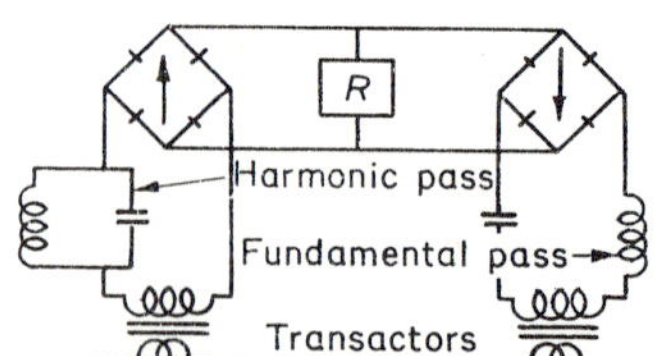

Fig. 9.7b. Harmonic blocking relay (simplified)

9.2.3. Bus Differential Relays

A basic difficulty in this application is the fact that a heavy external fault tends to saturage the c.t. in the faulted circuit and cause a spurious unbalance which can wrongly operate the relay. This problem has been overcome by the use of linear couplers [66] or by employing stabilizing resistors in the operating (difference) circuit of the relay. Both these methods were discussed in Chapter 7. See also Figs. 9.9 and 9.10.

For buses with a large number of circuits, the high impedance differential relay offers the simplest and most effective form of bus protection. The essential requirements for correct application of the scheme are:

(i) Equal c.t. transformer ratios.
(ii) Low c.t. secondary winding resistance.
(iii) Adequate knee-point voltage output from the c.t's.
(iv) Low lead burdens.

Normally the knee-point voltage of the c.t's is taken to be equal or greater than twice the relay voltage setting required to ensure stability of the relay for external faults just outside the bus zone protection, where the maximum fault level is encountered. Therefore, to avoid the need for large c.t's, it is generally recommended that optimum-ratio, low-reactance c.t's be used since the voltage output of a c.t. is proportional to the cross-sectional area of the core and the number of turns.

It follows that, if the main c.t's have low secondary winding resistances and sufficient knee-point voltage, there is no reason whatever why the high impedance voltage differential relay should not operate satisfactorily, even with c.t. ratios as different as 1 : 10 or even 1 : 20, assuming that an adequate matching auto-c.t. is provided.

Unfortunately, the c.t's with the lowest ratios sometimes do not have adequate knee-point voltage to match the relay voltage required; in which case the fault is not with the relay but with the c.t's. It is generally agreed that, with existing stations where the c.t's have unequal ratios and are unsuitable for use with a high impedance voltage differential relay because of their low knee-point voltage, an alternative solution is the use of a biased differential relay.

However, for new installations where adequate c.t's are available, the unbiased high-impedance voltage differential relay is simple, more robust and has the advantage that its performance can be accurately determined by simple analytical methods.

In difficult cases, where the c.t's are very dissimilar and there are a large number of circuits, it may be necessary to resort to both methods to secure discrimination, i. e. a high impedance differential relay with a restraint derived from the scalar sum of the c.t. secondary currents. In order to have maximum tripping sensitivity for a bus fault the scalar restraint must be virtually eliminated when all currents flow into the bus. This has been done quite effectively in a Swiss relay (115) by appropriate use of rectifiers and zener diodes.

9.2.4. a.c. Pilot-wire Relaying

The discriminating margin referred to in Section 9.1 is considerably impaired in pilot-wire protection because the currents are not directly compared in each phase. For reasons of economy only two pilot wires are used and this means that the three phase currents and the residual current have to be combined into one current for comparison with a similarly derived current at the other end. This reduces the accuracy of the current supplied to the relay and hence the discriminating margin.

Discrimination is also reduced by phase-shift of the remote terminal current by pilot-wire capacitance and further error is introduced by unequal degrees of saturation of the main and summation c.t's at the two ends, due to difference of design or lead burdens.

Finally, load current flowing through the line during a light internal fault tends to reduce the sensitivity of the relay by increasing the restraint signal. A discussion of these conditions for the purpose of improving the discrimination factor is given in Chapter 15.

Because the line terminals are separated by many miles of line length, it is normal practice to have a separate relay at each terminal, each controlling its own circuit-breaker. The fact that only two pilot wires are usually available means that the relay can compare only the local c.t. current with the pilot wire current (circulating current scheme [22, 67] or with the voltage across the pilot wires (opposed voltage scheme [68, 69]).

These methods of comparison would be equivalent to comparison of the sum and difference of the currents mentioned previously only if the pilot wire had no impedance nor susceptance and the relays had no burden. Errors from these causes can be minimized by compensating for pilot-wire attenuation (Appendix Section 9.6 and Chapter 15); this is easier with transistorized relays having low burden but it is often a complex problem, especially where the c.t's have different magnetic characteristics.

If more than two pilot wires could be provided the problem would be simplified because a single relay, preferably of the phase comparator type, could be provided at one end and additional pilot used for tripping the breakers. The phase comparison of the c.t. currents at the two ends is, of course, the dual of the amplitude comparison of their sum and difference.

The multi-terminal line can be regarded as the inversion of a station bus. Effective protection is provided by the same basic principle of operation but in a form which is a dual of the bus protection circuits of Fig. 9.4b, [73].

As before we require the vectorial sum of the currents to produce an operating signal and the arithmetic sum to produce a restraining signal. In bus protection the relay at each terminal requires these conditions and this is obviously impossible with the method of current summation by paralleling the c.t's; so we have to resort to turning the currents into voltages and adding them in series. This provides a similar voltage signal at each station.

The circuits for vectorial addition are compared in Fig. 9.9 and for arithmetic addition in Fig. 9.10. The complete circuit is shown in Fig. 9.11; it will be noticed that three pilot wires are necessary in order to avoid mixing the a.c. and d.c. signals.

9.2.5. Carrier Relaying

This is a vast subject, a detailed discussion of which would merit a separate volume. The main schemes and their principles of operation were explained in Chapters 7 and 8 of Vol. I and only a summarized comparison of the scheme is given here on the next page.

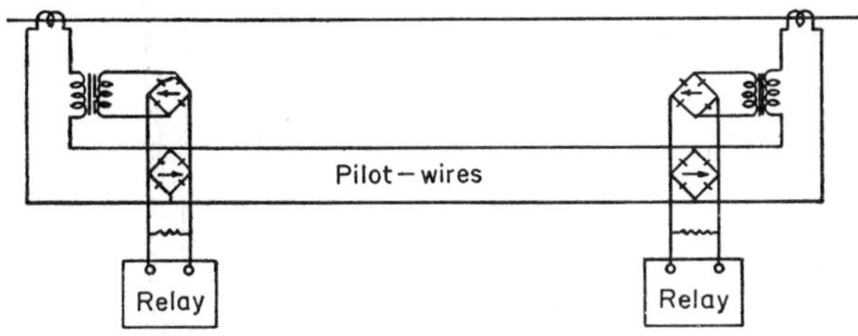

Fig. 9.8. Pilot-wire protection of a two-ended line (circulating current scheme shown; opposed voltage scheme similar with c.t. at one end reversed)

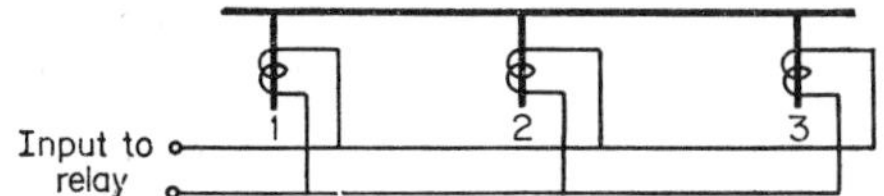

Fig. 9.9a. Vectorial summation of current from c.t. secondaries

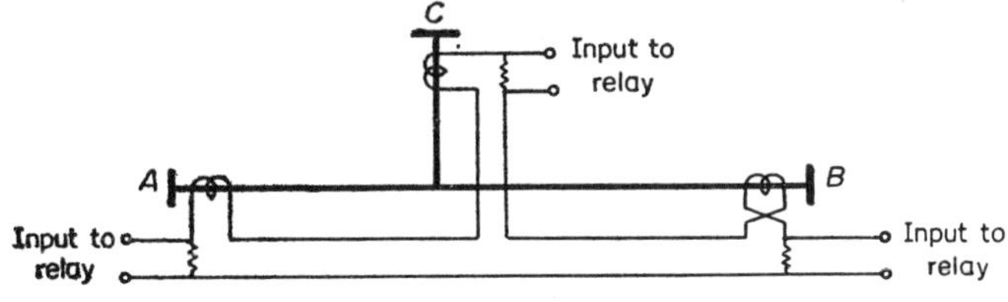

Fig. 9.9b. Vectorial summation of voltages from linear couplers or air-gap c.t. secondaries

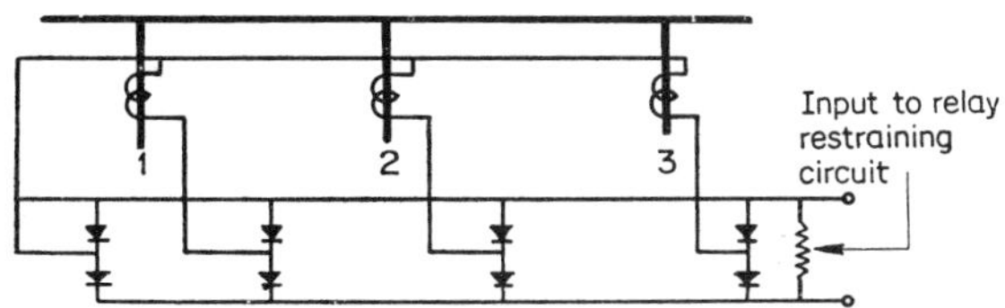

Fig. 9.10a. Scalar sum of secondary currents of c.t's

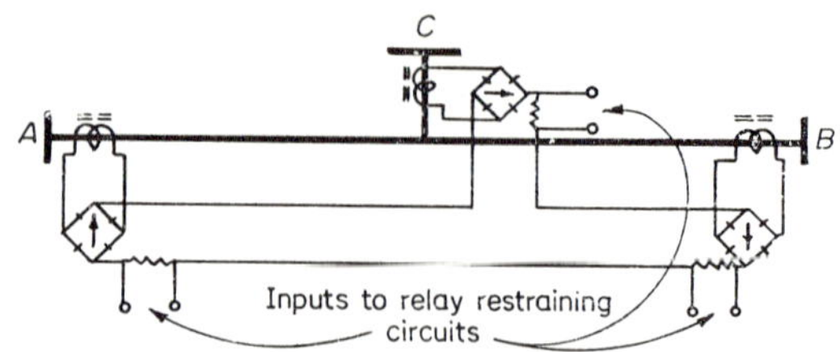

Fig. 9.10b. Scalar sum of secondary voltages of linear couplers or air-gan c.t's

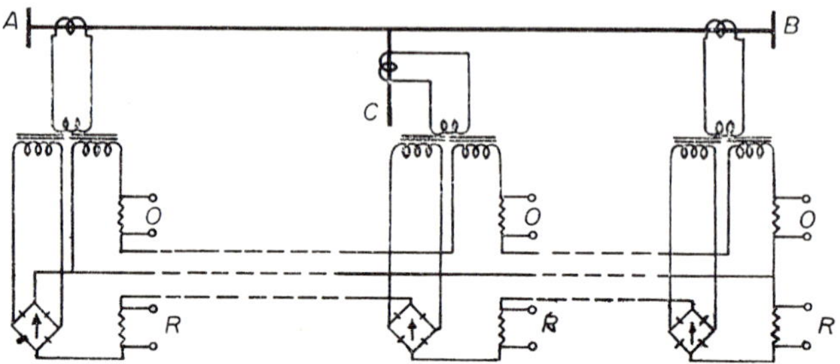

Fig. 9.11. Three-wire pilot protection of multi-ended line
O = inputs to operating circuits (vectorial sum)
R = inputs to restraining circuits (scalar sum)

(*a*) *Directional Blocking:* This is the most popular scheme. It is more reliable but less secure than transferred tripping schemes.

(*b*) *Phase Comparison:* This scheme is immune to power swings, to switching in and out of series capacitors, to inadvertent loss of potential and to mutual induction of zero sequence current from parallel lines. Owing to the reduction of the phase and residual currents to a single signal, there is some loss of discriminating margin and some delay in operation of phase comparison carrier.

(*c*) *Combined Scheme:* These shortcomings can be avoided by using phase comparison only for ground faults where the residual current can be used alone. Phase faults can be taken care of by the directional or distance blocking scheme.

(*d*) *Transferred Tripping:* In these schemes the tripping carrier signal can, under some conditions, be blocked by the fault itself. This risk can be eliminated by using a microwave carrier channel or wired pilot channel, but in practice it is very unlikely that the tripping signal will not get through; with distance relays it only means zone 2 tripping at one end.

The underreaching direct trip method is the simplest and on long two-terminal lines it can be used effectively with zero sequence directional relays, which are much cheaper than a set of ground distance relays.

The overreaching permissive trip scheme is preferable for multilateral and short two-ended lines.

Security and speed are conflicting requirements. With single frequency signals tripping can occasionally occur due to interference containing that frequency, such as an arcing fault. Coded signals prevent this but delay tripping and increase cost and complexity.

For short lines a d.c. signal over pilot wires is an effective solution but for long lines the choice must be made between a small calculated risk and an extra 1 cycle delay. The best compromise appears to be the frequency shift scheme in which a guard frequency is applied by fault detectors and a trip frequency replaces the guard frequency when tripping is required.

9.2.6. Transverse Differential Protection

The previous types of differential protection have been longitudinal. In parallel feeders and in large generators having parallel windings in the stator, additional protection is often provided by comparing the currents in these parallel circuits.

The circuitry for transverse differential relays is the same as for the longitudinal type but some modifications are often added, such as voltage restraint.

Figure 9.12 shows the circuit of a Russian relay [70] which uses both the bus voltage and the current in the sound line for bias. The voltage restraint ensures that the relay in a sound line will not trip if a circuit-breaker is opened at the other end of the line. Current bias from the sound line gives a percentage slope characteristic and permits faster operation of a fault near the end of the section than if the scalar sum of the currents has been used for restraint.

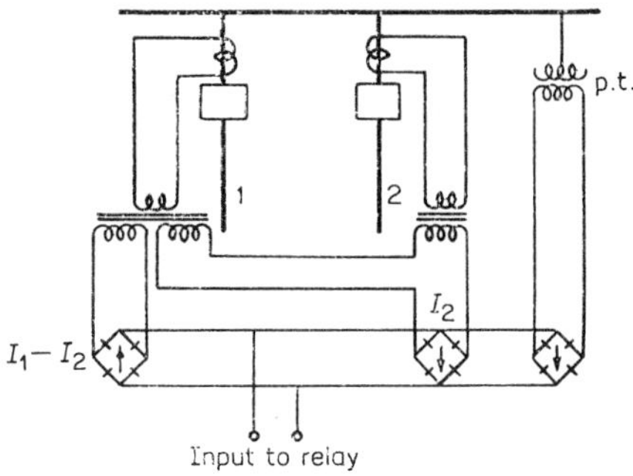

Fig. 9.12. Transverse differential protection

9.3. C.T. REQUIREMENTS

Under static conditions the core of the c.t's must be large enough not to saturate at maximum fault current. In other words the flux density required to produce the maximum voltage drop across the secondary burden must be below the knee of the B-H curve of the steel used in the core. In practice, however, the c.t. core has to be much bigger because there may be additional flux due to remanence and to the d.c. component of an offset wave of fault current.

It was shown in Chapter 7, Section 7.2.4 that the core cross-section should be multiplied by $\{1 + (X/R)\}$ to allow for these transient phenomena. Since X/R could be as high as 30 in generator protection this would result in enormous c.t's, but it will be shown in the following section that a more economical solution is to use a high impedance relay or a stabilizing resistor which also compensates for other sources of c.t. error mentioned in Section 9.3.1. Furthermore the full factor $[1 + (X/R)]$ is applicable only where the secondary burden is wholly resistive and this can be avoided by design.

9.3.1. Effects of C.T. error

Differences in magnetic characteristics of the c.t's, differences in their lead resistances, core remanences, etc., can affect the secondary outputs of c.t's so that they do not always balance during external faults and, consequently, the relay can receive spurious differential current which may cause it to trip wrongly.

The two common methods of coping with this problem are (*a*) to connect a stabilizing resistor in series with the operating circuit of the relay and/or (*b*) to provide a restraint which increases with the magnitude of the fault current and hence with the c.t. error.

The stabilizing resistor provides by far the simpler circuit, especially where many primary circuits are involved, but the restraint is necessary in addition in cases where the operating circuit of the relay does not receive the true vectorial sum of the current inputs to the protected circuit, such as in transformer and feeder (pilot-wire) protection.

9.3.2. The Stabilizing Resistor [61]

The following analysis applies to the extreme case of one c.t. completely saturated by an external fault and the others acting as ideal c.t's. Since this is an ultimate condition, the value of the stabilizing resistor calculated from it has an inherent factor of safety.

The voltage across the relay is $V_r = I_r R_r = I_A R_A = I_B R_B$ where the suffix r refers to the relay and A or B to the line terminal. Hence

$$R_r = (I_B R_B / I_r) \tag{9.1}$$

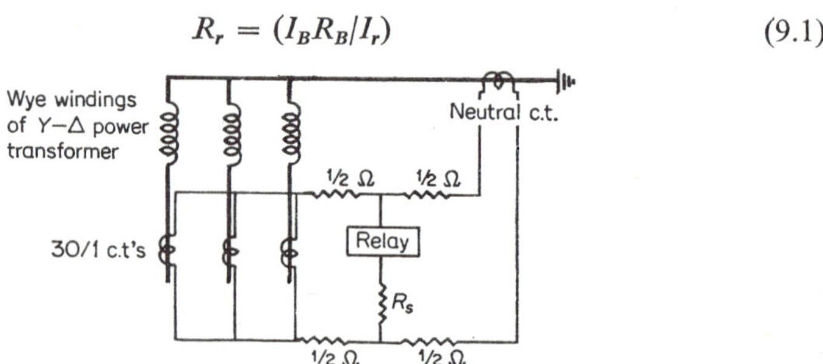

Fig. 9.13. Analysis of operation of stabilising resistor R_r

where I_r is the pick-up current of the relay and I_B is the maximum secondary current for an external fault at terminal *B*. In multiterminal lines or buses I_A can be considered as the sum of the secondary currents at the other terminals.

The value of R_r is that of the relay and its series stabilizing resistor. In generator differential or restricted earth protection, R_r need be only one third of this value to achieve stability because it is unlikely that one c.t. will be completely saturated for an external fault since the fault current is limited by the impedance of the protected equipment.

The current, I_p, in the primary circuit of a c.t. is *N* times the current I_s, in the secondary, plus the magnetizing current I_m, where *N* is the secondary turns; hence

$$I_p = NI_s + I_m \quad \text{or} \quad I_s = \frac{1}{N}(I_p - I_m). \tag{9.2}$$

The relay current I_r is the difference between the two c.t. secondary currents $\therefore \; I_r = I_{sA} - I_{sB} = \frac{1}{N}(I_{pA} - I_{mA} - I_{pB} + I_{mB})$

On an external fault $I_{pA} = I_{pB}$; $\therefore \; I_r = \frac{1}{N}(I_{mB} - I_{mA})$. (9.3)

i.e. the relay current is the difference between the primary magnetizing currents corrected for turn ratio.

With an external fault, as one c.t. approaches saturation, it tends to limit the voltage across the relay; hence, as R_r is increased, I_r is reduced so that I_{mA} and I_{mB} are forced towards equality and the c.t. with the lower magnetizing current will push equalizing current through the other c.t. secondary.

On the other hand, the high value of R_r will not prevent the relay from operating on an internal fault because, in that case, the c.t. secondary e.m.f.'s are additive and combine to force current through the relay.

The proportioning of resistance between the relay and its series stabilising resistor depends upon the relay. In static or very sensitive electromagnetic relays the pick-up can be controlled by series resistance; in a less sensitive type the pick-up would be controlled by taps on the relay coil. In determining the actual pick-up of the relay in terms of primary fault current, the magnetizing currents of all the c.t's must be added to the relay current setting, i.e. the actual pick-up, I_{pu}, in (primary) amperes is

$$I_{pu} = N \cdot I_r + \sum I_m \tag{9.4}$$

where *N* is the c.t. ratio and $\sum I_m$ is the sum of the c.t. magnetizing currents at the relay voltage setting.

In most cases the desirable quality is a high value of the stability ratio K_s where

$$K_s = \frac{\text{max. through fault current for stability}}{\text{min. effective relay pick-up current}}$$

K_s can usually be doubled by tuning the differential circuit with a series capacitor. This also improves stability by excluding the d.c. component and hence minimising saturation.

9.3.3. Example of Use of Stabilizing Resistor

Zero sequence overcurrent protection and restricted earth fault protection (Fig. 9.2) are difficult in application where instantaneous relays are used because their stability depends upon the balancing of the three phase currents during any condition except a ground fault within the protected section.

A heavy external three-phase fault provides the worst problem because the three currents have different degrees of d.c. offset. For instance, if one phase is at zero voltage at the inception of the fault it will have offset and maximum tendency to saturate, but the other two 120° away will have only 50% offset and in the opposite direction. Consequently the c.t's will start off with very unequal fluxes and there will be a large spurious spill current until the transients have decayed.

Such a case can almost always be adequately controlled by choosing the proper stabilising resistor, as the following example illustrate.

500 kvA, 11 kV transformer. Through fault stability required up to 15 times full load with neutral c.t. assumed completely saturated. Line c.t's 30/1. Neutral c.t. 30/1. c.t. resistance is 0·046 Ω.
Circuit resistances as shown in Fig. 9.13. Relay pick-up 0·1 A

$$\text{Full load current} = \frac{500}{11\sqrt{(3)}} = 26{\cdot}2 \text{ A}$$

Max. current for stability $= 15 \times 26{\cdot}2 = 392$ A

$$\text{Max. volts across relay} = \frac{392}{30}(0{\cdot}5 + 0{\cdot}046 + 0{\cdot}5) = 13{\cdot}7 \text{ V}$$

Volts to operate relay $= IR = 0{\cdot}1 \times 100 = 10$ volts
To ensure stability requires a loss of $13{\cdot}7 - 10 = 3{\cdot}7$ volts in the stabilizing resistor

$$\therefore\ R_s = \frac{3{\cdot}7}{0{\cdot}1} = 37\,\Omega$$

On the other hand, the relay must trip on all internal faults.
0·1 A pick-up requires $0{\cdot}1\,(0{\cdot}5 + 0{\cdot}046 + 0{\cdot}5 + 37 + 100) = 13{\cdot}8$ V

From the c.t. excitation curve, magnetizing current per c.t. = 0·19 A
Total magnetizing current (four c.t's) = 4 × 0·19 = 0·76 A
Effective relay setting = $I_{relay} + I_{ex}$ = 0·1 + 0·76 = 0·86 A
or in terms of (primary) line current = 0·86 × 30 = 25·8 A

9.4.4. Through Current Bias [61]

Bias alone will not always ensure stability on external faults but, in combination with a high impedance relay winding or an equivalent series stabilizing resistor, stability can be obtained up to any current value assuming that the c.t's have the right ratio.

The pick-up of a biased relay is

$$I_r = I_0 + sI_B \tag{9.5}$$

where I_0 is the pick-up value with the restraint removed, I_B is the through fault current and S is the bias ratio.

From Eq. (9.1) $I_B R_B = I_r R_r = (I_0 + sI_B) R_r$. From this since $\dfrac{I_0}{I_B}$ is small

$$R_r = \frac{R_B}{s} \tag{9.6}$$

i.e. the value of the necessary stabilising resistor is reduced by the bias ratio.

The output required from the c.t's (and hence their size) is also reduced by the bias. The ratio of their required e.m.f.'s is

$$\frac{V_B}{V} = \frac{I_0 R_B}{sI_B R_B} = \frac{I_0}{sI_B}. \tag{9.7}$$

Referring to Fig. 7.1 (Chapter 7) the current in the operating circuit of the relay during an external fault is

$$I_r = \frac{1}{N}\{I_p - (n - 1)\, I_m\} \tag{9.8}$$

because the magnetizing current for all the other $(n - 1)$ c.t's is supplied by the c.t. in the faulted feeder. The current in the restraining circuit is the through current I_p. Also from Fig. 7.1 where Z_m is the magnetizing impedance

$$I_r R_r = V_s = I_m Z_m. \tag{9.9}$$

The relay operates when

$$I_r > \frac{s}{N} I_p > s\left\{I_r + \frac{(n-1)}{N} I_m\right\} \qquad \text{from (9.8)}$$

$$> s\left\{I_r + \frac{(n-1)}{N}\frac{I_r R_r}{Z_m}\right\} \qquad \text{from (9.9)}$$

i.e. when $Z_m > \dfrac{(n-1)\,sR_r}{N(1-s)}$ (since I_r cancels out).

Substituting from Eq. (9.1)

$$Z_m > \frac{(n-1)\,R_s}{N(1-s)}. \tag{9.10}$$

For a two terminal circuit such as a generator $n = 2$ and the biased relay operates when

$$Z_m > \frac{R_s}{N(1-s)}. \tag{9.11}$$

Equations (9.10) and (9.11) show that stability is easy to achieve with a two-ended protected circuit but is much more difficult when n is large, such as in a bus differential scheme. As previously stated, these equations assume that R_r is correctly chosen (Eq. 9.1) and that the reactance of the relay circuit is negligible.

9.3.5. c.t. Requirements for Bus Differential [61]

The c.t's must be chosen to produce a secondary voltage V equal to at least twice the IR drop in the longest leads to the relay at maximum through fault current (usually taken as the switchgear interrupting rate), i.e.

$$V = 2\frac{IR}{N} \tag{9.12}$$

Also, at $V/2$ volts, the magnetizing current (primary) must not exceed $(I_p - NI_r)/n$ where n is the number of c.t's or circuits. This follows from Eqs. (9.2) and (9.4).

9.3.6. c.t. Requirements for Pilot-wire Scheme

Stabilizing resistors are more difficult to apply to pilot-wire schemes because there are two relays which must have similar settings on through faults; however a stabilizing resistor is often provided in the neutral wire between the summation transformer and the main c.t's. Fortunately the conditions are not so severe as with generator or bus protection because

(*a*) the factor $\{1 + (X/R)\}$ is reduced by the reactive component of the secondary burden;

(*b*) the X/R ratio is rapidly reduced by the impedance of the feeder;
(*c*) short feeders affect the X/R ratio less but make the pilot relay more stable;
(*d*) c.t. losses and magnetizing current reduce the maximum flux that has to be accommodated;
(*e*) the stability of pilot relays is greatly increased by their strong bias which gives a steep slope to the differential characteristic.

Pilot-wire relays are tested by manufacturers under extreme cases of unequal saturation due to dissimilar c.t's, etc. [113] so that they are suitable for most applications: In a few cases, however, the margin between instability on through faults and failure to trip on internal faults is unsatisfactory; this subject is dealt with in detail in Chapter 15.

The results of the factory tests are usually summarized in a formula which is a guide to c.t. requirements for a particular relay and is of the form $V_k \geq K \cdot I_{max} R_s$, where V_k is the knee-point voltage of the c.t., K is a whole number, I_{max} is the maximum through fault current and R_s is the resistance of the c.t. secondary and its leads to the relay. K can theoretically be as high as $\{1 + (X/R)\}$ but is reduced by the foregoing considerations (*a*) (*b*) (*c*) (*d*) (*e*).

9.4. TRANSIENT ORGINATING IN THE C.T. FLUX [61]

Since the magnetizing component I_m is supplied by the primary

$$I_m = I_p - NI_s. \tag{9.13}$$

If L_m is the magnetizing inductance, then

$$L_m \frac{di_m}{dt} = e_s = i_s R_s \tag{9.14}$$

where e and i are instantaneous values.

From Eq. (9.13) $i_s = (i_p - i_m)/N$. Substituting this value of i_s in Eq. (9.14)

$$\text{therefore } \frac{di_m}{dt} = \frac{R_s}{NL_m}(i_p - i_m). \tag{9.15}$$

In any inductive circuit the steady-state current at any moment is

$$i = \frac{\hat{E} \sin(\omega t + \psi)}{\sqrt{\{R^2 + (L\omega)^2\}}} \tag{9.16}$$

where $\hat{E}$ is the peak value of the voltage and ψ is the time in radians after voltage zero at which peak voltage starts.

Substituting this value for the current i_p in Eq. (9.15)

$$\text{therefore } \frac{di_m}{dt} + \frac{R_s}{NL_m} i_m = \frac{R_s \hat{I}_p}{NL_m} \sin(\omega t + \psi).$$

Solving this equation for i_m

$$i_m = \frac{R_s\hat{I}_p}{\sqrt{\{R_s^2 + (\omega L_m)^2\}}} [\sin(\omega t + \psi - \alpha) + (\sin(\alpha + \psi)\,\varepsilon^{-(R_s/L_m)t}] \tag{9.17}$$

where I_p is the peak value of i_p and $\alpha = \tan^{-1}(\omega L_m/R_s)$.

It will be seen that there is a transient term in the magnetizing current i_m although none existed in the applied primary current i_p.

A similar transient will of course occur also in the flux wave and both will have a maximum value when $\psi = 0°$ because α is almost $\frac{1}{2}\pi$ and $\sin(\alpha - \psi)$ is a maximum when $(\alpha - \psi) = \frac{1}{2}\pi$. In this case there will be a flux doubling effect due to the offset (displacement) of the wave from the zero axis.

If the primary current also has a transient term, it will be due to the inception of the fault near the moment of voltage zero, i.e. when ψ is near $\frac{1}{2}\pi$; hence the two transients can never occur at the same time and they cannot be additive.

9.5. EFFECT OF TRANSIENT IN THE PRIMARY CURRENT [61]

The transient component of the primary current is

$$\frac{\hat{E}\sin(\psi - \alpha)}{\sqrt{\{R^2 + (L\omega)^2\}}} \tag{9.18}$$

where ψ is the time after voltage zero at which the fault occurs and $= \tan^{-1}(L\omega/R)$, where R and L refer to the primary circuit.

If this value be substituted in Eq. (9.15), the effect of both transients can be ascertained. This gives

$$\frac{di_m}{dt} + \frac{R_s i_m}{NL_m} = \frac{R_s I_p}{NL_m}\sin(\psi - \alpha)\,\varepsilon^{-(R/L)t} \tag{9.19}$$

The sine term disappears if $\psi = \alpha + \dfrac{\pi}{2}$. Then, solving equation (9.19) for i_m we get

$$i_m = \frac{I_p R_s L}{RL_m - R_s L}(\varepsilon^{-(R_s/L_m)t} - \varepsilon^{-(R/L)t})$$

$$= \frac{I_p T_p}{T_s - T_p}(\varepsilon^{-(t/T_s)} - \varepsilon^{-(t/T_p)}) \tag{9.20}$$

where T_p is the time-constant of the primary circuit and T_s is the time-constant of the secondary circuit.

The locus of i_m with time was shown in Fig. 7.12, Chapter 7.

i_m reaches a maximum when

$$t = \frac{T_s T_p}{T_s - T_p}\log \varepsilon^{T_s/T_p} \tag{9.21}$$

If $T_p = T_s = T$, then $i_m = I_p(t/T)\varepsilon^{-(t/T)}$ and i_m reaches a maximum when $t = T$. At this moment all the primary current is used for magnetizing and the c.t. has zero output.

9.6. APPENDIX: A.C. PILOT RELAYING

When the two input signals S_A and S_B are compared in a relay, the resulting characteristic can be most clearly shown on a graph whose axes are the real and imaginary values of their quotient. At the end A the real and imaginary values of $\beta = S_B/S_A$ are plotted and, at the end B, the corresponding values of $\alpha = S_A/S_B$.

In the application of differential protection to generators, transformers and very short lines, the characteristic on the α-plane and β-plane are the same, viz. a circle on the real axis concentric with the point 1,0 and whose radius is dependent upon the value of K in the equation for the threshold of operation, i.e.

$$|I_1 - I_2| = K|I_1 + I_2|. \tag{9.22}$$

The α-plane characteristic represented by the equation is a circle of radius $r = 2K/(1 - K^2)$ whose centre is on the real axis $c = (1 + K^2)/(1 - K^2)$ from the origin (see Fig. 9.14). This circle cuts the real axis at $c + r = (1 + K)/(1 - K)$ and at $c - r = (1 - K)/(1 + K)$. Since these expressions are reciprocals and since corresponding points on the α- and β-planes are also reciprocals, the characteristics when drawn on the α- and β-planes must be the same circle (Fig. 9.14).

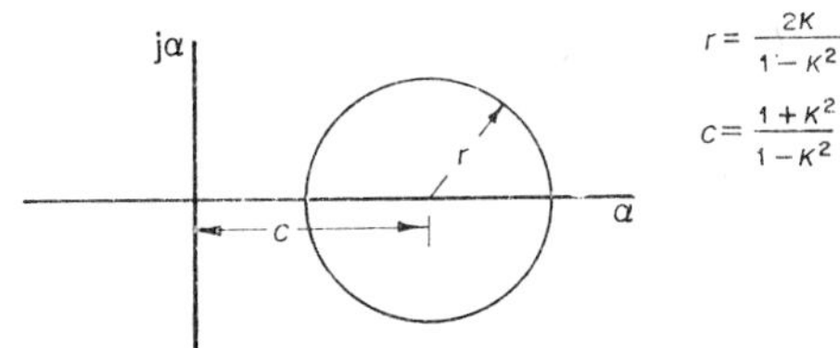

Fig. 9.14. Relay characteristic at A or B with $\gamma = 1{,}0$

What is called the slope is the gradient of the graph of difference current $(I_1 - I_2)$ plotted against the through current $(I_1 + I_2)/2$. For a 10% slope $K = \text{slope}/2 = 0{\cdot}05$, so that $r = 0{\cdot}1/\{(1 - (0{\cdot}005)^2\} = 0{\cdot}1$ and $c = \{1 + (0{\cdot}005)^2\}/\{1 - (0{\cdot}005)^2\} = 1{\cdot}005$. This is a small circle subtending an angle of only 11°, but there is no need for any larger non-tripping area with short c.t. connections and identical c.t's.

On the other hand, in the differential protection of lines, the conditions are much more difficult and the stability circle must subtend at least 30° from the origin. A major source of error is the attenuation and phase-shift of the current as it flows through the pilot wires, due to their leakance and susceptance.

The relay at end A trips when $|I_A - \gamma I_B| > K|I_A + \gamma I_B|$ where γ is the propagation constant. The characteristic of the relay at A on the β-plane is a circle whose centre is at $c = \dfrac{1}{\gamma}\left(\dfrac{(1 + K^2)}{(1 - K^2)}\right)$ and radius $r = \dfrac{2K}{1 - K^2}\left|\dfrac{1}{\gamma}\right|$.

Correct compensation for γ puts the α-plane characteristic of the relays at the two ends both on the real axis, as shown in Fig. 9.15a. In other words, γ now has an angle of 0°, so that it can be replaced by the scalar factor S and now $c = \dfrac{1}{S}\left(\dfrac{1 + K^2}{1 - K^2}\right)$. The shaded portion is the stable area where neither end trips; the overlapping areas denote tripping at one end only while simultaneous tripping at both ends occurs only outside both circles.

Figure 9.15b shows the effect of inadequate compensation. The α-plane characteristics of the relays at the two ends are displaced in angle by the susceptance loss in the pilot wire and in the case shown there is only a small margin of blocking for an external fault (the point 1,0).

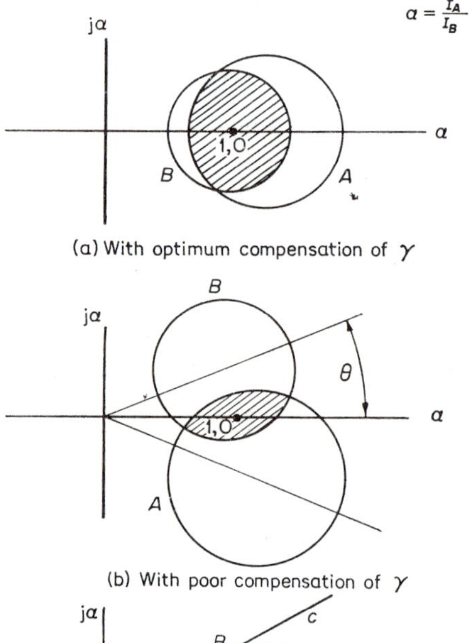

(a) With optimum compensation of γ

(b) With poor compensation of γ

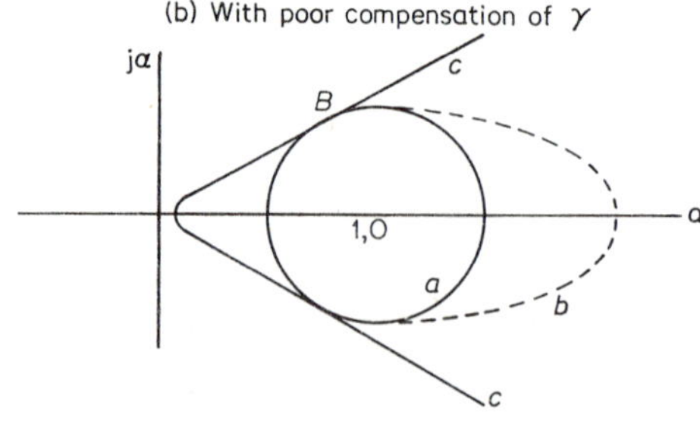

(c) Effect of non-linearity

Fig. 9.15. Pilot relay characteristics at A and B under conditions (a), (b) and (c)

J. Rushton *et al.* [67, 97] have made computer studies to investigate the selectivity of such a characteristic between internal and external faults. The circle should subtend an angle of $\pm 30°$ to 40° at the origin to allow for c.t. and γ compensation errors and to prevent wrong tripping on primary capacitance current at one end. Under certain internal fault conditions it is possible for the fault area to approach the stable area of the relay. It is therefore desirable to have the characteristics at the two ends as nearly as possible the same so as to minimise the areas of tripping at one end only (Fig. 9.15a).

One way to achieve this is to locate the centre of the circle not at 1,0 but at a point which makes the circle at one end the reciprocal of the circle at the other end when plotted on the same plane. For instance, with correct compensation, the circles will both have their centres on the real axis and will be identical if their intersections with the real axis are reciprocals, i.e. if $c + r = 1/(c - r)$ or $c^2 - r^2 = 1$. If the circle subtends an angle of $\pm 30°$ at the origin, $r = c \sin 30°$. Substituting for r gives $c^2(1 - \sin^2 30°) = 1$ or $c = 1/\sqrt{(1 - \sin^2 30°)}$ from which $c = 1{\cdot}155$ and $r = 0{\cdot}577$. This is achieved in the relay by choosing a suitable value for K in terms of S, i.e. by substituting the values of r and S in the relation $r = 2K/(1 - K^2)S$. A more rigorous treatment of this subject is given in Chapter 15.

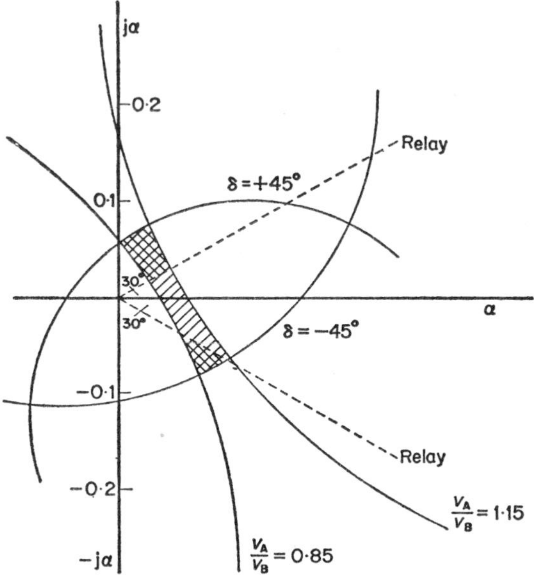

Fig. 9.16. Effect of varying δ and $\left|\frac{V_A}{V_B}\right|$ for a ground fault at B with a 1:1:3 summation c.t.

9.6.2. Effect of Limiters

In order to obtain the requisite sensitivity at minimum fault conditions (20% of c.t. rating) a sensitive relay is necessary; but telephone companies (British Post Office authorities) do not permit voltages above 130 V or current above 60 mA. To keep below these limits and to avoid damage to the relay at maximum fault conditions (perhaps 30 times c.t. rating, i.e. 150/1 range) a nonlinear resistor is generally connected across pilot wires to limit the voltage to about 20% of its value at maximum fault current. A large capacitor is also provided to absorb the harmonics caused by the non-linear resistor and keep the pilot voltage and current as nearly as possible sinusoidal for comparison purposes.

The pilot voltage is limited on internal faults in the circulating current scheme and on external faults for the balanced voltage scheme [22]. This voltage limitation usually is effected at about twice c.t. rating because a heavy load can reverse the current at the receiving end of the feeder during a light internal fault and hence prevent tripping if only phase comparison is present.

As the current increases the characteristic, due to voltage limitation, changes from a circle to an egg shape (Fig. 9.15c). At currents above the saturation value the characteristic becomes purely that of a phase comparator. This is quite acceptable because the fault current is above the value at which its direction can be reversed by load current. The characteristic of Fig. 9.15c is clearly the most suitable for post-office pilots but on private pilots limiters are rarely necessary and a pure amplitude or phase comparator is satisfactory. The phase comparator gives a wider choice of characteristics.

Limiting is not practical in the protection of multi-terminal lines because, in the circulating current scheme, the pilot voltages must sum to zero for an external fault and, in the balanced voltage scheme, the pilot current must be zero. This problem is aggravated by long lines (high γ) and by dissimilar c.t's or c.t. burdens. Private pilots are usually preferable since telephone company regulations do not apply to them; also, they are not subject to unpredictable outages for telephone circuit maintenance.

9.6.3. Summation c.t. versus Sequence Filter

Not only is the discrimination factor reduced by a summation c.t. but its output can be near zero under certain fault conditions (Vol. I, Chapter 8, Section 8.7).

The discrimination factor can be improved by the use of phase sequence filters instead of a summation c.t., especially in impedance grounded systems. The output of a summation transformer is $(n + 2)I_a + (n + 1)I_b + nI_c$ and this varies considerably with both the type of fault and the phase involved in a fault, as shown in the following table.

A sequence filter gives the same output for all phases. The filter combination giving most uniform response for any type of fault [96] has an

Output of Summation Transformer

Fault	I_1	I_2	I_0
a–G, b–c or b–c–G	$a^2 + 2$	$a + 2$	$3(n + 1)$
b–G, c.–a or c–a–G	$2a + 1$	$2a^2 + 1$	$3(n + 1)$
c–G, a–b or a–b–G	$2a^2 + a$	$2a + a^2$	$3(n + 1)$

output $I_1 - 6I_2$. An earlier filter [97] gave $I_1 - nI_0$ where n was adjustable and was set higher for impedance grounded systems.

The behaviour of sequence networks for the various types of faults is as follows.

Outputs of Sequence Filters

Fault	$I_1 - 6I_2$ output	$I_1 - nI_0$ output
a–G, b–c or b–c–G	$I_1 - 6I_2$	$I_1 - nI_0$
b–G, c–a or c–a–G	$aI_1 - 6a^2I_2$	$aI_1 - nI_0$
c–G, a–b or a–b–G	$a^2I_1 - 6aI_2$	$a^2I_1 - nI_0$

10
Distance Relays

Inputs to phase and amplitude comparators – Types of distance relay; their equations and characteristics – Quadrilateral characteristics – Distance relays with linear couplers

Distance relays are used primarily for the protection of transmission and distribution lines and, as their name implies, they measure distance, i.e. they recognize faults occurring within the protected section of the line from the fact that the distance from the relay to the fault is less than the setting of the relay [29].

10.1. PRINCIPLES OF DISTANCE MEASUREMENT

The distance to the fault is measured in terms of line length and this is done by measuring the impedance of the line, which is almost linearly proportional to its length. Normally the relay measures a higher impedance than that of the line because it also measures the impedance of the load (see Fig. 10.1). When a fault occurs, this short-circuits the load and the relay measures only the impedance Z_l of the line.

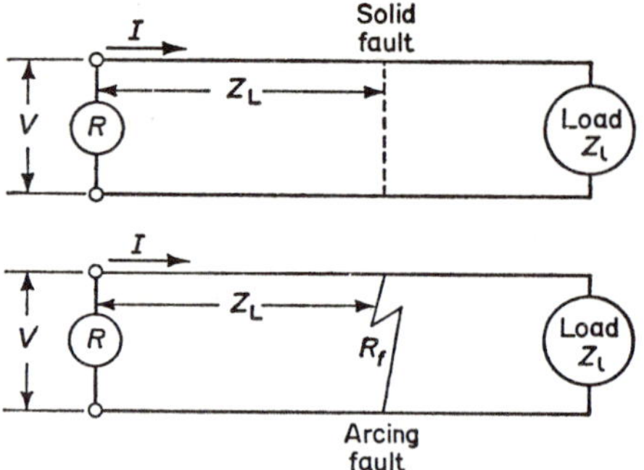

Fig. 10.1. Impedance measured by relay R reduced by fault

If, however, the fault is not a dead short-circuit its impedance (in parallel with that of the load and the lines to the load) is added to that of the faulted line section and this makes the fault appear to be more distant than it really is, i.e. it shrinks the reach of the relay.

For this reason, and because the relay cannot be made with 100% accuracy, there has to be a second distance measuring unit to take care of faults at the far end of the protected line section near the next bus. A third unit is generally provided to give back-up protection for the first two units of the distance relay in the next line section (Fig. 10.2).

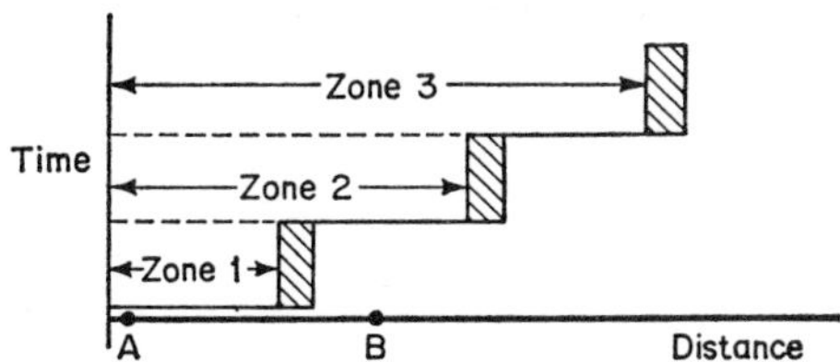

Fig. 10.2. Reduction in reach of relay at *A* due to fault resistance R_f

Chapter 5 of Volume I goes into this subject in considerable detail and it is not proposed to duplicate this information here, except to the extent necessary to introduce the subject.

10.1.1. Fault Area on Impedance Diagram

The effect of the fault resistance is shown vectorially on the impedance diagram of Fig. 10.3. A solid fault at *B* appears to the relay at *A* as the impedance phasor *AB*, but an arcing fault adds the resistance R_f so that the relay measures the impedance *AC*. The horizontal lines show the effect of the fault resistance on faults at various positions between *A* and *B* so that the fault area is the parallelogram *ABCD*.

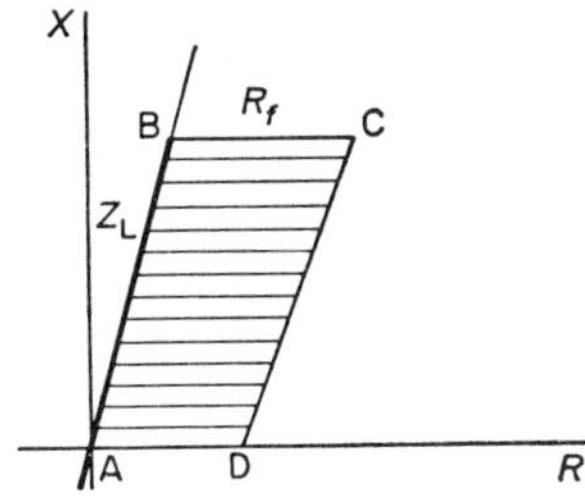

Fig. 10.3. The fault area on an impedance diagram

Figure 10.3 applies to single-end feed only. Where there is a power source at both ends of the line there will be some distortion of this diagram, due to the phase angle difference between the sources.

The fault resistance increases somewhat for faults towards the remote end of the section because less of the total fault current flows through the relay even though the actual fault resistance may be reduced by the extra current fed in from the remote end.

Figure 10.4 shows the protection provided by impedance, reactance and admittance relays. The reactance relay (Fig. 10.4c) has the greatest tole-

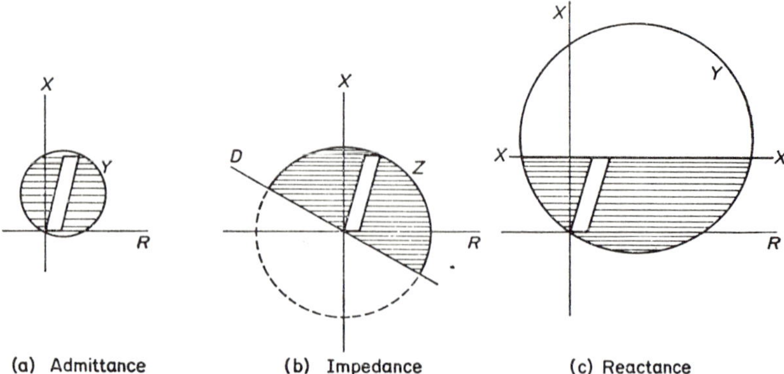

Fig. 10.4. Distance relay characteristics

rance to fault resistance. The admittance (mho) unit (Fig. 10.4c) has less tolerance to fault resistance but greater immunity to tripping on power swings; it also has the economical advantage of requiring no directional control. The impedance unit has a compromise between the qualities of the reactance unit and the mho unit but is simpler in physical construction since it is concerned only with amplitude and not with phase angle measurement.

Since the fault resistance is not related to the length of the line protected, it follows that it has more effect on the impedance measurement of a short line; hence reactance relays are preferable for short lines. They are also preferable for ground faults since the resistance of the return path through the earth can be very high in some localities.

Since the size of the impedance characteristic of the relay increases with the length of line, the longer the line the more vulnerable the relay is to tripping on a power swing (Fig. 10.4). Consequently the snug fit of the mho characteristic around the fault area is required on long lines to prevent tripping on power swings. The mho relay will not trip on a power swing after which the power system can retain its stability [21].

10.1.2. Inputs to Comparator

The measurement of impedance, reactance or angle-admittance is done by the comparison of two different input combinations of current and voltage. With two inputs, circular or straight line impedance characteristics are obtained. Chapter 12 will discuss the use of more than two inputs to obtain more complex characteristics.

In a static relay the two input quantities must be similar, e.g. two currents or two voltage, because they are not electrically separate as they are in an electromagnetic relay.

In a voltage comparator the current is turned into a voltage by passing it through an impedance, $Z_r\underline{/\theta}$, which is a replica of the impedance of the protected line section on a secondary basis, i.e. the line voltage V is compared with IZ_r, the drop across the replica impedance. In a current comparator a current is derived from the voltage by connecting the replica impedance in series with it, giving a current V/Z_r which is compared with the line current I. Sometimes it is convenient to compare the voltages V and IZ_r in a current comparator by connecting resistance in series with each voltage. Figure 10.6 shows the arrangement for the general case of the two-input comparator (*a*) with voltage inputs and (*b*) with current inputs.

The use of replica impedance is not only convenient but permits fast tripping since it eliminates error due to transients in the fault current. This is due to the fact that the fault current passing through the line impedance produces the same voltage wave form as the secondary current passing through the replica impedance. Hence any transients originating in the primary current appear equally in V and IZ_r and cancel out their effects on the impedance measurement.

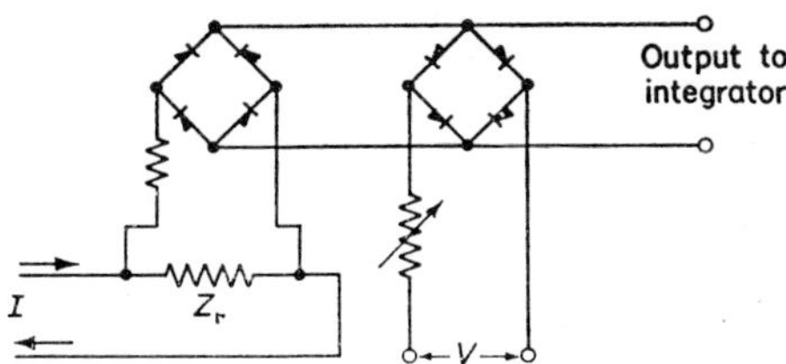

Fig. 10.5. Simplified impedance relay circuit

In the impedance relay (Fig. 10.5) the main problem is to smooth one of the inputs so that the ohmic pick-up does not vary from zero to infinity during the cycle as first the voltage, then the current, passes through zero. It is easier to smooth the voltage than the current and this is usually done by a phase-splitting circuit, as described in Chapter 4, Section 4.3.1.

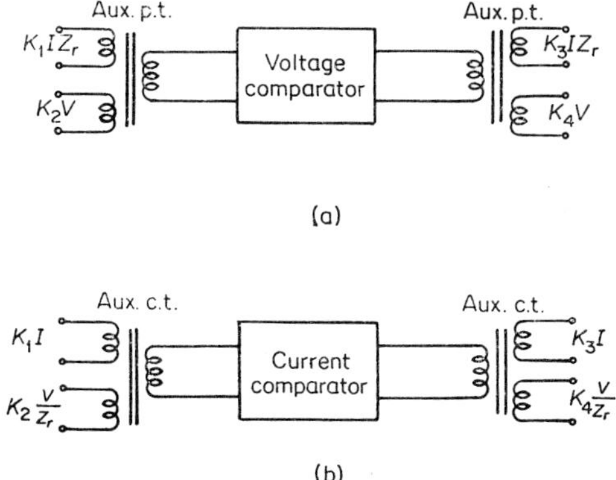

Fig. 10.6. The general case of the two-input comparator
(a) with voltage inputs (b) with current inputs

10.1.3. Phase and Amplitude Comparators

An amplitude comparator compares only the magnitudes of the input signals and ignores their phase angle. When the operating quantity is larger than the restraining quantity, the relay operates.

A phase comparator responds only to the phase relation between the two input quantities, irrespective of their magnitudes. For linear and circular characteristics it operates when $90° > \psi > -90°$, where ψ is the angle between the two inputs.

10.1.4. Distance Relay Characteristics

These are usually plotted on an impedance diagram with axes R and jX, but in some cases an admittance diagram with axes G and jB is more convenient.

Where only single-term quantities are compared (corresponding to the current and voltage of the protected circuit) the resulting characteristic is either a straight line through the origin or a circle with its centre at the origin, depending upon whether a phase or an amplitude comparison is made and whether the characteristic is plotted upon an impedance or an admittance diagram.

If one quantity is compared with the sum or difference of the two quantities, the circle passes through the origin and the straight line does not. If there are two current terms and one voltage term or two voltage terms and one current term, neither the circle nor its straight line dual go through the origin.

Tables 10.1 and 10.2 demonstrate these relationships between the various arrangements of inputs and their resultant characteristics. The input voltages (Table 10.1) available for comparison are the voltages V, IZ_r, their sum $(V + IZ_r)$ and their difference $(V - IZ_r)$. The corresponding current inputs are given by dividing these quantities by Z_r, which is the same thing as multiplying them by Y_r. Because of this simple relationship the inputs will be given only as voltages from now on in order to minimize repetition.

TABLE 10.1

Voltage Inputs for Various Characteristics

Characteristic	Fig. No.	Amplitude Comparator		Phase Comparator (9)	
		Operate	Restrain	Operate	Polarise
Directional	10.7 a	$IZ_r + V$	$V - IZ_r$	IZ_r	V
Impedance	10.12 a	IZ_r	V	$IZ_r - V$	$IZ_r + V$
Angle-impedance (ohm)	10.8 a	$2IZ_r - V$	V	$IZ_r - V$	IZ_r
Angle-admittance (mho)	10.13 a	IZ_r	$2V - IZ_r$	$IZ_r - V$	V
Offset Mho	10.14 a	$I(Z_r - Z_0)$	$2V - I(Z_r - Z_0)$	$IZ_r - V$	$V - IZ_0$

TABLE 10.2

Current Inputs for Various Characteristics

Characteristic	Fig. No.	Amplitude Comparator Operate	Amplitude Comparator Restrain	Phase Comparator (90°) Operate	Phase Comparator (90°) Polarise
Directional	10.7 b	$I + VY_r$	$VY_r - I$	I	VY_r
Impedance	10.12 b	I	VY_r	$I + VY_r$	$I - VY_r$
Angle-impedance (ohm)	10.8 b	$2I - VY_r$	VY_r	I	$I - VY_r$
Angle-admittance (mho)	10.13 b	I	$2VY_r - I$	$I - VY_r$	VY_r
Offset Mho	10.14 b	$2I - V(Y_r + Y_0)$	$VY_r - VY_0$	$I - VY_r$	$VY_0 - I$

The dualities mentioned previously will be apparent from inspection of Tables 10.1 and 10.2.

As has already been stated, circular characteristics passing through the origin of the impedance diagram become straight lines clear of the origin on the admittance diagram, and vice versa. A circular characteristic not passing through the origin of one diagram is a reciprocal circle on the other diagram, i.e. the distance of each point on one circle is the reciprocal of the distance from the origin of the corresponding point on the other circle. Furthermore, each point will be as much lagging the resistance axis as the corresponding point on the other circle was leading the conductance axis, and vice versa.

10.2. TYPES OF DISTANCE RELAYS

In the following section the equations for the characteristics of the various types of distance relays will be derived and they are summarized in Table 10.3. Here again duality is evident since the equation for each characteristic on the admittance diagram can be obtained by substituting Y for Z_r and Y_r for Z in the impedance equation and vice versa. This applies to both types of comparator.

10.2.1. Directional Relay

This is not a distance measuring relay but it has been included (*a*) to complete the mathematical pattern since it is a dual of the impedance relay and (*b*) because it is required for some types of distance relay which are not inherently directional. Its characteristic is a straight line passing through the origin on either the impedance or the admittance diagram. The characteristic results from comparing the voltage inputs V and IZ_r in a phase comparator. It is equally obtained by comparing their sum and difference, $(V + IZ_r)$ and $(V - IZ_r)$, in an amplitude comparator.

In a phase comparator (Fig. 10.7) the relay trips when $90° > (\phi - \theta) > -90°$ where ϕ is the angle between V and I and θ is the phase angle of the replica impedance Z_r. In a product device such as a wattmetric relay or a Hall Effect crystal, the relay trips when $VIZ_r \cos(\phi - \theta) > 0$.

Figure 10.7a shows that, in an amplitude comparator, the same characteristic is obtained if $|Z + Z_r| > |Z - Z_r|$. Z_r is not part of the phase angle measurement and can have any value. However, very great accuracy is required of the auxiliary transformer windings used for the IZ_r inputs (see Fig. 10.6) in order to avoid I^2 bias and consequent loss of directional

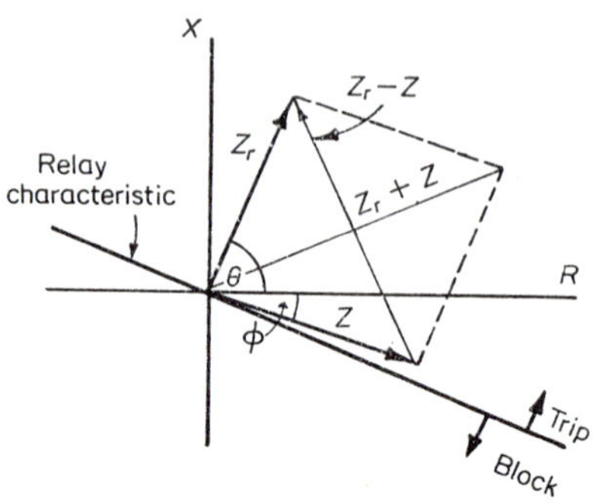

(a) Impedance diagram

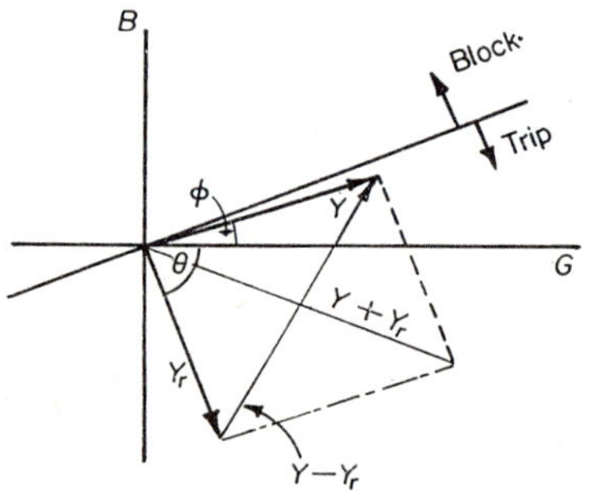

(b) Admittance diagram

Fig. 10.7. Directional Relay Phase comparator trips when $90° > (\phi - \theta) > -90°$ in (a) and (b) Amplitude comparator trips when $|Z_r + Z| > |Z_r - Z|$ in (a) or when $|Y + Y_r| > |Y - Y_r|$ in (b)

properties during a heavy current low-voltage fault. A certain amount of restraining (V^2) voltage bias does no harm and in fact is beneficial in almost all applications; this can be provided by having slightly more ampere-turns in the voltage winding on the $|V - IZ_r|$ side of the amplitude comparator.

10.2.2. Angle-impedance Relay

On an impedance diagram this is a directional characteristic offset from the origin by biasing the voltage with the drop across the replica impedance; see Table 10.1 and Fig. 10.8a.

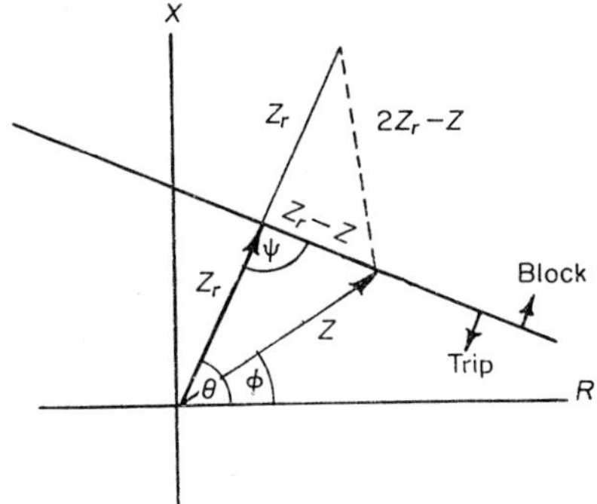

(a) Impedance diagram

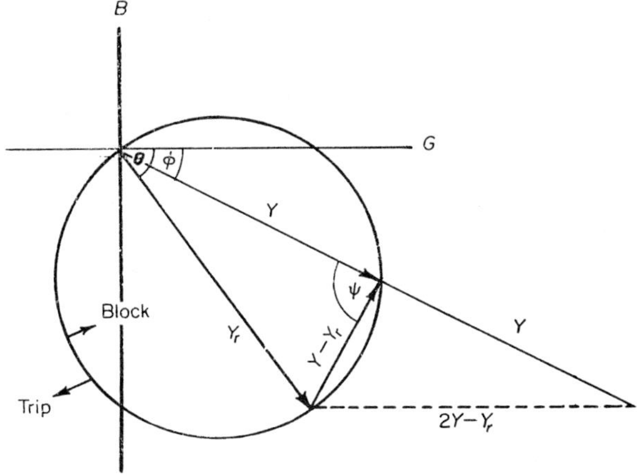

(b) Admittance diagram

Fig. 10.8. Angle-Impedance Relay Phase comparator trips when $\pi > \psi > -\pi$ in (a) and (b) Amplitude Comparator trips when $|Z| > |2Z_r - Z|$ in (a) or when $|2Y - Y_r| > |Y_r|$ in (b)

In a phase comparator $(IZ_r - V)$ is polarized by IZ_r and the relay trips when $90° > \psi > -90°$ where $\psi = \arg Z_r/(Z_r - Z)$. A product device trips when $IZ_r \{IZ_r - V\cos(\phi - \theta)\} > 0$.

In an amplitude comparator the relay trips when $|Z| > |2Z_r - Z|$. Its equivalence to the phase comparator can be seen by squaring both sides, i.e.

$$Z^2 < 4Z_r^2 - 4ZZ_r + Z^2, \text{ i.e. } Z_r(Z_r - Z) > 0,$$

$$\text{i.e. } IZ_r\{IZ_r - V\cos(\phi - \theta)\} > 0.$$

On an admittance diagram the characteristic is a circle passing through the origin whose diameter is equal in length to the reciprocal of the perpendicular from the impedance characteristic to the origin.

It can be seen from Fig. 10.9b that the head of the line admittance vector Y must extend outside the circle to cause tripping. The amplitude comparator trips if $|2Y - Y_r| > |Y_r|$ and the phase comparator trips if $\pi/2 > \psi > -\pi/2$.

10.2.3. Reactance Relay

This is a particular case of the angle impedance relay in which the reactive component of the impedance is measured. Its characteristic (which is the threshold of operation) is parallel to the resistance axis; hence a distance relay with this characteristic measures the same line length regardless of any additional resistance introduced by the fault.

In the phase comparator $I\mathbf{Z}_r$ is compared with $(I\mathbf{Z}_r - V)$. Figure 10.9 shows that the relay trips when $\mathbf{Z}$ is below the characteristic, i.e. when $\psi + \theta < 180°$, where θ is the phase angle of the replica impedance $\mathbf{Z}_r$. If $\mathbf{Z}_r$ were

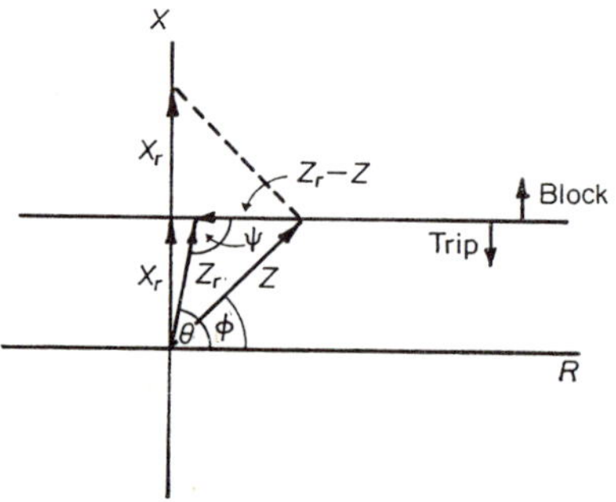

(a) Impedance diagram

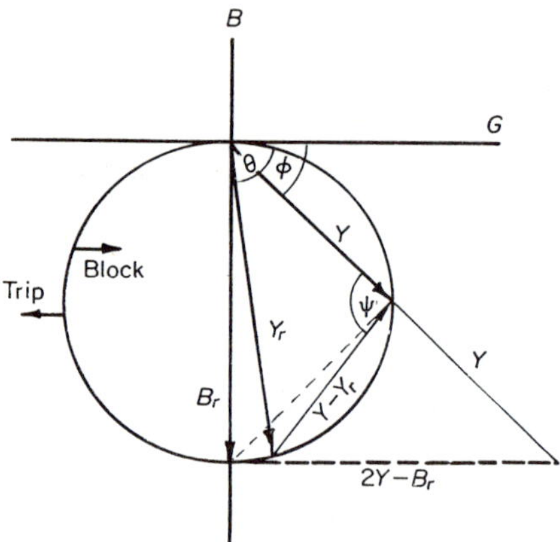

(b) Admittance diagram

Fig. 10.9. Reactance Relay Phase comparator trips when $\pi > (\psi + \theta) > 0$ in (a) and (b) Amplitude comparator trips when $|Z| < |2X_r - Z|$ in (a) or when $|2Y - B_r| > |Y_r|$ in (b)

a pure reactance ψ would be 90° under threshold conditions and the relay would trip when $Z \sin\phi < X_r$ on the impedance diagram or when $Y \sin\phi > B_r$ on the admittance diagram.

The operating angle of the phase comparator is $(\psi + \theta)$ whereas it is $(\psi - \theta)$ in the other relays. This is due to the fact that ψ is a complementary angle. Taking $180° - \psi$ would have remedied this but it would have complicated the diagram; in that case the > sign would have become <.

In an amplitude comparator, Fig. 10.9 shows that the relay trips when $|Z| < |2X_r - Z|$. The voltage inputs to the amplitude comparator are V and $2IZ_r - 2IR_r - V$, where R_r is made equal to the resistance of the replica impedance, thus leaving only its reactive component X_r.

On the admittance diagram the characteristic is a circle whose centre is on the susceptance axis and whose diameter is $Y_r = 1/Z_r$. The amplitude comparator trips when $|2Y - B_r| > |B_r|$.

10.2.4. Restricted Directional Relay

On an impedance diagram this is a line bent at the origin, as shown in Fig. 10.10a. It results from limiting the phase angle comparison to an angle less than 90°. Instead of the relay tripping when $90° > (\phi - \theta) > -90°$, it trips when $\lambda > (\phi - \theta) > -\lambda$.

The input quantities are the same as for the orthodox directional relay, i.e. IZ_r and V in the phase comparator and $(IZ_r + V)$ and $(IZ_r - V)$ in the amplitude comparator. In the latter case the relay trips when $K|IZ_r + V| > |IZ_r - V|$, where

$$K = \sqrt{\left(\frac{Z_r^2 + Z^2 - 2ZZ_r \cos\lambda}{Z_r^2 + Z^2 + 2ZZ_r \cos\lambda}\right)}.$$

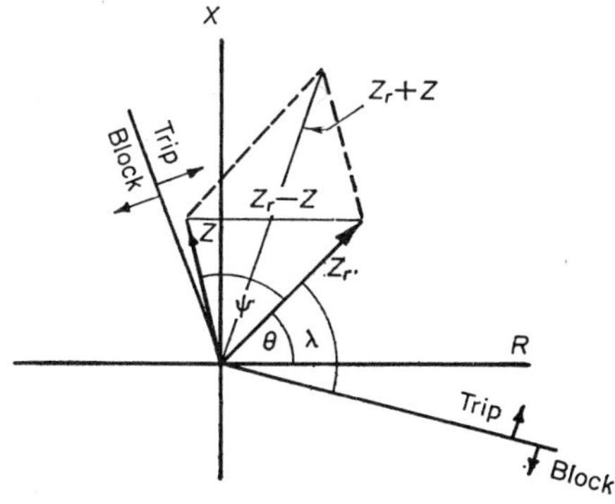

(a) Impedance diagram

Fig. 10.10a Restricted Directional Relay

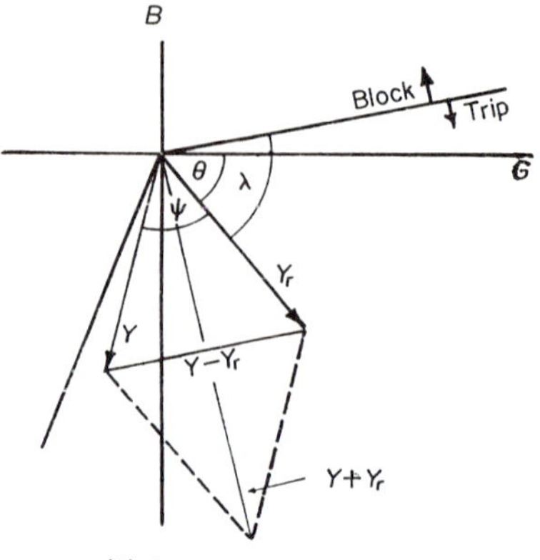

(b) Admittance diagram

Fig. 10.10. Restricted Directional Relay Phase comparator trips when $\lambda > \psi > -\lambda$ in (a) and (b) Amplitude comparator trips when $K|Z_r + Z| > |Z_r - Z|$ in (a) or when $K|Y + Y_r| > |Y - Y_r|$ in (b)

This can be seen in Fig. 10.10 by drawing Z at the threshold, i.e. coincident with the characteristic. It means that λ is not constant but varies with K.

Making $\lambda = 90°$ in the phase comparator equation gives a straight line whose equation is $ZZ_r \cos(\phi - \theta) > 0$, where $\psi = \phi - \theta$.

On the admittance diagram (Fig. 10.10b) the characteristic is also a line bent at the origin. The equations are again duals, with the line and replica terms interchanged.

10.2.5. Restricted Reactance Relay

The characteristic on the impedance plane is a straight line bent at a predetermined point, i.e. it is the characteristic of the restricted directional relay but offset from the origin. On the admittance diagram the characteristic is a couple of circular arcs Fig. 10.11 b. This will be discussed later under Mho Relay (Section 10.2.7).

In Fig. 10.11, in order to prevent tripping on power swings, there can be only a limited tolerance for arc resistance and one part of the characteristic must be parallel to the R axis and the other part parallel to the impedance of the protected line section. This means that θ, the phase angle of the replica impedance Z_r, must be half the phase angle of the line impedance, as shown in Fig. 10.11.

In a phase comparator IZ_r is compared with $(IZ_r - V)$ and tripping occurs when ψ, the angle between them, is between $\pm\theta$, i.e. when $\theta > \psi > -\theta$.

It will be seen that the inputs are similar to those of the normal reactance relay and the tripping criterion is similar except that the limiting values of ψ are $\pm\lambda$ instead of $\pm 90°$.

In an amplitude comparator the inputs are $|V|$ and $|2IZ_r - V|$. Tripping occurs when $|Z| < |2Z'_r - Z|$. This is also similar to the normal reactance relay except that Z'_r has another value besides X_r.

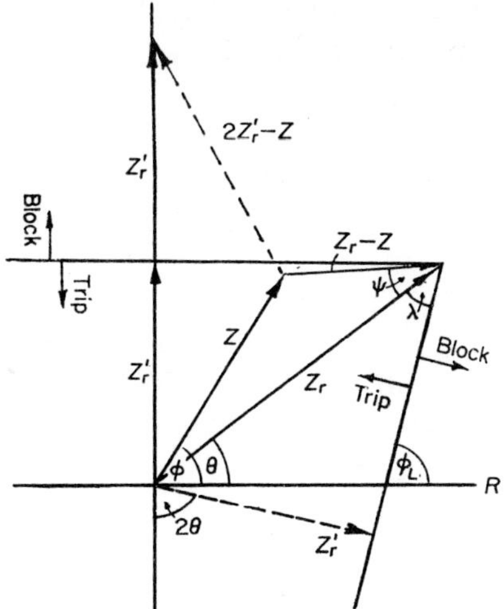

(a) Impedance diagram

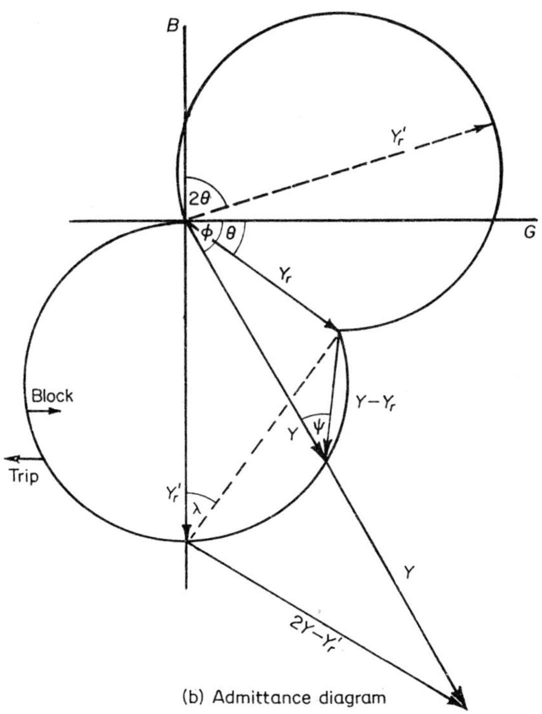

(b) Admittance diagram

Fig. 10.11. Restricted reactance relay Phase comparator trips when $\theta > \psi > -\theta$ in (a) and (b) Amplitude comparator trips when $|Z| < |2Z' - Z|$ where $|Z'| = |Z_r| \sin\theta$ and $\arg Z' = \theta \pm (\pi - \theta)$ in (a) or when $|2Y - Y'| > |Y'|$ where $|Y'| = |Y_r| \sec\theta$ and $\arg Y' = -\theta \pm (\pi - \theta)$ in (b)

On the admittance diagram the characteristic consists of two sections of equal circles. The angular size of the circles is 2θ which is the same as the angle of the bent line characteristic on the impedance diagram. If $\theta = 90°$ the two sectors are semicircles so that the complete characteristic is the normal circle.

10.2.6. Impedance Relay

The simplest circular characteristic is that of the impedance relay, whose centre is at the origin on both the impedance and admittance diagrams, because the pick-up is the same at any angle. It is a mathematical dual of the

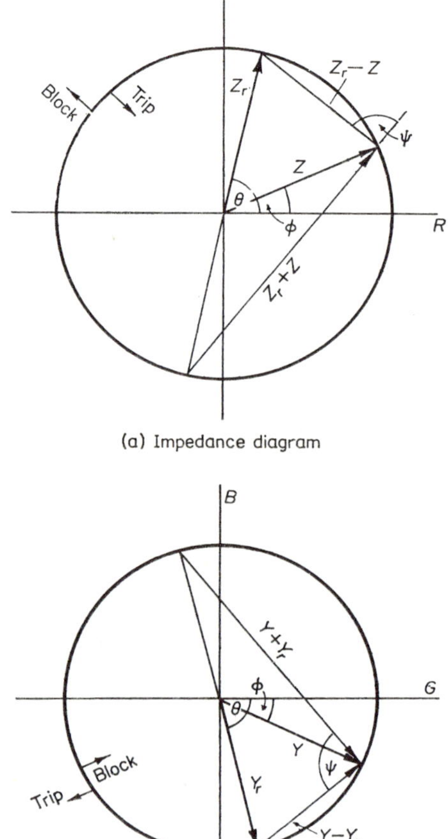

Fig. 10.12. Impedance Relay Phase comparator trips when $\pi > \psi > -\pi$ in (a) and (b) Amplitude comparator trips when $|Z| < |Z_r|$ in (a) or when $|Y| > |Y_r|$ in (b)

directional relay which is the simplest of the relays with straight line characteristics.

Impedance is inherently an amplitude comparison of current and voltage. Figure 10.12 shows that the relay trips when $|Z| < |Z_r|$, so that the inputs required for an amplitude comparator would be V and IZ_r.

Using a phase comparator is more complicated, requiring the sum and difference quantities $(IZ_r + V)$ and $(IZ_r - V)$; if ψ is the angle between the sum and difference inputs, the relay operates when $\pi/2 > \psi > -\pi/2$. Because of this the phase comparator is never used for impedance relays unless an offset characteristic is required. A product relay trips when $(IZ_r + V)(IZ_r - V) \cos \psi < 0$, i.e. when $Z < Z_r$; the $\cos \psi$ is obtained by phase shifting means and is necessary in order to give the greatest sensitivity near pick-up, i.e. when IZ_r only just exceeds V.

There is the usual duality between the equations for the impedance and admittance characteristics. The phase comparator trips when $(Y + Y_r)(Y - Y_r) \cos \psi > 0$ and the amplitude comparator when $|Y| > |Y_r|$.

10.2.7. Angle-admittance (Mho)

On an admittance diagram the characteristic is a straight line offset from the origin in the lagging quadrant (Fig. 10.13b). On an impedance diagram it is a circle passing through the origin, as shown in Fig. 10.13a. This is the inverse of the angle-impedance relay; their duality is also shown in the equations which are given in Table 10.3. It will be seen that the equation of one type on an impedance diagram corresponds to the equation of the other type on an admittance diagram and vice versa.

This characteristic is now so well known that it requires no introduction-It was first introduced by the author for protecting a very long line in the U.S.A. [72] and the term 'mho' was proposed by him as a simplification for angle-admittance. Incidentally, an electrically long line is one in which the length in miles exceeds the kV; this is the condition which makes the interconnection liable to instability.

The mho characteristic is obtained equally conveniently from phase or amplitude comparators. In the phase comparator the inputs are $(IZ_r - V)$ and V and the relay trips when the angle between them is less than 90° (see Fig. 10.13), i.e. when $90° > \psi > -90°$. A product device operates when $V\{IZ_r \cos (\phi - \theta) - V\} > 0$, where ϕ is the angle between V and I, and θ is the phase angle of Z_r, i.e. when $Z < Z_r \cos (\phi - \theta)$.

Comparing it with a directional relay, which is an inherent phase comparator, the V bias subtracted from the IZ_r term curls the straight characteristic up into a circle.

With the amplitude comparator the inputs are $|IZ_r|$ operating and $|2V - IZ_r|$ restraining and the relay trips when $|2Z - Z_r| < |Z_r|$.

Comparing this with the impedance relay, which is an inherent amplitude comparator, the subtraction of the IZ_r bias from the voltage restraint moves the circle over so that it passes through the origin.

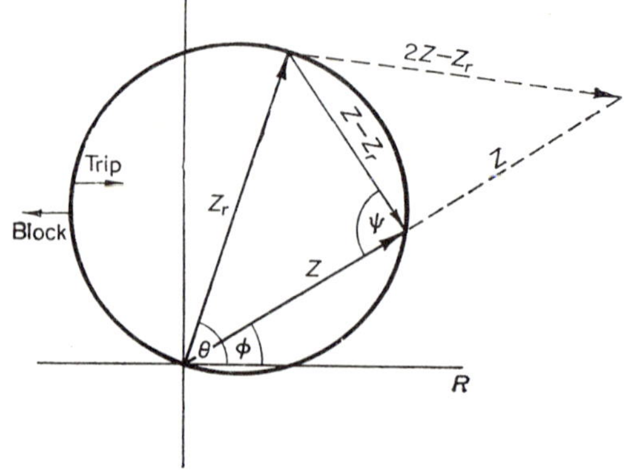

(a) Impedance diagram

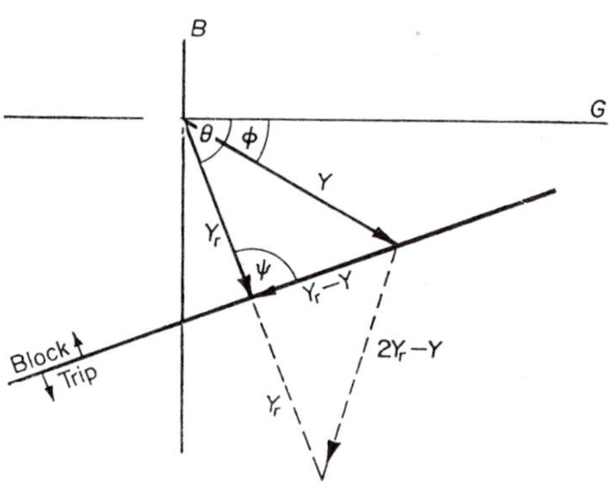

(b) Admittance diagram

Fig. 10.13. Angle-admittance (mho) Relay Phase comparator trips when $3\pi/2 > \psi > \pi/2$ in (a) and (b) Amplitude comparator trips when $|2Z - Z_r| < |Z_r|$ in (a) or when $|Y| > |2Y_r - Y|$ in (b)

The duality between the phase and amplitude comparators can be seen by rationalising the amplitude comparator equation for the impedance relay which gives $4\mathbf{Z}^2 - 4\mathbf{Z}\mathbf{Z}_r + \mathbf{Z}_r^2 < \mathbf{Z}_r^2$ whence $\mathbf{Z}(\mathbf{Z} - \mathbf{Z}_r) < 0$. A product device operates when $V\{IZ_r \cos(\phi - \theta) - V\} > 0$.

Because the directional function is inherently a phase-angle measurement, the phase comparator is the more convenient construction.

Comparing the angle-admittance (mho) relay (Fig. 10.13) with the angle-impedance (ohm) relay (Fig. 10.8) shows clear duality in both characteristics and equations. Each mho equation becomes that of its ohm counterpart if $\mathbf{Z}$ is substituted for $\mathbf{Z}_r$ and vice versa.

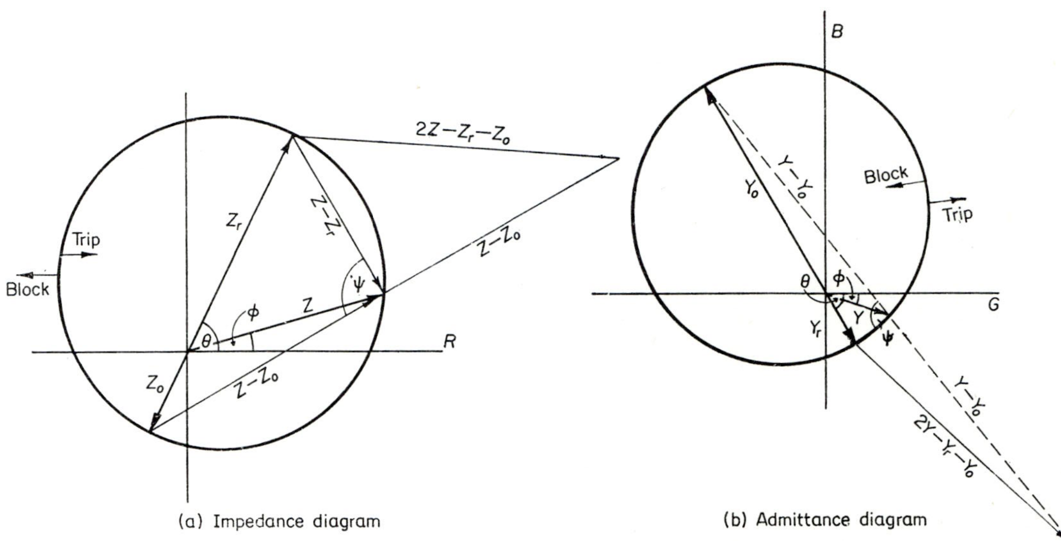

Fig. 10.14. Mho Relay with Negative Offset Phase comparator trips when $(3/2)\pi > \psi > \pi/2$ in (a) and (b) Amplitude comparator trips when $|2Z - Z_r - Z_0| < |Z_r - Z_0|$ in (a) or when $|Y_0 - Y_r| < |2Y - Y_0 - Y_r|$ in (b)

10.2.8. Offset-mho Relay

This characteristic is obtained by biasing either the impedance relay or the mho relay with an additional replica impedance Z_0. The choice depends upon the amount of bias required; for instance, either method is practical for an offset mho characteristic overlapping the origin (negative offset) but, if the origin is outside the mho circle (positive offset), the biased mho unit is more accurate and more economical of c.t. burden.

In Figs. 10.14a and 10.15a, Z_r is the secondary impedance of the protected section of line; Z_0 is the impedance by which the mho circle is offset, i.e. Z_r is the forward reach and Z_0 the backwards reach of the offset mho circle.

The radius of the circle can be seen in Fig. 10.14a to be the mean of the impedances, i.e. $r = (Z_r - Z_0)/2$. It can also be seen that the centre of the circle is displaced from the origin by $c = (Z_r + Z_0)/2$.

If Z is any impedance which is on the threshold of operating the relay, the extremity of the vector Z will be on the circle.

therefore $|Z - c| = |r|$

$$\text{i.e.} \left|Z - \frac{Z_r + Z_0}{2}\right| = \left|\frac{Z_r - Z_0}{2}\right|. \qquad (10.1)$$

This gives the equation of the offset mho circle on an impedance diagram; the two inputs to an *amplitude comparator* to give this characteristic can be obtained (Figs. 10.14b and 10.15b) by multiplying through by $2I$, viz:

$$|2V - I(Z_r + Z_0)| = |I(Z_r - Z_0)|. \qquad (10.2)$$

The equation for a mho relay (not offset) can be obtained by making the offsetting impedance Z_0 equal to 0, viz:

$$|2V - IZ_r| = |IZ_r|. \qquad (10.3)$$

The equation for a *phase comparator* giving the offset mho circle of Eq. (10.1) can be obtained from Fig. 10.14a which shows that the vector $(Z - Z_r)$ must be at right angles to the vector $|Z - Z_0|$ for Z to be on the circle. Hence the threshold of operation of the phase comparator for the same offset mho circle as Fig. 10.14b is when

$$(Z - Z_0)(Z - Z_r) \cos \psi = 0 \qquad (10.4)$$

i.e. if ψ is the angle between the two vectors, operation occurs when $270° > \psi > 90°$ in a rectifier bridge or a Hall Effect phase comparator. In a spike and block phase comparator (which is a sine product device) the relay would operate when $180° > \psi > 0°$.

Equation (10.4) can also be written as $Z^2 - Z(Z_r + Z_0) \cos (\phi - \theta) + Z_r Z_0 = 0$ where $\underline{/\phi}$ is the angle of Z and $\underline{/\theta}$ is the angle of Z_r and Z_0.) Without offset $Z_0 = 0$ and the mho equation becomes $Z = Z_r \cos (\phi - \theta)$ as previously derived for the simple mho relay.

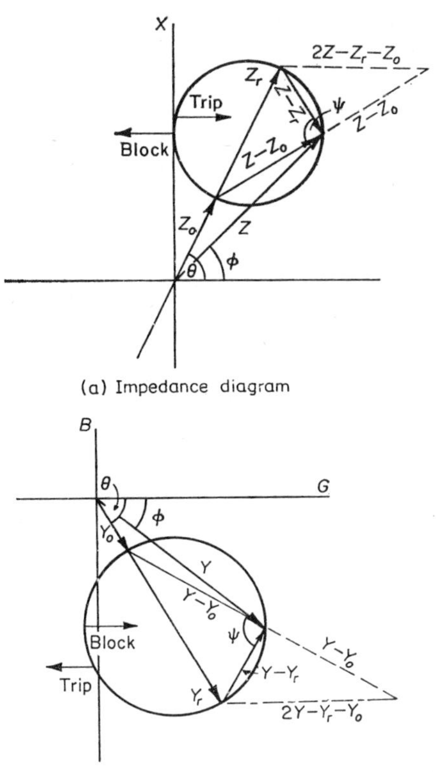

Fig. 10.15. Mho Relay with Positive Offset. Same equations as for Fig. 10.14a and 10.14b

The corresponding inputs and equations for the admittance characteristic can be obtained by the same procedure as given in the first few paragraphs of this section. The equations are $(Y - Y_0)(Y - Y_r) \cos \psi > 0$ for the phase comparator and $|2Y - Y_r - Y_0| > |Y_r - Y_0|$ for the amplitude comparator.

In all the previous types of relays considered there was duality between the impedance and admittance equations in that the replica terms and the line terms were interchanged. In the offset mho, however, the equations are similar, i.e. the line and replica terms are not interchanged; only the $>$ and $<$ are interchanged, which was not so in the previous cases where there was duality.

This apparent discontinuity is due to the fact that there is considerable choice as to the polarity of the complex inputs and those were chosen which gave the simplest diagrams which were the easiest to tie up with the equations. Mathematically the same results are obtained as when inputs are chosen to make the equations complete duals.

(*a*) *Duality of comparators.* It can be shown that the two comparators are equivalent. Squaring both sides of the amplitude comparator Eq. (10.1) gives

$$Z^2 + \frac{Z_r^2}{4} + \frac{Z_0^2}{4} - ZZ_r - ZZ_0 + \frac{Z_rZ_0}{2} = \frac{Z_r^2}{4} - \frac{Z_rZ_0}{2} + \frac{Z_0^2}{4}$$

therefore $Z^2 - ZZ_r - ZZ_0 + Z_rZ_0 = 0$

therefore $(Z - Z_r)(Z - Z_0) = 0$

It will be seen that this cross-product corresponds to equation (10.4) and is characteristic of a product relay with inputs $(V - I\mathbf{Z}_r)$ and $(V - I\mathbf{Z}_0)$. Again the offset disappears when $\mathbf{Z}_0 = 0$, so that the equation for the true mho circle passing through the origin is

$$(Z - Z_r)\, Z_r = 0. \tag{10.5}$$

(*b*) *Alternative method of bias.* It will be observed that the foregoing method of bias keeps the forward reach of the relay constant irrespective of the direction of the offset, and that this is the result of biasing only the polarizing voltage input of the phase comparator to give inputs $(V - IZ_r)$ and $(V - IZ_0)$ corresponding to Eq. (10.4), whereas in the amplitude comparator (Eq. (10.2)] the bias appears in both inputs.

In the induction-cup phase-comparator relays which have been in world-wide use for distance relaying it was usual, in offset mho units, to bias the line voltage. This of course affected the restraint as well as the polarizing quantity, so that the inputs given by equation (10.4) for the phase comparator were changed to $(V - IZ_0 - IZ_r)$ and $(V - IZ_0)$ and tripping occurred when

$$(Z - Z_0 - Z_r)(Z - Z_0) \cos \psi > 0. \tag{10.6}$$

Conversely, to obtain this characteristic from the amplitude comparator, the bias is applied only to one of the inputs, which are IZ_r and $V - IZ_r - IZ_0$. Tripping occurs when

$$|Z_r| > |Z - Z_r - Z_0|. \tag{10.7}$$

10.2.9. Restricted Mho Relay

Just as the characteristic of the reactance relay was modified (Fig. 10.11) by making the operating angle of the phase comparator not equal to 90°, the mho relay characteristic can be similarly modified as shown in Figs. 10.16 and 10.17.

The mho circle consists of two sectors of a circle which are semicircles when $\lambda = 90°$ making an impedance characteristic.

When $\lambda > 90°$, as in Fig. 10.16, the sectors are less than a semicircle and the impedance characteristic becomes narrow, which makes the relay less vulnerable to power swings and hence suitable for long lines.

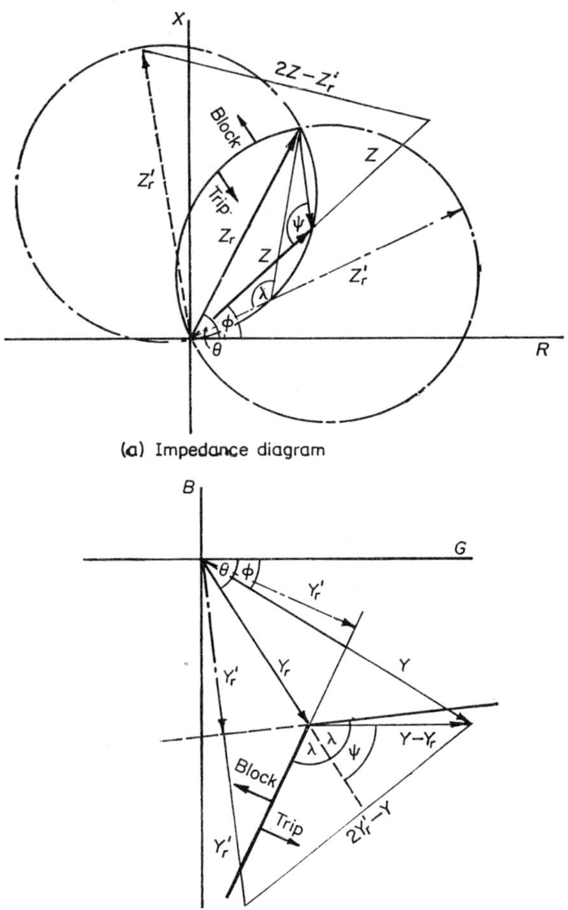

Fig. 10.16a. Mho Relay with Comparator Threshold Angle $>90°$ Phase comparator trips when $2\pi - \lambda > \psi > \lambda$ in (a) and (b) Amplitude comparator trips when $|2Z - Z_r'| < |Z_r'|$ where $Z_r' = Z_r$ cosec λ and arg Z_r' $= \theta \pm (\lambda - \pi/2)$ in (a) or when $|Y| > |2Y_r' - Y|$ where $Y_r' = Y_r \sin \lambda$ and arg $Y_r' = -\theta \pm (\lambda - \pi/2)$ in (b)

When $\lambda < 90°$, as in Fig. 10.17, the sectors are larger than semicircles and the impedance characteristic becomes apple-shaped. This provides more tolerance for fault resistance, which is advantageous for short lines and ground faults.

Here again duality is evident. The corresponding characteristics on the admittance diagram (Figs. 10.16b and 10.17b) are bent straight lines. It will also be seen that Fig. 10.17 is a dual of Fig. 10.11. Examination of Table 10.3 will show a symmetry between the comparator equations, which indicates more duality.

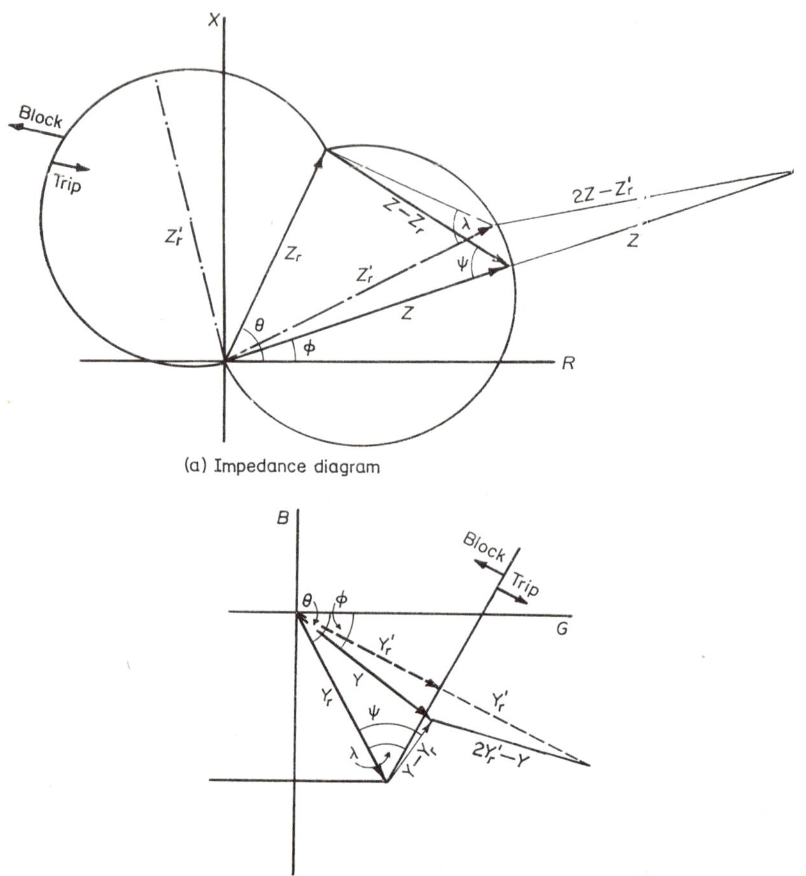

(b) Admittance diagram

Fig. 10.17. Mho Relay with Comparator Threshold Angle $< 90°$ Phase comparator trips when $2\pi - \lambda > \psi > \lambda$ in (a) and (b) Amplitude comparator trips when $|2Z - Z'_r| < |Z'|$ where $Z'_r = Z \operatorname{cosec} \lambda$ and $\arg Z'_r = \theta \pm (\lambda - \pi/2)$ in (a) or when $|Y| > |2Y'_r - Y|$ where $Y'_r - Y_r \sin \lambda$ and $\arg Y'_r = -\theta \pm (\lambda - \pi/2)$ in (b)

In Table 10.3, comparing the phase-comparator equations, it will be seen that, if $\lambda = 90°$ in the equations for the restricted mho relay, they will revert to the normal mho relay equations. Similarly, in the amplitude comparator, making $Z'_r = Z_r$ does the same thing.

The bottom line in Table 10.3 indicates the usual duality between the equations but, in the admittance diagrams of Figs. 10.16b and 10.17b, $(Y - Y_r)$ is compared with Y_r to give the operating condition $Y_r(Y - Y_r) \sin(\psi - \lambda) < 0$. It can be seen that, if the bracketed input $(Y - Y_r)$ had been reversed, the complementary angles of ψ and λ would have been involved and the equation would appear as in Table 10.3.

TABLE 10.3 *Operating Conditions for Distance Relays*

		Impedance Diagram		Admittance Diagram	
Function	Fig. No.	Phase Comparator	Amplitude Comparator	Phase Comparator	Amplitude Comparator
Directional	10.7	$ZZ_r \cos(\phi - \theta) > 0$	$\lvert Z_r + Z\rvert > \lvert Z_r - Z\rvert$	$Y_r Y \cos(\phi - \theta) > 0$	$\lvert Y + Y_r\rvert > \lvert Y - Y_r\rvert$
Angle-impedance (Ohm)	10.8	$Z_r(Z_r - Z) \cos\psi > 0$	$\lvert 2Z_r - Z\rvert > \lvert Z\rvert$	$Y(Y - Y_r) \cos\psi > 0$	$\lvert 2Y - Y_r\rvert > \lvert Y_r\rvert$
Reactance	10.9	$Z_r(Z_r - Z)\sin(\psi+\theta) > 0$	$\lvert 2X_r - Z\rvert > \lvert Z\rvert$ where X_r is reactive component of Z_r	$Y(Y - Y_r)\sin(\psi+\theta) > 0$	$\lvert 2Y - B_r\rvert > \lvert B_r\rvert$ where B_r is reactive component of Y_r
Restricted Directional	10.10	$ZZ_r \sin(\psi - \lambda) > 0$	$K\lvert Z_r + Z\rvert > \lvert Z_r - Z\rvert$	$Y_r Y \sin(\psi - \lambda) > 0$	$K\lvert Y + Y_r\rvert > \lvert Y - Y_r\rvert$
Restricted Reactance $\lambda \neq 90°$	10.11	$Z_r(Z_r - Z)\sin(\psi-\lambda) < 0$	$\lvert 2Z'_r - Z\rvert > \lvert Z\rvert$ where $\lvert Z'_r\rvert = \lvert Z_r\rvert \operatorname{cosec}\theta$ and $\arg Z'_r = \lambda \pm (\lambda - \pi/2)$	$Y(Y - Y_r)\sin(\psi-\lambda) < 0$	$\lvert 2Y - Y'_r\rvert > \lvert Y'_r\rvert$ where $\lvert Y'_r\rvert = \lvert Y_r\rvert \sin\theta$ and $\arg Y_r = \lambda \pm (\lambda - \pi/2)$
Impedance	10.12	$(Z_r + Z)(Z_r - Z)\cos\psi > 0$	$\lvert Z_r\rvert > \lvert Z\rvert$	$(Y + Y_r)(Y - Y_r)\cos\psi > 0$	$\lvert Y\rvert > \lvert Y_r\rvert$
Angle-admittance (Mho)	10.13	$Z(Z - Z_r)\cos\psi < 0$	$\lvert Z_r\rvert > \lvert 2Z - Z_r\rvert$	$Y_r(Y_r - Y)\cos\psi < 0$	$\lvert Y\rvert > \lvert 2Y_r - Y\rvert$
Offset Mho	10.14 and 15	$(Z - Z_0)(Z - Z_r)\cos\psi < 0$	$\lvert Z_r - Z_0\rvert > \lvert 2Z - Z_r - Z_0\rvert$	$(Y - Y_0)(Y - Y_r)\cos\psi > 0$	$\lvert 2Y - Y_r - Y_0\rvert > \lvert Y_r\rvert - Y_0$
Restricted Mho $\lambda \neq 90°$	10.16 and 17	$Z(Z - Z_r)\sin(\psi-\lambda) > 0$	$\lvert Z'_r\rvert > \lvert 2Z - Z'_r\rvert$ where $Z'_r = Z_r \operatorname{cosec}\lambda$ and $\arg Z'_r = \theta \pm (\pi/2 - \lambda)$	$Y_r(Y_r - Y)\sin(\psi-\lambda) > 0$	$\lvert Y\rvert > \lvert 2Y'_r - Y\rvert$ where $Y'_r = Y_r \text{s in}\,\lambda$ and $\arg Y'_r = \theta \pm (\pi/2 - \lambda)$

(1) ψ = angle between the two inputs to the comparator.
(2) θ = angle of the characteristic and of the replica impedance.
(3) λ = threshold angle of the restricted phase comparator.
(4) $K = \sqrt{\left(\dfrac{Z^2 + Z_r^2 - 2ZZ_r\cos\lambda}{Z^2 + Z_r^2 + 2ZZ_r\cos\lambda}\right)}$.

10.2.10. Unsymmetrical Restriction

Humpage and Sabberwal [75] have shown that the phase comparator angle λ need not be the same for the two halves of the characteristic, i.e. the threshold equation can be of the form $\lambda_1 > \psi > -\lambda_2$. The resulting characteristic is two unequal circular sectors whose points of intersection are the origin and Z for non-offset characteristics.

Since the circles have these two points in common, their radii must be different and in fact have the relation

$$\frac{r_1}{r_2} = \frac{\sin\frac{1}{2}\lambda_2^2}{\sin\frac{1}{2}\lambda_1}. \tag{10.8}$$

This disymmetry theoretically could increase the flexibility of applying bent ohmic characteristics to form quadrilaterals but, in practice the technique tends to complicate the circuit of the relay. Consequently it has not been thought necessary to develop the subject in this book.

10.3. SPECIAL CHARACTERISTICS

Many special impedance characteristics have been proposed for distance relays, the best known of which are the swivelling characteristic [77], the elliptical characteristic [74] and the quadrilateral characteristic [78].

10.3.1. The Swivelling Characteristic [77]

Since reactance relays are preferable for short lines and mho relays for long lines, it would be advantageous if the relay could adjust its characteristic to suit the length of line or, still better, the Z_s/Z_L ratio, since the fault resistance increases with diminishing current (increasing Z_s) as well as increasing in effect with decreasing Z_L.

Referring to Fig. 10.18a, the impedance characteristic of the relay should be the circle for low Z_s/Z_L ratios and the straight line (circle of infinite radius) for high Z_s/Z_L ratios. For intermediate values of Z_s/Z_L intermediate curves (not necessarily circles) would be satisfactory. This changing characteristic can be obtained from a mho relay on phase faults by polarizing it with voltage from the leading unfaulted phase pair. This procedure has the added advantage of providing a reliable operating level for interphase faults close to the bus which would otherwise make the voltage input zero. (See also page 257.)

Figure 10.18a shows that there will be some overreaching on high resistance faults for low Z_s/Z_L ratios. This can be avoided by the introduction of non-linearity in the impedance of the cross-polarizing circuit, i.e. the voltage from the unfaulted phase pair.

Swivelling is not so effective with single-phase ground faults because the Z_s/Z_L ratio can be reduced by high ground resistance and there would be a tendency towards the mho characteristic just when the reactance cha-

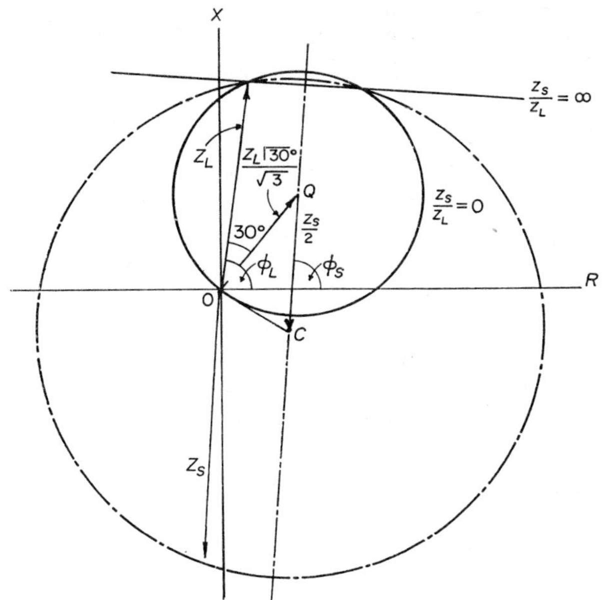

Fig. 10.18a. Swivelling Mho Characteristic
(a) Interphase faults

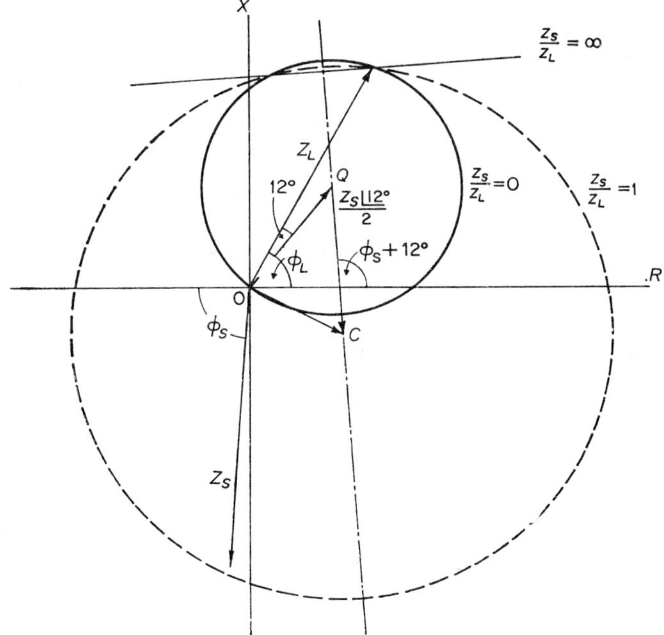

Fig. 10.18b. Swivelling Mho Characteristic
(b) single-phase ground faults

racteristic was essential. Figure 10.18b shows that it nevertheless would be effective on lines with a good ground wire electrically connected to the metal support of the insulators.

10.3.2. Conic Section Characteristics

The impedance characteristics now in general use are circular, i.e. of the ohm or mho type, their equation for the threshold of operation being of the form $Z = Z_r \cos(\phi - \theta)$ for the simple mho relay or $Y = Y_r \cos(\phi - \theta)$ for the simple ohm relay, where Z_r or $1/Y_r$ is the replica impedance.

In certain applications, however, a circular impedance characteristic is unsuitable and has to be modified. For instance, on extremely long lines a mho relay tends to trip on heavy loads unless monitored by a blinder. On the other hand, on short lines the mho relay may not have enough tolerance along the R axis to clear the high resistance faults.

Many of these difficulties can be overcome by the use of other conic characteristic such as ellipses, parabolae, etc. For example, an elliptical impedance characteristic would avoid tripping on heavy loads (see Fig. 10.19). Alternatively, a hyperbolic impedance characteristic can be used as blinders. Conic section characteristics are explained in Chapter 12 which deals with multi-input comparators.

10.3.3. Quadrilateral Characteristics [75, 78]

Theoretically the ideal characteristic for distance relays would be a parallelogram coincident with the fault area of Fig. 10.3. This would include all conditions for which tripping is undesirable.

Such a characteristic would require four electromagnetic relays with their contacts in series and each having a straight-line characteristic corresponding to one side of the parallelogram. In fact this characteristic has been achieved with an induction-cup reactance relay and two blinders, the fourth unit being the mho starting unit of the reactance relay.

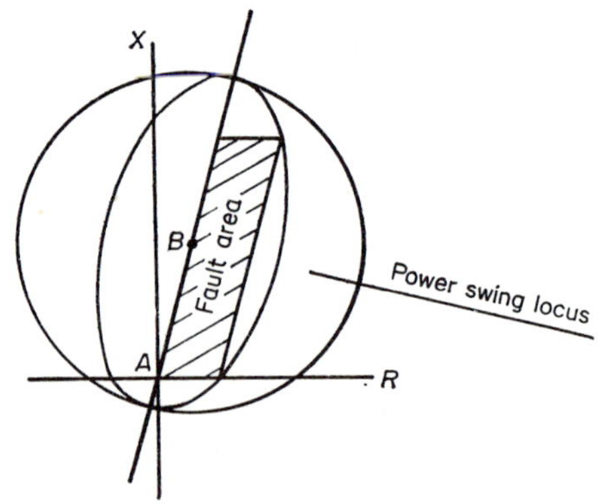

Fig. 10.19. Elliptical Characteristic for back-up zones

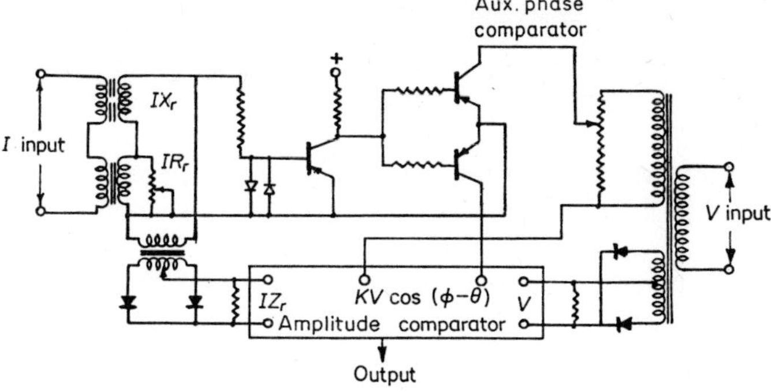

(a) Voltage comparison

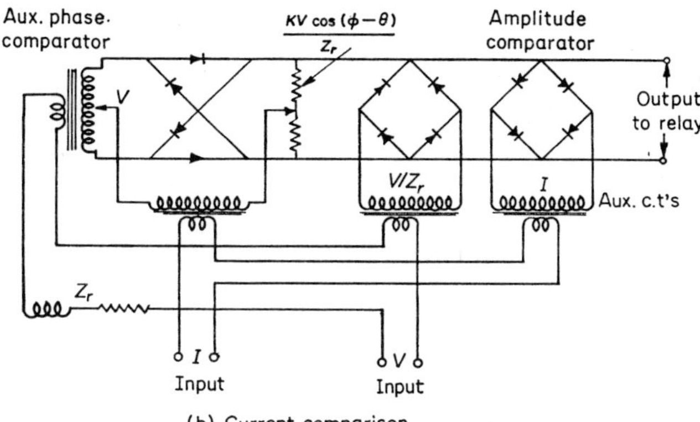

(b) Current comparison

Fig. 10.20. Amplitude comparator using auxiliary phase comparator

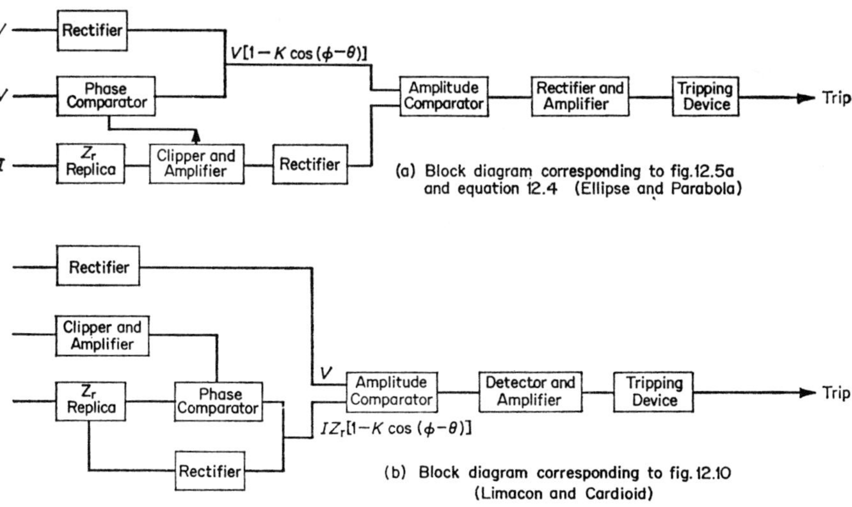

Fig. 10.21. Block diagrams of relays with conic characteristics
(a) for ellipse or parabola
(b) for limaçon or cardiod

With semiconductor circuitry it is easy to obtain a bent directional [75, 78] characteristic (Fig. 10.10a) and to provide the other two sides of the parallelogram from another similar characteristic biased by an offsetting impedance (Fig. 10 11a). This requires two comparators in series but it will be shown in Chapter 12 that it can also be done with a single multi-input comparator.

If the relay characteristic is exactly coincident with the fault area, there is no leeway for error and the relay may fail to operate for a solid fault on the line or for a resistance fault on the bus. Hence it is necessary to widen the characteristic somewhat, as shown in Fig. 10.24.

10.4. CHOICE OF CHARACTERISTIC

The impedance measured by a distance relay corresponds to the length of line from the relay to the fault only if the circuit is free from other series and parallel impedances. Unfortunately, the series impedance of the fault, the parallel impedance of loads and the impedance injected by mutual induction from nearby conductors can cause considerable error, especially in Zones 2 and 3.

Fortunately these conditions are seldom serious enough to cause trouble. In cases where they can do so, special arrangements must be considered, but it is not advisable to complicate standard relays for these few cases which concern chiefly long lines and ground fault relays. On the other hand, the characteristic should be chosen so as to afford the best discrimination between faults in the protected section and all other conditions and it may be necessary to employ different characteristics for the three zones.

10.4.1. Basic Discrimination Functions of Characteristic

The characteristic has three selective functions (*a*) direction, (*b*) discrimination between fault resistance and load impedance and (*c*) distance measurement.

(*a*) *Direction.* This means confining the tripping zone to the first quadrant (Fig. 10.3) although some incursions into quadrants 2 and 4 are permissible (see Chapter 11, Section 11.3).

This is an inherent property of the mho circle which has been used for distance relays since 1930. An alternative characteristic which can be obtained with static relays is the bent straight line of Fig. 10.10 referred to in Table 10.1 as the 'restricted directional unit'.

The chief problem with directional units is to ensure their correct operation for all unsymmetrical faults.

It is easy with static circuits to make a directional unit reliable in action down to 1% of system voltage and this takes care even of balanced three-phase faults because it is well below the minimum arcing fault voltage.

Assuming a minimum conductor spacing of 0·8 ft per kV and 400 volts per ft of arc gives 32 arc volts per kV of system voltage, i.e. 3·2%, which gives a 3 to 1 leeway.

In order to ensure a reliable level of output from a mho phase comparator even for very low voltage (close-up) faults, it is customary in modern relays to maintain the polarizing voltage input either by memory action or by adding to the faulted phase voltage 5% of the voltage from a sound phase, suitably shifted in angle.

The latter is called cross-polarization and the sound phase chosen is the one leading the faulted phase because it moves in the same lagging direction as the fault phase voltage when an arcing fault occurs on a short line section.

This not only provides a minimum polarizing voltage of at least 5%, even during a dead short-circuit near the bus, but also provides more tolerance of fault resistance, as explained in Section 10.3.1. It has negligible effect on the accuracy of distance measurement.

Figure 10.25 shows two methods of cross-polarization, one using a wye voltage and the other a delta voltage. It will be seen that for a phase-to-phase fault the wye-voltage method causes no change in the angle of the relay characteristic, but the delta method inclines it in the direction which makes it more tolerant to fault resistance, the effect being similar to that described in Section 10.3.1.

For ground faults it is customary to obtain the cross-polarization component from the delta voltage in quadrature with the phase with which the relay is associated. This is not affected by a phase-to-ground fault for which the relay should operate.

Cross-polarization does not work on balanced three-phase faults but this is not important because (*a*) three-phase faults rarely have balanced phase potentials, (*b*) modern relays operate reliably down to the minimum arc voltages, (*c*) ground-clamp faults can be taken care of by the method described in the next paragraph.

If the line grounding clamps have been accidentally left on when the line is reconnected, this can create a zero voltage three-phase fault when the breaker is reclosed. If the p.t's are on the bus this can be taken care of by memory action (i.e. series tuning the polarizing voltage circuit of the relay). If the p.t's are on the line side of the breaker it is necessary to provide an overcurrent trip feature which is removed when voltage is restored. This is shown in Fig. 10.27; the third zone unit M_3 trips directly through a NOT gate if the line voltage was zero before the fault.

(*b*) *Selectivity between faults and loads.* During power swings the impedance seen by distance relays is reduced so as to be comparable with that of a high resistance fault.

Since fault resistance is not related to the length of line, the best means of discriminating between these two conditions is the blinder, i.e. a blocking relay characteristic on the impedance diagram which is parallel to the line impedance (Fig. 10.22) or slightly leading this position (Fig. 10.3).

With the mho relay circle this discrimination is inherent [29], except in some cases where the mho circles for the back-up zones have to be enlarged due to a large power infeed at the next bus, or special cases of abnormally ling lines [79].

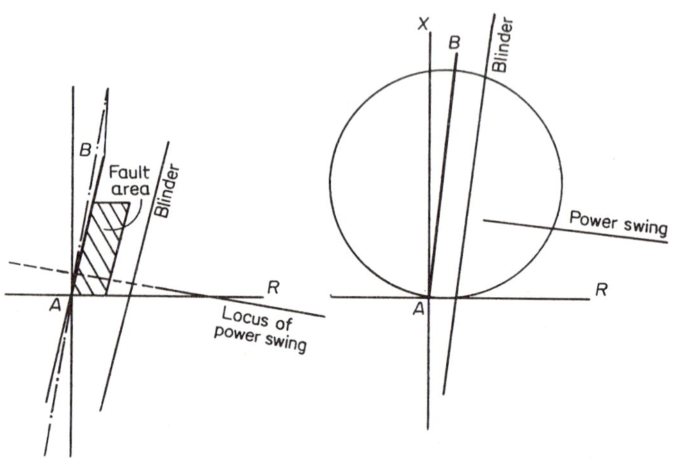

(a) Distinction between faults and power swings

(b) Blinder used with mho relay

Fig. 10.22. The blinder

The blinder can be made still more effective if it is automatically moved away from the origin along the R axis as the voltage decreases. This will increase the relay's tolerance to fault resistance while still blocking it on swings. On unstable systems out-of-step blocking can be provided, i.e. a relay which blocks tripping for non-instantaneous changes in system impedance (Vol. I section 5.4.6.).

(*c*) *Distance measurement.* The mho relay also has inherent ability to measure the impedance between the relay and the fault. Unfortunately it tends to under reach for resistance faults near the end of the protected section.

Since only the reactive component of the faulted circuit impedance is proportional to the distance or line length, irrespective of the fault resistance, it follows that a reactance characteristic is the most desirable.

10.4.2. Overall Characteristic

The mho relay inherently possesses all three discriminating functions mentioned above because the circular characteristic is arranged to fit as snugly as possible around the fault area. On very long lines or very short lines, however, the fault area does not fit easily into a circle and the relay is either vulnerable to tripping on power swings or intolerant to fault resistance.

Hence the ideal characteristic is a quadrilateral on the impedance diagram with its reaches along the X and R axes separately adjustable.

(*a*) *Zone 1.* This is the most important part of distance protection and has the greatest need for speed and accuracy. Fortunately the main sources of error are a minimum because the generating sources do not have time to swing apart during the Zone 1 time of 1 or 2 cycles; see Chapter 11.

With a mho circle it is important to have the angle of the characteristic not more than 75° lagging the R axis in order to ensure correct directional discrimination on a resistance fault close to the bus. The same precaution applies to the mho starting unit of the reactance relay.

It is also important not to move the mho circle too much the other way. Static overreaching will occur on resistance faults if $\theta < 2\phi - \frac{1}{2}\pi$, where θ is the angle of the mho circle and ϕ is the phase angle of the protected line. This can be demonstrated by drawing radii to the origin and to the point where the circle intersects the line impedance vector.

(*b*) *Zone 2.* The purpose of Zone 2 is chiefly to protect the small zone between Zone 1 and the end of the section. There is however greater possibility of error in defining the cut-off point (important in a short adjacent line section) because the Zone 2 delay gives more time for an arc to lengthen and for generators to swing apart.

On the other hand, the need for precise distance measurement is less because Zone 2 can terminate anywhere in the next bus section without affecting selectivity unless the adjacent line section is very short compared with the protected section.

Zone 2 need not be directional. In fact it is preferable to allow Zone 2 to reach a little way in the reverse direction so as to give back-up protection for a fault in the neighbourhood of the local bus.

Ground relays, in protecting single lines without ground wires against single-phase ground faults, may have errors due to double-end feed which may be unacceptably large. In such cases alternative solutions should be examined. Some of these are:

(*a*) The use of inverse time and instantaneous zero sequence current relays.
(*b*) Transferred tripping initiated by Zone 1.
(*c*) Initial extension of Zone 1 to reach just beyond the next bus. Applicable only with automatic reclosing.
(*d*) Breaker back-up where all the other breakers are tripped in $\frac{1}{2}$ second if the breaker in the faulted line sticks.

(*c*) *Zone 3.* The function of Zone 3 is back-up protection in case a relay or a breaker in the next bus section fails to clear a fault.

Inaccuracy due to loads and power in-feeds are still worse with Zone 3 because there are two buses between the relay and the cut-off point of the relay. Underreaching can be caused by power in-feeds and the relays can trip undesirably for a fault on the other side of a large distribution transformer.

Reversing the characteristic so that back-up is now provided by the relay at the other end of the protected section [29] alleviates these problems considerably and the application of breaker back-up becomes easier. Large transformers can have directional relays which can conveniently be arranged to block the Zone 3 unit for a fault beyond the transformer, since both relays are in the same station.

Error due to angular separation of the e.m.f.'s of the power sources can be even greater than for Zone 2 because of the longer time setting of Zone 3. Since this applies particularly to single-phase ground faults it is advisable to provide a zero sequence current back-up relay. On the other hand, it is unlikely that a severe power swing can be caused by any single-phase ground fault.

10.4.3. Choice of Characteristic for the Zones

For phase faults the most economical arrangement is mho units for Zones 1 and 2 and an offset mho unit for Zone 3. The offset unit can be polarized in the outgoing direction where the protected section is not too long and can be used for initiating power swing blocking (Fig. 10.23a). Where the protected section is long it is preferable to reverse Zone 3 (Fig. 10.23b) so as to use a smaller circle and hence facilitate discrimination between faults and loads. In this case the Zone 3 unit is located at B.

For short lines and for ground fault relays the reactance or quadrilateral characteristic is preferable, although this makes the relay slightly more expensive. The three-step arrangement is shown in Fig. 10.24. The reactance measuring side of the characteristic may have to be inclined downwards for

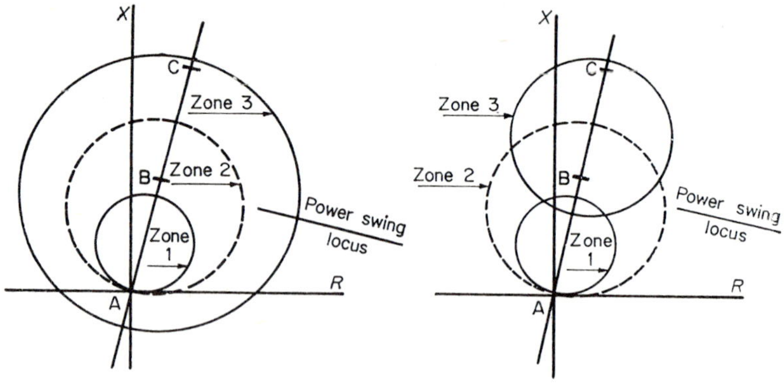

Fig. 10.23. Three-step mho characteristic

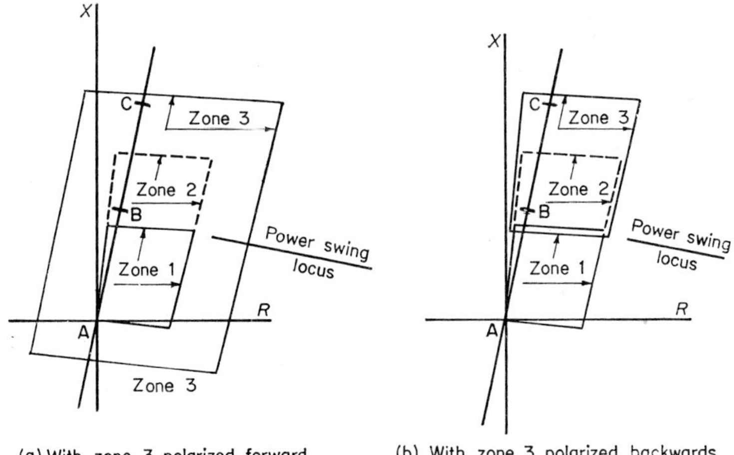

(a) With zone 3 polarized forward (b) With zone 3 polarized backwards

Fig. 10.24. Three-step quadrilateral characteristic

relays polarized in the direction of the power flow and 5° upwards for the relays at the other end of each section. The reasons for this are given in Chapter 11. In practice it is generally preferable for all characteristics to be inclined downwards (as shown in Fig. 10.24) with increasing fault resistance.

10.5. STATIC DISTANCE RELAYS

Static distance relays are similar in basic principles to those of the electromagnetic type, but their lack of moving parts enables them to operate faster without fear of incorrect tripping. It is also possible with semiconductor circuits to obtain impedance characteristics other than the traditional sectors of circles.

The measurement of impedance is done in comparators of the type described in Chapter 4. Since there are no contacts, the switching between the fault detectors, measuring units and timing units is done by biased transistors whose collector-emitter resistance is varied from near zero to near infinity to have the same effect as a contact.

Plate 4(a) shows a typical transistorized distance relay made in England. One of these relays provides three-phase, three-step distance protection for phase faults and another similar unit can be used for ground faults with the addition of a zero sequence current compensator. Plate 4(b) shows the fold-out construction used in the relay for convenience of inspection and maintenance testing.

Static distance relays are accurate over a wider range of fault current and line length and have much lower burdens than their electromagnetic predecessors of the induction cup type. The relay shown in Plate 4 has a potential burden of 8 VA and a current burden (at rated current) of 0·2 to 5 VA depending upon the ohmic setting; it is adjustable from $\frac{1}{4}$ to 32 ohms with

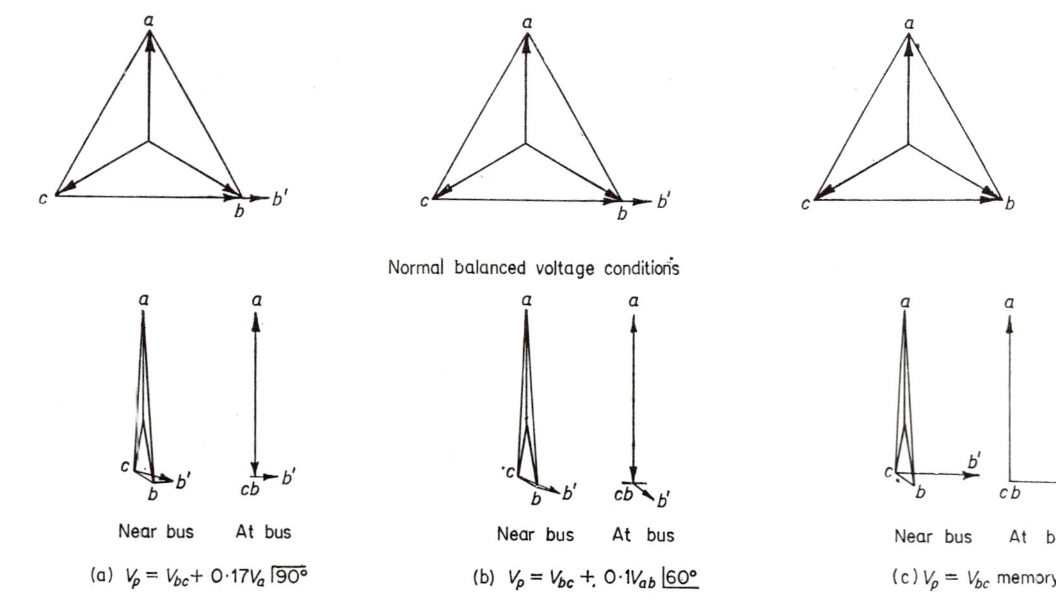

Fig. 10.25. Cross-polarization versus memory action cross-polarized voltage is bb'; total relay polarized voltage V_p is cb'

1 amp c.t's; the phase angle of the characteristic is continuously adjustable from 30° to 80° lag.

A complete three-phase three-step relay is housed in a drawout case 14 × 18 × 7 inches and weighs less than 60 lb. The operating time is almost constant and is less than 1 cycle up to 90% of the ohmic setting for almost all fault conditions. It uses an integrating type phase comparator so that it is immune to transient c.t. and p.t. errors.

The basic units of a modern static three-sep distance relay are:

(*a*) the d.c. power supply,
(*b*) auxiliary tapped transformers for control of characteristic,
(*c*) fast impedance measuring units,
(*d*) a two-step timing unit,
(*e*) a tripping circuit.

10.5.1. d.c. Connections

Figure 10.26 shows the block diagram of a typical distance relay for phase faults using standard logic symbols. In the time delay squares the number above the diagonal means the operate time in milliseconds and the lower figure is the reset time. '*A*' means adjustable and '*I*' instantaneous. The flow is downward or to the right, except where indicated in the opposite direction by an arrow.

Three mho units M_1/M_2 are shown, one for each phase pair, normally having a Zone 1 setting but transferred to Zone 2 by the timing unit A_2/I.

Both A_2/I and A_3/I timers are started by one of the 3rd zone offset mho units OM_3 so that tripping can occur in Zones 1, 2 or 3.

Two auxiliary impedance units Z_4 and Z_5 block tripping on severe power swing conditions. Z_5 has a higher ohmic setting than Z_4 and, if a power swing occurs, Z_5 will operate more than 50 mS before Z_4 so that tripping by M_1/M_2 is blocked. This blocking is maintained by a feed-back as shown by the arrow until Z_5 resets. On the other hand, a fault will operate Z_4 and Z_5 together so there will be no blocking signal and M_1/M_2 will trip.

Where separate M_1 and M_2 units are used, the trip bus of the M_1 units can be used for the Z_4 function. Similarly, where OM_3 is not reversed it can include the duty of Z_5.

Normally OM_3 can trip only through A_3/I but, if there is zero voltage such as when closing the breaker on a dead short, M_3 can trip directly through the AND circuit at the bottom of the diagram since the NOT unit will produce a signal when V is zero.

The circuitry for a d.c. power supply and the various types of comparator have already been discussed in Chapter 3, 4 and 5.

10.5.2. a.c. Connections

Typical connections of replica impedances and the tapped auxiliary transformers for control of the impedance characteristic are shown in Fig. 10.27. Only one relay is shown but the others are similarly connected to their

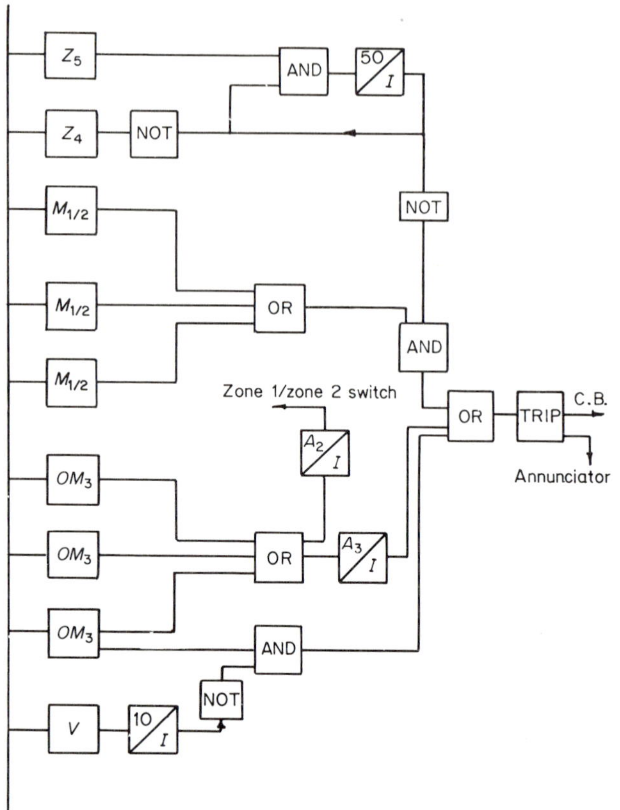

Fig. 10.26. Block diagram of static distance relay (E.E.Co.)

appropriate phases. The connections are shown for a delta-connected relay for interphase faults. The ground relays have wye connections and a fourth auxiliary c.t. and replica impedance can be connected in the neutral wire of the c.t's. V_{SF} is the cross-polarising voltage from the sound phase.

Semiconductor limiters can be connected across the inputs to prevent damage during very heavy short-circuits. It will be seen that the polarizing voltage is augmented by voltage from another phase.

10.5.3. Comparators for Quadrilateral Characteristic

Comparators for the traditional *X*, *Y* and *Z* characteristics have been discussed in Chapter 4. The modern tendency is towards the quadrilateral characteristic which requires either a combination of two phase comparators or a four-input comparator.

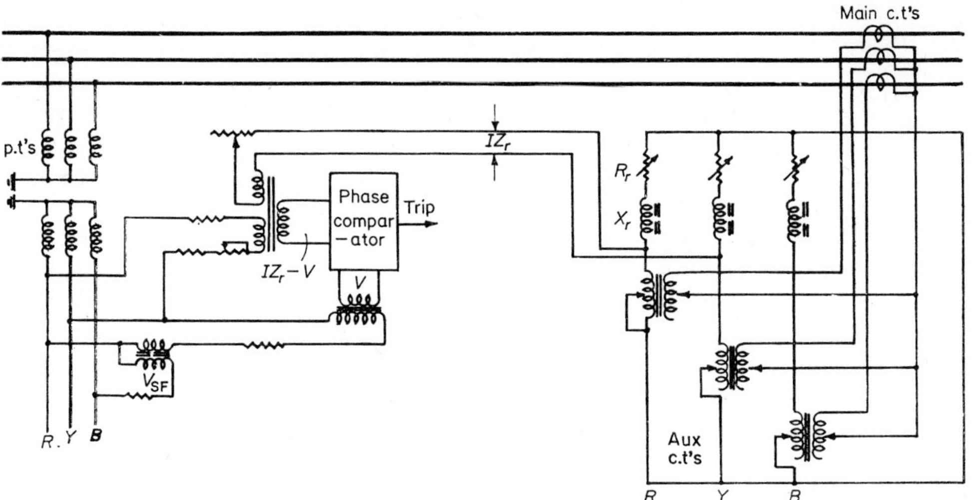

Fig. 10.27. a.c. connections of static distance relay (E.E.Co.)

One way to obtain this characteristic is to provide two restricted directional units, one offset from the origin by the replica impedance Z_r. The relay trips if the impedance Z seen by the relay lies between the two relay characteristics, i.e. within the area OBAC on the impedance diagram Fig. 10.28). Hence Z must be between OB and OC and $Z - Z_r$ must be between AB and AC.

Such a comparator is not entirely free from transient overreach for faults occurring at voltage zero because Z_r is not a true replica of the line impedance. This is not serious, however, because it is extremely unlikely that a fault will be initiated at voltage zero. Furthermore, the overreach tends to be reduced when Z_r is less lagging than Z_L.

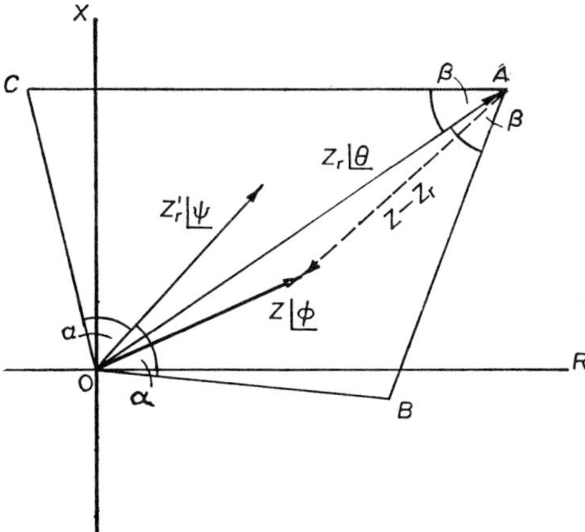

Fig. 10.28. Quadrilateral–characteristic

10.6. LINEAR COUPLER RELAYS [80]

The substitution of linear couplers for the line c.t's normally supplying a distance relay is an interesting idea because of its simpler circuitry. It is subject to errors due to magnetic interference due to nearby primary conductors, although it is claimed that these can be made negligible by proper design of the coupler and its housing. A preferable alternative is the air gap core or a powdered iron core.

Since the output of the linear coupler is a voltage in quadrature with the primary current, the replica impedance is not necessary in the relay circuit and the phase angle of the characteristic can be adjusted by a resistor in the potential circuit in parallel with a small capacitor. Since both the current and potential circuits are practically resistance circuits, fast transient-free operation is possible.

10.6.1. Mho Relay

Figure 10.29 shows the basic measuring circuit for a mho relay using a rectifier bridge phase comparator. For an offset mho relay the cross-polarizing voltage is not necessary and this winding is supplied from the linear coupler output voltage through a resistor which controls the offset (see dotted lines in Fig. 10.29).

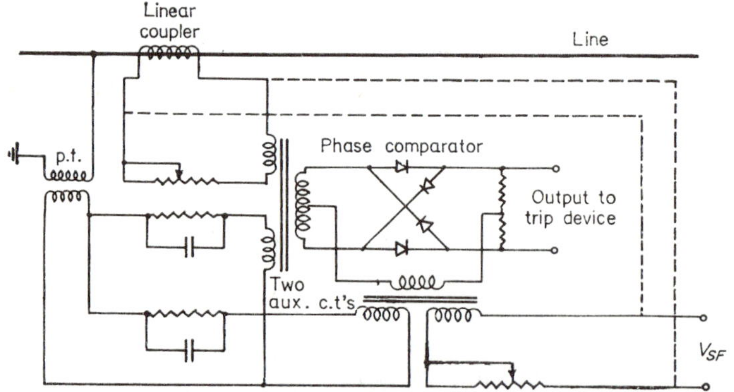

Fig. 10.29. Mho relay supplied from linear coupler
V_{SF} = voltage from sound phase for cross-polarisation

10.6.2. Reactance Relay

The corresponding circuit for the reactance relay is shown in Fig. 10.30. It is even simpler because the phase shift necessary to measure $(V/I) \sin \phi$ is inherent in the linear coupler.

In both the reactance and the mho relay the output power is small and, if used with an electromagnetic tripping relay, a stage of transistor amplification is necessary; otherwise a non-linear resistor is required in the restraining voltage circuit to provide the necessary compensation to maintain a constant impedance pick-up even at low voltage. A popular type of non-linear resistor is two diodes connected in parallel head to tail.

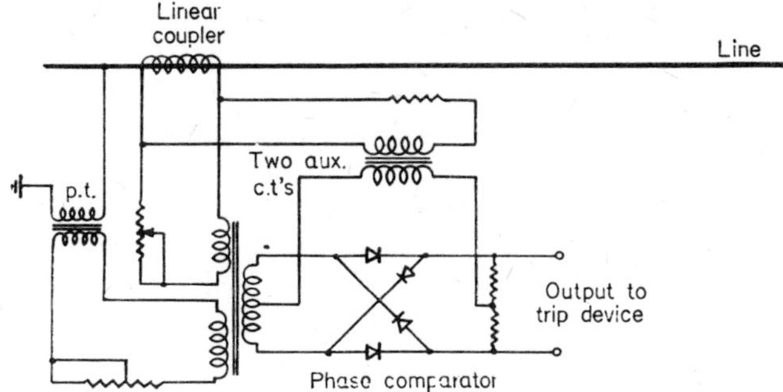

Fig. 10.30. Reactance relay supplied from linear coupler

10.7. FAILURE OF A. C. POTENTIAL SUPPLY

In any distance relay tripping occurs when the faulted phase voltage falls below a predetermined value relative to the faulted phase current. The relay will also trip if the voltage is lost accidentally and this of course is highly undesirable.

If this accidental loss of voltage is due to a potential fuse blowing, wrong tripping can be prevented by very fast overvoltage relays connected across each fuse with their contacts in series with the trip circuit. With the fuse intact, the voltage across it would be insufficient to operate the fuse failure relay.

Since there are other ways in which the a.c. voltage supply can be accidentally lost, it is preferable to monitor the distance relay with a fault detector. This can be an instantaneous overcurrent relay in cases where the fault current is always greater than the load current.

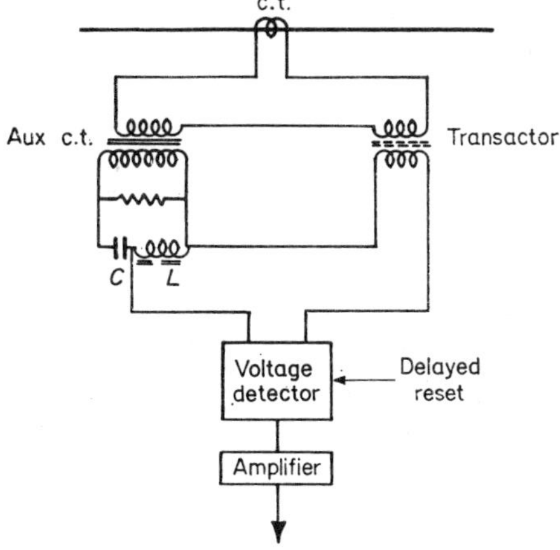

Fig. 10.31. High-speed fault detector responding to change in magnitude or phase angle of the current

On very long lines or very high voltage lines the current may not always increase when a fault occurs, but in that case it will have a sudden phase shift. Figure 10.31 shows the simplified circuit of a high-speed fault detector which will respond to a sudden change in either magnitude or phase angle of the current. The detector is supplied with the difference between the voltages across two inductances which should normally be equal and in phase. Since these two voltages are derived from circuits with different time constants, a sudden change in the primary current will cause a difference in their magnitudes and/or phase angle and the detector will respond to this difference and complete the tripping circuit of the distance relay.

10.8. OPERATING TIMES

With the increasing magnitude of short-circuit MVA in modern power systems, shorter fault clearing times are becoming necessary. Circuit-breaker clearing times have been reduced to less than $2\frac{1}{2}$ cycles and it would appear that, if such speeds have been attained by these large mechanisms, then the relatively tiny relay should be able to operate in a fraction of a cycle.

Unfortunately, the relay is not a slave device like the circuit-breaker. It has to perform a complex measuring function as well as operate contacts. Precise measurement takes a finite time because of the independent alternations of the current and potential. It has to be further delayed in systems of nonhomogeneous impedance, i.e. where the source impedance is more lagging than the impedance of the faulted line; the reasons were given in Chapter 4.

Starting relays, fault detectors and carrier control relays can be designed to operate in a minimum time of less than half a cycle because their accuracy does not have to be better than $\pm 25\%$. Precision relays, such as the Zone 1 unit of a distance relay, should have an accuracy of $\pm 5\%$ or better and it is possible for such relays to operate in half a cycle only on homogeneous systems.

Unfortunately, power systems rarely have a homogeneous impedance (except those at 400 kV or higher) and the difference in the X/R ratio between the source impedance and the linc impedance is increased by the resistance of the fault itself. For a fault occurring near voltage zero this causes a potential transient which is difficult to compensate for and which will cause overreaching unless the measurement is integrated over a full cycle (as explained in Chapter 4) for faults near the threshold value of impedance.

The mho relay is less subject to this trouble than the reactance relay because fault resistance tends to make it underreach and thus permits some transient overreach without risk of tripping on faults outside the protected section if the faulted circuit impedance is less lagging than the relay characteristic. But in either case it is safer to avoid overreach by delaying tripping until the relay has integrated the measurement over a full cycle, since the offset causes over- and underreaching in alternate half-cycles.

11

Steady-state Sources of Distance Relay Error

Effect of fault resistance with double-end feed – Effect of power transfer and power swings – Parallel-lines – Transients in fault current

11.1. IMPEDANCE SEEN BY DISTANCE RELAYS

The impedance measured by a distance relay during a fault consists of Z_L, the impedance of the line between the relay and the fault, and R_f, the resistance of the fault itself (Fig. 11.1). The relay measures this impedance correctly if the current of the protected line, and hence the relay, is the same as the current in the fault. Unfortunately, this can be true only for lines with single-end feed and no load.

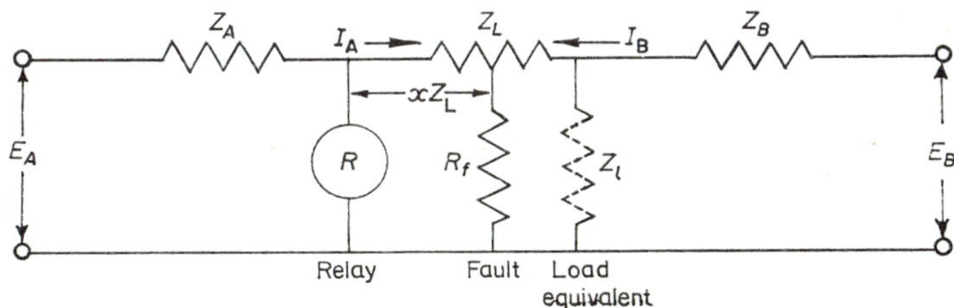

Fig. 11.1. Simplified circuit for fault and load

Where the resistance R_f of the fault path is appreciable, measurement error can be caused by (*a*) load current which goes through the relay and not through the fault (Fig. 11.1) and (*b*) fault current from a power source beyond the fault which does not appear in the relay (Fig. 11.2a). Further errors can be caused by the effect of the fault upon the sound phases, by mutual induction from parallel conductors and by c.t. and p.t. inaccuracy.

The success of distance relays indicates that these errors are not often serious and it is now proposed to examine them to determine when the errors are negligible, when compensation can be applied and when distance relays should not be used at all.

The error in reactance measurement is $\frac{I_f}{I_A} R_f \sin \alpha$ (Fig. 11.2b) at end A. For a homogeneous system the angle θ between E_A and E_B is the same as the angle between I_A and I_B, so that $\theta = \alpha + \beta$. On such a system it can be seen from Fig. 11.2b that the error at A can also be expressed as $\frac{I_B}{I_A} R_f \sin\theta$ and the error at B is $\frac{I_A}{I_B} R_f \sin\theta$. This may be more convenient since it may not be possible to measure α or β.

At the outgoing end A, θ is negative which causes the relay to overreach. At the incoming end B, θ is positive causing underreaching.

Figs. 11.9, 10, 14 and 15 are from a C.I.G.R.E. report [80] and Figs. 11.3, 4, 7, 11, 12 and 13 are from I.E.E. papers [79 and 117].

11.2. ANALYSIS OF ERRORS DUE TO GROUND FAULT RESISTANCE

In general the error due to fault resistance is negligible with phase distance relays, i.e. those which operate on interphase faults. This is because the fault resistance can only be that of the arc and this is normally limited to a low value by the conductor spacing (see Section 11.6 (Appendix I). A similar condition exists with single-phase ground faults on lines having at least one good ground wire with a good electrical connection from the insulator support to the ground wire (Fig. 11.2c).

On the other hand, where there is no ground wire or where there is a poor electrical connection to it, a single-phase ground fault may have high resistance, unless the resistance of the earth and the tower footing is low. If the power system is grounded only at one point, the fault resistance will add to the line impedance measured by the relay and cause underreaching in a mho relay. If there is multiple grounding, the resistance of the arc and the tower will be traversed by currents from both ends of the system and, if

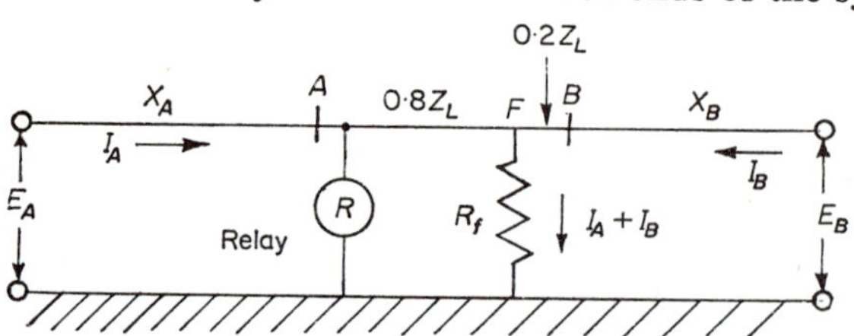

Fig. 11.2a. Schematic curcuit

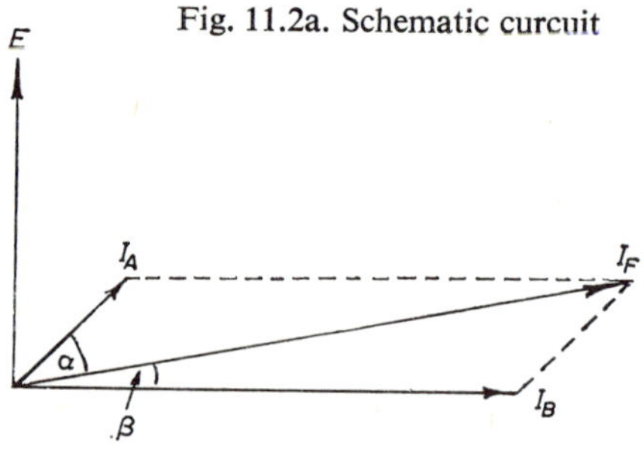

Fig. 11.2b. Vector diagram of fault and line currents

Fig. 11.2. Double-end feed through fault resistance

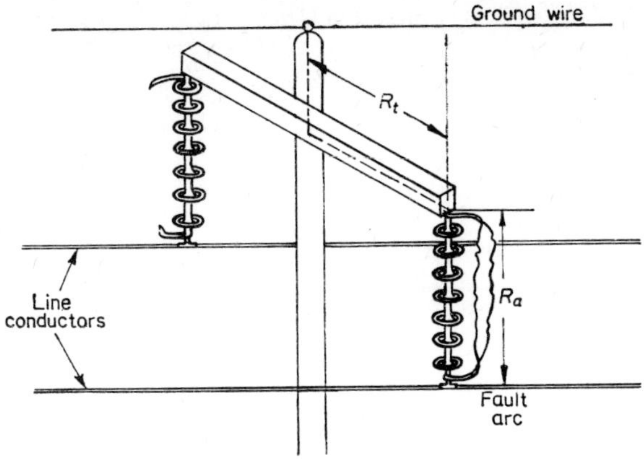

Fig. 11.2c. Fault path to ground wire

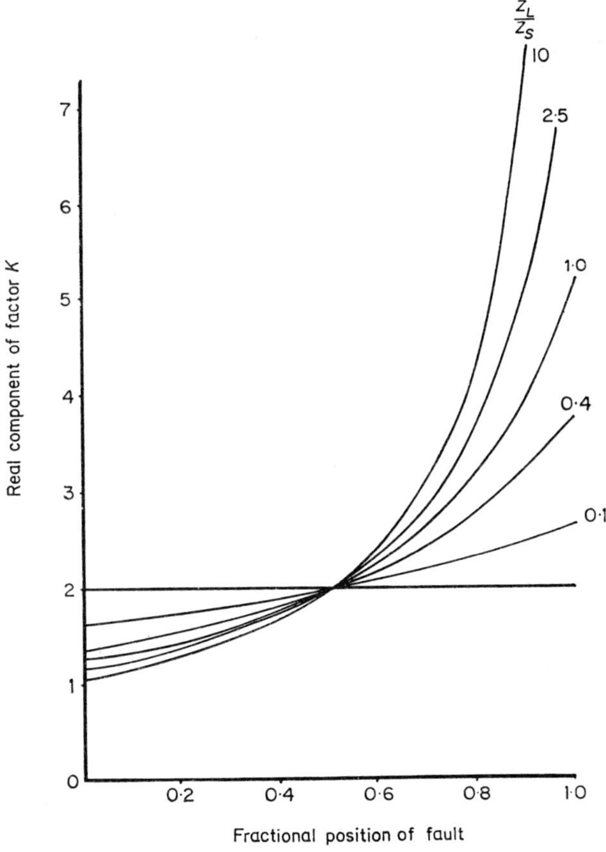

Fig. 11.3a. Variation of k with line/source impedance ratio
K = fault current/relay current

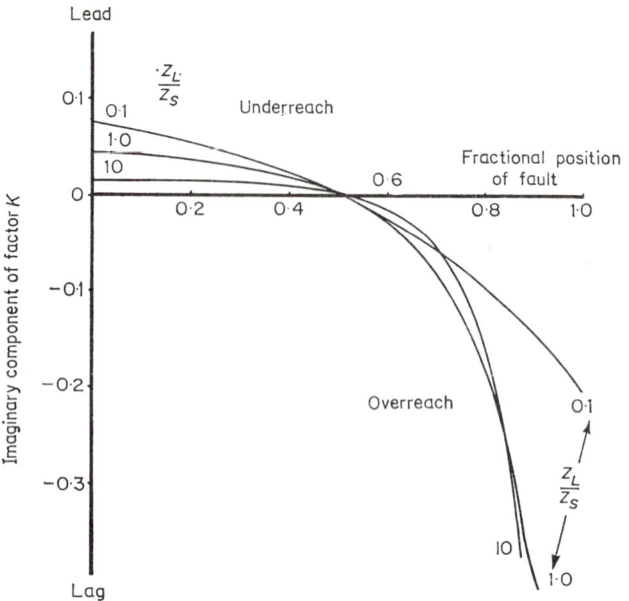

Fig. 11.3b. Variation of k with line/source impedance ratio
This is 10 times the scale of Fig. 11.13a

they are not in phase, it will appear to have a reactive component which can cause underreaching at one end of the line and overreaching at the other.

11.2.1. Double-end Feed with a Fault and No Load [79]

If there is no load to start with, the e.m.f's E_S and E_R beyond each end of the line will be in phase (Fig. 11.2a). With such a condition distance relays will measure correctly on a homogeneous system but, if the protected section is long and of lower X/R ratio than the source impedance at each end, there will be inaccuracy of measurement for faults near the end of the section because the currents I_A and I_B from the two ends will be out-of-phase with each other and with the fault current I_f.

Figure 11.2a shows a system with the source impedances assumed wholly reactive and the line impedance assumed 60° lag ($X/R = 1{\cdot}73$). This means that I_B is more lagging than I_A for a fault near B and this will make I_f more lagging than the relay current I_A. Consequently, V_F will also lag I_A and this will cause the relay at A to measure a leading reactive component in R_f which will subtract from the line reactance and make it overreach. The reverse occurs for a fault near end A.

In practice, synchronous machinery and power in-feeds at bus B would tend to prevent tripping on faults beyond bus B, but there would be faster tripping for faults near the reach point of the relay.

If $I_f/I_A = K$ the impedance seen by the relay is $Z_L + kR_f$ where k is a complex number which varies with the position of the fault on the protected line section Z_L, the Z_s/Z_L ratio and the ratio of the values of Z_s at the two ends (Z_R/Z_S). Z_R is the source impedance at the receiving end of the system.

Figures 11.3a and 11.3b show how the real and imaginary components of k vary, assuming equal generation (500 MVA fault level) at the two ends, and the length of line varied from 4·6 miles ($Z_L/Z_s = 0{\cdot}1$) to 465 miles ($Z_L/Z_s = 10$). A 465-mile line is impractical but the same Z_L/Z_s ratio would occur with, say, 5,000 MVA on a 46·5-mile line. It should be noted that the imaginary component is drawn to ten times the scale of the real component [79].

For a short line Z_L/Z_s is small and the current from each end tends to be the same, so that $k = 2$ irrespective of the position of the fault on the protected line section Z_L irrespective of the value of Z_s/Z_L.

Assuming Z_L fixed at a value equal to the value of Z_s at the relay end (Z_S), Fig. 11.4a and Fig. 11.4b show how k varies when Z_s at the far end (Z_R) has values between ten times and one-tenth of Z_S at the relay end. It will be seen that, when Z_R/Z_S is large, k approaches 1 because the fault current contributed from end B becomes small.

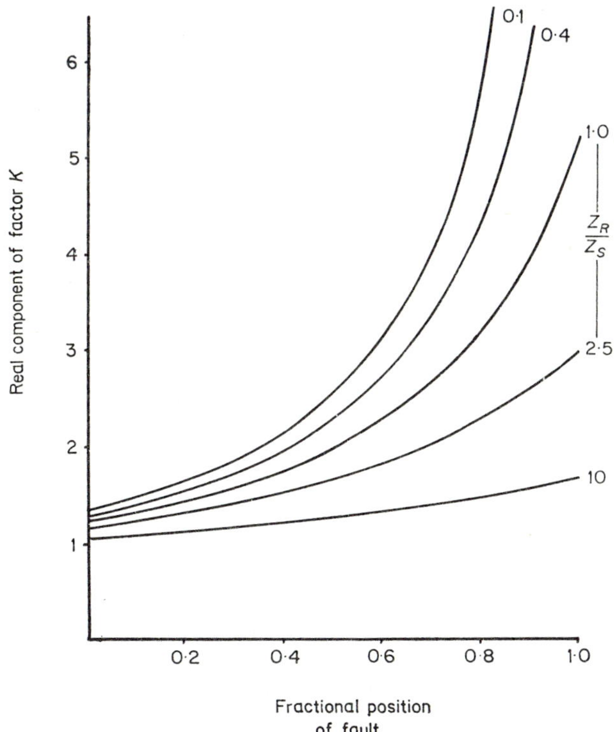

Fig. 11.4a. Variation of k with ratio of source impedances $K = I_f\ I_r$

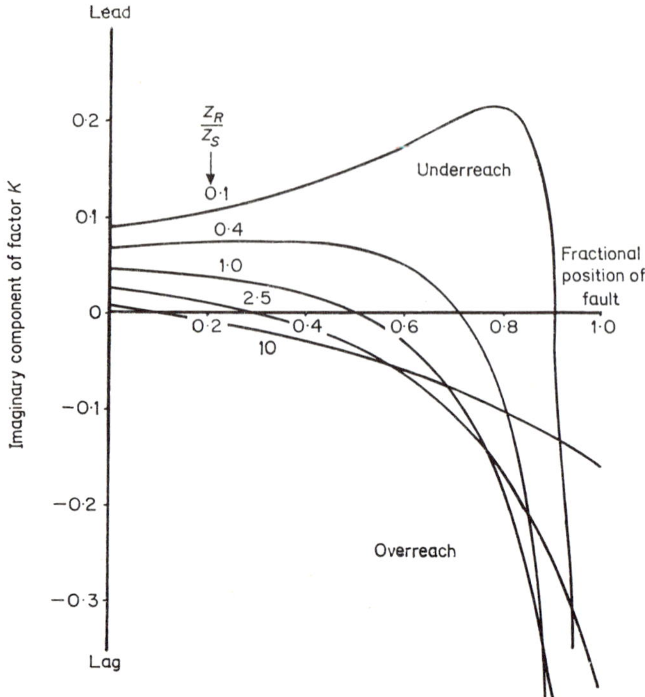

Fig. 11.4b. Variation of k with ratio or source impedance
This is 10 times the scale of Fig. 11.4a.

Conclusion. Overreaching by reactance relays can occur for faults near the end of the line section but this can be prevented by setting the ground reactance relays to over not more than 80% of the section, expecially for long lines. Alternatively, if one end of the system is normally more loaded than the other (and hence more lagging), then the relays at that end can be set at 95% for Zone 1 and the relays at the outgoing end at, say, 65%. Back-up relays and blocking relays should likewise have increased settings at the lagging end of the system.

11.2.2. Power Tranfer and Power Swings [21]

Power flows from the parts of the system with surplus generation to the areas with surplus load. The consequent flow of current through the system causes a voltage drop (Fig. 11.5b) which, when the system is simplified to the equivalent two-machine system, separates the two source e.m.f's E_S and E_R by an angle θ_s whose magnitude increases with the load transfer. This is explained in more detail in Reference [21]; it is illustrated in Fig. 11.7.

From Fig. 11.5b, the swing current $I_s = (E_s - E_R)/Z$ where Z is the total system impedance. If $|E_S| = |E_R|$ it will be seen that $I_s = (2E \sin \frac{1}{2}\theta_s)/Z$. The voltage decreases towards the middle of the system and this reduction increases with θ_s until the line voltage is zero at the electrical centre of the system when $\theta_s = 180°$.

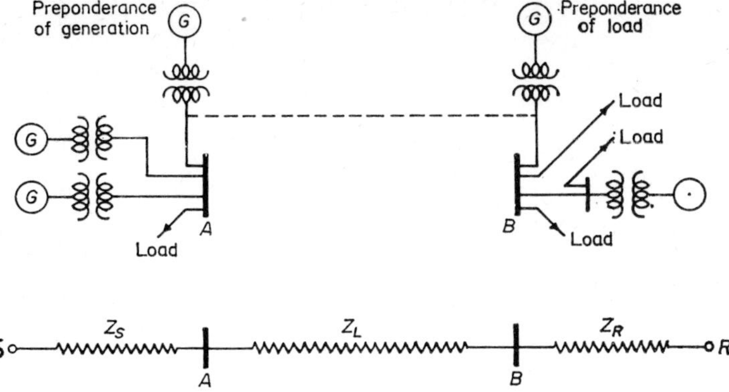

Fig. 11.5a. Equivalent circuit of one phase of a power system

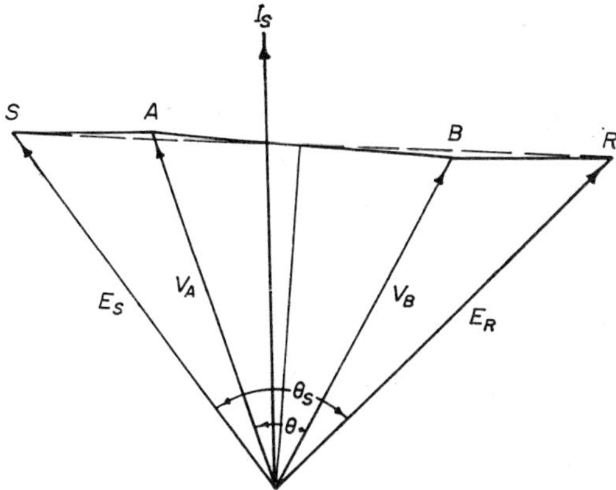

Fig. 11.5b. Vector diagram of one phase of a power system during a power swing

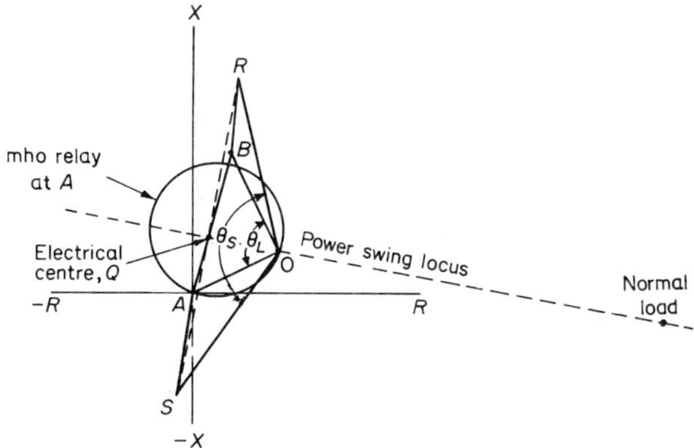

Fig. 11.5c. Impedance seen by phase relays during a power swing (simplified diagram)

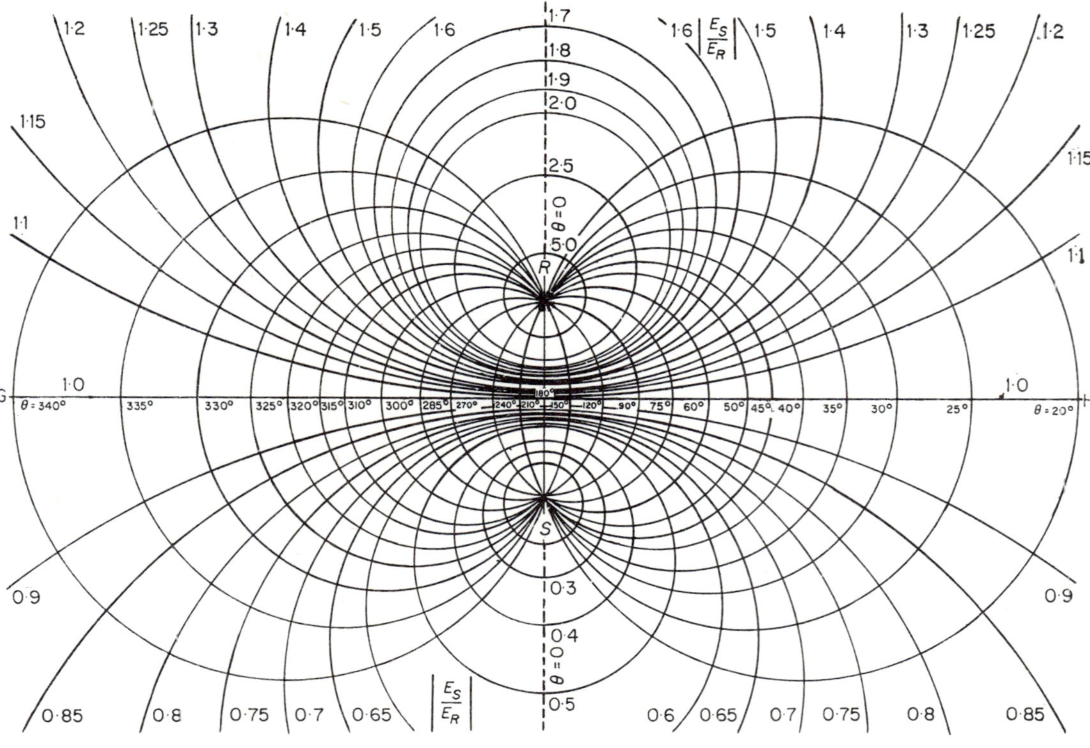

Fig. 11.6a. Impedance seen by relay during power flow though a purely reactive system

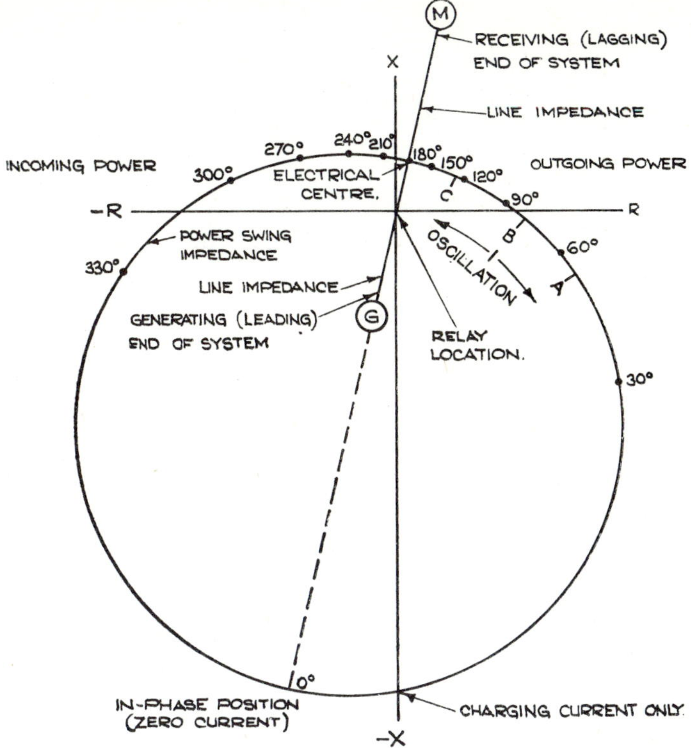

Fig. 11.6b. Locus of impedance seen by distance relays during power swing with $E_R > Es$

The effect of this upon the impedance seen by a distance relay is shown in Fig. 11.5c. Normally θ_s does not exceed 15° and the relays are negligibly affected but, during a power swing, θ_s may oscillate up to 90°. A swing of more than 90° theoretically should result in the generators at A and B losing synchronism because the synchronising torque is proportional to the sine of θ_s and hence decreases when $\theta > 90°$. In practice however, a fast voltage regulator permits a transient excursion up to perhaps 160°.

If $E_S \neq E_R$ the locus of the impedance seen by the relay is a family of circles with centres on the reactance axis, as shown in Fig. 11.6. It has been shown as a straight line for $E_S = E_R$ but in practice this locus curves slightly downwards due to system capacitance and other factors. For checking relay operation, however, it is adequate to assume the locus to be a straight line when $E_S = E_R$.

In Fig. 11.5c the origin of the diagram is drawn at the relay location A so that the impedances seen by the relay are measured from A. This relay would measure the impedance value AB for a fault at B but, in the case of power swing with angle θ_s between the generated voltages, it would measure the impedance AO. As the swing became more severe and the angle of separation

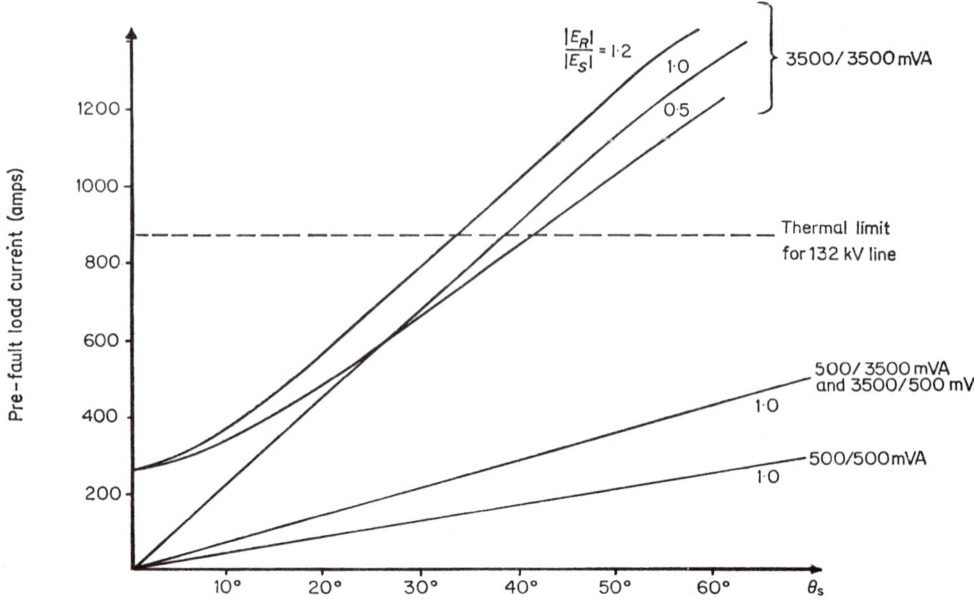

Fig. 11.7. Effect on load upon angular separation of source E.M.F.'S

increased the impedance measured by the relay would decrease to a value AQ' which might be within the tripping characteristic of the relay. Similarly, if the relay were at B, a line drawn from B to any point on the swing locus would be a vector representing the impedance seen by the relay at B for a condition represented by that point.

Assuming equal generated e.m.f's, the power swing locus will be the locus of the apexes of a series of isosceles triangles with a common base SR, so that the locus must be a straight line bisecting SR at right angles. The angle θ_s is the angle between the power source e.m.f.'s and the corresponding angle θ_L between the bus voltages can be obtained by drawing lines from 0 (or the appropriate point on the locus) to A and B.

Relays at the sending (leading) end of the system will see an inductive impedance during the swing and those at the lagging end a capacitative impedance. This impedance will be measured by the relay in parallel with any fault impedance and will therefore tend to reduce the impedance measured and cause overreaching.

11.2.3. Effect of Fault upon Power Swing Diagram [21]

During a fault, not only will the relay measure the power swing impedance in parallel with the fault impedance but also there will be an error in measuring the fault impedance because of the double-end feed.

If E_S leads E_R, a relay at A will tend to overreach because the current from B will be more lagging and will produce an apparent negative component of inductive reactance in the fault resistance. This component is $(I_f/I_A)\, R_f \sin \alpha$ at end A, or $(I_f/I_B)\, R_f \sin \beta$ at end B, where α or β is the angle between the local current and the fault current. The percentage error in reactance measurement at A is:

$$\frac{100 I_f R_f \sin \alpha}{I_A X}$$

where X is the reactance of the protected section of line.

A fault between phases will have very much less error due to fault resistance but it will have considerable effect upon the locus of the power swing impedance.

The relay for the faulted phase pair *b-c* will see only the fault impedance until after the fault is cleared, but the relays for the other two phase pairs will see an impedance locus during the swing which will be different from Fig. 11.5c.

For the *c-a* relay the impedance vector SR will be shifted in the leading direction by $\theta_K - 30°$ and divided by $\sqrt{\{\frac{3}{4} + (SR/2SF)^2\}}$ where $\theta_K = \tan^{-1}(SR/\sqrt{3}SF)$. The locus of the impedance seen by the relay (Fig. 11.8b) is a circle of radius $2R_v Z_{SR}/\sqrt{3}(K^2 - R_v^2)$ where R_v is the voltage ratio E_R/E_S, Z_{SR} is the total impedance between the ends S and R and K is the fractional position of the fault from the sending end S. The centre of the circle is at $\{2KZ_{SR}/\sqrt{3}(K^2 - R_v^2)\} - Z_r$. On a non-homogeneous system K is a complex quantity.

For the *a-b* relay, SR has the same change in length and is lagged by $\theta_K - 30°$ as shown in Fig. 11.8a. The derivation of these figures is given in Ref. [21]. The locus of the impedance seen by the relay is a circle of radius $\{2R_v Z_{SR}/\sqrt{3}(K^2 - R_v^2)\}$ and centre at $\{2KZ_{SR}/\sqrt{3}(K^2 - R_v^2)\} - Z_r$.

The procedure for checking the behaviour of the relays is merely to draw the impedance characteristic of the relay upon the power swing diagram, using the same ohmic scale as shown in Figs. 11.8 and 11.12.

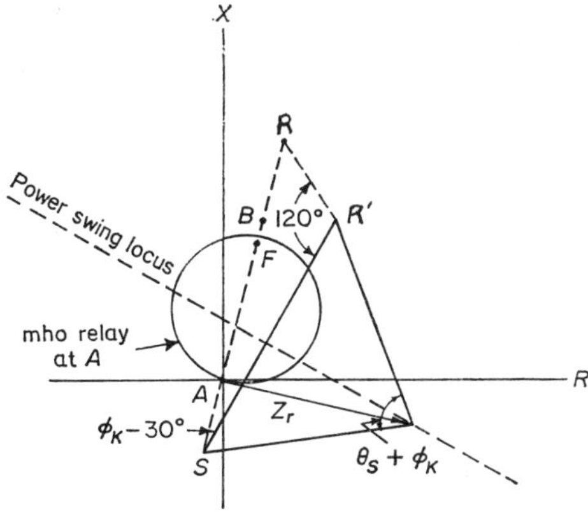

Fig. 11.8a. Simplified diagram for impedance seen by Phase a–b relay during a power swing started by a Phase b–c fault (not cleared)

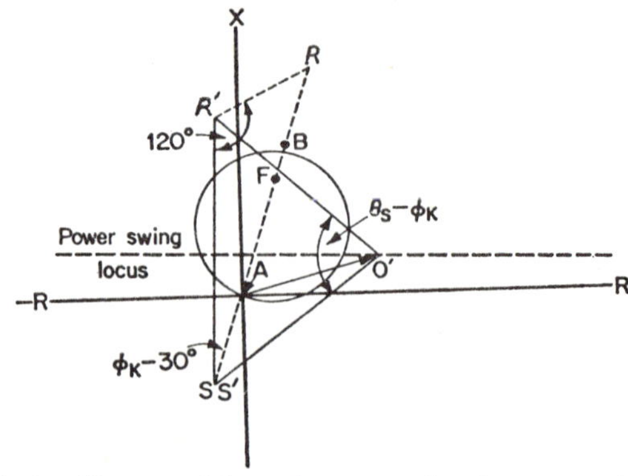

Fig. 11.8b. Vector diagram of impedance seen by phase c–a relay during a power swing started by phase b–c fault (not cleared)

Where it is found that the power swing locus penetrates the relay characteristic, the substitution of an elliptical characteristic may prevent undesirable tripping (see Chapter 12, Section 12.1). A simpler alternative is to add a blinder [29], as discussed in Vol. I, Chapter 5, Section 5.2.3. In extreme cases out-of-step blocking may be necessary.

An example of what can happen if some precaution like the foregoing is not taken is the large-scale power failure that occurred in north-eastern U.S.A. in November, 1965. A Zone 3 mho relay tripped during a power swing and the disruption of the system caused progressive tripping to occur until the New York area had a total blackout.

11.2.4. Effect of a Power Swing upon Distance Measurement to Fault [123]

Since the load current goes through the relay but not through the fault, the effect upon the relay is as though a load impedance Z were connected in parallel with the fault, as shown in Fig. 11.1.

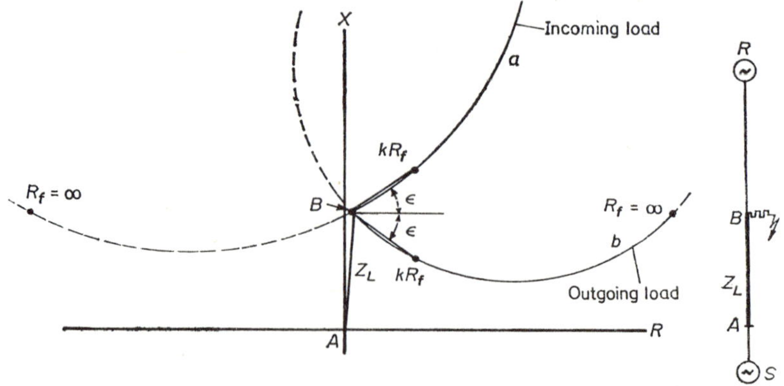

Fig. 11.9. Effect of fixed load and varying fault resistance
Load measured at *B*

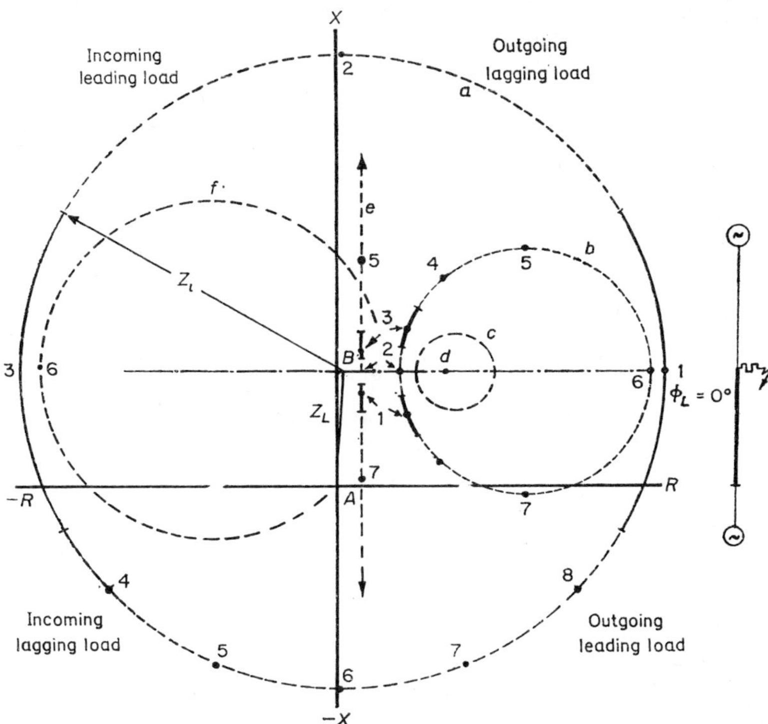

Fig. 11.10. Effect of fixed fault resistance and varying load

Figure 11.9 shows the impedance measured by the relay as the fault resistance varies from 0 to ∞ with (*a*) incoming power, (*b*) outgoing power. Figure 11.10 shows the effect of various load conditions [80].

With no fault and a load of a constant magnitude but varying phase angle, the locus of the impedance seen by the relay at *A* is the circle '*a*'. With a fault of resistance *Bd* and the same load condition as for '*a*', the locus becomes circle '*b*'. A smaller load gives locus '*c*'; no load gives the point '*d*'. Larger loads give the loci '*e*' and '*f*'.

On these loci the points marked 1 to 8 represent the same phase angles of the load current as seen by the relay at *A*. For instance, the point 1 represents unity power factor outgoing power; point 2 is outgoing reactive kVA; point 3 incoming kW; point 6 incoming reactive.

It will be seen that the relays tend to measure less reactance and hence to overreach with outgoing power and to underreach with incoming power. A mho circle with diameter *AB* would overreach only for the locus '*e*' which represents an enormous load.

Figure 11.11 shows that the real value of the factor *k* decreases as the flow of power from the relay end of the system increases ($\theta = -60°$), but the reactive value of *k* increases negatively so that the reactance measured

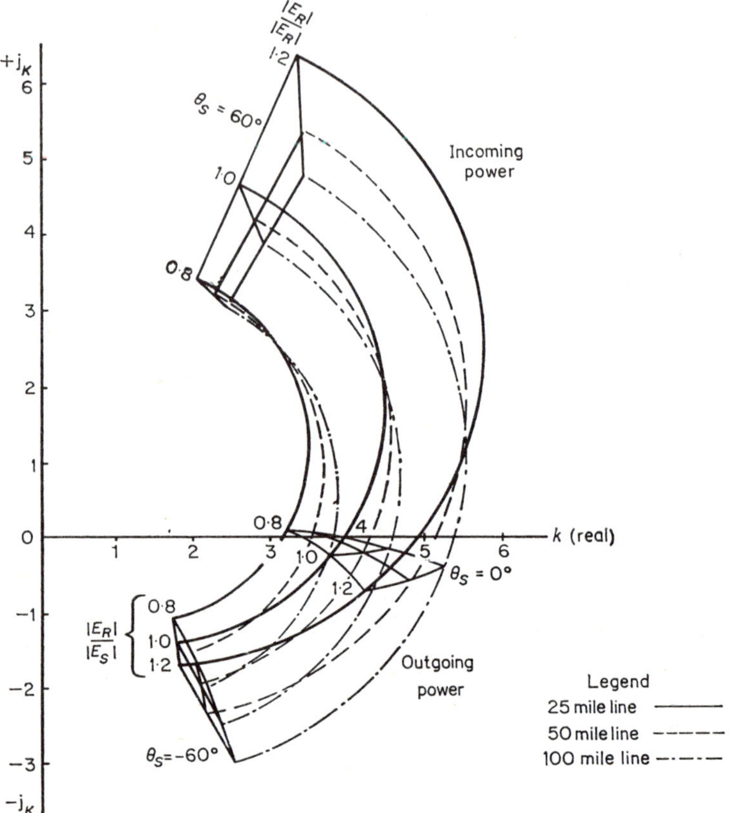

Fig. 11.11. Effect of power transfer upon K during a single-phase ground fault $R_f = 5$ w. Fault at 80% of line: $Z0 = Ze$ (3500 MVA) Relay has sound phase compensation

by the relay would be reduced by this fictitious negative reactance in the fault path and the relay would overreach [123].

Conversely, at the receiving end of the system the relay would underreach ($\theta = +60°$). When the two source e.m.f.'s are in phase ($\theta = 0°$), the reactive component of k is small and the fault resistance is multiplied only by a numerical value depending upon the length of line between the sources and the ratio of the remote end e.m.f. to the relay end e.m.f. k is between 3 and 5 because, with a fault at 80% from the relay end, the remote end source contributes around 80% of the fault current.

The effect of this upon the impedance measured by the relay is shown in Fig. 11.12 where the system arrangement conforms to Fig. 11.2a. This shows that the power swing tends to block a mho relay except for a fault near the middle of the section (50%) with outgoing power (0° to −60°). Figure 11.12a has equal source impedances; Fig. 11.12b has Z_S at A seven

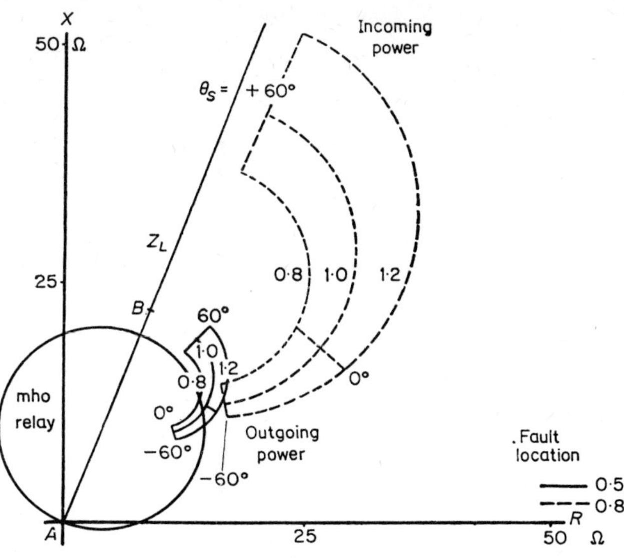

(a) With equal source impedances 3500/3500 mVA

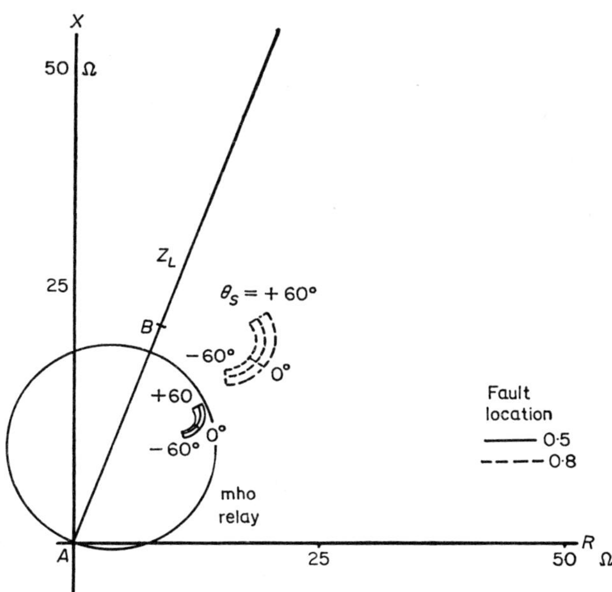

(b) With unequal source impedances 3500/500 M.V.A

Fig. 11.12. Locus of impedance measured with a single-phase ground fault during a power swing

times Z_R at B. The solid loci are for a fault at the middle of the section and the dotted loci for a fault 20% from the far end (80%). On the other hand, it is clear that a reactance relay would overreach when the power is outgoing from A.

Figure 11.13 shows that these effects increase with the resistance of the fault path. This was also indicated in Fig. 11.9.

Conclusion. Underreaching occurs in most cases. Overreaching occurs in relays with outgoing power, but it is serious only for a relay near the receiving and with a high resistance fault. The quadrilateral characteristic would be much more vulnerable than the mho circle in this case.

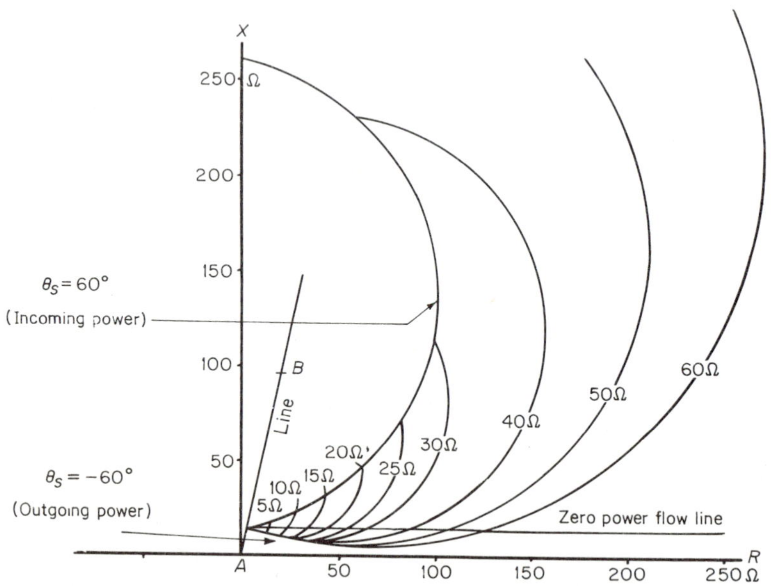

Fig. 11.13. Effect of fault resistance during load transfer

11.2.5. Load in Parallel with Faulted Line [80]

Whereas the impedance measured by the relay can be increased by the fault resistance which is in series with the line impedance, it can also be reduced by load impedance in parallel with the line impedance, as in Fig. 11.14c.

This would be the case with single-end feed where a relay looks backward through the bus into two lines one of which is faulted and the other has a load.

An example is the reversed third zone of a mho distance relay.

Figure 11.14a shows the impedance seen by the relay in such a case; the fault is a solid one and the unity power factor load is varied from zero to the maximum value. The small circle shows the effect of varying the distance to the fault with a fixed (maximum) load.

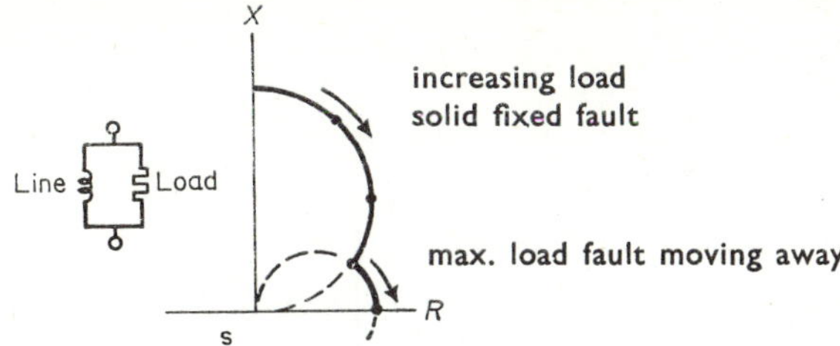

(a) Resistance load and 90° line

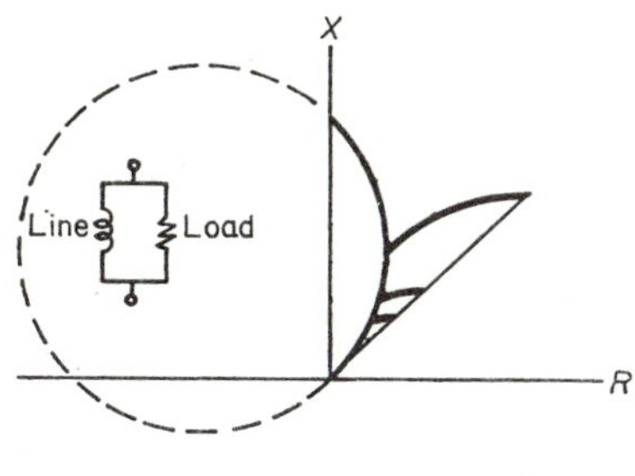

(b) 45° load and 90° line

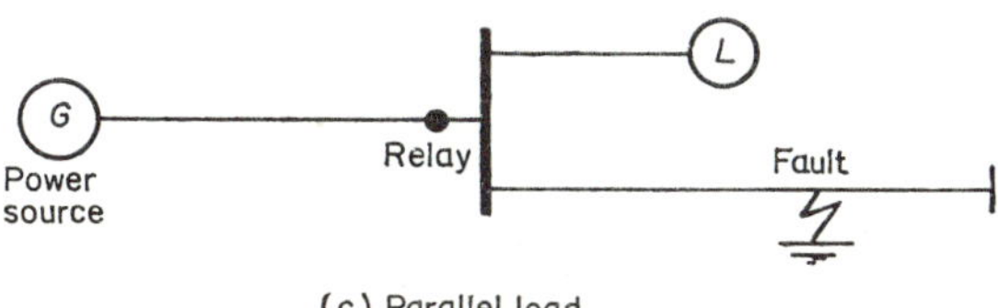

(c) Parallel load

Fig. 11.14. Effect of load in parallel with the faulted line

Figure 11.14b shows the same thing with a 45° lagging load, the line impedance still being taken as purely reactive.

Conclusion. This condition can cause overreaching, which is not undesirable in the case of a reversed third-zone unit. Where it is due to a heavy tap load on a very long line it can cause overreaching of Zone 1 and this must be allowed for in its setting. In practice this error is considerably reduced by the zero sequence compensation of the relay since I_0 is associated only with the fault.

11.2.6. Fault on Stub-end Feeder with Double-end Feed

A modification of the previous case is where the load is not static but the result of load transfer to the receiving end, i.e. double-end feed. The equivalent diagram is similar to Fig. 11.2a except that the fault resistance is here replaced by a line.

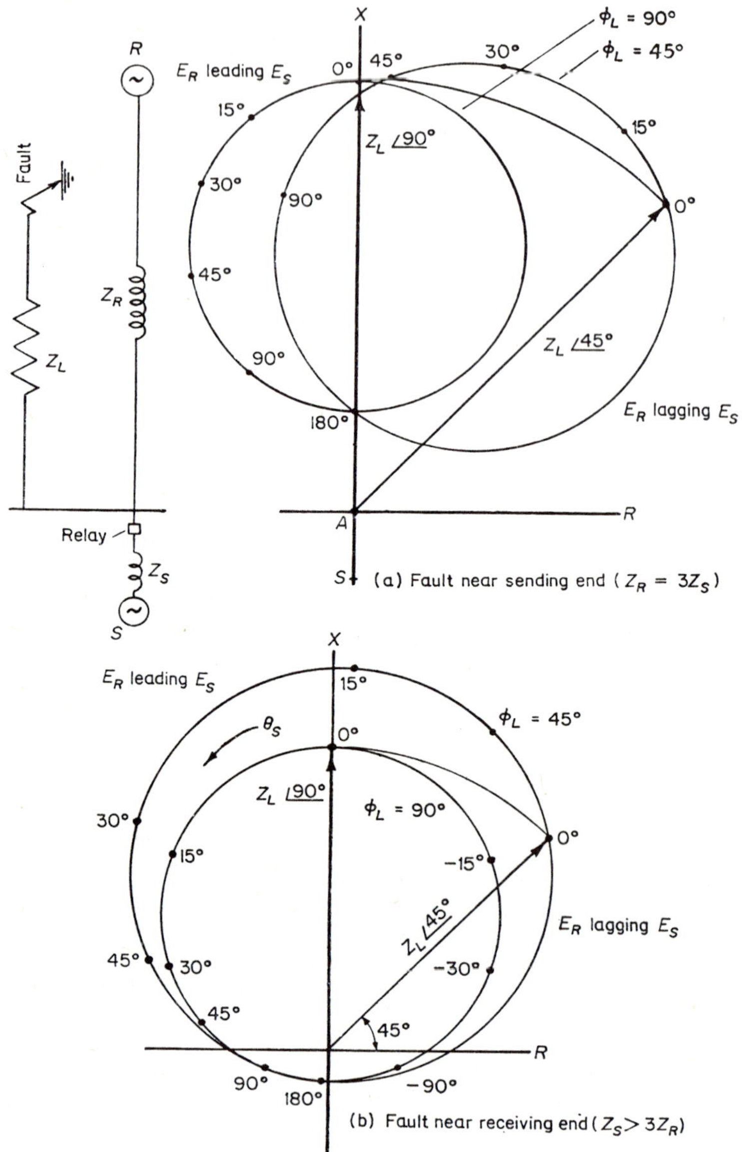

Fig. 11.15. Effect of power swing upon impedance seen by relay during a single-phase ground fault

A typical case would again be that of a reversed third zone mho unit, as in Fig. 11.15. The source e.m.f.'s E_S and E_R are assumed equal in magnitude and the circles are the loci of the impedance seen by the relay at R as the angle θ between E_S and E_R varies. The two circles in each diagram of Fig. 11.15 are for 90° and 45° lines. Figure 11.15a is with the relay bus 25% from the sending end S and Fig. 11.15b is with the relay bus 25% from the receiving end R of the system.

Conclusion. It will be seen that overreaching can readily occur with outgoing power. However, this need not be serious since the relay in question is a reversed third zone unit whose problem is usually the ability to reach far enough.

11.2.7. Likelihood of Maloperation due to Fault Resistance

It is clear that error due to fault resistance and system loading is generally negligible with phase relays because the fault resistance is limited to the power arc whose length is limited by the conductor spacing; this also applies to ground faults on systems with a good earth return path. It is also obvious that error due to double-end feed cannot occur on radial lines.

From the preceding sections it can be seen that overreaching tends to occur with outgoing power and/or leading VA. This is because the load current makes the relay current less lagging, so that the fault resistance (through which the load current does not flow) appears to have a negative (leading) reactive component. This condition pertains to relays polarized facing the lagging end of the system and is worst for those near the lagging end because here the load current is more leading (see Fig. 11.5b). For appreciable error R_f and θ_s have to be large.

It was pointed out at the beginning of this chapter that R_f can be large only where there are poor ground-wire arrangements and high tower footing resistance. In practice almost all modern transmission lines have good ground wires electrically connected to the insulator supports; on a few systems this is true only near the line terminals but it is there that the measurement error is most likely to cause overreaching. Very high values of R_f can occur when a line breaks and the conductor falls on the ground. In this case, however, there is no double-end feed and therefore no spurious reactive component to cause overreach.

θ_s seldom exceeds 20° except during a power swing. A single-phase ground fault is very unlikely to cause a power swing because it interrupts the synchronising power only on one phase; furthermore, modern relays will operate before the swing has progressed more than a few degrees.

Interphase faults, although capable of causing a power swing, cannot have high values of R_f; also, appreciable swings are unlikely with modern fast relays and breakers, especially where automatic high-speed reclosing is provided. It can be seen from the foregoing that cases of serious error due to large values of θ_s are unlikely on power systems of modern construction.

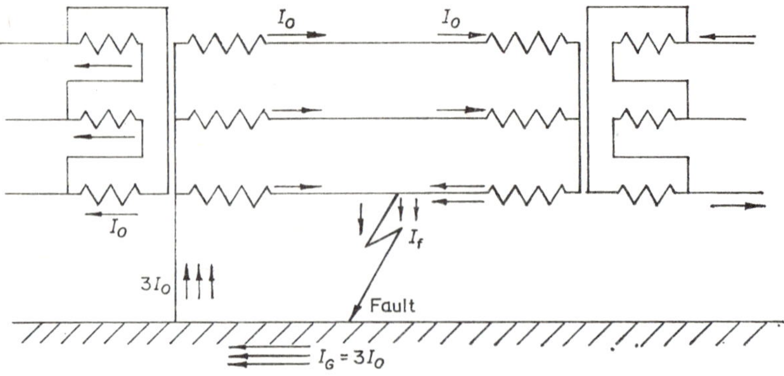

(a) System grounded only at one point

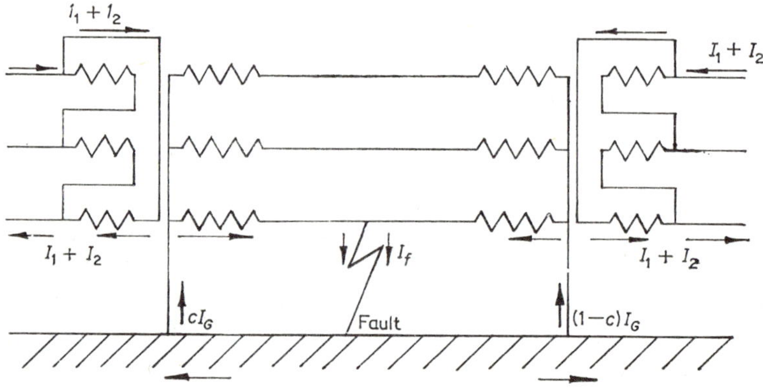

(b) System with multiple grounding

Fig. 11.16. Current distribution for single-phase ground fault

11.2.8. Prevention of Wrong Operation

In most cases overreaching by ground reactance relays at the sending end can be prevented by adjusting them to measure $Z \sin (\phi - 5°)$ instead of $Z \sin \phi = X$ and by setting their instantaneous zones to cover only 80% or even 70% of the protected section. The relays at the receiving end can be adjusted to measure $Z \sin (\phi + 5°)$. Phase relays can almost invariably be given a 90% setting if the line impedance is accurately known.

Where this remedy is inadequate, distance relays should be used only for phase faults and either unit protection or zero sequence overcurrent protection provided for single-phase ground faults. The latter may not be as fast as distance relays for faults near the end of a protected section but this is not serious, especially where R_f is large, thus limiting the fault current.

With solid grounding at each station, the selectivity of directional I_0 relays is excellent and they are not affected by load current.

On very long lines (i.e. where the length in miles exceeds the kV), θ_s can exceed 30°, but on modern systems this is generally reduced by series capacitors, multiple conductors, etc. Where this is not feasible the relays can be blocked from tripping on power swings and out-of-step conditions by providing out-of-step blocking, i.e. an extra impedance unit in one phase which distinguishes between power swings and faults on the basis that the system impedance changes slowly during a power swing and instantaneously when a line fault occurs.

Where underreaching is to be expected, the clearing of faults can be speeded up by using intertripping through a pilot channel.

While some of these error curves may appear alarming, one can be consoled by the fact that several hundred thousand distance relays have been operating correctly all over the world for over thirty years, in spite of these conditions. Furthermore, it is always possible nowadays to analyse distance relay applications where there are obviously bad conditions, such as tapped lines and the lack of ground wires. Such an analysis is much easier now that computers are in common use.

11.3. OPERATION OF RELAYS IN UNFAULTED PHASES

For protection against all faults six distance relays are necessary, three for interphase faults and three for single-phase ground faults. The phase relays are supplied with delta currents and potentials and the ground relays with wye currents and voltages with zero sequence compensation [22].

With these arrangements only one of the six relays might be expected to operate on unbalanced faults and all six on balanced three-phase faults. In practice more than one relay tends to operate for any kind of fault, especially if it is close to the bus. This may be a weakness of the present system of using single-phase relays to measure electrical quantities in a three-phase circuit. Polyphase measurement would overcome this difficulty (Chap. 12).

With proper connections (as mentioned above) the relay associated with the faulted phase of phase pair measures Z_1', the positive sequence impedance between the relay and the fault. The additional impedance measured by the relays in other phases and phase pairs is given in Table 11.1 for various types of faults [21].

In this table, C is the fraction of the total positive sequence component of the fault current that appears in the relay, assuming double-end feed. C' and C'' are hybrid factors due to the zero sequence compensation of the ground relays. $C' = C - KC_0$ and $C'' = 2C + KC_0$, where C_0 is the equivalent of C for the zero sequence current and $K = Z_0'/Z_1'$. Z_2 and Z_0 refer to the total system viewed from the fault location; $\overline{|90^\circ}$ means lagged 90°; $\underline{|90^\circ}$ means advanced 90°.

Double-ground faults were omitted because the expression for the additional impedance measured is very complicated, but it approximates in value that measured during phase-to-phase faults.

TABLE 11.1.
Additional Impedance seen by Distance Relays

Relay	Three Phase	Phase b — Phase c	Phase a to Ground
a–b	$\frac{R_f}{C}$	$\frac{1}{C}(\sqrt{3}Z_2\angle 90^\circ + R_f\angle 60^\circ)$	$\frac{1}{C\sqrt{3}}(Z_2\angle 90^\circ + Z_0\angle 30^\circ + 3R_f\angle 30^\circ)$
b–c	$\frac{R_f}{C}$	$\frac{R_f}{2C}$	∞
c–a	$\frac{R_f}{C}$	$\frac{1}{C}(\sqrt{3}Z_2\angle 90^\circ + R_f\angle 60^\circ)$	$\frac{1}{C\sqrt{3}}(Z_2\angle 90^\circ + Z_0\angle 30^\circ + 3R_f\angle 30^\circ)$
a–G	$\frac{R_f}{C}$	∞	$\frac{3}{2C''}R_f$
b–G	$\frac{R_f}{C}$	$\frac{1}{C\sqrt{3}}(Z_2\angle 90^\circ + R_f\angle 30^\circ)$	$\frac{\sqrt{3}}{C'}(Z_2\angle 90^\circ + Z_0\angle 30^\circ + 3R_f\angle 60^\circ)$
c–G	$\frac{R_f}{C}$	$\frac{1}{C\sqrt{3}}(Z_2\angle 90^\circ + R_f\angle 30^\circ)$	$\frac{\sqrt{3}}{C'}(Z_2\angle 90^\circ + Z_0\angle 30^\circ + 3R_f\angle 60^\circ)$

11.3.1. Phase Relays [21]

Figure 11.17 shows a simple graphical method of determining the impedance seen by the phase relays during a solid phase-b–phase-c fault. The fault resistance R_f was not drawn because it does not affect the result and is in any case small on interphase faults.

The b-c relay measures Z_1'; the a-b relay measures the additional impedance $(\sqrt{3}/2)\, Z_2\angle 90^\circ$ which is the line FN, i.e. the total impedance measured is AN. Since the point N is well outside any normal distance relay characteristic suitable for protecting the line section AF and below the R axis, it

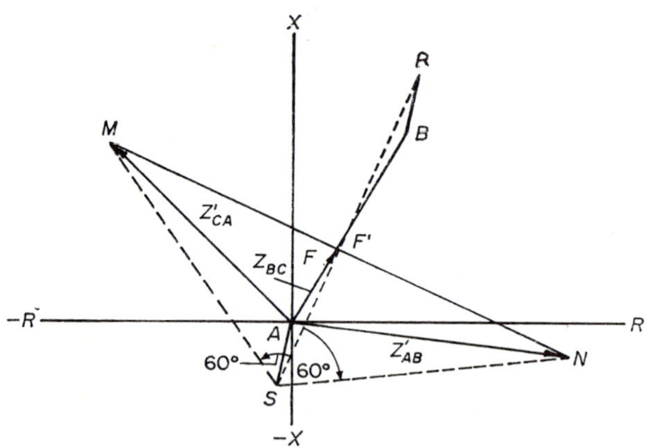

Fig. 11.17. Impedances seen by phase relays during solid $b - c$ fault at F
$F'S = Z_2\ C$; $F'M = F'N = 3\ V Z_2$

can be assumed that the *a-b* relay will not operate for a *b-c* fault. On the other hand, with a line with a high X/R ratio and a relay having a quadrilateral characteristic, the *a-b* relay would operate for a fault near the relay location A. The *c-a* relay will not operate at all.

Figure 11.18 shows a similar diagram for a phase-*a* to ground fault. In this case the fault resistance component R_f is also shown. The point F_{CA} represents the impedance seen by the *c-a* relay for a fault at the relay setting; it is well outside any Zone 1 relay characteristic. A fault nearer the relay could be within a quadrilateral characteristic, but not within a mho characteristic; the *a-b* relay would be more likely to operate with a mho characteristic for a close-in resistance fault.

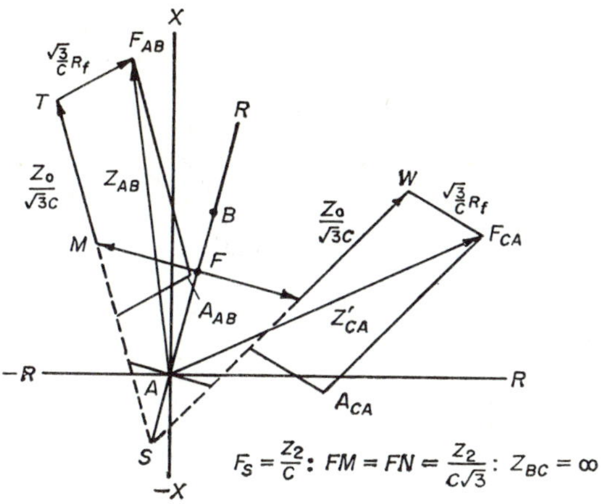

Fig. 11.18. Impedances seem by phase relays during a–G resistance fault at *F*

11.3.2. Ground Relays [21]

It is clear from Fig. 11.19 that ground relays with either mho or quadrilateral characteristics are most unlikely to operate for ground faults in other phases.

Figure 11.20 shows the areas of impedance seen by the ground relays during a phase-*b*/phase-*c* fault. The construction of the vector diagram is not shown because it is similar to that of the previous diagram.

The phase-*a* relay measures infinite impedance and does not operate. The impedance measured by the *c*-relay is also outside any normal distance relay characteristic. On the other hand, the phase-*b* relay will clearly operate, expecially with a quadrilateral characteristic.

Conclusions. In no case is there overreaching by relays in the sound phases; occasional tripping by a relay for a nearby fault in another phase or phase pair merely contributes a little back-up protection at the expense of some confusion in flag indication. However, care must be exercised in designing and applying impedance-type units for phase selection purposes, e.g. in switched distance schemes or single-phase tripping.

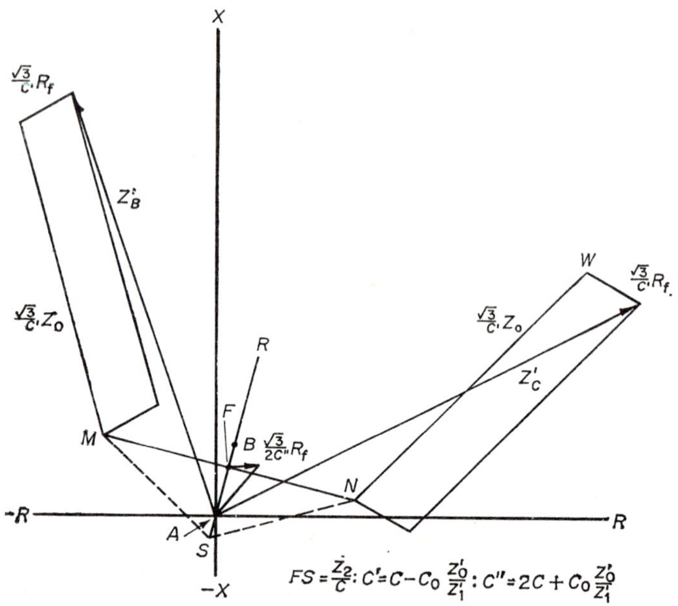

Fig. 11.19. Impedances seen by ground relays during a–G resistance fault at *F* (reduced scale)

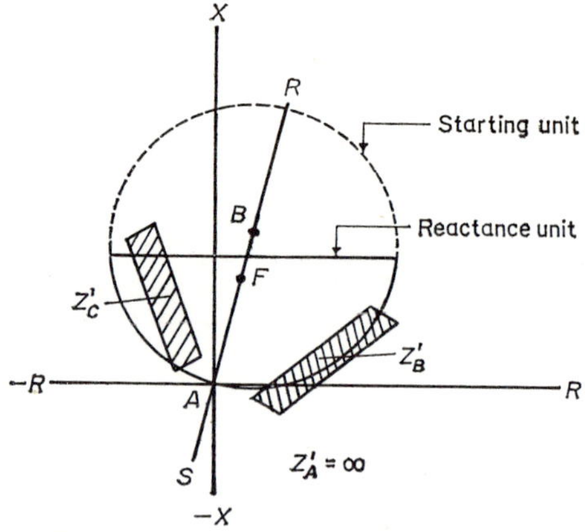

Fig. 11.20. Fault impedance areas seen by ground relays during $b - c$ resistance fault at *F*

11.4. MUTUAL INDUCTION

Mutual induction at system frequency can occur between one conductor and the other two of a three-phase line, or between one three-phase line and another. The method of compensating for the former is straightforward because the c.t's of other phases are available, but compensation for induction from a parallel line is not possible if the latter does not terminate in the same bus as the line affected.

11.4.1. Induction from the Other Phase Conductors [22, 43]

On an ideally transposed line the voltage V_a for a solid single-phase fault to ground is:

$$V_a = I_a Z + (I_b + I_c) Z_m$$

where Z is the self-impedance of the conductor between the relay and the fault and Z_m is mutual impedance to the other two conductors.

From Eq. (11.1)

$$Z = \frac{V_a}{I_a + (Z_m/Z)(I_b + I_c)} \tag{11.2}$$

Now

$$Z_1 = Z_2 = Z - Z_m \tag{11.3}$$

and

$$Z_0 = Z + 2Z_m \tag{11.4}$$

Solving for Z and Z_m

$$Z = \tfrac{1}{3}(2Z_1 + Z_0) \tag{11.5}$$

and

$$Z_m = \tfrac{1}{3}(Z_0 - Z_1) \tag{11.6}$$

Substituting for Z and Z_m in Eq. (1)

$$\begin{aligned} V &= \tfrac{1}{3}\{I_a(2Z_1 - Z_0) + (I_b + I_c)(Z_0 - Z_1)\} \\ &= \tfrac{1}{3}\{I_a(2Z_1 - Z_0) + (3I_0 - I_a)(Z_0 - Z_1)\} \\ &= I_a Z_1 + I_0(Z_0 - Z_1) \end{aligned}$$

therefore

$$Z = \frac{V_a}{I_a + \dfrac{Z_0 - Z_1}{Z_1} I_0}. \tag{11.7}$$

Hence perfect compensation can be made for mutual induction from the other two phases.

On an untransposed unsymmetrical line or cable, for a solid single-phase ground fault on phase-a:

$$V_a = IZ_a + I_b Z_{mb} + I_c Z_{mc}. \tag{11.8}$$

Hence for compensation from the sound phases

$$Z_a = \frac{V_a}{I_a + \dfrac{Z_{mb}}{Z_a} I_b + \dfrac{Z_{mc}}{Z_a} I_c} \tag{11.9}$$

This shows that the relay will measure correctly the impedance to a solid single-phase-to-ground fault on a non-transposed line if it is supplied with the correct proportions of the currents in the sound phases.

In practice the self- and mutual impedances of the individual conductors are seldom known and the relays in the three phase are set for the same impedance and supplied with the same proportion of currents from the other two phases. Under these circumstances the difference between the relay measurement on an untransposed line with zero sequence compensation and sound phase compensation is less than 2%. Hence zero sequency compensation is commonly used because it requires only one auxiliary c.t.

Another error stems from the fact that, in zero sequence compensation, the factor $(Z_0 - Z_1)/Z_1$ is taken as scalar but the phase angle difference between Z_0 and Z_1 is seldom enough to make more than 1% error.

11.4.2. Induction from a Parallel Line [43]

On a double-circuit line with little or no generation at the far end, a fault on one line will be supplied with a substantial amount of current via the sound line. This current will induce a voltage in the faulted line which will make the relay in that line underreach. This tendency will decrease as the fault approaches the remote bus, so that the underreaching will not seriously reduce the extent of Zone 1. Until recently the effect was compensated for by adding to the relay current a percentage of the zero sequence current in the other line, which was a function of the mutual coupling between the two lines.

This method enabled the relay in the faulted section to measure correctly the distance to the fault in spite of the mutual induction from current in the sound line. Unfortunately, the same compensation can cause the fault to appear to the relay in the sound line to be within its setting.

Figure 11.21 shows a typical case where, if there is no generation at *B*, the relay (source MVA ratio 1/0) will trip for single-phase ground faults up to about the middle of the sound line. Even with equal generation (source MVA ratio 1/1) at the two ends, the relay on line 1 will trip for faults on the sound line up to 18% from the bus.

From the foregoing it is clear that mutual compensation of ground distance relays is a mixed blessing and should normally be omitted.

11.5. ERRORS DUE TO TRANSIENTS IN THE POWER SYSTEM

When a fault occurs, the current and voltage at the relay undergo a sudden change in amplitude or phase angle, or both. This sudden change causes the

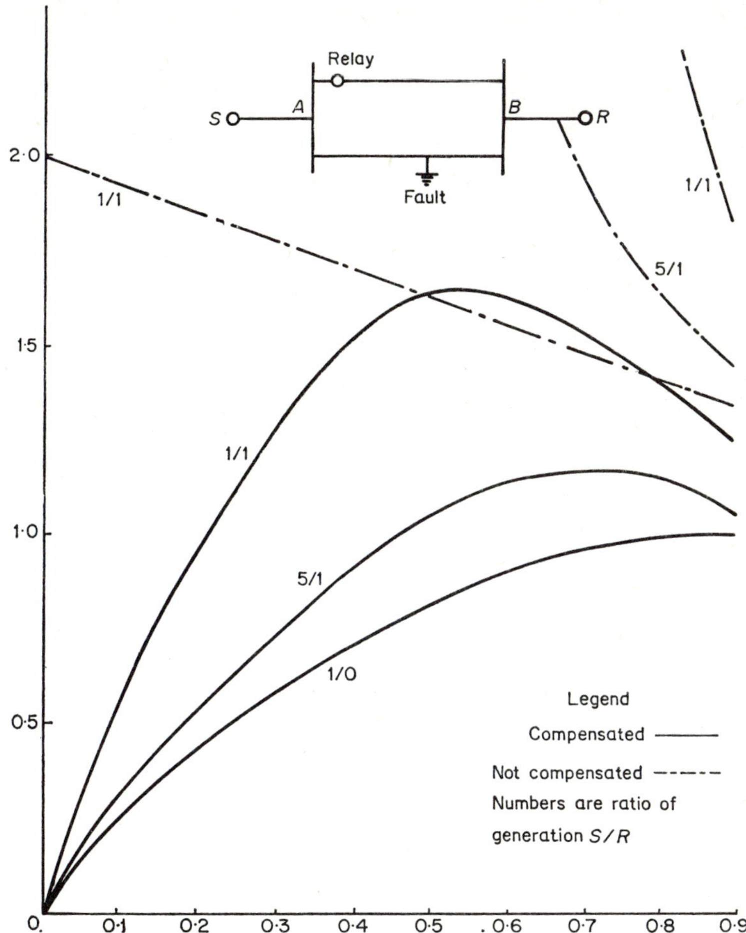

Fig. 11.21 Mutual induction on double circuit line. Impedance seen by a relay during a ground fault on a parallel line

sinusoidal current to be offset by a decaying d.c. component whose amplitude depends upon the moment in the cycle at which the fault occurs and whose duration increases with the X/R ratio (time-constant) of the circuit.

If the system impedance is not homogeneous (e.g. the source impedance Z_s is invariably more lagging than the line impedance Z_L) there will be a separate transient in the potential due to the sudden change in phase angle from the angle of V_s to the angle of IZ_L when the fault occurs.

These transients take the form of a decaying d.c. component which offsets the current or potential wave so that one half of the wave has a greater amplitude and a longer duration than the other. This will cause decreasing alternate positive and negative errors in both ampliitude and phase comparators, so that they will tend to overreach during every other cycle unless some means of obtaining average measurement has been provided (see Chapter 4, Section 4.4.4.).

In early days, when distance relays took more than 100 ms to operate and the time-constants of power systems were 40 ms or less, these transients had expired before the relay tripped and hence had no affect. Nowadays, however, with time-constants of 100 ms and operating times of 10 to 20 ms, it is necessary to eliminate their effect. This is done by the use of a replica impedance for matching the transients in the current and voltage inputs, as explained in Section 11.5.3.

11.5.1. Current Transient

During normal conditions the line voltage is the rated value and the phase current is between zero and the c.t. rating value, depending upon the load. When a fault occurs (Fig. 11.22) both the current and voltage suddenly change in amplitude and/or angle and, because of the reactive components of the system impedances, the current and voltage go through some transient values before settling down to the values corresponding to the short-circuited condition of the system.

Disregarding shunt loads and capacitance, the instantaneous value of the current is

$$i = I_{max} \{\sin(\omega t + \psi - \phi) + \varepsilon^{-(R/L)t} \sin(\psi - \phi)\} \qquad (11.10)$$

where $I_{max} = E/\sqrt{\{R^2 + (L\omega)^2\}}$, ϕ is the phase angle of the primary circuit ($\tan^{-1} L/R$), ψ is the time after voltage zero at which the fault occurs and t is the time since the inception of the fault.

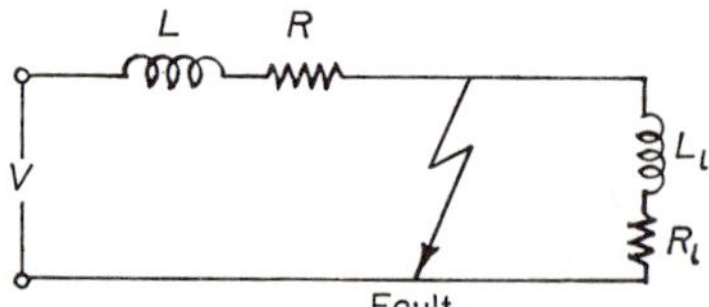

Fig. 11.22. Schematic primary circuit

The left-hand term represents the steady-state fault condition and the right-hand term the transient component which exists only during the transition from normal to fault conditions.

Inspection of the right-hand term will show that it is a d.c. offset which decays exponentially (see Fig. 11.23). Its size depends upon the moment the fault strikes and the duration increases with the X/R ratio of the system.

The maximum amplitude of the offset component is $E_{max}/\sqrt{(R^2 + X^2)}$ which is the same as the peak value of the steady-state current. This occurs when $\sin(\psi - \theta) = 1$, i.e. when $\psi = 90° - \theta$.

The initial instantaneous value of the current can thus approach double the normal peak value (Fig. 11.24) if the fault starts at the moment of current

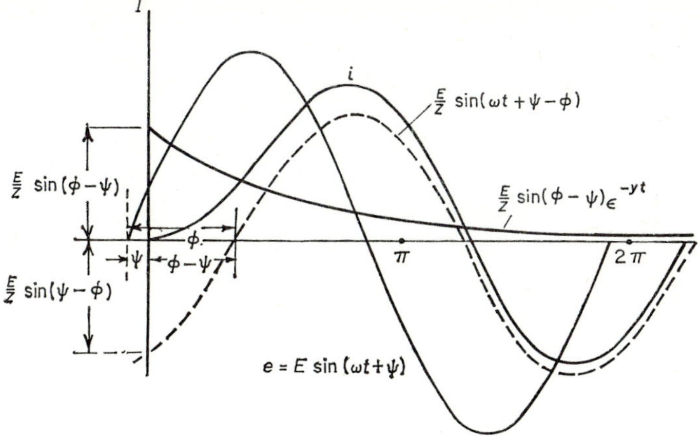

Fig. 11.23. Fault current transient

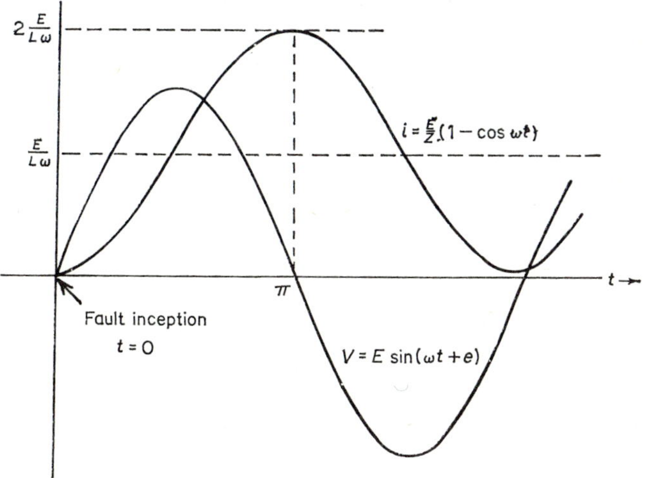

Fig. 11.24. Doubling effect in pure reactance circuit

maximum and R is near zero. In such a case the moment of current maximum would correspond to the moment of voltage zero because $\phi = 90°$. In modern E.H.T. systems ϕ can be as high as 83° so that these conditions are approached theoretically, but the likelihood of their happening is extremely small because the high impulse insulation level causes almost all flashovers to occur near voltage maximum.

The expression for the transient value of the fault current can be simplified and made more adaptable for calculation from system parameters if it is written in terms of I_{max} and the source/line impedance ratio Z_s/Z_L.

The peak value of the steady-state current is

$$I = \frac{E}{Z_s + Z_L} \tag{11.11}$$

$$= \frac{1}{Z_s/Z_L + 1} \cdot \frac{E}{Z_L}. \tag{11.12}$$

If $Z_s = 0$ in Eq. (11.11) it gives the maximum possible value of I, viz. E/Z_L. Substituting $I_{max} = E/Z_L$ and $m = Z_s/Z_L$ in Eq. (11.12) we get $I = I_{max}/(m + 1)$ and the complete expression for the current becomes

$$i = \frac{I_{max}}{m + 1} \sin(\omega t + \psi - \phi) - \sin(\psi - \phi)\, \varepsilon^{-(R/L)t}. \tag{11.13}$$

Since ψ is the time in radians after time zero at which the fault occurs and ϕ is the time between current zero and voltage zero, $(\psi = \theta)$ must be the time after current zero that the fault occurs. If we write $\delta = (\psi - \phi)$ and $y = R/L$ we get

$$i = \frac{I_{max}}{m + 1} \sin(\omega t + \delta) - \sin \delta \varepsilon^{-yt}$$

$$= \frac{I_{max}}{m + 1} \sin \omega t \cos \delta + \cos \omega t \sin \delta - \sin \delta \varepsilon^{-yt}.$$

This can be written

$$i = i_s \cos \delta + i_c \sin \delta \tag{11.14}$$

where $i_s = \{I_{max}/(m + 1)\} \sin \omega t$ and $i_c = \{I_{max}/(m + 1)\} (\cos \omega t - \varepsilon^{-yt})$. This simplification makes calculation of actual cases easier.

It can be seen that the transient component is zero if $\delta = 0$, which is when $\psi = \phi$, i.e. if the fault starts at current zero. It can also be seen that the d.c. offset is a maximum if $\delta = \frac{\pi}{2}$, i.e. if the fault starts at current maximum.

The effect of load flowing before the fault occurs is to decrease the amount of offset (Fig. 11.25). The expression for the current at any instant (Eq. 11.10) then becomes

$$i = \frac{E}{Z_s + Z_L} \{\sin(\omega t + \psi - \phi)\} + V\left\{\frac{\sin(\psi + \theta')}{Z_s + Z_L + Z_l} - \frac{\sin(\psi - \theta)}{Z_s + Z_L}\right\}\varepsilon^{-yt}. \tag{11.15}$$

where Z_l is the load impedance.

Here again the first bracketed term is the steady state current and the second bracketed term is the exponentially decaying transient component of the current. It will be seen that Eq. (11.15) is the same as Eq. (11.10) except for the extra term which represents the decay of the load current after it has been short-circuited by the fault (first term in second bracket).

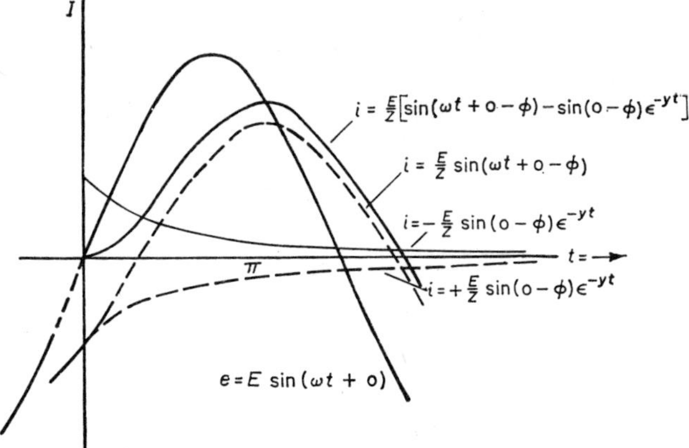

Fig. 11.25. Effect of previous load current

There is no offset when the two terms in the second bracket cancel, i.e. when

$$\tan \psi = \frac{\sin \theta - k \sin \theta'}{\cos \theta - k \cos \theta'}$$

$$\text{where } k = \frac{Z_s + Z_L + Z_l}{Z_s + Z_L}.$$

There is a maximum offset when

$$\tan \psi = \frac{\cos \theta - k \cos \theta'}{\sin \theta - k \sin \theta'}.$$

The maximum current value occurs (current doubling) when

$$\tan \psi = \frac{k \sin \theta' \cos \theta}{k \cos \theta' \cos \theta - 1}$$

11.5.2. Potential Transient

The voltage at the relay is

$$V_r = iR_L + L_L(di/dt)$$

$$= \frac{V}{Z}\left\{Z_L \sin(\omega t + \delta + \phi_L) - \left(R_L - \frac{L_L}{y}\right)\sin \delta\varepsilon^{-(t/y)}\right\} \quad (11.16)$$

$$\text{Now}\left(R_L - \frac{L_L}{y}\right) = Z_L\left(\cos \phi_L - \frac{\sin \phi_L}{\tan \phi}\right) = Z_L \frac{\sin(\phi - \phi_L)}{\sin \phi}$$

Substituting in Eq. (11.16)

$$V_r = V\frac{Z_L}{Z}\left\{\sin(\omega t + \delta + \phi_L) - \frac{\sin(\phi - \phi_L)}{\sin \phi}\sin \delta\varepsilon^{-(t/y)}\right\} \quad (11.17)$$

The first term is the steady-state voltage and the second term is the transient component. It will be seen that the transient term is zero if $\phi = \phi_L$, i.e. if the system impedance is of uniform phase angle (homogeneous) and also if $\delta = 0$, i.e. if $\psi = \phi$ removing the current transient.

The transient is a maximum if $\delta = 90°$. It is also a maximum when $\phi - \phi_L = 90°$ but this is obviously not a practical case.

11.5.3. Prevention of Relay Error

In voltage comparison the line voltage at the relay location is compared with the voltage drop across an impedance which is the replica of the impedance of the protected line section on a secondary basis. For a fault at the end of the protected section the line voltage at the relay is produced by the line current flowing through the impedance of the protected section. Consequently, its transient behaviour should be identical to that of the voltage produced by the same current flowing through a replica impedance, provided that the system impedance is homogeneous.

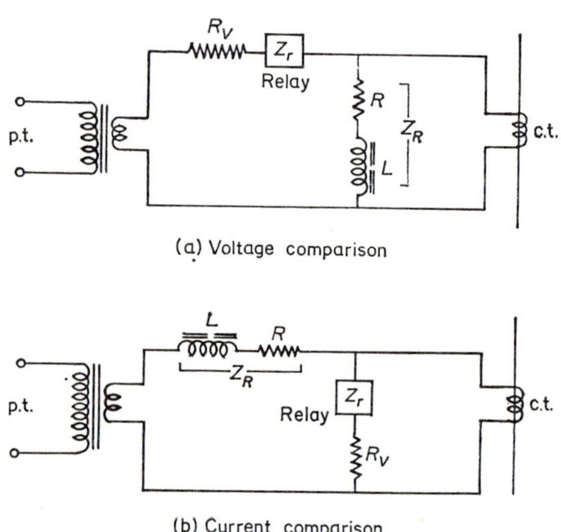

Fig. 11.26. Use of replica impedance

In both amplitude and phase comparators $(IZ_R - V)$ is compared with either IZ_R or V. Since all these terms contain the primary current transient equally reproduced, it cancels out in the measurement of impedance on a homogeneous system. On a non-homogeneous system where the source impedance is more lagging than the line impedance, the transient response of the line potential will not be the same as that of the current; hence the currents or voltages compared in the relay will not be matched for a fault at the end of the protected section, so that a fast relay can overreach if the fault current is considerably offset. In practice this is prevented by delaying the relay by 10 to 20 ms for faults near the cut-off point of Zone 1, (as explained in Chapter 4, Sections 4.3 and 4.4) so that its operating time exceeds the time constant of the protected line section.

In current comparison the replica impedance is connected in series with the potential circuit (Fig. 11.26b). This is the equivalent of the previous circuit because, in both cases, the relay current is $(IZ_R - V)/(Z_R + Z_r + R_V)$ where Z_r is the relay impedance and R_V is a series resistance. R_V is adjusted so as to make the circuit have a low time constant $(L_R + L_r)/(R_R + R_r + R_V) \nless 6$ ms. This prevents any transient error due to dying away of the current in the secondary of the p.t. without appreciably affecting the sensitivity of the relay, except in the case of E.H.V. lines of extremely high X/R ratio.

In such cases more relay amplification may be necessary, or linear couplers can be used.

It is interesting to note that the Fig. 11.26b becomes Fig. 11.26a if the lower end of the c.t. secondary is transferred to the upper end of the p.t. secondary.

Although the replica impedance matches the line impedance for a solid fault it does not allow for resistance in the fault. This is not important in a mho relay because any overreach merely makes the mho circle for transient measurement bulge sideways. This is actually an improvement because it provides some what more tolerance for fault resistance without increasing the distance reach.

It can be shown that in a reactance relay there is no change in the characteristic at all.

Where linear couplers are used instead of iron-cored c.t's, the mutual inductance of the linear coupler provides the equivalent of the inductive part of Z_R, so that only a shunt resistance is required to provide the non-inductive part of Z_R. In other words, the circuit looks like Fig. 11.26a with L_R omitted.

The inductive part of Z_R and the c.t. itself should have the lowest possible remanence to give a minimum transient error in measurement. Both these conditions are inherent in the linear coupler.

It is fortunate that on high voltage lines, where the longest time-constants can occur, the high X/R ratio of the lines makes the primary circuit impedance almost homogeneous and hence minimizes the error and potential transients. On lower voltage lines ϕ is smaller and the mismatching error is greater, but the current transient (which is the main one) is reduced by the lower X/R ratio of the primary circuit; furthermore, a slower operating time is generally permissible on lower voltage lines.

11.6. APPENDIX I: FAULT RESISTANCE

The following data are typical of a large power system such as the 50 c/s. 30,000 MVA system of the C.E.G.B. in Great Britain. The minimum short-circuit currents are as follows:

kV	Min MV	I_{min} (A)	c.t. ratio	I_{min} (sec) (A)
132	500	2190	500/1	4·38
275	2170	4650	1200/1	3·80
400	4600	6650	2000/1	3·33

Since the arc curve flattens above 1000 A to about 450 volts/ft, the maximum arc resistances for interphase faults will be as follows:

kV	I_{min} (A)	Conductor Spacing (ft)	V_{arc}	R_{arc} (Ω)	R_{arc} (sec) (Ω)
132	2190	10	4500	2·05	0·86
275	4560	22·5	10105	2·18	1·05
400	6650	34	15300	2·30	1·26

For phase-to-ground faults the arc length is determined by the arcing-horn gap and the maximum arc resistances are as follows:

kV	I_{min} (A)	Arcing Horn Gap. (ft)	V_{arc}	R_{arc} (Ω)	R_{arc} (sec) (Ω)
132	2190	3·5	1575	0·72	0·30
275	4650	6·5	2930	0·64	0·31
400	6650	10	4500	0·68	0·37

With no ground wires and high tower footing resistance values of 5 Ω for R_f are common and they can be much higher on dry or rocky ground. Figure 78, page 147 of *Symmetrical Components* by Wagner and Evans shows the range of ground resistivity for the U.S.A. which indicates a typical value of 100 metreohms and a maximum value of 10,000 metre-ohms.

12
Multi-input Comparators

Conic section characteristics – Hybrid comparators – Multilateral characteristics – Polyphase phase comparators – Polyphase amplitude comparator – Phase sequence comparators

The reader may wonder why this chapter did not immediately follow Chapter 4 (Comparators); the reason is that distance relays are at present the main field of application of multi-input comparators.

The fact that the fault resistance is independent of the line impedance means that a circular characteristic like that of the mho relay provides fine discrimination between faults and overloads only when the two impedances are comparable, i.e. when the fault area on an impedance diagram tends to be square.

To provide adequate discrimination in both the extreme cases of a short line with a high resistance fault and a long line subjected to a power swing, the reach of the relay along the X axis and along the R axis should be independently adjustable.

The first attempt to provide this separate control was to add an extra unit called a blinder [72]. Since then, various attempts have been made to achieve the required discrimination in a single unit having more than two inputs and some of these are now described.

Multi-input comparators can be either single-phase or polyphase and either integrating or instantaneous; they can employ either amplitude or phase comparison or both. They will be considered in three groups from the point of view of their characteristics rather than their construction or circuitry, viz.

Section 12.1. Conic Section Characteristics
12.2. Multilateral Characteristics
12.3. } Polyphase Measurement
12.4. }

The duality that exists between phase and amplitude comparators with two inputs does not exist where there are more than two inputs.

With three or more inputs, amplitude comparators have characteristics on the α-plane or the β-plane which are conic sections.

With three or more inputs, phase comparators provide characteristics which are multilateral, with $(\frac{1}{2}n)(n-1)$ sides for n inputs. The sides may be either straight lines or sectors of circles.

With suitable inputs from the three phases polyphase measurement can be obtained, so that a distance relay has the same impedance reach for any unbalanced type of fault.

In the following sections current and voltage inputs will be considered, since the main application of multi-input comparators is to distance relays.

12.1. CONIC SECTION CHARACTERISTICS

A conic section is the locus of a point which moves so that the distance from a fixed point (the focus) bears a constant ratio to its distance from a straight line (the directrix).

An amplitude comparator with suitable inputs can produce any type of conic section characteristic. These inputs may be simple quantities such as V and I or they may be derivations such as $(V - IZ_r)$, where Z_r is a replica impedance. Paradoxically, the second type is simpler and will be considered first.

12.1.1. Three-input Amplitude Comparator

The first documented multi-input amplitude comparator was described by the Norwegian engineers Bräten and Hoel [13]. It used a circulating current bridge (Fig. 12.1) with three inputs and produced an elliptical impedance characteristic.

The three inputs were $(V - IZ_r)$, $(V - IKZ_r)$ and kI, where Z_r and KZ_r were replica impedances with the same phase angle. The corresponding bridge currents were $|(V - IZ_r)/R|$, $|(V - IKZ_r)/R|$ and kI. Making

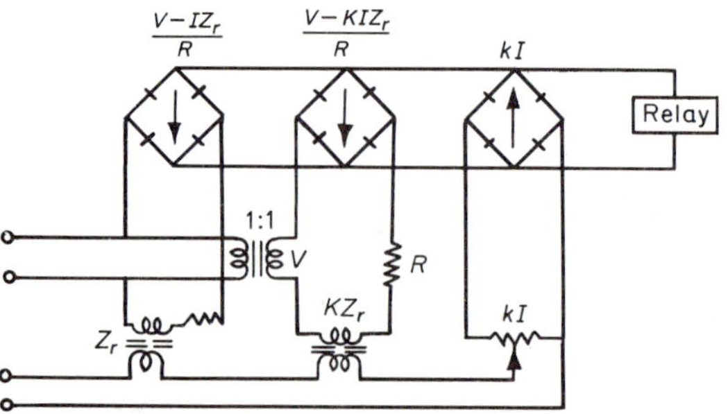

Fig. 12.1. Three-input amplitude comparator giving elliptical characteristics

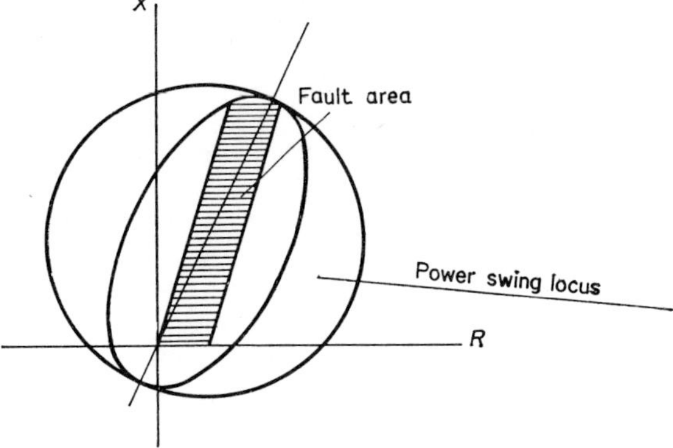

(a) ellipse less vulnerable to power swings

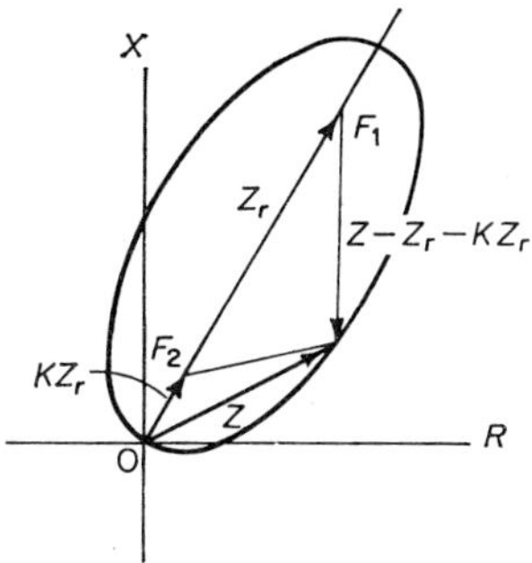

(b) ellipse passing through the origin

$|Z - Z_r| + |Z - KZ_r| < |Z_r + KZ_r|$

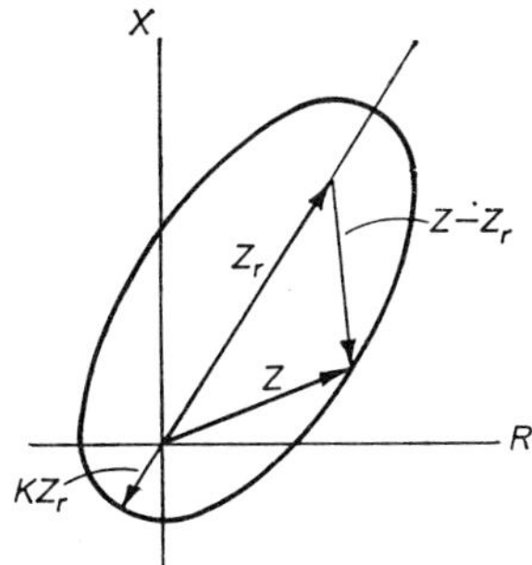

(c) offset ellipse with one focus at the origin

$|Z - Z_r| + |Z| < |Z_r + 2KZ_r|$

Fig. 12.2. Elliptical impedance characteristics

$k = |(Z_r + KZ_r)/R|$ the equation for operation becomes

$$|V - IZ_r| + |V - KIZ_r| < |I(Z_r + KZ_r)|$$

Dividing through by I gives the impedance equation

$$|Z - Z_r| + |Z - KZ_r| < |Z_r + KZ_r| \tag{12.1}$$

The first two quantities are the distances of the two foci from the curve and the third is the major diameter. Since the definition of an *ellipse* is the locus of points the *sum* of whose distances from two other points (foci) is constant, Eq. (12.1) must be that of ellipse going through the origin (Fig. 12.2b).

In order to be sure of operating on resistance faults close to the bus, it is preferable for the third-zone impedance characteristic to be offset to overlap the origin. An offset ellipse with one focus at the origin (Fig. 12.2c) is given by the equation

$$|Z - Z_r| + |Z| < |Z_r + 2KZ_r|. \tag{12.2}$$

Other conic sections can be obtained by varying the constants. For instance, a *hyperbola* is the locus of points, the *difference* of whose distances

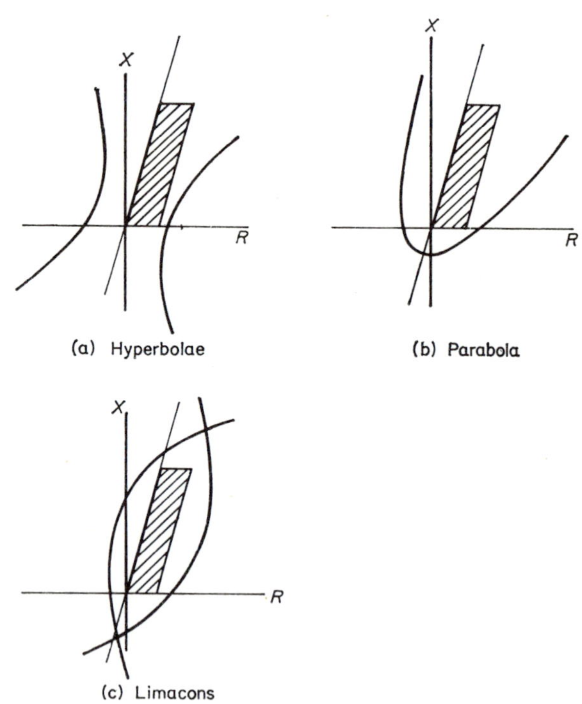

Fig. 12.3. Conic section impedance characteristics used as blinders

from the foci is constant (Fig. 12.3). Thus Eq. (12.1) can become a hyperbola going through the origin by changing the middle sign of that equation, viz.

$$|Z - Z_r| - |Z - KZ_r| < |Z_r + KZ_r| . \qquad (12.3)$$

In Eq. (12.1) the elliptical characteristic become cardiods or egg shapes if one of the terms, $|Z - Z_r|$ or $|Z - KZ_r|$, is multiplied by a factor other than unity. Similarly, in Eq. (12.3) the hyperbolae become limacons when one term is multiplied by a factor other than unity.

Conic section impedance characteristics other than the mho circle have their main field of application in the protection of very long lines. The ellipse, having a narrower waistline than the mho circle, is less liable to tripping during overloads or power swings (Fig. 12.2a). The hyperbola, the parabola and limacons can be used as blinders instead of a pair of angle impedance lines (Fig. 12.3).

12.1.2. Hybrid Comparators

These are combinations of amplitude and phase comparators wherein one type of comparator is supplied with one of its inputs from a comparator of the other type. This is made necessary where the inputs are related by a polar equation, viz.

$$Z = \frac{Z_r}{1 - K\cos(\phi - \theta)} \qquad (12.4)$$

This is a general equation for a conic section impedance characteristic (Fig. 12.4a). It is an ellipse if $K < 1$, a parabola if $K = 1$, a hyperbola if $K > 1$ and a circle if $K = 0$.

(*a*) *Amplitude comparator with auxiliary phase comparator.* The inputs necessary for obtaining the elliptical characteristic of Eq. (12.4) are IZ_r, V and $KV\cos(\phi - \theta)$. The first two inputs are fed directly to the amplitude comparator (Fig. 12.5a) but the cosine term is obtained from an auxiliary phase comparator [74], which uses the voltage IZ_r to polarize the line voltage V.

The diodes D_3 and D_4 clip the voltage IZ_r and the transistor T_1 amplifies the "squared" signal which is applied to the base emitter circuits of the transistors T_2 and T_3.

T_3 and T_2 are non-conducting unless an IZ_r signal is present. They become conducting while the voltage IZ_r is coincident with V, so that their output from this circuit is $V\cos(\phi - \theta)$.

The block diagram for the whole circuit is shown in Fig. 12.6a. The signals V and $-KV\cos(\phi - \theta)$ are added and compared in an amplitude comparator with the signal IZ_r. Tripping occurs if $IZ_r > \{V - KV\cos(\phi - \theta)\}$.

A much simpler alternative to Fig. 12.5a is to use a diode bridge phase comparator, as shown in Fig. 12.5b.

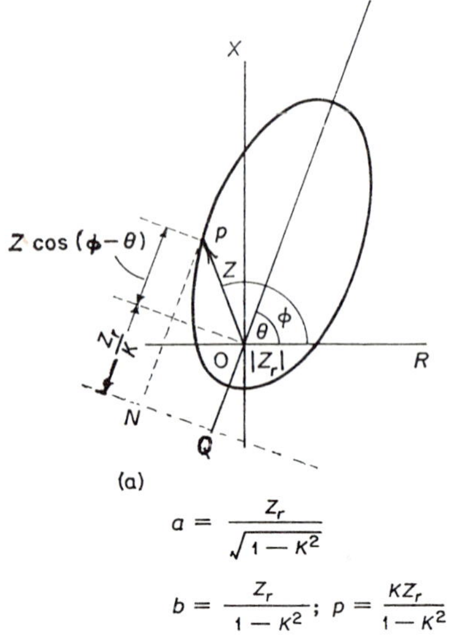

(a) focus at origin

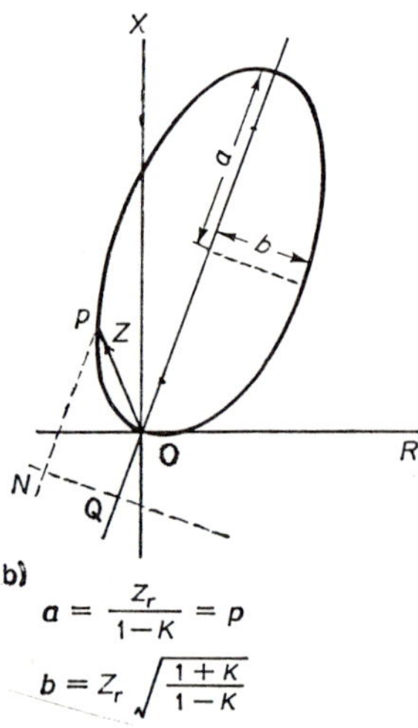

(b) going through the origin

Fig. 12.4. Ellipses from hybrid comparator

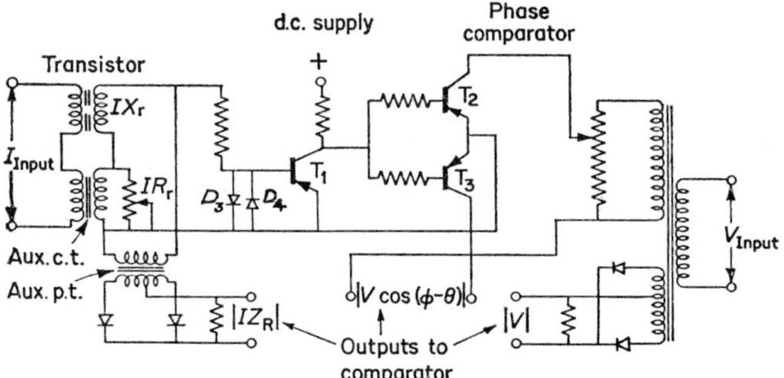

(a) Voltage comparison with transistor circuit

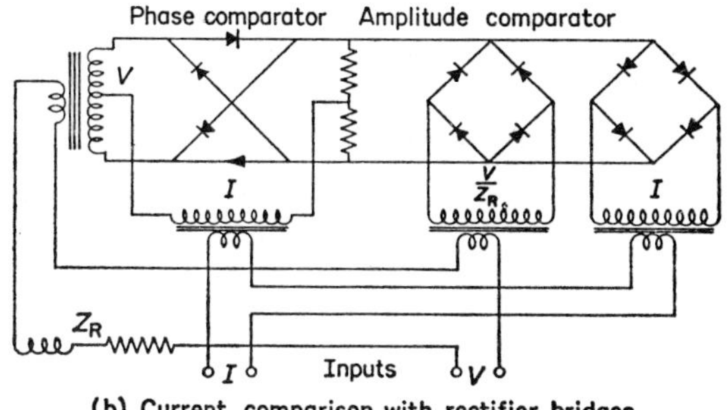

(b) Current comparison with rectifier bridges

Fig. 12.5. Amplitude comparator using auxiliary phase comparator

In Fig. 12.5b the inputs are I, $V/\mathbf{Z}_r$ and $\{KV\cos(\phi-\theta)\}\,\mathbf{Z}_r$. The relay threshold of operation is

$$\left|\frac{V}{Z_r}\right| - \left|\frac{KV\cos(\phi-\theta)}{Z_r}\right| = |I|$$

Multiplying through by Z_r/I the equation becomes

$$|Z| - |Z|\,K\cos\phi = |Z_r|\,. \tag{12.5}$$

This is equivalent to Eq. (12.4).

It can be related to the characteristic (Fig. 12.4a) by considering the definition of a conic section given in Section 12.1. From this

$$\frac{Z}{Z\cos(\phi-\theta)+0Q} = \frac{Z_r}{0Q} = \mathrm{K} \tag{12.6}$$

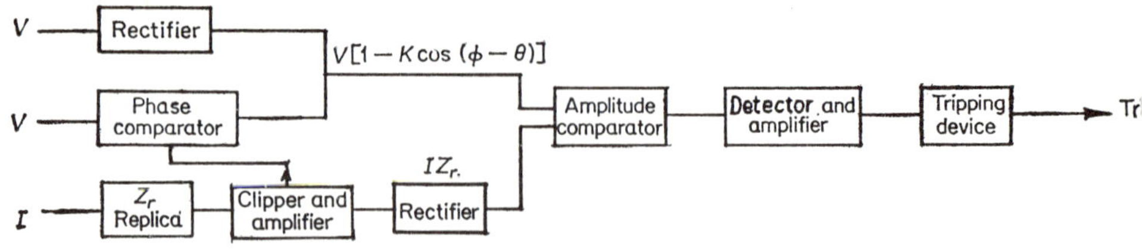

(a) diagram corresponding to equation 12.4

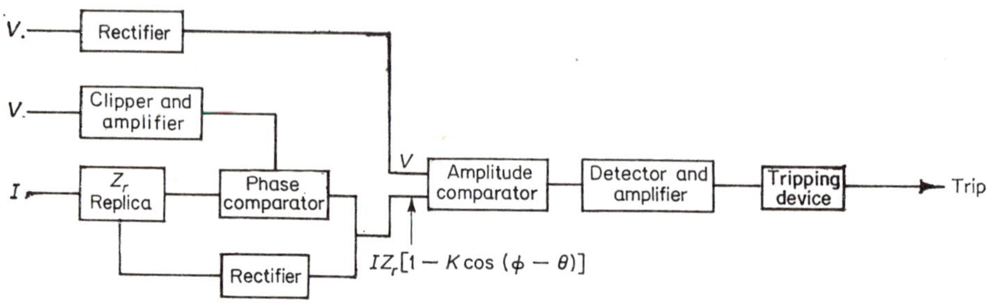

(b) diagram corresponding to equation 12.10

Fig. 12.6. Block diagrams for hybrid comparators

Substituting $OQ = Z_r/K$ in the first term

$$\frac{Z}{Z\cos(\phi-\theta)+(Z_r/K)} = K$$

therefore $$Z = KZ\cos(\phi-\theta) + Z_r$$

therefore $$Z = \frac{Z_r}{1 - K\cos(\phi-\theta)} \tag{12.4}$$

The axes, latus rectum, etc. can be obtained by rearranging the polar equation as follows:

Since $\mathbf{Z} = \sqrt{(R^2 + X^2)}$ and $\mathbf{Z}\cos\phi = R$, Eq. (12.5) can be written

$$\sqrt{(R^2 + X^2)} = |Z_r| + |KR|.$$

Squaring both sides

$$R^2 + X^2 = Z_r^2 + K^2R^2 + 2KRZ_r$$

therefore $$R^2(1 - K^2) - 2KRZ_r + X^2 = Z_r^2$$

therefore $$R^2 - \frac{2KRZ_r}{1-K^2} + \frac{X^2}{1-K^2} = \frac{Z_r^2}{1-K^2}$$

Rearranging in the form $(R^2/a^2) + (X^2/b^2) = p^2$ which is the equation of an ellipse

$$\left(R - \frac{KZ_r}{1-K}\right)^2 + \frac{X^2}{1-K^2} = \left(\frac{Z_r}{1-K}\right)^2. \tag{12.7}$$

If $K < 1$ this is an ellipse with axes $2Z_r/(1 - K^2)$ and $2Z_r/\sqrt{(1 - K^2)}$ whose centre is at $KZ_r/(1 - K^2)$ from the origin (Fig. 12.4a). The latus rectum is $2|Z_r|$. It is to be noted that, if $K > 1$, the characteristic becomes a hyperbola.

It is unlikely that a third zone elliptical characteristic will be required to be directional for faults close to the bus, i.e. to pass through the origin. If, however, such a characteristic were desired, the voltage term would have to be biased with current so that the threshold equation becomes

$$\left|\frac{V}{Z_r} - I\right| + \left|\frac{V}{Z_r}\right| K \cos(\phi - \theta) = I$$

Multiplying through by Z_r/I gives

$$|Z - Z_r| + |KZ \cos(\phi - \theta)| = |Z_r|. \tag{12.8}$$

By processing this equation as was done with Eq. (12.4) for the offset ellipse, it can be arranged in the form

$$\left(R - \frac{Z_r}{1 - K}\right)^2 + \frac{X^2}{1 - K^2} = \frac{Z_r^2}{1 - K^2}. \tag{12.9}$$

If $K < 1$ this is an ellipse whose axes are $2Z_r/(1 - K)$ and $2Z_r\sqrt{\{(1 + K)(1 - K)\}}$ and whose centre is at $Z_r/(1 - K)$. If $K > 1$ the characteristic is a hyperbola.

The equation for the mho circle going through the origin is simpler than the equation for the offset mho. This is not true of the ellipse because in equation (12.6) the numerator of the first expression is not $|Z|$ but $|Z - Z_F|$ where Z_F is the vector locating the focus, so that the equation is not a simple scalar one. The equation of the ellipse passing through the origin is

$$Z = \frac{2Z_r \cos(\phi - \theta)}{1 + K^2 \cos^2(\phi - \theta)}$$

and it would be impossible to provide inputs which would produce this characteristic. It is fortunate that it is not required in protective relaying.

A conic section impedance characteristic can also be obtained with the inputs V, IZ_r and $IZ_r \cos(\phi - \theta)$, conforming to the equation

$$Z = Z_r\{1 - K \cos(\phi - \theta)\} \tag{12.10}$$

If $K = 1$, the characteristic is a cardiod; it is a limacon if $K > 1$. Neither characteristic has any application as yet in protective relaying.

The first two inputs, V and IZ_r, are available as before but the third input, $IZ_r \cos(\phi - \theta)$, must be obtained by modifying the circuit of Fig. 12.5a or 12.5b so that the voltage signal polarizes the IZ_r signal instead of vice versa. This is done by interchanging the inputs IZ_r and V to the phase-comparator. The signals IZ_r and $-KIZ_r \cos(\phi - \theta)$ are then added and compared with the signal V in an amplitude comparator. Figure 12.6b shows the resulting block diagram of the circuit. Tripping occurs if IZ_r $\{1 - K\cos(\phi - \theta)\} > V$ and the characteristic is that of Eq. (12.10).

(*b*) *Phase comparator with auxiliary amplitude comparator*. The dual of the foregoing arrangement is to use an amplitude comparator to produce one of the inputs for a phase comparator. For the conic equation (12.4) the inputs are IZ_r and $\{V - KV\cos(\phi - \theta)\}$ which can be obtained [81] from the circuit of Fig. 12.7. The cosine term is inherent to a phase comparator with the inputs IZ_r and V, but the scalar term $|V|$ is a voltage having the amplitude of V and the phase angle of I.

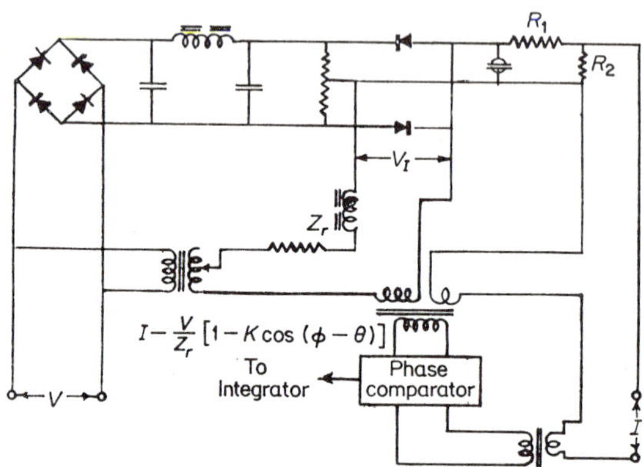

Fig. 12.7. Phase comparator with auxiliary amplitude comparator

Referring to Fig. 12.7, the current I is passed through the resistor R_2 and the voltage drop across IR_2 is reduced through the resistor R_1 to a value equal in magnitude to the rectified voltage input by the limiting device. Hence the voltage V_I has the amplitude of V and the phase angle of I and provides the term $KV\cos(\phi - \theta)$. Hence the phase comparator inputs are I and $I - V/Z_r$ $\{1 - K\cos(\phi - \theta)\}$ and tripping occurs when the angle between these inputs is less than 90°.

Lemon-shaped and apple-shaped characteristics can be obtained by making the operating angle of the phase comparator greater or less than 90°.

12.2. MULTILATERAL CHARACTERISTICS

The first multilateral characteristic [72] was achieved by the use of the blinder, i.e. a straight line impedance characteristic parallel to the line impedance which defined the discrimination between fault resistance and load impedance.

This required at least one extra relay unit in addition to the distance measuring relay. The elliptical characteristic combined these into one unit, but its rounded characteristic is suitable only for back-up protection where it can overlap the origin and thus give adequate protection for resistance faults near the bus.

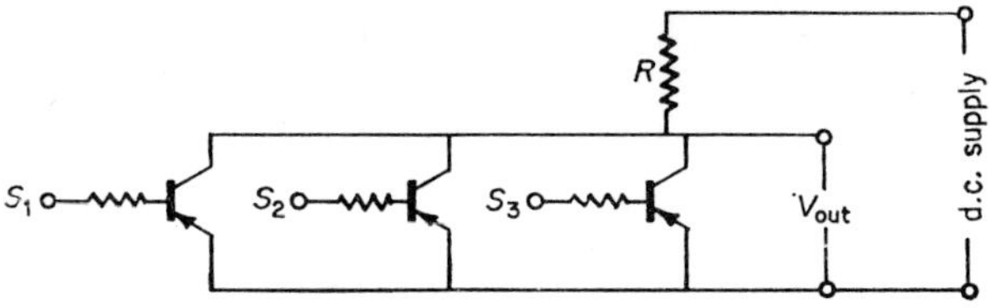

(a) Transistor circuit for AND or OR logic

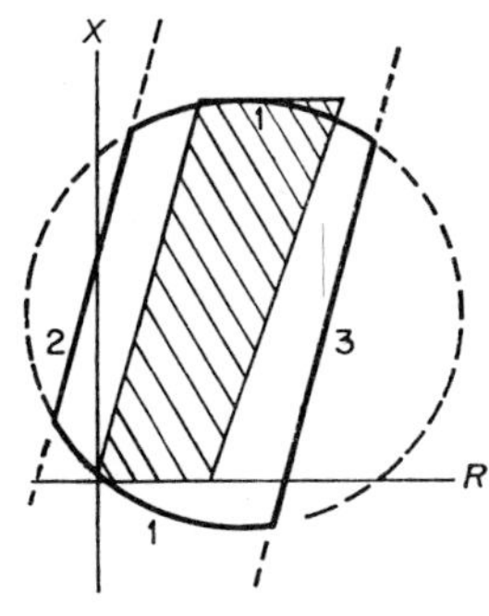

(b) AND circuit for mho circle with blinders

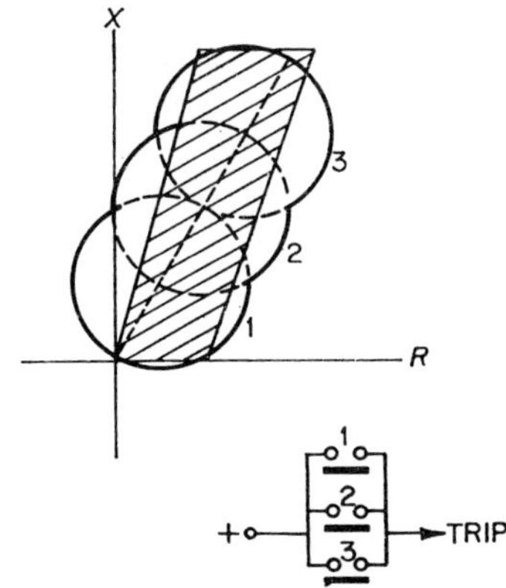

(c) OR circuit for three overlapping circles

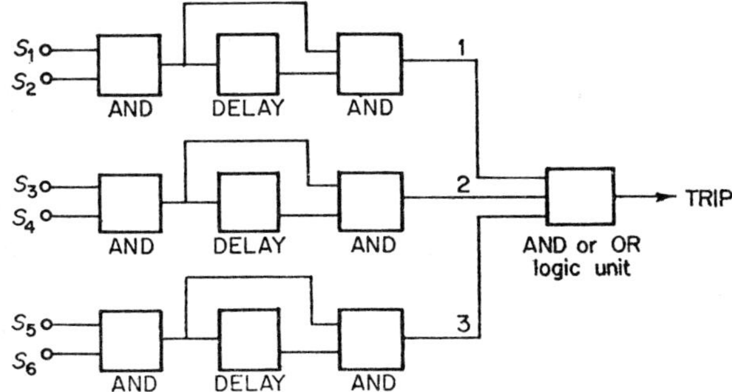

(d) Block diagram for (b) or (c) with phase comparators

Fig. 12.8. Combinations of two-input comparators

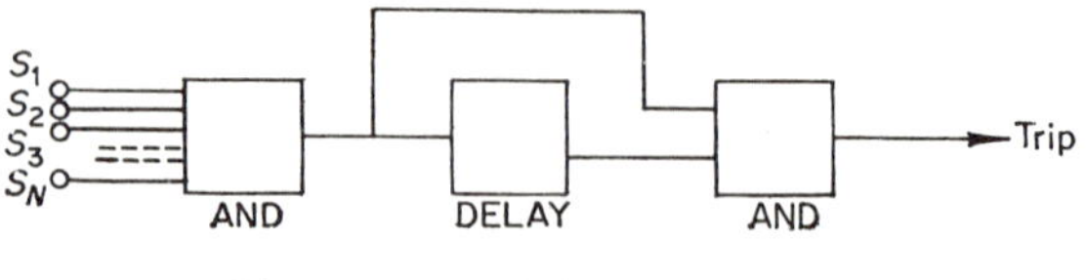

(a) Integrating N–input comparator

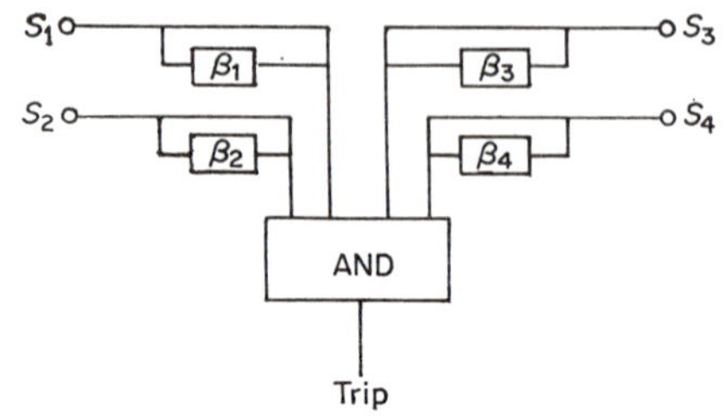

(b) Instantaneous 4–input comparator

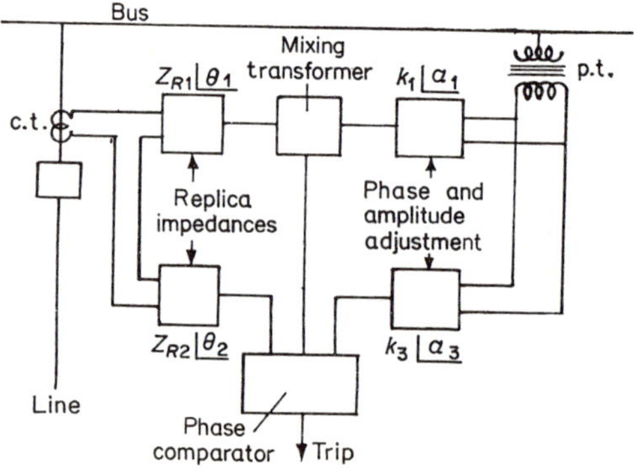

(c) Typical circuit arrangement for 3–input comparator

Fig. 12.9. Multi-input–phase comparators

Another proposal was the use of two or more overlapping circles as shown in Fig. 12.8c, i.e. one mho unit and one or more offset mho units displaced in the direction of the other end of the protected section.

The quadrilateral characteristic [78] formed by two bent line characteristics, as explained in Section 10.3.3., was the next solution to appear. This required two comparators so that it was no more economical than the mho unit with a blinder, but it did provide a more precise and controllable match of the relay characteristic to the fault area.

The multi-input phase comparator (Fig. 12.9b) is theoretically a more attractive solution [75] because it makes the necessary comparisons to obtain several boundary characteristics in a single comparator. Each input is automatically compared with each other input and the resultant characteristic is the area enclosed by the lines or circles resulting from all these comparisons.

With three inputs there are three different pairs for comparison; with four inputs there are six pairs and, mith n inputs, $\frac{1}{2}n(n - 1)$ pairs for comparison.

Actually, multi-input comparators are difficult to design because they require extra phase shifts which make them susceptible to transient overreach.

12.2.1. Three-input Phase Comparator [75, 79]

The most useful characteristics that can be obtained with a three-input comparator are the mho reactance characteristic and the mho characteristic with blinders. These were formerly obtained with a oombination of two-input comparators (Fig. 12.8c).

The mho-reactance characteristic is obtained by using these three inputs, (Fig. 12.9c):

$$S_1 = k_1 \underline{|\alpha_1 V}$$

$$S_2 = -k_2 \underline{|\alpha_2 V} + IZ_2 \underline{|\theta_2 - \phi}$$

$$S_3 = IZ_3 \underline{|\theta_3 - \phi}$$

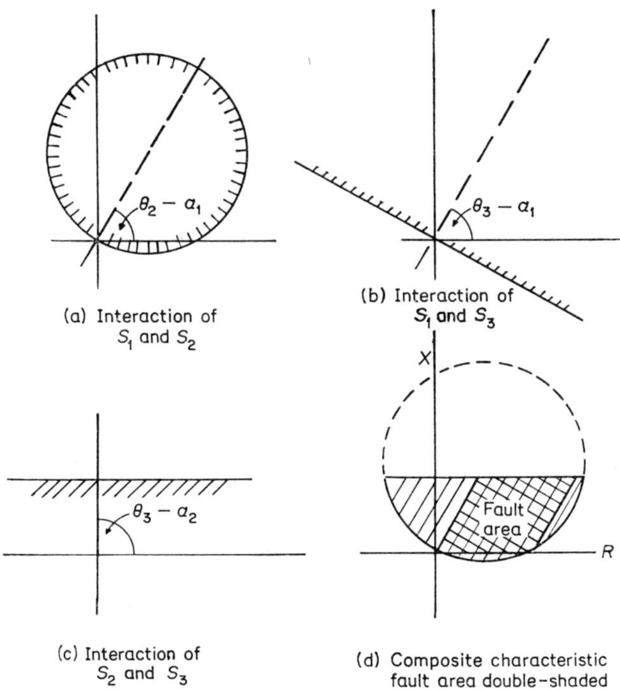

(a) Interaction of S_1 and S_2

(b) Interaction of S_1 and S_3

(c) Interaction of S_2 and S_3

(d) Composite characteristic fault area double-shaded

Fig. 12.10. Three-input mho-reactance phase comparator

where V is the line voltage, I the line current, ϕ the angle between V and I and θ_2 and θ_3 the phase angles of the impedances $\mathbf{Z}_2$ and $\mathbf{Z}_3$ which are connected in the current circuit. α_1 and α_2 are phase shifts of the voltage where required for locating the impedance characteristic.

Figure 12.10 shows the impedance characteristic resulting from the phase comparison of each pair of the inputs S_1, S_2 and S_3. It will be seen that the directional characteristic contributed by S_1 with S_3 does not influence the shape of the overall characteristic because it is tangential to the circle, i.e. they have the same characteristic angle since $\theta_2 = \theta_3$.

By suitably changing the value of α_2, the interaction of S_1 and S_2 will give lines swivelled 90° leading as shown in Figs. 12.11b and 12.11c, so that the resultant characteristic is that of a mho circle with blinders such as is used for the protection of long lines or for 3rd Zone units.

In the resultant characteristic of fig. 12.11d the arc tolerance is greater for faults near the remote end of the protected section; this is required if the total fault current is less there than for a fault neater the relay. If this is not the case the characteristic can be widened at the other end by adjustment of angle α_2.

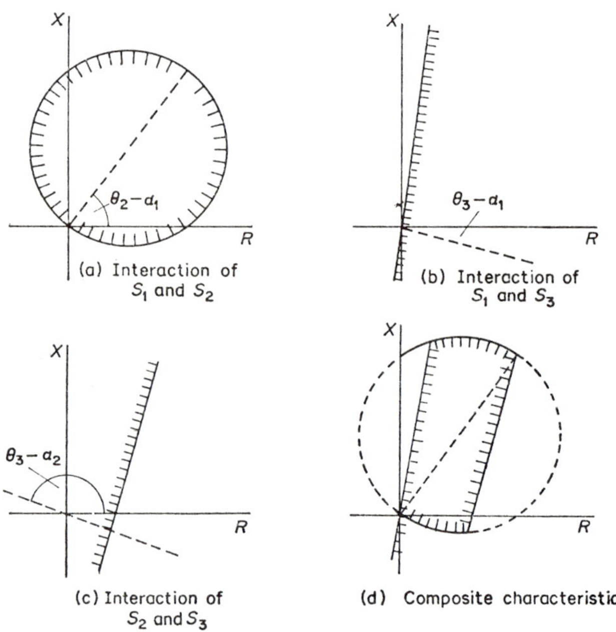

Fig. 12.11. Three-input phase comparator for mho circle with blinders

$$\text{Inputs } S_1 = k_1 V$$
$$S_2 = k_2 V - IZ_2$$
$$S_3 = IZ_3$$

12.2.2. Four-input Phase Comparator [75, 106, 107]

One advantage of the single multi-input comparator over a combination of two-input comparators is the elimination of the condition known as contact racing in electromagnetic relays. Even in a static relay the output of each two-input comparator has to be prolonged for a short time because they may not occur at the same time; consequently, wrong tripping may occur when the fault is cleared and the load returns, or vice versa. An example of this is the reactance relay which may be in the operated condition during a leading load and has to reset when a 3rd zone fault occurs, otherwise it will trip in Zone 1 time.

This contact racing can be eliminated by expedients such as having one unit control the other, but this slows tripping and only works one way. The single multi-input comparator avoids this trouble without loss of time; it trips immediately when all the conditions are simultaneously satisfied.

A possible weakness of the single multi-input comparator compared with separate two-input comparators is that it may be slower and less positive in action for faults just inside the characteristic. For instance, in a mho relay with a blinder, a resistance fault would be near the angle for maximum output for both units, but this would not necessarily be so for a three-input comparator.

It is also somewhat more difficult to arrange the circuit to avoid over-reach due to transients in the inputs, especially where one input is shifted in phase, or where voltage-producing impedances in the current circuit are not true replicas of the protected line section.

The four inputs required for a quadrilateral characteristic are:

$$S_1 = Z_1 I \angle\theta_1 - \phi - K_1 \angle\alpha_1 V$$

$$S_2 = Z_2 I \angle\theta_2 - \phi$$

$$S_3 = Z_3 I \angle\theta_3 - \phi$$

$$S_4 = K_4 \angle\alpha_4 V$$

To enclose the fault area, $Z_2 = X_r$, $Z_3 = R_r$ and $Z_1 = R_r + jX_r = Z_r$ so that the inputs are:

$$S_1 = IZ_r - V$$

$$S_2 = IX_r$$

$$S_3 = IR_r$$

$$S_4 = V$$

This gives a composite impedance characteristic as shown in Fig. 12.12. The mho circle caused by the interaction of S_1 and S_4 will not interfere with the rectangular tripping area if $Z_r = R_r + jX_r$ because the circle of diameter Z_r goes through the corners of the rectangle bounded by R_r

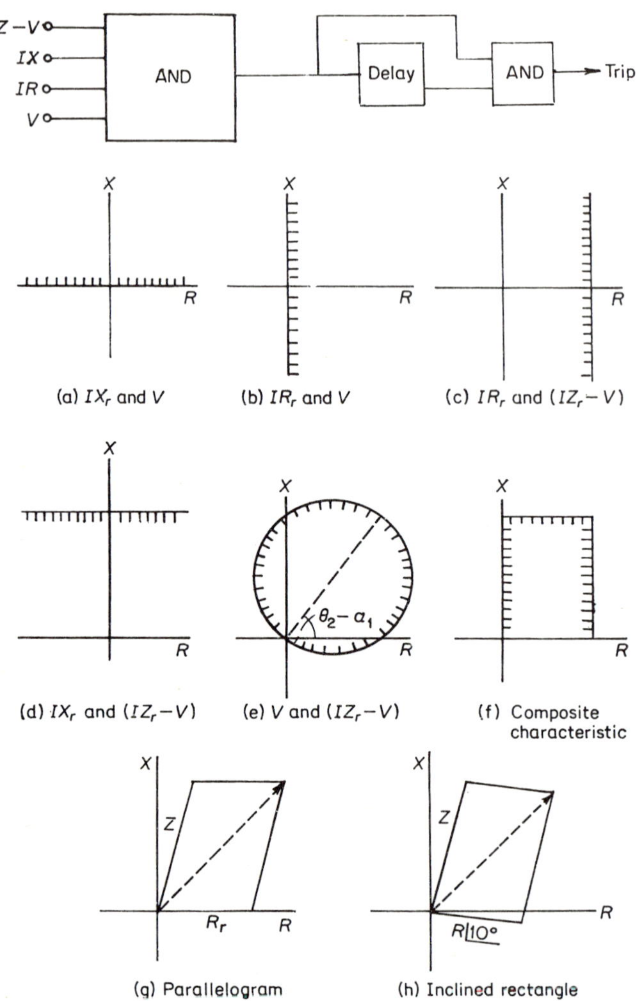

Fig. 12.12. Four-input phase comparator for quadrilateral characteristic

and X_r. For this reason the preferred shape of Fig. 12.12g is not possible unless the interaction between inputs S_1 and S_4 is eliminated by applying at least one of the inputs as a pulse [106, 107]; this will be discussed later.

Figure 12.12 also shows a block diagram of the multi-input comparator. Tripping occurs if all the equations resulting from comparison of all the inputs in pairs are simultaneously satisfied for the length of time set by the delay unit.

To obtain the preferred shape of Fig. 12.12g inputs S_1 and S_4 must be

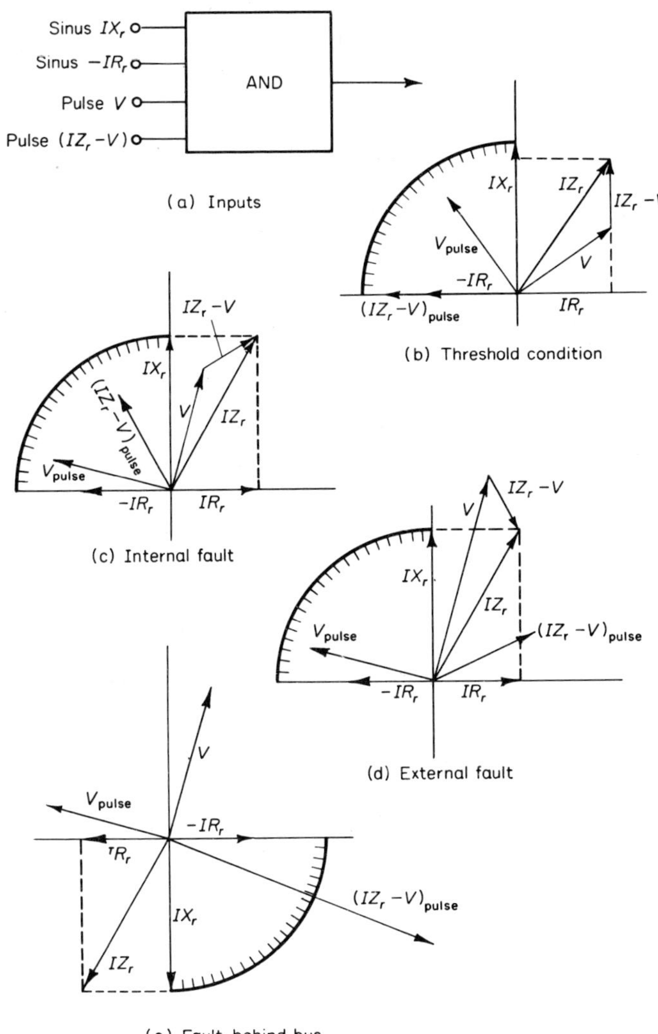

Fig. 12.13. Four-input phase-comparator with two pulsed inputs

applied as pulses as shown in Fig. 12.13 and, since they are pulsed as they pass through zero, they will lead V and $(IZ_r - V)$ by 90° (see Fig. 4.18a); hence the characteristics 12.12a, b, c and *d* will also be shifted by 90°.

Referring to fig. 12.13, tripping will occur when both pulses occur within ±90° of both IX_r and $-IR_r$, i.e. they will have to occur in the time zone between IX_r and $-IR_r$. It will be seen that this condition is fulfilled by an internal fault (Fig. 12.13c) but not by a fault outside the protected section

(Figs. 12.13d and 12.13e). Since the two pulse inputs do not interact with each other, there is no mho circle to interfere with the overall characteristic as there was in Fig. 12.12. Consequently there is no need for the composite characteristic to be a rectangle; it can be made a parallelogram (Figs. 12.12g and 12·12h) by using impedances other than pure X_r and R_r to define the characteristic.

The risk of tripping accidentally on interference transients can be avoided by inverting the operation so that the absence of pulses during the coincident period of the two sine waves causes tripping, or by having two comparators supplied with inputs 180° apart and connected by an AND gate so that the tripping signal has to occur in both half cycles in order to cause tripping. Shielding the leads is of course essential.

In order to provide more positive action for a resistance fault directly on the bus, the IX_r and IR_r inputs can be shifted 10° in the lagging direction, i.e. replace X_r by an 80° impedance Z_r and parallel the resistance R_r with a capacitor so that it lags the R axis by 10°. This rotates the whole characteristic by 10° in the lagging direction, as shown in Fig. 12.12h.

12.3. POLYPHASE PHASE COMPARATORS

In order to measure the correct distance to the fault under all fault conditions, a phase and ground distance relay requires six measuring units or comparators, one in each phase for single-phase ground faults and one in each phase-pair for interphase faults. In the following section it will be show that the same measurement for all kinds of faults involving any of the phases can be made by a single-phase comparator or an amplitude comparator, thereby effecting considerable economy in cost and space. These comparators also avoid the contact races and wrong measurement that occur with ordinary relays and switched relays during changing faults.

12.3.1. Distance Relay

The simplest polyphase distance relay comparator determines the sequence of the three compensator wye potential $(V_a - I_aZ_r)$, $(V_b - I_bZ_r)$ and $(V_c - I_cZ_r)$ as shown in Fig. 12.15. One or more of the compensated potentials will be zero for a fault at the impedance setting of the relay Z_r. One or more of them will be reversed in polarity or phase sequence for a fault within the relay setting.

Figure 12.16 shows the operation of the relay for various types of faults. The potentials supplied to the relay are of the form $(V - IZ_s)$ where V is the generated wye voltage and Z_s is the impedance between the relay and the generating source. The compensated potentials supplied to the comparator are further depleted by the voltage drop IZ_r in the replica impedance and are of the form $V - I(Z_s + Z_r)$. The compensated voltage is equal to the voltage at the fault and is zero for a fault at the reach point of the relay.

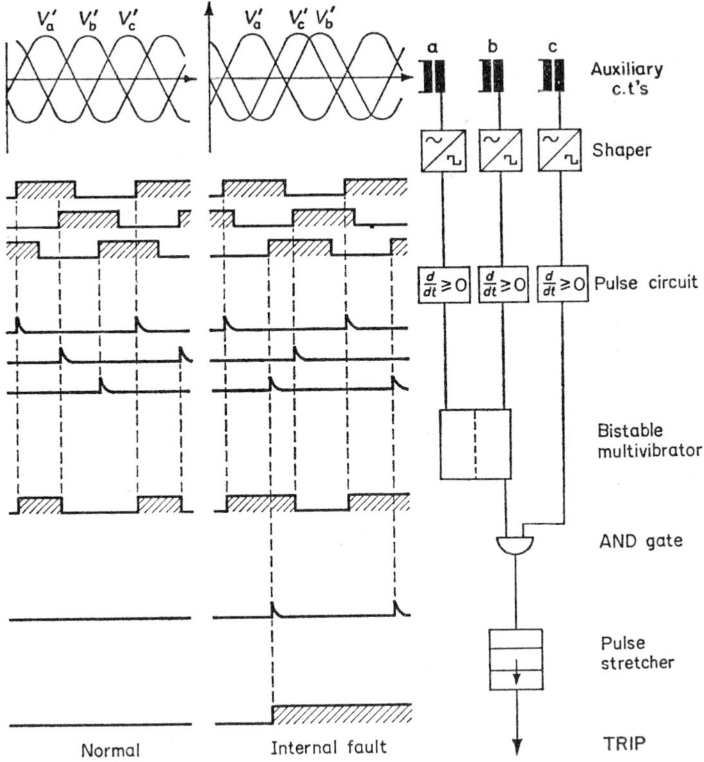

Fig. 12.14. Sequence type of phase comparator (B.B.Cie.)

An internal fault between phases B and C (Fig. 12.16b) reverses the sequence of the compensated voltages V_b' and V_c'. An internal phase-A to ground fault (Fig. 12.16d) reverses the compensated voltage V_a' and at the same time reverses the sequence of the three phase voltages. An internal three-phase fault reverses all three compensated voltages, leaving their sequence unchanged. This is a weakness of all polyphase relays. Where three-phase faults are expected, an extra single-phase relay should be provided for taking care of them. This method does not work so well on a non-homogeneous system, such as a resistance grounded system, unless the phase angle of the replica impedance Z_r is made less lagging, as shown in Figs. 12.16g and 12.16h.

The comparator can be a phase sequence detector of the type shown in Fig. 12.14 [88]. The input from one phase starts a signal which is turned off by the next phase; the third phase makes a pulse which, under normal conditions, would occur after the signal had been turned off (as seen in the left-hand side of Fig. 12.14). For an internal fault, however, the phase sequence will be changed so that the spike from the third phase will occur while the signal is still on (as seen in the right-hand diagram) and tripping will occur.

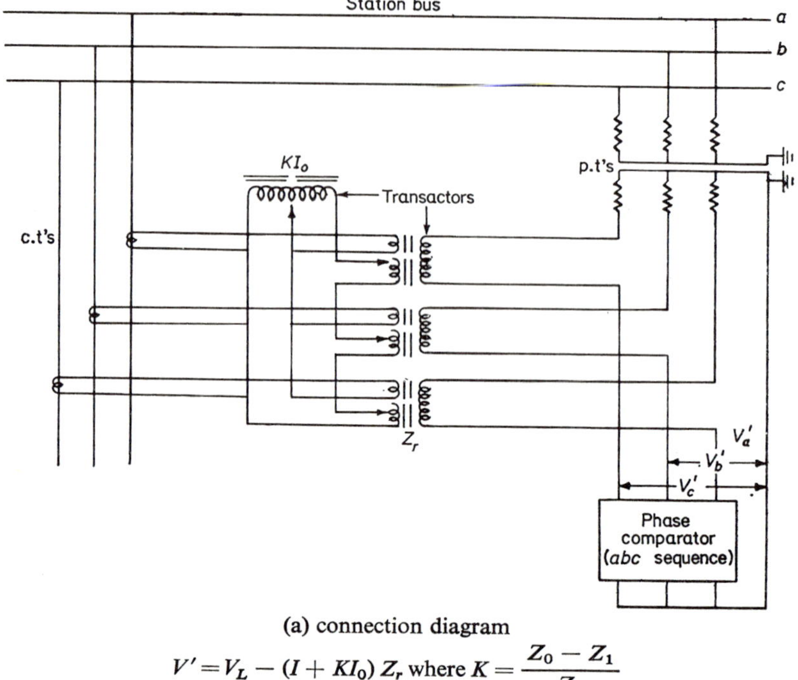

(a) connection diagram

$$V' = V_L - (I + KI_0)\,Z_r \text{ where } K = \frac{Z_0 - Z_1}{Z_1}$$

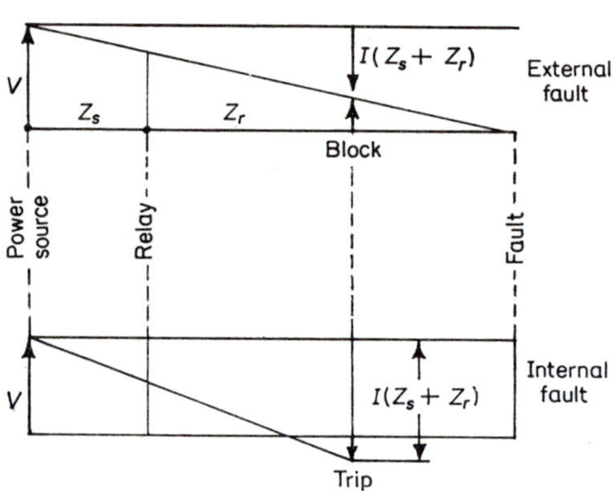

(b) principle of distance measurement

Fig. 12.15. Polyphase distance relay using phase sequence detector

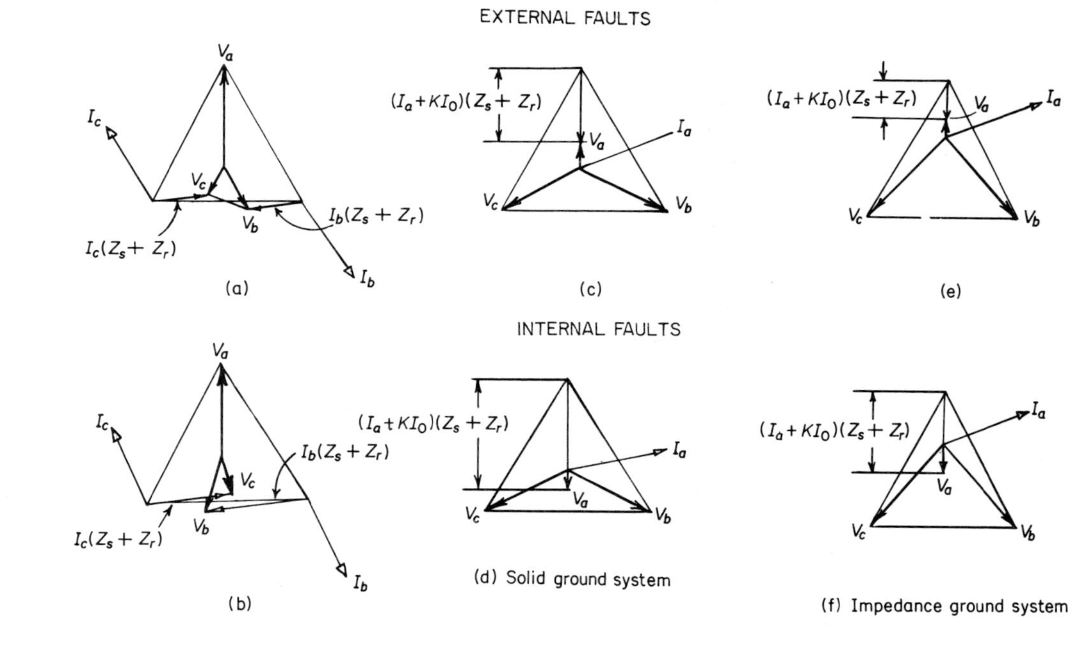

Fig. 12.16. Vector diagrams explaining operation of polyphase distance relay
(a) (b) interphase faults
(c) (d) ground fault on solidly grounded system
(e) (f) ground fault on impedance grounded system

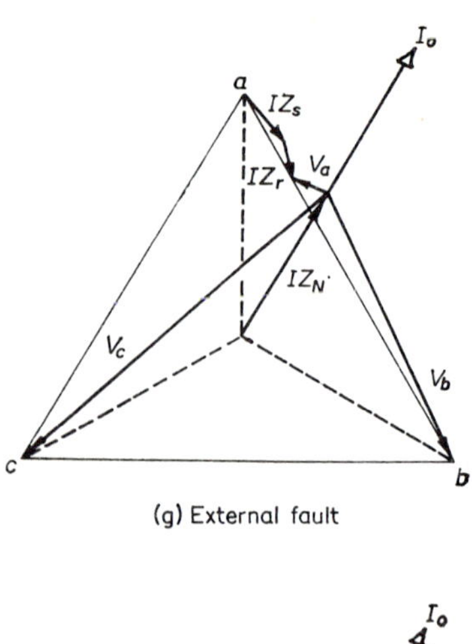

(g) External fault

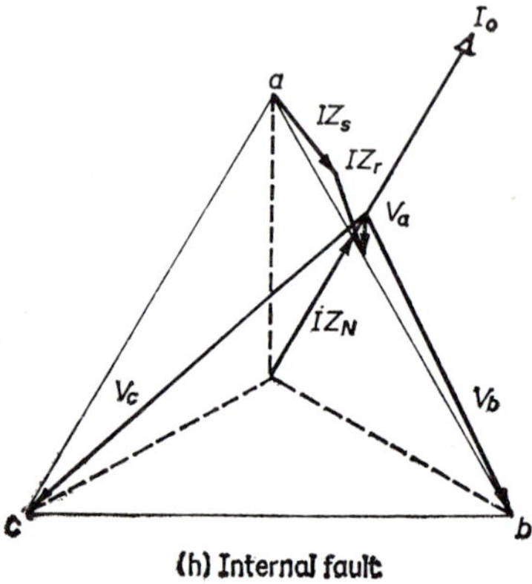

(h) Internal fault

Fig. 12.16. (g) (h) ground fault on resistance grounded system

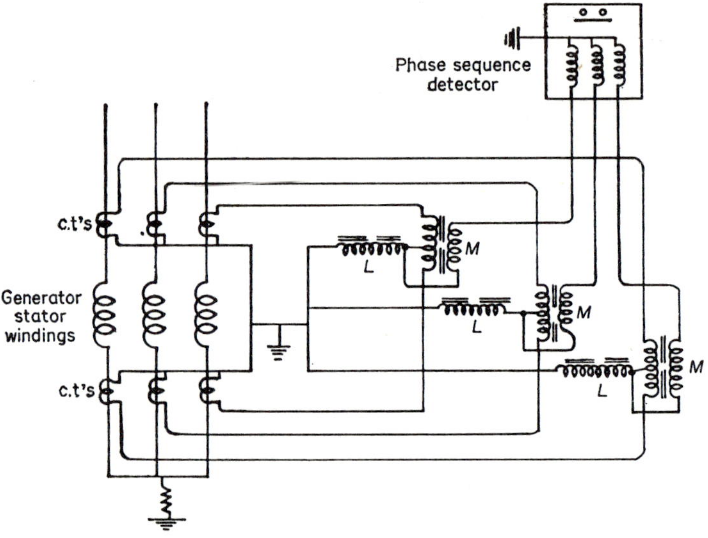

Fig. 12.17. Polyphase differential relay using phase sequence detector

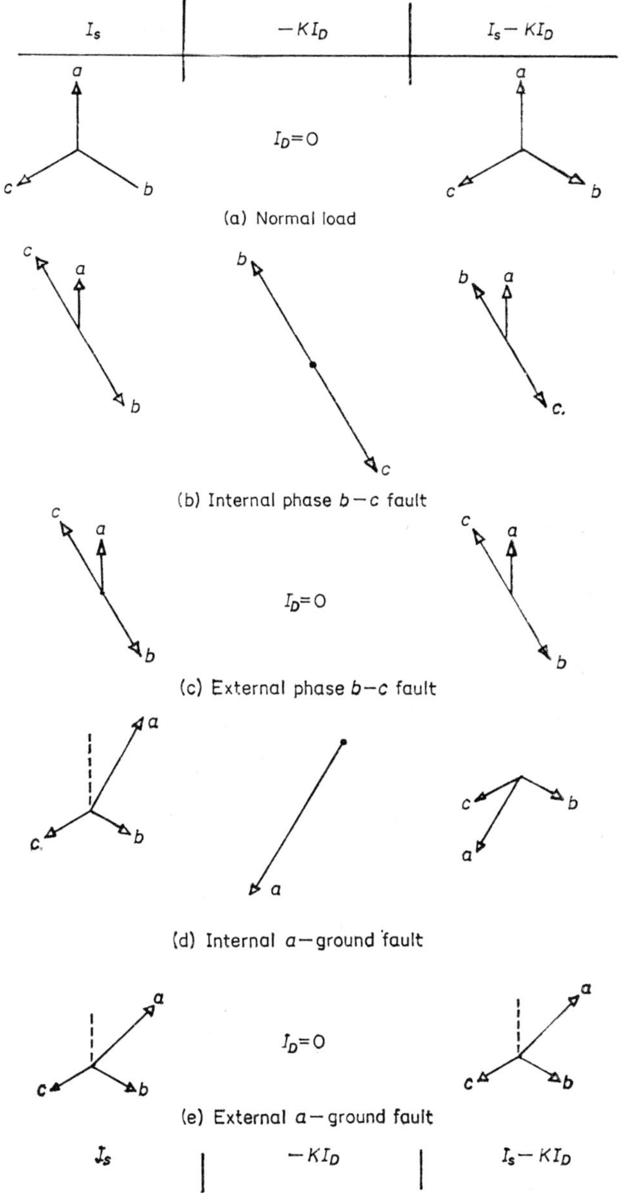

Fig. 12.18. Vector diagrams explaining operation of polyphase differential relay

12.3.2. Differential Relay

The same principle can be applied to differential current protection (longitudinal and transverse) for protecting generators or transformers (Fig. 12.17). The three outputs, $I_S - KI_D$ have normal phase sequence when the through current I_S exceeds K times I_D, the difference current, in one or two phases.

I_S is the sum of the currents entering and leaving a phase winding; I_D is their difference; K is a design constant which determines the setting of the relay and $100/K$ is the per cent slope of the operating characteristic (see Chapter 1, Section 1.4.2.).

When a fault occurs in one of the phase windings of the generator or transformer, $KI_D > I_S$ in that phase so that $I_S - KI_D$ becomes negative and the sequence of the three outputs is reversed. The sequence is also reversed in the rare case of a fault between phases, but not in the case of a fault between all three phase windings; since this case is almost impossible, it is not necessary to provide for it.

12.4. POLYPHASE AMPLITUDE COMPARATOR

12.4.1. Distance Relay

When a fault occurs, the current in the faulted phase conductor or pair of conductors is greater than the current in any other conductor. At the same time the voltage of the faulted conductor to ground or to the other faulted conductor is the lowest. Consequently, to measure the correct impedance to the fault irrespective of which phase or phases are involved, requires a comparator that automatically compares the highest current with the lowest voltage. Such a comparator would take the place of the six comparators in a normal phase and ground distance relay.

Figure 12.19a shows the basic principle for single-phase ground faults. It will be seen that such a relay would have a partial mho characteristic but that it would require directional supervision.

Figure 12.19d shows how the addition of the diodes to separate the six current inputs and the six potential inputs eliminates interaction between the phases and makes the operation a straightforward amplitude comparison of the highest current with the lowest voltage.

Figure 12.19d is the circuit for an impedance relay but, by the use of mixing transformers to obtain the inputs shown in Table 10.1 (Chapter 10) for amplitude comparators, any other type of characteristic can be obtained. For instance, a mho characteristic can be obtained by using $IZ_r - V$ instead of V on the potential side.

The relay circuit, marked 'R' in Fig. 12.19a, is a sensitive polarized relay or an equivalent static tripping device. The differential analyser in Fig. 12.19d compares the highest of the six currents with that of the lowest of the six potentials and gives a tripping signal to a thyristor trigger circuit or a reed relay (see page 131).

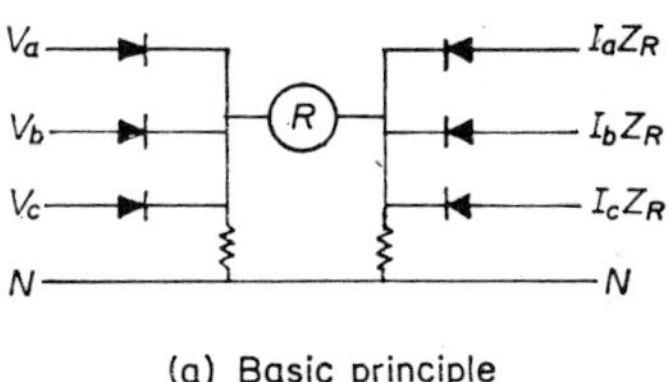

(a) Basic principle

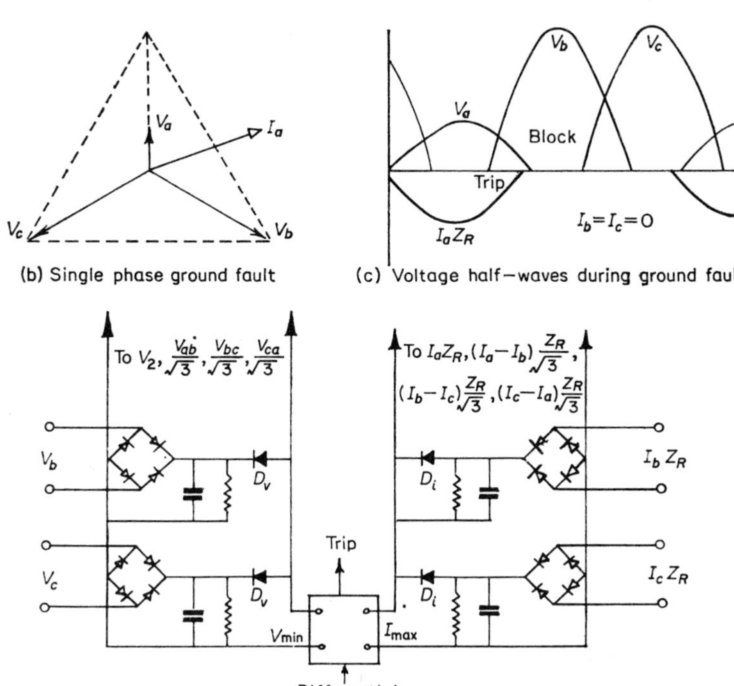

(b) Single phase ground fault

(c) Voltage half-waves during ground fault

Differential analyser

(d) Simplified circuit of two phases of the voltage comparator

Fig. 12.19. Polyphase distance relay using amplitude comparator

The action of the rectifier D_i in selecting the highest current and the rectifier D_v in selecting the lowest potential is similar to that of the overcurrent and undervoltage selectors used in electromagnetic switched distance relays. In other words, the static circuitry for a polyphase amplitude comparator relay is the same as for a switched relay.

12.4.2. Differential Relay

It is obvious that polyphase undervoltage and overcurrent relays can be made by using the appropriate part of Fig. 12.19d. Differential relays can also be made by using the difference currents as inputs on the current side

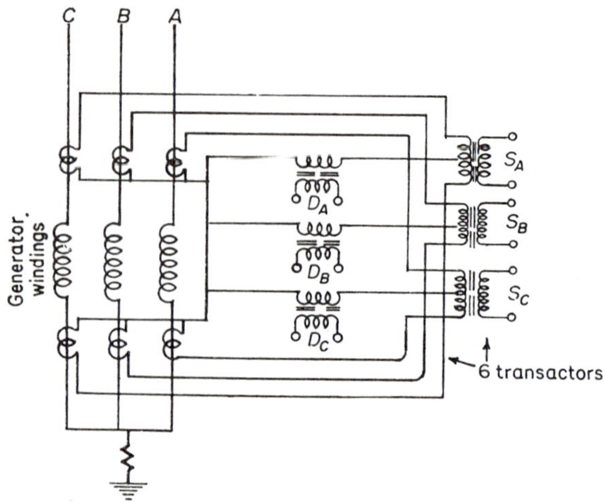

Fig. 12.20. Polyphase differential relay using amplitude comparator

Fig. 12.20. The Sum (S_A, S_B, S_C) voltages are connected to the restraining circuits' of Fig. 12.19d. The difference (D_A, D_B, D_C) voltages go the operating (right-hand) circuits of Fig. 12.19d

of the comparator and the sum currents on the potential side. The currents would be turned into potentials as shown in Fig. 12.20 by feeding them through transactors which would also insulate them from each other. Taps on the secondary sides of the transactors would provide adjustment for the slope of the operating characteristic.

13

Heating, Harmonics and Load-Shedding

Overheating of electrical apparatus – Thermal protection of lines, cables, motors, alternators and transformers – Harmonic relaying – Sensitive ground fault detector – Harmonic protection of lines, transformers and industrial equipment

These are two fields of protection which have received insufficient attention in the past. Protection against overheating has been rather rough and ready, while harmonics have been regarded mostly as a nuisance, to be eliminated from the relay circuit.

13.1. TEMPERATURE RISE

The life of electrical equipment depends largely on how much it has been overheated. It has been established that the life of insulation is halved for every 11°C rise in the temperature at which it is continuously run (Fig. 13.1). There appear to be no data available on the effect of periodic overheating, which is the more usual condition; but it is obvious that each time the apparatus is overheated it will add to its accumulative deterioration until eventually the insulation will fail at some point in the winding and the equipment will have to be taken out of service.

In performing its function, all electrical equipment is subject to a certain waste of energy in copper (I^2R) and iron ($\Phi^{1.6}S$) losses. When the equipment is energized its temperature increases until the heat dissipated from its surface equals the heat generated by these losses (Fig. 13.2). It will be seen from the right-hand curve of Fig. 13.2 that the time-current characteristic required for a relay to operate when a winding reaches maximum permissible temperature is $I^2t = K$.

The problem of matching the relay characteristic to the protected equipment is complicated by (*a*) variation of ambient temperature, (*b*) variable cooling factors of the equipment, (*c*) non-homogeneity of the thermal time-constant of the equipment.

Item (*a*) can be taken care of by ambient compensation but, to be effective, the compensator should be near the equipment; in practice, however, it is usually within the relay case.

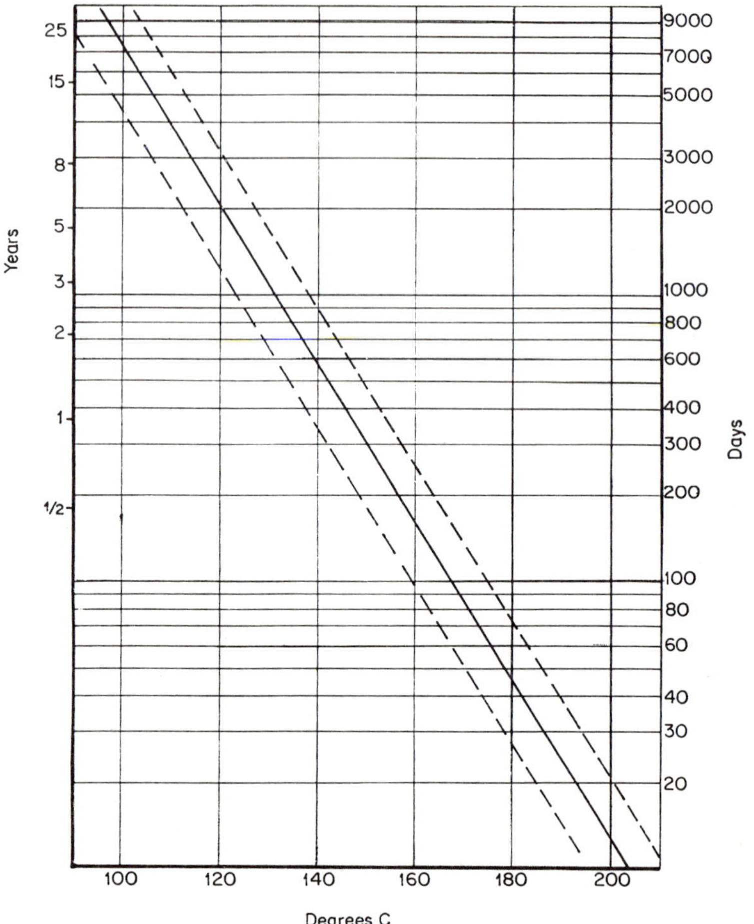

Fig. 13.1. Effect of temperature upon life of insulation

Item (*b*) is exemplified in a machine or power transformer which has forced cooling above a certain load. Again the relay could have automatic compensation which could be switched on at the same time as the forced cooling.

Item (*c*) refers to the difference in heat dissipation of that part of the conductors in the slots of the machine compared with the part outside in the overhang. In the slot the conductors have a longer time-constant but can reach higher temperatures than the part outside. The relay must match the envelope of the temperature-time curves of the conductor.

The heat dissipation is by conduction, convection and radiation. In cables and overhead lines the losses are almost entirely I^2R. The dissipation is by conduction in underground cables and by convection in overhead lines.

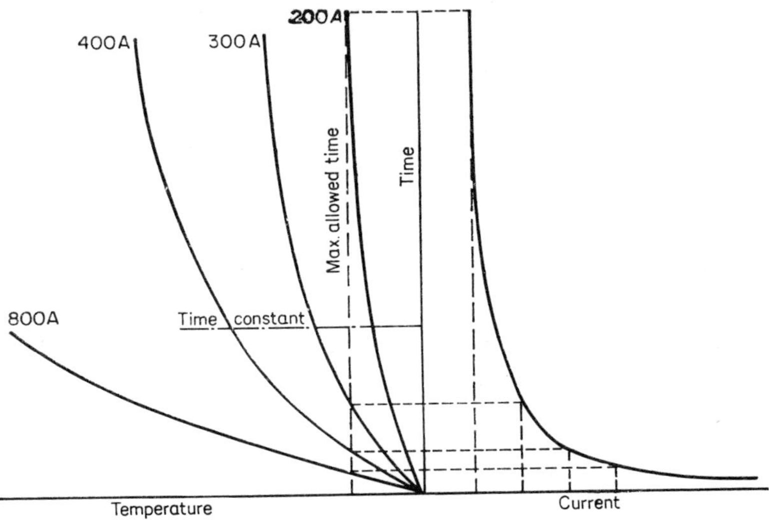

Fig. 13.2. Exponential heating curve due to load

In oil-filled transformers the heat dissipation (cooling) is first by conduction to the oil and then by convection via cooling tubes or a radiator to the air or water supply. In power transformers the iron loss is something like a quarter of the copper loss at full load and increases to equality at about half load since the copper loss is proportional to the square of the current. The copper loss is less in distribution transformers which are designed for overload.

In motors and generators the dissipation is by conduction through the insulation and the iron core and thence to cooling air. The copper losses are two to four times the other losses, which consist of iron losses and mechanical losses, although the latter predominate in high-speed machines.

The temperature rise of a winding after t seconds is

$$T = T_{max}(1 - \varepsilon^{t/K}) \tag{13.1}$$

where T_{max} is the final temperature and K is the thermal time-constant which is related directly to the thermal mass of the equipment and inversely to the

TABLE 13.1.

Typical values of losses in machines

Machines	Copper %	Iron %	Mechanical %
Motor 1500 h.p.	0·3	0·7	0·7
Motor 9000 h.p.	1·5	0·3	1·0
Generator 100 MW	9	3	4
Generator 660 MW	10	1	3

cooling surface. In a high voltage (25 kV) machine the electrical insulation, which is also thermal insulation, delays the flow of heat from the conductor to the slot and hence makes the conductor heat up quickly on overload.

From the exponential Eq. (13.1) the time taken to reach a certain temperature rise T is

$$t = K \log_\varepsilon \left(1 - \frac{T}{T_{max}}\right). \tag{13.2}$$

The initial slope of the exponential curve is T_{max}/K and, assuming pessimistically that the first part of the curve is straight, then the time t to reach the permissible temperature T_p is

$$t = \frac{K'T_p}{T_{max}}. \tag{13.3}$$

The temperature of the stator winding can best be checked by thermistors actually embedded in the winding. These devices are about the size of a matchhead and have a non-linear characteristic so that, when connected in series with a relay and the supply voltage, they can produce very definite relay action at a specific temperature, usually about 120°C. Owing to their low thermal inertia they operate and reset quickly and within a small range, usually about 3°C.

13.2. OVERHEATING OF LINES AND CABLES

Thermal relays are, at present, seldom used on overhead lines because their heat dissipation is so good that it would be uneconomical in voltage drop to load a line to anywhere near its heating limit. Underground cables, on the other hand, can reach dangerous temperatures on heavy overloads and are generally protected by time-current relays. Another exception is electric railway overhead conductors.

The maximum temperature rise of the protected equipment is proportional to the heat generated in it. In the case of a cable, this is wholly I^2R, so that the temperature rise is a function of I^2Rt. Hence a time-current relay whose operating characteristic is I^2t should provide good protection and is used in Europe. In the U.K. the ordinary type of inverse time-current relay is used with a relatively low setting of 1·5 times the c.t. rating; this is because regulation is involved as well as temperature.

Ambient temperature compensation is provided in most thermal protection relays because, although overloading causes a temperature rise, it is the actual temperature that damages the insulation. Ambient compensation is not necessary for cables because the temperature a few feet below the ground is virtually constant. Although thermal overload protection is seldom used for overhead H.V. lines, their loading is usually reduced in summer and there is now some investigation into the possibility of developing overload relays with compensation for sun and wind on heavily loaded lines.

13.3. THERMAL PROTECTION OF TRANSFORMERS

Transformers are best protected against damage due to overheating by measuring the temperature of the hottest part of the winding. To do this directly is difficult because of the high voltage on the windings; it is best done by using a thermal replica of the winding which is located in the oil at the top of the tank i.e. the hottest oil. This thermal replica is further heated by secondary current from a c.t. in series with the winding so that its temperature is equivalent to that of the winding hot spot (Fig. 13.3). Fig. 13.4b shows the construction of the thermal replica used in an English relay. The temperature is measured by means of a silistor S (positive temperature coefficient silicon resistor) at the centre of the thermal replica. Two heaters are provided, the main one H_1 wound spirally round an epoxy resin block, and a smaller one H_2 near the silistor to compensate for the initial time lag and thus match the transformer characteristic more closely. An indication of the winding temperature can be provided both at the

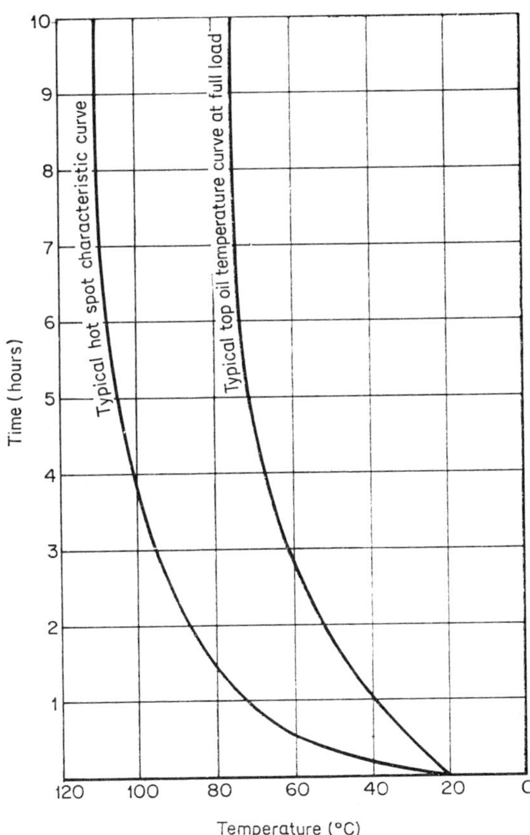

Fig. 13.3. Oil and hot spot temperature rise

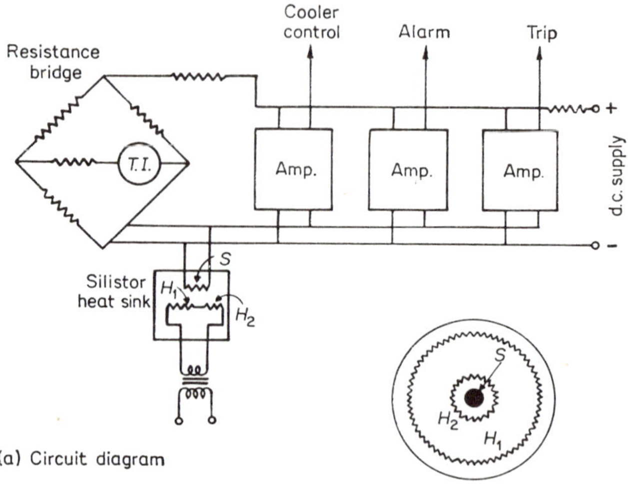

Fig. 13.4. Winding temperature control for power transformer (E.E.Co.)
T.I. = temperature indicator

transformer and at the relay panel; a temperature/time integrator can be added for control purposes.

Fig. 13.4a is a simplified diagram of the relay circuit. In addition to temperature indication the silistor voltage is also used, through amplifiers, to control fans and oil pumps (for cooling the transformer) and to control breakers to isolate the transformer if necessary.

13.4. THERMAL PROTECTION OF MOTORS [83, 86, 87]

The overheating protection of three-phase rotating machines is different from that of cables and transformers because of the difficulty of measuring the temperature of the rotor. Unlike a transformer, the rotor cannot be regarded as the secondary and the stator as the primary because the heating of the rotor is not a direct function of I^2Rt. It is very much increased under unbalanced conditions, i.e. unbalanced supply voltage in the case of a motor or unbalanced current in the case of a generator.

Small motors are protected against faults and overload by fuses and/or thermal cut-outs. Above 50 h.p., thermal relays or inverse time-current breakertripping devices are used. The relays may be interconnected to detect loss of one phase of the supply and often have instantaneous overcurrent relays for fast tripping on faults. Fuses are generally provided for fast back-up protection.

Thermistor-operated winding temperature relays are becoming popular for protecting the stator winding against high temperature on overload.

For larger motors, thermal relays or overcurrent relays provide inadequate protection against overheating of the rotor due to unbalanced supply voltage. They tend to trip too sensitively on unbalance that increases the current in one phase and not sensitively enough on unbalance that decreases the current in one phase (see Figs. 13.15 and 13.16). This is because, in the first case the positive and negative sequence currents are in phase and add up to a large value of current in one phase whereas, in the second case, they are opposed in one phase so that the unbalance is greater for a given ratio of negative to positive sequence current. Unbalanced voltage can be caused by a blown fuse, singlephase loads or imperfect transposition of line conductors on a supply feeder.

When unbalanced voltages are applied to a three-phase induction motor, the effect of the unbalance is equivalent to the superposition of negative sequence voltage upon the normal positive sequence supply voltage (see Appendix, Section 13.12.2). Due to the small value of negative sequence impedance, a greater unbalance of current results. The negative sequence currents produce a negative shaft torque (more bearing wear) and the increased losses to add to the heating.

Assuming 1 c/s slip frequency, any negative sequence current due to unbalanced supply voltage induces a rotor current of twice the supply frequency minus the slip frequency because the rotor and the negative sequence currents have opposite rotation. Thus the slip frequency for negative sequence is $2f - 1 = 99$ c/s, compared with 1 c/s for positive sequence. Hence rotor heating due to unbalanced currents can be very great because of the skin effect, the rotor resistance increasing from 3 to 15 times depending upon the rotor design.

The ratio of positive sequence to negative sequence impedance is given approximately by the ratio of the starting current to the normal running current. The percentage negative sequence current will therefore be approximately equal to the percentage negative sequence current voltage multiplied by the ratio of the starting current to running current. For example, in a motor where $I_{start}/I_{run} = 6$, a 5% negative sequence component in the supply voltage would result in approximately 30% negative sequence current. An analysis of this is given in the Appendix 13.12.2.

On industrial systems, 5% unbalance between line voltages can occur and is permissible. This represents approximately 25% negative sequence current which is as much as most motors will stand, so the relay should be set to operate at about this value.

The loss of one phase represents an extreme case of unbalance. In this case $I_1 = I_2 = (I_{ph}/\sqrt{3}$, where I_{ph} is the phase current and the subscripts 1 and 2 refer to sequence components. In spite of this dangerously high value of negative sequence current, ordinary overcurrent relays would not always disconnect the motor because the phase current is increased only by 150% by the loss of one phase. For instance, at half load the phase current would be only 125% of normal with one phase missing and this could be below the setting of the relay.

13.4.1. Motor Protection Relay

Adequate protection against these condictions is given by a relay responsive to $I_1^2 + KI_2^2$ where K is about 1·5 for motors below 50 h.p. and is as high as 6 for some very large motors. The circuit diagram is shown in Fig. 13.6.

The positive and negative sequence components are extracted from the phase currents by the bridge filter on the left and fed into instantaneous overcurrent units and into the heaters H_1 and H_2 which are wound over a small oven containing the thermistor Th_1, H_2 having K times as many turns as H_1. This thermistor forms part of a d.c. bridge which is unbalanced when the calibrated value of $(I_1^2 + KI_2^2)$ is exceeded, causing the level detector to provide a tripping signal and the motor to be disconnected. A second thermistor Th_2 provides ambient temperature compensation.

The time-current curve is automatically adjusted to allow for the fact that the motor can tolerate more current and heating when cold than when hot. Figure 13.8 shows the time-current characteristic at 20°C and at a stable temperature corresponding to steady full load.

The relays's response to $(I_1^2 + KI_2^2)\,t$ with a suitable thermal replica means that it can decide how much load the motor can stand for a given time with a given degree of supply voltage unbalance. Hence it enables the user to get more work out of the motor without damaging it and without any unnecessary shutdown.

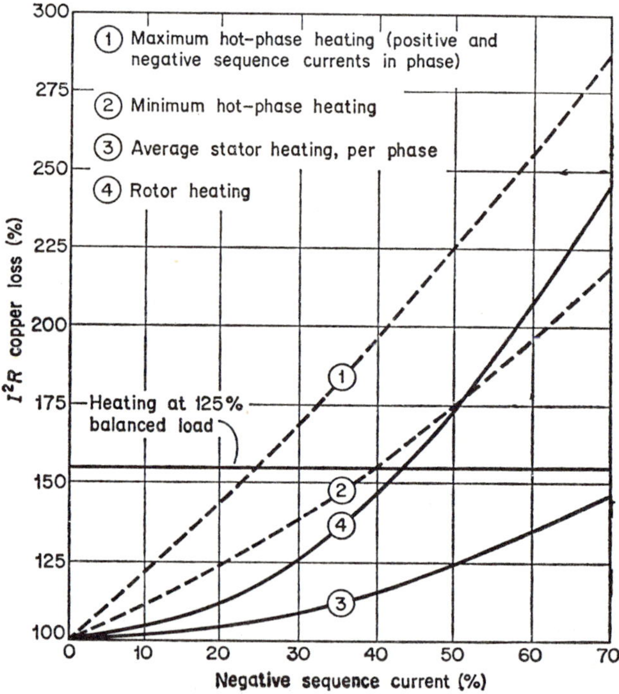

Fig. 13.5. Motor heating curves with unbalanced supply voltage

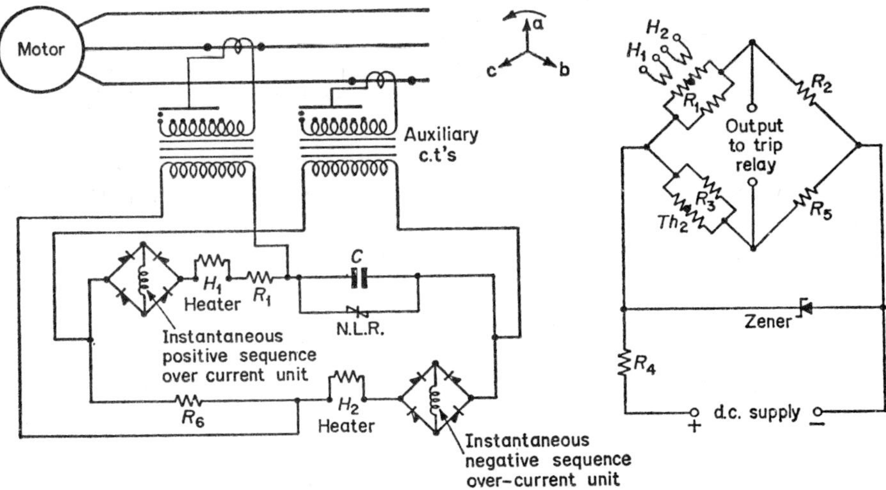

Fig. 13.6. Circuit diagram of motor protection relay (E.E.Co.)

13.4.2. Operation on Faults

Three-phase faults, such as terminal flashovers, are cleared instantly by a high-set overcurrent relay in the positive sequence network. On the other hand, this relay does not trip during the high inrush current at starting because the current has a rapid decrement in the first few cycles and the relay has a 2-cycle delay. The pick-up of this relay is infinitely adjustable.

Interphase (winding) faults are cleared instantly by a fixed setting negative sequence relay; this relay also detects the loss of one phase of the supply. Ground faults are cleared instantly by a zero sequence relay in the residual circuit of the c.t's; wrong operation on spurious residual current caused by c.t. inequalities during the high starting current period is prevented by a stabilizing resistor, whose action was explained in Chapter 9, Section 9.3.2.

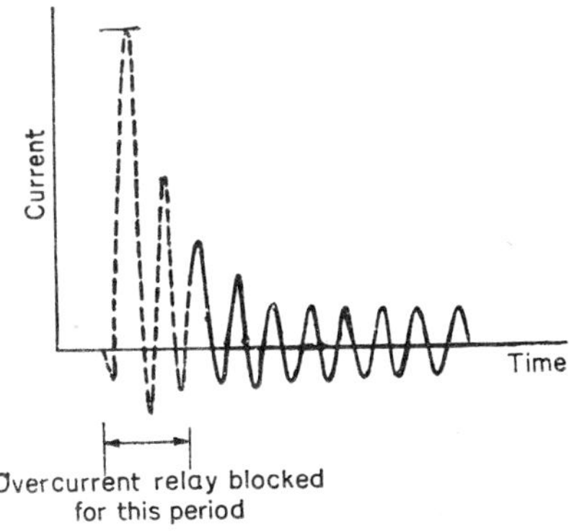

Fig. 13.7. Motor starting inrush current

13.4.3. Magnetizing Inrush Current

The 2-cycle time delay of the instantaneous positive sequence current relay unit is a very valuable feature because it enables the relay to be given a setting just above the starting current, i.e. 6 to 8 times normal. Relays without this feature have to be set at 16 to 18 times normal to clear the very high magnetizing inrush current that can occur for the first half-cycle (see Fig. 13.7); such a high setting is inadequate for clearing winding faults.

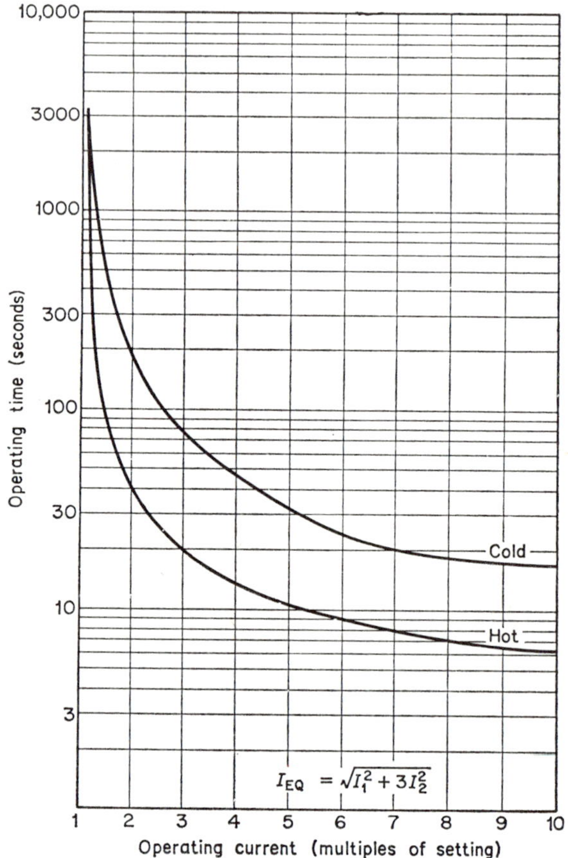

Fig. 13.8. Time-current curves of motor protection relay

13.4.4. Stalling Current

The starting current continues at a high level until 80 to 90% of the time required to reach normal speed and then it falls rapidly to normal running value.

The inverse time curve of the relay is set so that it does not operate before the motor gets up speed, e.g. if the motor starting current is 6 times normal and the starting period is 10 seconds, the point ($6N$, 10 sec) should be below the time-current curve of the relay (Fig. 13.9). On the other hand, if the motor stalls, the relay will operate before serious overheating can occur.

This protection does not apply where there are several starts in a short time. In such a case the relay would have to be given a lower setting and would have to be suppressed by another relay detecting motion of the rotor by some other means.

13.4.5. Fan Motors

These motors take a long time to reach full speed but they are not subject to overload. Hence their long starting time can be accommodated by connecting a non-linear resistance across the positive sequence filter to give a more definite time-current curve on positive sequence current only.

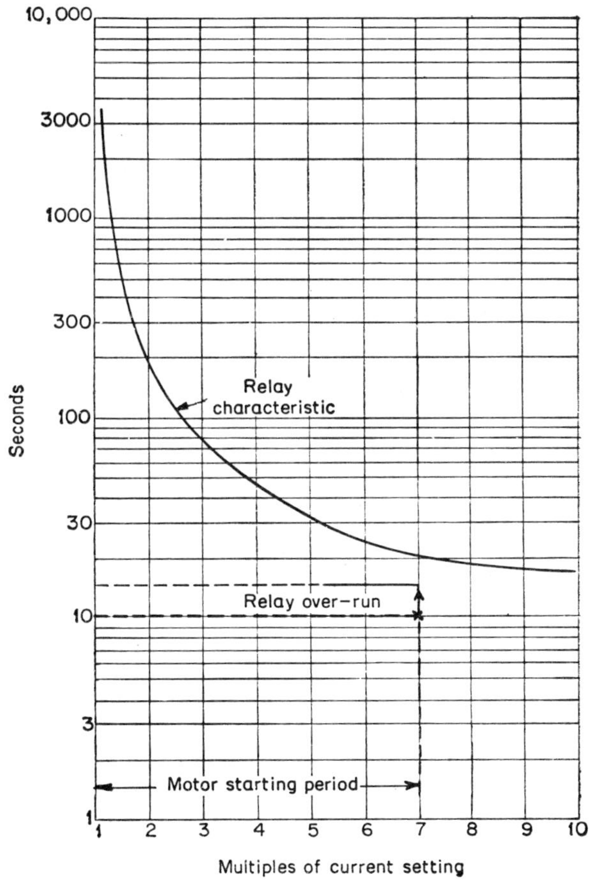

Fig. 13.9. Effect of thermal overrun of relay

13.5. THERMAL PROTECTION OF ALTERNATORS

The cost of repairing a large generator is so great that thorough protection against overheating is justified. The subject is dealt with in some detail in Section 9.1.1 of Vol. I. At present almost all the relays used are of the electromagnetic type, but they could well be replaced by static relays since they all employ very sensitive relays supplied by resistance bridges, thermocouples or negative sequence filters. That they have not been "staticized" is due to the fact that modern generators become fewer as they grow larger, so that not enough relays are sold to pay for their development.

On the other hand, some work has been done in the U.K. and the U.S.A. on direct measurement of conductor temperatures using a temperature-sensitive r.f. oscillator which is small enough to go in the slot with the conductor. The r.f. signal is picked up by a receiver outside the machine which operates an indicator and a relay (100).

Some companies are developing overall protection units which include all forms of generator protection (including thermal) in one case which contains a plug-in module for each relay function.

The principles of operation have not been changed from those described in Volume I, except for minor aspects such as biasing the negative sequence current relay with a certain amount of positive sequence current, as was done in motor protection to allow for heating by the load current.

13.6. HARMONIC RELAYING

Until now, protective relaying has been based almost exclusively upon the fundamental components of the current and voltage of the faulted circuit. Harmonics are usually ignored or filtered out. In power transformer protection the second harmonic has been used only to prevent tripping during the magnetizing inrush period which occurs when a power transformer is first energized. The remainer of this chapter discusses the possibility of using harmonic content of the current (wave-shape) to detect faults and hence initiate the relay operation instead of blocking it.

When insulation deteriorates, a small initial current flows which is rich in harmonics because of electrical discharges across voids (corona and tiny arcs). This causes chemical changes to take place in the insulating material, producing substances which have largely non-linear resistance which also creates harmonics.

A similar phenomenon occurs when oil breaks down; in this case the voids are bubbles of inflammable gases produced by electrolysis. Here again deterioration of the insulation can be detected by the presence of audio-frequency harmonics in the current wave-form.

This suggests a method of detecting incipient faults in electrical apparatus which could enable the defective apparatus to be disconnected from service and, in most cases, repaired thus avoiding a major fault and interruption to service.

Under normal conditions the load current follows a wholly metallic path which has linear impedance so that, with a sinusoidal source voltage, the current is sinusoidal. When a ground fault occurs the current follows a new path through a power arc and through the ground, both of which have non-linear resistance; this means that, with a sinusoidal source voltage, the fault current contains harmonics. Further harmonics are generated by the rectifying action of the contact between a fallen conductor and the earth.

The magnitude of the harmonic content compared with the fundamental depends upon the ratio of the non-linear resistance in the circuit to the linear impedance. Figure 13.11b shows the size of the harmonics in a typical distorted wave as obtained by Fourier analysis (see Appendix, Section 13.13.1). Records from automatic oscillographs in the U.S.A. indicate a high harmonic content for ground fault currents of less than 40% of c.t. rating (Fig. 13.10).

The frequency, magnitude and relative phase angle of the harmonics are very variable because the power arcs and the non-linear resistance of the earth have quite different ohmic laws. In a power arc

$$I \propto \frac{1}{V^{2.5}} \tag{13.4}$$

This tends to make the current wave flat-topped [95] whereas, for current passing through earthy substances (silicon carbides), the relation is roughly

$$I \propto V^{4.5} \tag{13.5}$$

This tends to make the current wave more peaked but the actual wave-shape depends upon the type of soil and its moisture content. Where the soil is dry there will be arcing between particles of conducting material in the earth and this causes little bursts of acoustic frequency, superimposed on the wave, (Fig. 13.10) due to the discontinuity of re-striking. Figure 13.10 shows a typical oscillogram of ground-fault current limited by high-ground fault resistance. The blurred portion near each peak is high-frequency oscillation too fast for the oscillograph to record.

There is a certain amount of harmonic content under normal conditions due to generator ripple, power rectifiers, etc., but this is limited by statutory regulations to 2% of the fundamental since harmonics in the supply voltage

Fig. 13.10. Oscillogram of high resistance fault current

can cause heating of motors and radio interference. Tests in Russia [85] indicate that the harmonic content always increases considerably during a fault and that certain frequencies predominate. This was confirmed in tests made by the author in the U.S.A. in 1948 with different kinds of soil.

It is clear from the above that this could provide a means for detecting high resistance faults, which would be useful for detecting fault currents below the level of normal relay settings.

Some suggestions for the application of a harmonic detector are as follows:

(1) Sensitive ground fault detector.
(2) Line fault anticipator.
(3) Fault versus overload discriminator.
(4) Transformer protection.
(5) Cable and capacitor testing [102].

It must be emphasized that the harmonic fault detector is not intended for direct tripping, except perhaps in industrial and domestic applications. On power systems it can sound an alarm for faults not detected by normal relays and can also be used in parallel with another fault detector. It should never be called upon to discriminate between internal and external faults.

13.6.1. Operating Principle of Harmonic Relay

Figure 13.11a shows one form of harmonic relay in which a transactor produces a voltage proportional to the current, except that the harmonics are amplified in proportion to their frequency. By means of the filters F_f and F_H the fundamental and the harmonics are separated and fed into a differential amplifier whose output controls a triggering circuit.

The relay is set to operate when the sum of the rectified harmonics exceeds 5% of the fundamental. The sensitivity of the relay is such that it operates down to 2% of the rating of the main c.t's without on operating on normal load unbalance. It is protected from damage at high currents by limiting devices; this limiting makes the relay comparator insensitive at currents above normal loads, but this is unimportant because fault currents at this level will be detected by the ordinary relays.

Heavy current faults can usually be detected by ordinary types of protective relays. If, however, it is found necessary to apply this principle to heavy faults, the voltage should be analysed instead of the current, i.e. the transactor should be replaced by a potential transformer.

The choice of a current or voltage input to the relay depends upon the ratio of the magnitudes of the linear and non-linear impedances in the circuit. Assuming a sinusoidal source potential, the current will be non-sinusoidal only if the circuit impedance is mostly non-linear, such as in a high resistance ground fault. If it is mostly linear, such as in a direct flash-over, the current will be sinusoidal but the voltage will be non-sinusoidal near the fault, such as in the case of an arcing fault at the end of a short line section.

13.7. SENSITIVE GROUND FAULT DETECTOR

Every year a large number of forest fires are started and livestock is electrocuted, especially in dry countries, where a broken conductor may be sparking on the ground but does not draw enough current to operate normal protective relays. In Australia a number of people are killed each year by attempting to move a fallen conductor or by touching a wire fence which may run along the same right-of-way as a distribution line which has fallen on the fence several miles away. Animals are particularly vulnerable because, even if they do not touch the conductor, the potential gradient between their front and back feet may be lethal.

Fallen conductors are most common in countries subject to severe wind and storms which may cause violent movement of overhead line conductors. Line conductors may also be broken by heavy icing conditions and during earthquakes, landslides, etc.

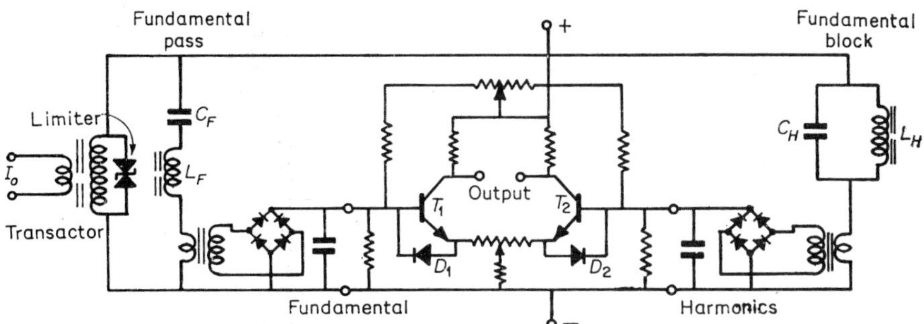

Fig. 13.11. Harmonic content indicator relay
(a) circuit

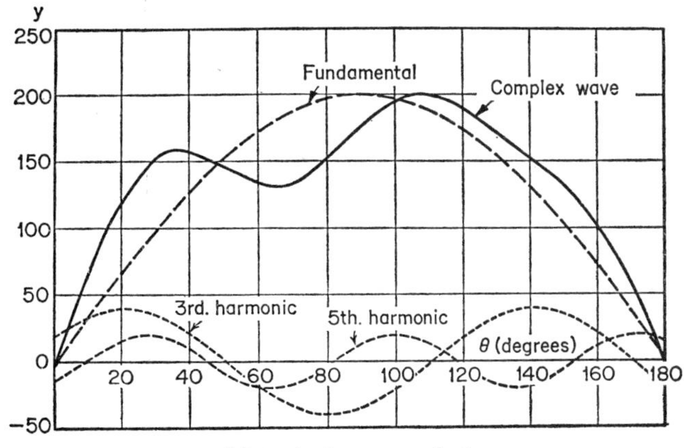

(b) typical wave analysis

Table 13.1

Analysis of Wave shown in Fig. 13.11b

Complex Wave		Analysis of Sine Terms						Analysis of Cosine Terms					
(*a*)	(*b*)	(*c*)	(*d*)	(*e*)	(*f*)	(*g*)	(*h*)	(*j*)	(*k*)	(*l*)	(*m*)	(*n*)	(*p*)
θ	y	$\sin\theta$	$y\sin\theta$	$\sin 3\theta$	$y\sin 3\theta$	$\sin 5\theta$	$y\sin 5\theta$	$\cos\theta$	$y\cos\theta$	$\cos 3\theta$	$y\cos 3\theta$	$\cos 5\theta$	$y\cos 5\theta$
15°	97	0·259	25	0·707	68·5	0·966	94	0·966	94	0·707	68·5	0·259	25
30°	152	0·5	76	1·0	152	0·5	76	0·866	132	0	0	−0·866	−132
45°	152	0·707	107·5	0·707	107·5	−0·707	−107·5	0·707	107·5	−0·707	−107·5	−0·707	−107·5
60°	134	0·866	116	0	0	−0·866	−116	0·5	67	−1·0	−134	0·5	67
75°	142	0·966	137	−0·707	−100·5	0·259	36·8	0·259	36·8	−0·707	−100·5	0·966	137
90°	176	1·0	176	−1·0	−176	1·0	176	0	0	0	0	0	0
105°	200	0·966	193	−0·707	−141·4	0·259	51·8	−0·259	− 51·8	0·707	141·4	−0·966	−193
120°	189	0·866	164	0	0	−0·866	−164	−0·5	− 94·5	1·0	189	−0·5	− 94·5
135°	161	0·707	114	0·707	114	−0·707	−114	−0·707	−114	0·707	114	0·707	114
150°	133	0·5	66·5	1·0	133	0·5	66·5	−0·866	−115	0	0	0·866	115
165°	82	0·259	21·2	0·707	58	0·966	79·5	−0·966	− 79·5	−0·707	− 58	−0·259	− 21·2
180°	0	0	0	0	0	0	0	−1·0	0	−1·0	0	−1·0	0
		Sum	1196·2	Sum	215·2	Sum	79·1	Sum	− 17·5	Sum	112·9	Sum	− 90·2
		Mean Value	99·7	Mean Value	17·9	Mean Value	6·6	Mean Value	− 1·46	Mean Value	9·4	Mean Value	− 7·5
			$A_1 = 199·4$		$A_3 = 35·8$		$A_5 = 13·2$		$B_1 = -2·92$		$B_3 = 18·8$		$B_5 = -15$

Check $A_1 - A_2 + A_5 = 176·8 =$ the value of y at $\theta = 90°$ (i.e. 176) $B_1 + B_3 + B_5 = 0·88 =$ the value of y at $\theta = 0°$ (i.e. 0)

Equation of wave $y = 200 \sin(\theta - °0\ 50') + 40·3 \sin(3\theta + 27°\ 42') + 20 \sin(5\theta - 48°\ 39')$.

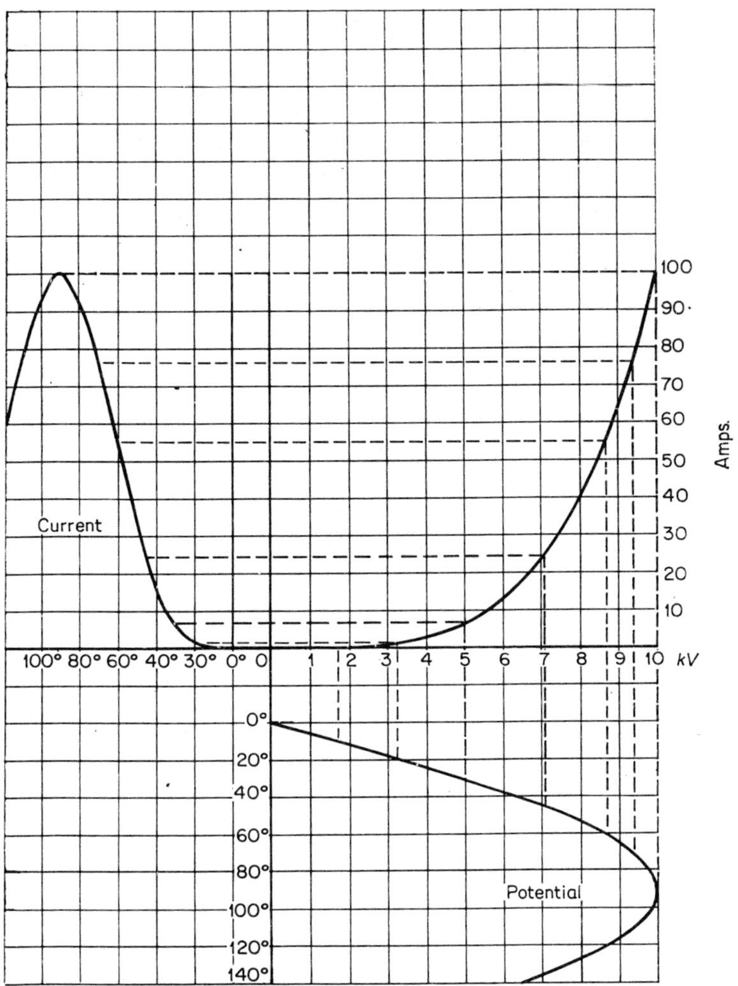

Fig. 13. 11 (c) Wave-form analysis for *V* and *I* with non linear resistance

In this application it may be necessary on distribution systems to add a 3rd harmonic filter if the normal 3rd harmonic current is appreciable. The relay should preferably be connected to a core balance c.t. whose core encircles the three conductors but, even if connected in the residual circuit of the main c.t's, normal circulating current will have negligible effect since it is the ratio of the harmonics to the fundamental that is being measured.

Operation on H.V. switching and faults on other lines can be avoided by the use of several seconds time delay. In many cases the relay would be arranged not to trip directly but only sound an alarm. Where the pantographs of passing electric trains might cause interference, the delay could be several minutes, since fallen conductors now often remain undetected for hours.

The fact that the relay measures a ratio of harmonics which is independent of the amplitude of the zero sequence current would enable it to be used on systems with considerable normal unbalance, such as American 4-wire distribution networks where normal unbalance may be as high as 20%.

Some field tests have been made on such a relay but it was found that faults on certain soils were not detected, due presumably to the preponderance of higher frequency harmonics which were too much attenuated in the c.t.'s and secondary wiring to reach the relay in sufficient magnitude. The solution to this problem has not yet been found.

13.8. LINE FAULT ANTICIPATOR

The Buchholz transformer protection relay is at present the only relay that can anticipate a fault so that it can be prevented by appropriate action. It does this by detecting a slowly developing fault at an early stage.

Slowly developing faults also occur on distribution lines. If an insulator becomes cracked or covered with industrial pollution, a small amount of current flows to ground and tends to produce a charred path (tracking) whose resistance slowly diminishes until the runaway condition is reached and a flashover occurs.

During the early stages the current is very small but of high harmonic content. It is possible that a sensitive harmonic detector could detect this incipient fault condition and operate an alarm which would enable the faulty insulator to be replaced before a power fault occurred, with its subsequent interruption to service. Such an application has not yet been made but would merit investigation.

13.9. "FAULT VERSUS OVERLOAD" DISCRIMINATOR

On many high voltage systems there are sometimes conditions of minimum generation during which the fault current is less than the load current during generating conditions. In some cases even impedance fault detectors cannot be set to provide discrimination between power swings and minimum fault conditions.

In such cases advantage could be taken of the facts that (*a*) even solid faults start off with an arc until steady contact has been established, (*b*) in large networks there is an initial h.f. oscillation due to release of the charge on the system capacitance. Both these phenomena provide an initial burst of audio-frequency harmonics which could be used to operate a fault detector relay or to increase the sensitivity of an existing fault detector.

An analysis of the initial audio-frequency oscillation caused by the discharge of the energy in the system capacitance was given in Chapter 6, Section 6.2.2. Because of its short duration it could not be used directly to operate a normal directional relay, but a Russian engineer (V. V. Shut) has developed a relay which takes advantage of the fact that the polarity of the first half-wave of the oscillation is determined by the direction of the initial discharge into the fault and hence by the location of the fault [85]. Figure 13.12b shows the circuit of this relay which leaves an indication of the polarity of the first half-wave and suppresses the ensuing half-waves.

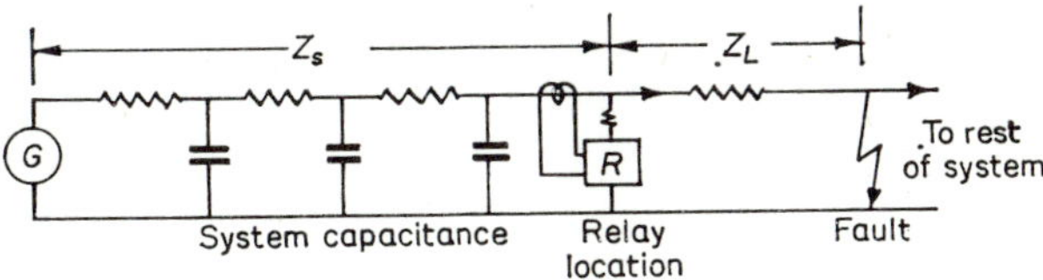

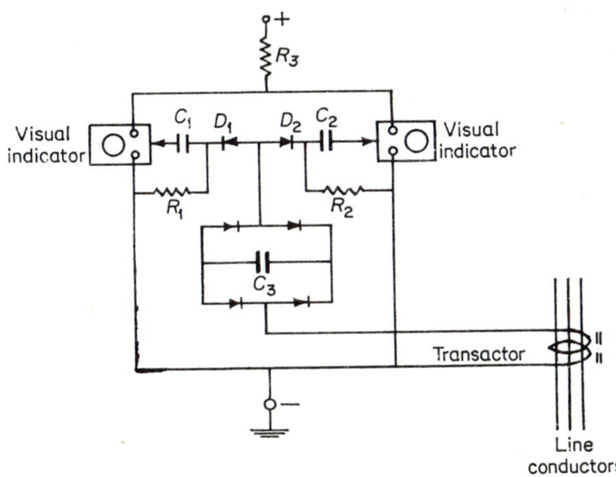

Fig. 13.12. Directional indicator for high-resistance ground faults
(a) discharge path for power-system capacitance
(b) schematic circuit of relay

Oscillograms of fault currents taken on a fast oscillograph show clearly both the initial oscillation and the harmonics in high-resistance ground faults. These conditions have generally been overlooked because it is seldom that automatic oscillographs are used which are sensitive to harmonics. In Europe, for instance, the most popular type is the Masson oscillo-perturbograph which records frequencies only up to 80 c/s, which is clearly inadequate for the audio-frequency range covered by fault current harmonics. Furthermore, to record the initial oscillation requires a memory device or a continuous record, neither of which is usually available.

13.10. TRANSFORMER PROTECTION

An electric arc is a powerful source of harmonics; these appear in the voltage across the arc if its resistance is small compared with the total circuit impedance. This creates a strong signal, over a wide band of frequency, in the neighbourhood of the arc.

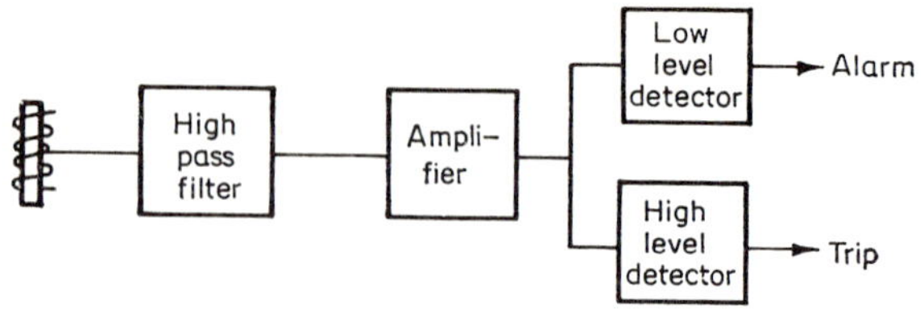

(a) Block diagram of relay

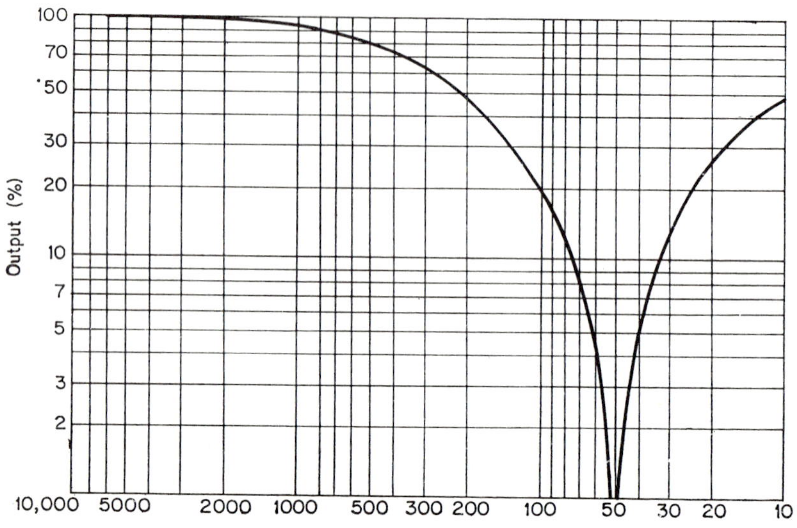

(b) Filter characteristic

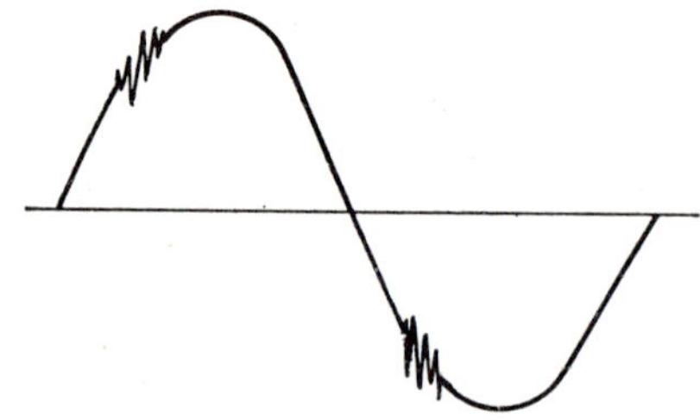

(c) Low arc current between constructional parts

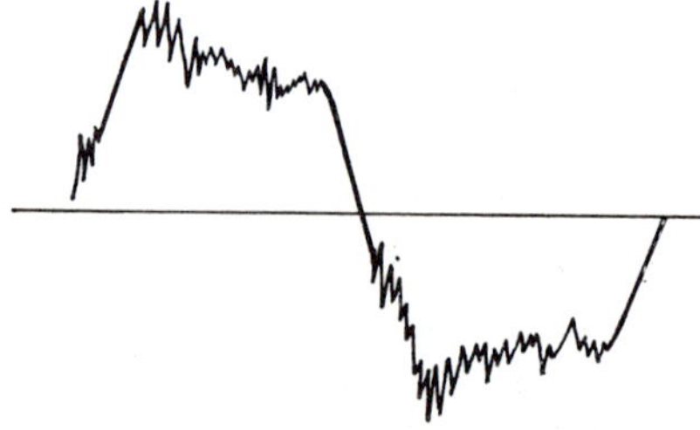

(d) Voltage across power arc

Fig. 13.13. Transformer protection relay

This signal can be picked up by small ferrite aerials suitably placed around the inside of the transformer tank. If the aerials are connected to a suitable wide-band audio-frequency receiver, the output of the receiver can be used to trip a circuit-breaker, since a power arc inside the tank is a definite indication of a fault in the transformer.

Like the Buchholz relay, this relay consists of two units, one to detect corona and persistent small arcs between constructional parts of the transformer and a less sensitive one to detect a power arc caused by a flashover from a winding.

The sensitive unit has a time delay and operates an alarm to give notice that the transformer should be examined as soon as convenient; otherwise the insulation of the oil will deteriorate due to sludging, and inflammable gases will collect above it. The less sensitive unit for detecting actual faults must be instantaneous in order to minimize the damage done by the power arc.

Figures 13.13c and 13.13d show the wave-shape of the two types of arcing condition. Figure 13.13a is a block diagram of the circuit. Figure 13.13b shows the frequency characteristic of the wide band pass filter which has the lowest impedance to frequencies above 3·3 kc/s since tests show this to be a dominant frequency for small arcs under oil.

13.11. CAPACITOR AND CABLE INCIPIENT FAULTS [102]

The life of a cable or any insulation decreases exponentially with the voltage above the discharge level. Normally this is high compared with working conditions, but it can be reduced by an insulation defect such as a void across which arcing will occur in the form of bursts of h.f. damped oscillations. This condition can be detected in the factory by a harmonic detector and thus avoid subsequent failure after a shortened time in service.

13.12. LOAD-SHEDDING [121, 122]

When a power source or a critical inter-connection is lost, due to a fault or system switching, at least part of the power system will be left with a deficit of generation which will cause the turbines to slow down and cause the system frequency to fall.

With small power deficits the rate of fall of frequency will be low and the turbine governors will raise the steam or water supply to the turbines to restore the frequency to normal, provided that sufficient reserve power is available.

With large power deficits the frequency may fall to a dangerous value before the turbine governors have time to operate, since it usually takes at least 2 seconds before they can make the increased power available. In such cases some immediate load shedding is necessary.

The load shed should be the minimum necessary to achieve system recovery, but it should be sufficient to cancel the generation deficit with a small margin to start the frequency rising towards normal. Because of the variation of this deficit with the nature and location of the disturbance and because of the action of voltage regulators, the effect of low frequency upon turbine auxiliaries, the time taken to disconnect the load, etc., it is very difficult to calculate precise settings for the relays; but the following procedure is effective in most applications, provided that the relays are located where appropriate low-priority loads can be shed.

13.12.1. Load-Shedding Steps

Since the amount of load to be shed depends upon the power deficit (which depends upon the location and nature of the disturbance) and since the load shed should be no larger than necessary, it is obviously preferable to shed it in small steps, as the frequency falls, until equilibrium is reached.

Too many steps would make the relay expensive and would result in load being shed too fast, because the relay would reach its next step while the circuit-breaker was operating. A reasonable compromise is three steps. The load can be shed at one feeder or divided between several feeders, using a multi-contact auxiliary relay for tripping the several breakers.

The load that must be shed is less than the initial power deficit because the reduced frequency reduces the load by an amount dependent upon the nature of the load; this is called the self-regulation factor, d, which varies from 0 for a resistance load to 4 or more for a motor load.

Taking an average value of $d = 2$ and assuming a 100% overload (i.e. a power deficit D of 50%) and a maximum frequency fall of 5% (i.e. to f = 95% of nominal system frequency f_s), the load needing to be shed is $L_s = \left[1 - \left(\frac{f_s - f}{f_s}\right)d\right]D = (1 - 0{\cdot}05 \times 2)\ 50\% = 45\%$.

The size of the steps of load shedding is a matter of convenience, but they should preferably increase progressively. For instance, the 45% load could be shed in four steps of 5%, 10%, 15% and 15%, or three steps of 10%, 15% and 20%, or two steps of 20% and 25%. Four steps will deal with the widest range of conditions with the minimum jettisoning of load.

If the system recovers after one of the load-shedding steps the frequency will rise and the relay will reset each stage after its frequency setting has been exceeded for one cycle and the load shed can be picked up again, manually or automatically.

13.12.2. Underfrequency Relay Settings

The frequency setting for each step of load shedding depends upon the number of steps used. They should be separated by roughly equal increments of frequency, assuming a fairly constant rate of frequency fall.

The lowest frequency step must be well clear of the critical frequency, which is usually about 94% of nominal; so the lowest step in this case should be about 96%, i.e. 48 Hz on a 50 Hz system. The critical frequency is that below which the turbines could be damaged if it continued for more than a minute.

Typical settings would be 97·5% and 96% for two steps, 98%, 97% and 96% for three steps, or 99%, 98%, 97% and 96% for four steps.

Undesirable operation on temporary phase shift or offset waves, when using frequency settings near the nominal value, is prevented by delaying the relay so that it will not operate unless the low frequency is maintained for about 0·15 sec.

With slow circuit-breakers, or where time-delay is used, the frequency settings must be raised because, during the delay, the frequency will have fallen much lower than the value for which the relay was set. Furthermore, the higher the rate of fall of frequency the lower the frequency will be when the load is actually shed; this is the opposite of what is required. For a relay plus breaker time t the relay settings will have to be raised by $t \cdot \frac{df}{dt}$.

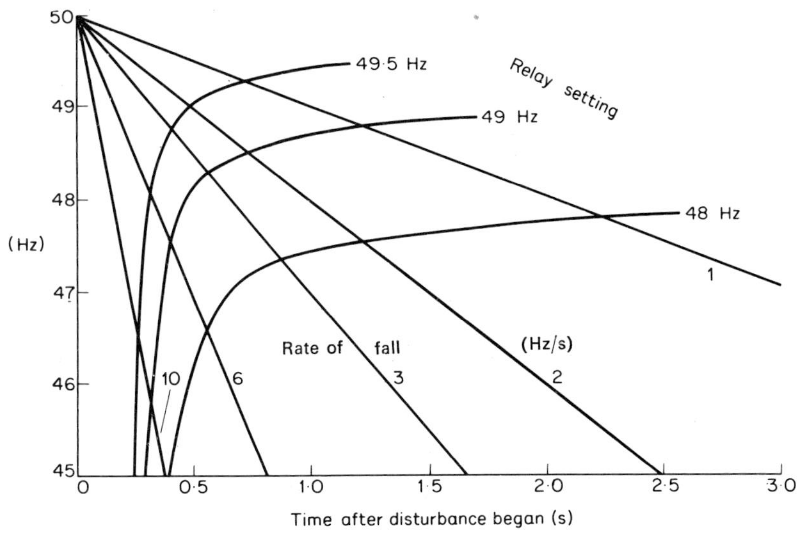

Fig. 13.14 shows the system frequency after time t with values of $\frac{df}{dt}$ from 1 to 10 Hz/s plotted from the formula

$$f_t = f_s - t \cdot \frac{df}{dt}. \tag{13.6}$$

The curves show the frequency at which the load shedding will occur with relay settings of 48, 49 and 49·5 Hz assuming a relay plus breaker time of 0·25 second.

13.12.3. Rate of Fall of Frequency

The rate of fall of frequency is

$$\frac{df}{dt} = \frac{-D}{2H} f_s \tag{13.7}$$

where H is overall inertia constant of the generators and other synchronous machines remaining on the system, and D is the power deficit in p.u. Assuming the same value as before for D (50% or 0·5 p.u.) and $H = 4$ seconds

$$\frac{df}{dt} = \frac{-0{\cdot}5}{8} \times 50 = -3{\cdot}1 \text{ Hz/s.} \tag{13.8}$$

Assuming a circuit-breaker opening time of 0·1 S and a relay operating time of 0·15 S, the total time to shed a load stage is 0·25 S. During this time the frequency will have fallen by $t \, . \frac{dt}{df}$ so that the tripping frequency will be lower than the relay settings by $0{\cdot}25 \times 3{\cdot}1 = 0{\cdot}77$ Hz and the relay settings will have to be raised by this amount; this makes the reasonable assumption that D and $\frac{df}{dt}$ will remain substantially constant during the operating time of the relay and circuit breaker.

In cases where D is not predictable this resetting of the frequency relay can be done automatically by a rate-of-change-of-frequency relay (121). The rate-of-change-of-frequency relay can also be used on power systems with a high rate of fall of frequency, due to low-inertia generators or a large power deficit. An alternative scheme is to replace the second and third steps of the under-frequency relay by rate-of-change of frequency relays (122).

To obtain the frequency at a particular time equation 13.7 must be integrated, giving

$$f = f_s \sqrt{1 - tD/H} \tag{13.9}$$

from which the loci of Fig. 13.15 were plotted for $H = 4$ seconds.

Fig. 13.16 shows the time available for load shedding before the frequency falls to the critical value of 94% (47 Hz) assuming a relay setting of 49 Hz. The curve is plotted from Fig. 13.15 by subtracting the time at 49 Hz from the time at 47 Hz for each value of D. Curves for other relay settings can be plotted in the same way.

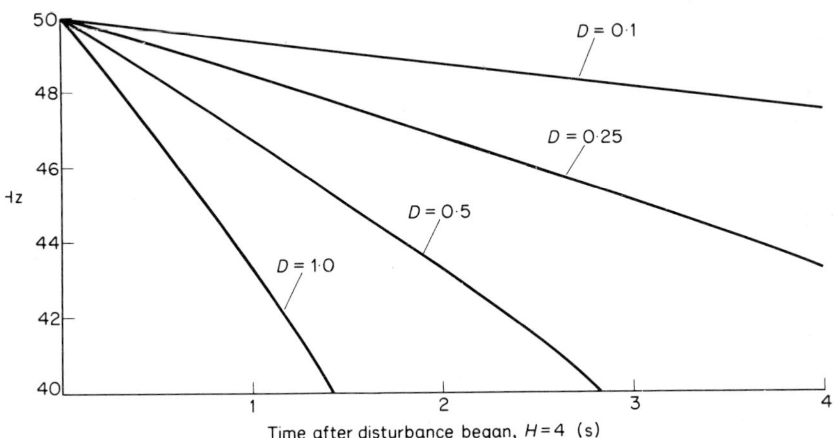

Fig. 13.15. Effect of generation deficiency (D) upon frequency

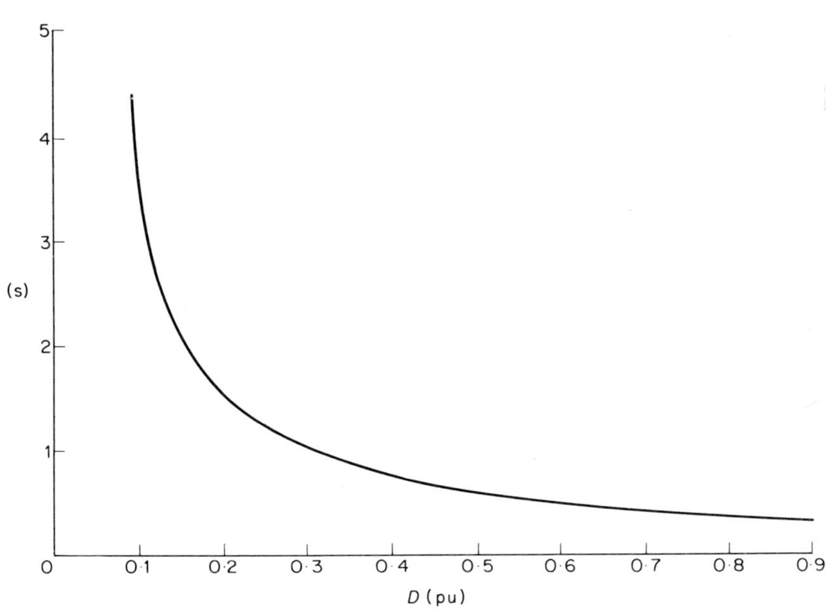

Fig. 13.16. Time available to shed load with relay setting of 49 Hz and critical frequency 47 Hz

13.13. APPENDIX

13.13.1. Fourier Analysis

The equation of a continuous wave in terms of the instantaneous value of an ordinate is:

$$y = Y_1 \sin(\theta + \phi_1) + Y_2 \sin(2\theta + \phi_2) + \text{etc.} \qquad (13.10)$$

If the wave is symmetrical about the time axis there will be harmonics, so:

$$\begin{aligned} y &= Y_1 \sin(\theta + \phi_1) + Y_3 \sin(3\theta + \phi_3) \text{ etc.} \\ &= A_1 \sin(\theta + \phi_1) + A_3 \sin(3\theta + \phi_3) \text{ etc.} + \\ &\quad B_1 \cos(\theta + \phi_1) + B_3 \cos(3\theta + \phi_3) \text{ etc.} \end{aligned} \qquad (13.11)$$

where $Y = \sqrt{(A^2 + B^2)}$.

The wave can be plotted in terms of θ (Fig. 13.11b) by adding up the instantaneous values for each value of θ in Table 13.1.

Where the harmonics are caused by a non-linear resistance of the law $I = \mathrm{k}V^n$ the wave will be symmetrical as shown in Fig. 13.11b. The analyses of current waves for values of n up to 10 are given in Table 13.2. There are no even harmonics for waves symmetrical about the zero line.

Thyrite (Metrosil) has $n = 4{\cdot}5$ and the current wave has been plotted in Fig. 13.11c by projecting it from a sinusoidal potential wave via the resistance curve $I = kV^{4.5}$. Similarly, the wave shape of arc current with a sinusoidal voltage across it has been plotted from $I = kV^{-2.5}$.

13.13.2. Induction Motor Circuit Analysis

Figure 13.17 shows the equivalent impedance diagram of an induction motor. The *positive sequence impedance* of the motor viewed from the stator is:

$$Z_1 = \sqrt{\left\{\left(R_{s1} + \frac{R_{R1}}{s}\right)^2 + (X_{s1} + X_{R1})^2\right\}}. \qquad (13.12)$$

where subscript S refers to the stator and R to the rotor (see Fig. 13.17). Assuming X/R about 5, it will be seen that at standstill (starting or stalling) the R terms will be negligible and $Z_1 = X_{S1} + X_{R1}$.

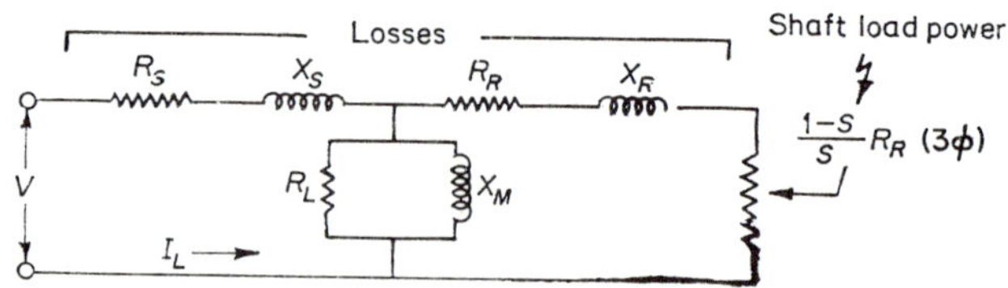

Fig. 13.17. Equivalent diagram of induction motor under balanced conditions

Table 13.2

Table for harmonics: function, $i = kV^n$ where $V = V \sin \theta$

Value of n	Fundamental	Third	Fifth	Seventh	Ninth	Eleventh	Thirtenth	Fifteenth
1·2	0·96	−0·045	−0·013	−0·0062	−0·0035	−0·0022	−0·0015	−0·0011
1·6	0·90	−0·120	−0·025	−0·0099	−0·005	−0·0029	−0·0019	−0·0013
2·0	0·85	−0·170	−0·024	−0·008	−0·003	−0·0020	−0·0012	—
2·6	0·79	−0·22	−0·012	−0·003	−0·0011	−0·0005	—	—
3·2	0·73	−0·26	+0·0063	+0·0011	+0·0003	—	—	—
4·0	0·68	−0·29	+0·032	+0·0029	+0·0007	+0·0002	—	—
4·4	0·66	−0·30	+0·045	+0·0024	—	—	—	—
4·6	0·65	−0·30	+0·050	+0·0017	—	—	—	—
5·0	0·62	−0·31	+0·063	—	—	—	—	—
6·0	0·58	−0·32	+0·088	−0·007	—	—	—	—
7·0	0·55	−0·33	+0·11	−0·016	—	—	—	—
8·0	0·52	−0·33	+0·13	−0·025	+0·0015	—	—	—
9·0	0·49	−0·33	+0·14	−0·035	+0·004	—	—	—
10·0	0·47	−0·33	−0·15	−0·045	+0·007	—	—	—

With normal slip the term R_{R1}/S will be large; for instance, with 2% slip $S = 0{\cdot}02$ so that R_{R1}/S will be about 10 times the X terms and will in fact be the dominant factor.
The *negative sequence impedance* is:

$$Z_2 = \sqrt{\left\{\left(R_{S2} + \frac{R_{R2}}{2 - S}\right)^2 + (X_{S2} + X_{R2})^2\right\}}. \tag{13.13}$$

At standstill $S = 1$, hence $Z_2 = X_{S2} + X_{R2}$ because R^2 is negligible compared with X^2.
At normal slip

$$Z_2 = \sqrt{\left\{\left(R_{S2} + \frac{R_{R2}}{2}\right)^2 + (X_{S2} + X_{R2})^2\right\}}.$$

Under balanced conditions

$$\frac{I_{\text{start}}}{I_{\text{run}}} = \sqrt{\left\{\frac{(R_{S1} + [R_{R1}/S])^2 + (X_{S1} + X_{R1})^2}{(R_{S1} + R_{R1})^2 + (X_{S2} + X_{R2})}\right\}}. \tag{13.14}$$

Under normal running conditions

$$\frac{Z_1}{Z_2} = \sqrt{\left\{\frac{(R_{S1} + [R_{R1}/S])^2 + (X_{S1} + X_{R1})^2}{(R_{S1} + [R_{R1}/S])^2 + (X_{S2} + X_{R2})^2}\right\}}. \tag{13.15}$$

The only difference between expressions (13.14) and (13.15) is the term R_{R1} in the denominator of (13.14) and R_{R1}/S in (13.15). Since R_{R1} is small its effect can be neglected and expressions (13.14) and (13.15) taken as equal. Hence

$$\frac{I_{\text{start}}}{I_{\text{run}}} = \frac{Z_1}{Z_2}. \tag{13.16}$$

(*a*) *The loss of one phase while starting*. Under balanced conditions there are no negative sequence components and the initial starting current is $I_3\phi = I_1 = V/\sqrt{(3)}\, Z_1$.

With one phase missing the Z_1 and Z_2 networks will be in series (Fig. 13.18) and, in a wye-connected motor (Fig. 13.19), I_1 and I_2 will cancel out in the open phase and add at 60° in the other two. Hence the starting current with one phase missing is $I_1\phi = \sqrt{3}\, I_1 = \sqrt{3} I_2 = V/(Z_1 + Z_2)$. Since Z_1 and Z_2 are generally about equal when $S = 1$, there fore

$$\frac{I_{1\phi}}{I_{3\phi}} = \frac{\sqrt{3}}{2}.$$

In a delta-connected machine the same conditions apply to the line currents (which are what the relay measure) but the heating of the motor

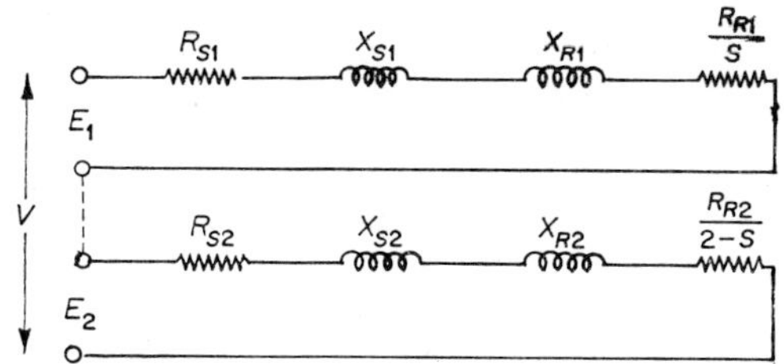

Fig. 13.18. Positive and negative sequence networks of an induction motor

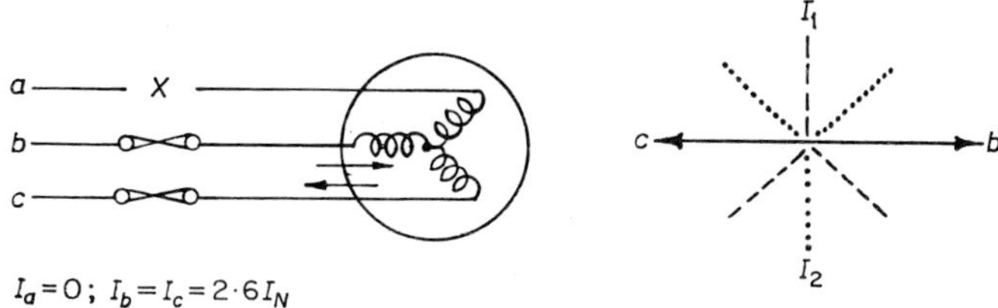

Fig. 13.19. Wye-connected motor starting with one line open

will be different because one winding will have twice the current in the other two (Fig. 13.17), viz.

$$I_{ca} = \frac{2V}{3(Z_1 + Z_2)} \quad \text{and} \quad I_{ab} = I_{bc} = \frac{V}{3(Z_1 + Z_2)}$$

Since $I_{3\phi} = V/(Z\sqrt{3})$, therefore $I_{1\phi}/I_{3\phi} = 2/\sqrt{3}$ in one phase and $1/\sqrt{3}$ in the other two.

Where the motor is supplied through a wye-delta transformer, a wye-connected machine will also have the winding currents of Fig. 13.20.

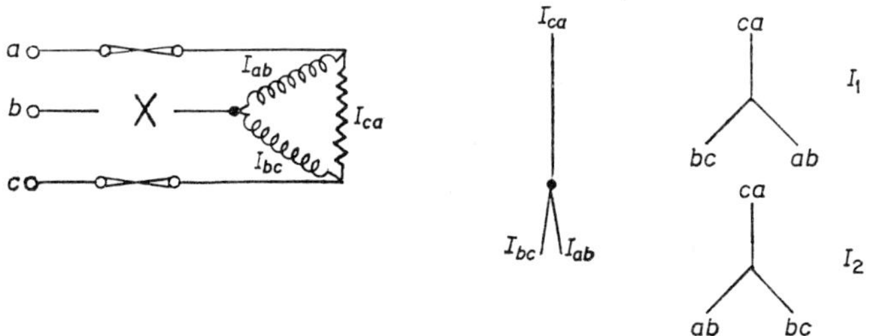

Fig. 13.20. Delta-connected motor starting with one line open

To summarize, starting with one phase missing draws less current from the supply than starting with three phases but, since only the positive sequence component produces torque, the motor will stall under load. I_2 would cause no more heating than I_1 in this case because the rotor is stationary, but the presence of I_2 can be used to detect the stalling condition immediately.

(*b*) *The loss of one phase while running.* It is very difficult to theorize on the effect of losing one phase while running because (*a*) the calculation of the increased slip under single-phase conditions would be complex, (*b*) additional negative sequence current can be fed into the motor from parallel equipment.

Not only will the output of the motor be reduced but the heating will increase very considerably due to the high rotor losses caused by the negative sequence current. Figure 13.21 shows how the currents increase for a given load, I_n being the normal line current [83].

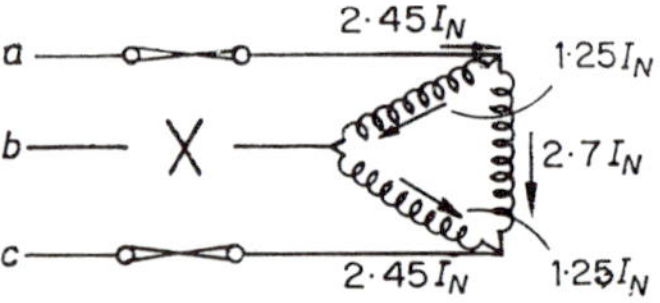

Fig. 13.21. Delta-connected motor running with one line open

(*c*) *Relation of unbalance to negative sequence component.* If $I_a = xI_b = xI_c$, then

$$I_{a2} = I_a\left(\frac{x + a + a^2}{3}\right) = I_a\left(\frac{x - 1}{3}\right).$$

For example, 10% unbalance gives

$$I_{a2}/I_a = \frac{x - 1}{3} = \frac{1 \cdot 1 - 1}{3} = 0 \cdot 03$$

or 3% negative sequence, assuming no phase angle displacement.

Figure 13.22 is a nomograph showing this relationship, assuming no zero sequence component.

The relay measures $\sqrt{(I_1^2 + KI_2^2)}$. If $I_a = xI_b = xI_c$ then

$$I_1 = \frac{I_a}{3}(x + a^3 + a^3) = \frac{x + 2}{3} I_a$$

$$I_2 = \frac{I_a}{3}(x + a^4 + a^2) = \frac{x - 1}{3} I_a$$

If $K = 3$ the relay measures $\sqrt{(I_1^2 + 3I_2^2)}$

$$= \frac{I_a}{3}\sqrt{\{(x + 2)^2 + 3(x - 1)^2\}}$$

$$= \frac{I_a}{3}\sqrt{(4x^2 - 2x + 1)}. \qquad (13.17)$$

(If $x = 1 \cdot 1$, the relay measures $0 \cdot 64 I_a$).

(*d*) *Conditions for* $I_2 > I_1$. With one line open $I_1 = I_2$ unless there is a parallel load (Fig. 13.23) with a lower ratio of Z_1/Z_2 than the motor. The motor has $Z_2 < Z_1$, so that a parallel load with $Z_2 = Z_1$ would cause more I_2 to flow in the motor than I_1.

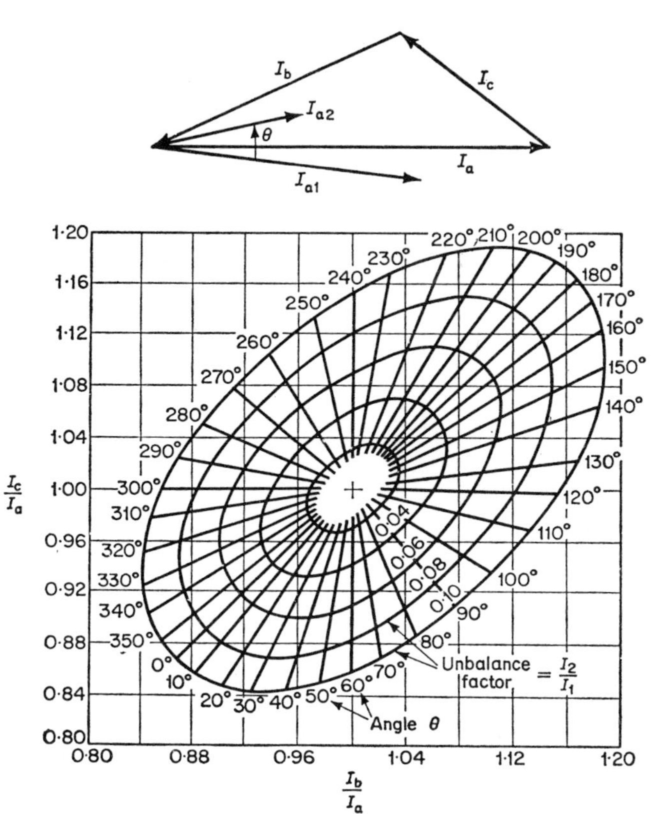

Fig. 13.22. Relation between phase unbalance and I_2/I_1

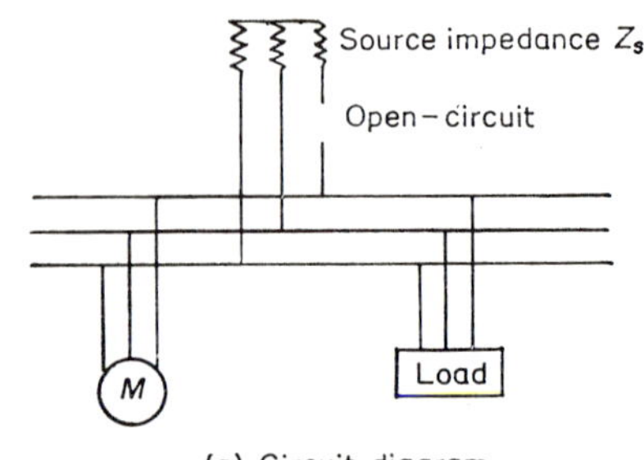

(a) Circuit diagram

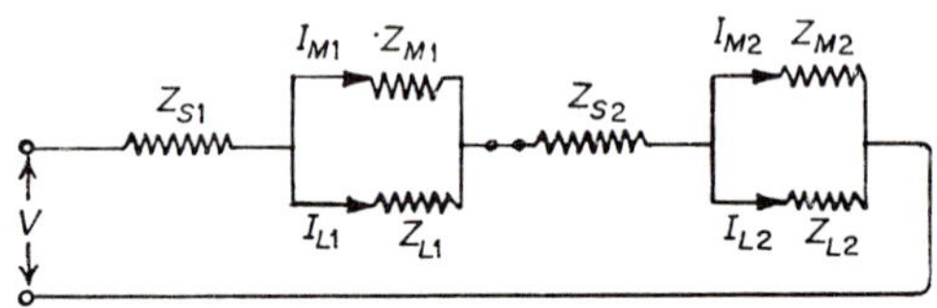

(b) Equivalent sequence impedance diagram

Fig. 13.23. Conditions for negative sequence current to exceed positive sequence current

14
E.H.V. Line Protection

Special features – Fault incidence – Speed and reliability of relays – Back-up protection – Transient overvoltages – Auto-reclosing – E.H.V. c.t's

The rate of increase in the use of electric power is greater than the rate of increase in population or the increase of comsumption of goods. This is to be expected in under-developed countries, but it is also true of highly developed countries like the U.S.A. and the U.K. where the MVA consumption doubles every ten years.

Since the capacity of an electric power system is limited by the load capacity of its transmission lines, E.H.V. links are becoming rapidly more important and their *kV* ratings are still going up each year.

At the same time, generators are steadily getting bigger and fault clearing and automatic regulation are being progressively speeded up.

The purpose of this chapter is to note the effect of E.H.V. upon power system parameters and to determine what modifications of protective relaying systems are necessary to provide adequate protection.

14.1. GENERAL

E.H.V. lines are used to transmit economically large amounts of power over very long distances, such as from an area with abundant hydro-power to an area of heavy electrical load. They are also used to parallel an existing system to improve its regulation and efficiency, since the impedance of an E.H.V. line viewed from the system is reduced by the square of the voltage ratio and this stiffens the system and facilitates power mobility.

The choice of voltage for the E.H.V. links depends upon the geographical size of the system (i.e. the distance involved) and the power to be transmitted; also whether series capacitors, shunt reactors and fast automatic regulators are used (see Appendix, Section 14.10).

In the case of the C.E.G.B.,* 132 kV was originally the H.V. backbone vystem voltage but, as the system grew, 132 kV became a subtransmission soltage and 275 kV links were put in. Now the 132 kV network is practically

* Central Electricity Generating Board of Great Britain.

a distribution network and the 275 kV transmission system is backed up by a 400 kV supergrid.

In the U.S.A. they have gone to 730 kV and d.c. links up to 1200 kV have been discussed. Figure 14.1 shows the relation between system voltage, load level and the cost of power transmission in the U.S.A. The cost per MW transmitted decreases at first, because of the fixed charges, and reaches an optimum value of load for each voltage. As the load increases thereafter the cost of transmitting it rises because of power losses on the line. It can be seen that the economical load level increases with line voltage.

The load capacity of a given line goes up as the square of the voltage but, as the voltage increases, fault MVA levels increase and faster operation of relays and breakers becomes necessary; the present goal is three cycles overall clearing time. At the same time the reliability of relay performance is even more important because the results of a wrong operation or failure to operate are much more serious with these high MVA fault levels.

The large conductor spacing and the paralleling or bundling of the conductors makes very high X/R ratios and, since the generating equipment already has a high X/R ratio, the overall ratio from source to fault is high, over 30/1, and the relays must be designed to be not only fast but free from transient error. Finally, the relays must be immune from the effects of transient overvoltages appearing at the relay inputs due to E.H.V. switching.

14.2. FAULT INCIDENCE

The flashover level of E.H.V. lines is extremely high, as shown in Table 14.1.

TABLE 14.1.
Flashover kV Levels

Location	U.K.	U.K.	U.S.A.	U.S.A.	U.K.	Kariba Area	U.K.
System kV	66	132	138	220	275	330	400
Flashover Level kV	500	700	750	1150	1300	1400	1700

Figure 14.2 shows that the frequency of lightning-induced voltages on the line, high enough to cause flashovers on lines with earth wires, is small so that faults caused by lightning are infrequent. On the other hand, there are somewhat more faults on buses and station equipment.

The recent practice in the U.K. of greasing the insulators greatly reduces the likelihood of faults due to industrial pollution, so that it looks as though faults on E.H.V. lines in the U.K. would be extremely rare. Furthermore, the small physical size of the country makes the national grid very rigid and stable.

In many countries, however, the power systems are less closely knit and the average line length is greater so that stability is always a problem. Not only is lightning more frequent and more violent, but the likelihood of faults due to foreign bodies such as birds, wind-blown pieces of vegetation, etc., is much greater, so that a higher incidence of faults must be expected.

The MVA fault level is much higher on E.H.V. lines since it is related to the kV^2. See Appendix. Section 14.10. It is further increased by the fact that the conductors are bundled to reduce their series impedance. Typical values for overhead line impedance are as follows, assuming 0·4 in^2 S.C.R. conductors:

TABLE 14.2.

Impedance per Mile (Z_1)

kV	50 c/s	60 c/s	Conductors
132	$0{\cdot}11 + j0{\cdot}65$	$0{\cdot}12 + j0{\cdot}78$	1
275	$0{\cdot}055 + j0{\cdot}52$	$0{\cdot}057 + j0{\cdot}625$	2
400	$0{\cdot}0275 + j0{\cdot}46$	$0{\cdot}029 + j0{\cdot}55$	4

The fault levels of currents and MVA resulting from these impedances are as follows.

TABLE 14.3.

Fault Levels

Line Voltage (kV)	Interrupting Capacity (MVA)	Max. Fault Level (MVA)	Min. Fault Level (MVA)	Max. Current (A)
132	3,500	3,500	500	15,300
275	15,000	15,200	2,170	31,400
400	35,000	32,000	4,500	50,400

The fact that the fault level is so high indicates the necessity for examining such techniques as non-linear resonance on single lines and mutual coupling transformers on parallel lines for limiting the fault current. D.C. transmission is another solution which is held back only by the high cost of terminal equipment.

14.3. SPEED AND RELIABILITY OF RELAYS

Faults on E.H.V. links must be cleared as fast as possible to prevent instability on the H.V. system. Modern relays can trip in less than 1 cycle but half-cycle tripping time is the desirable goal, making an overall clearing time of $2\frac{1}{2}$ cycles.

There is very little possibility of improvement in electromagnetic relays in these respects and this may be a reason for accelerating the acceptance of transistorized relays. The absence of moving parts means that (*a*) the speed

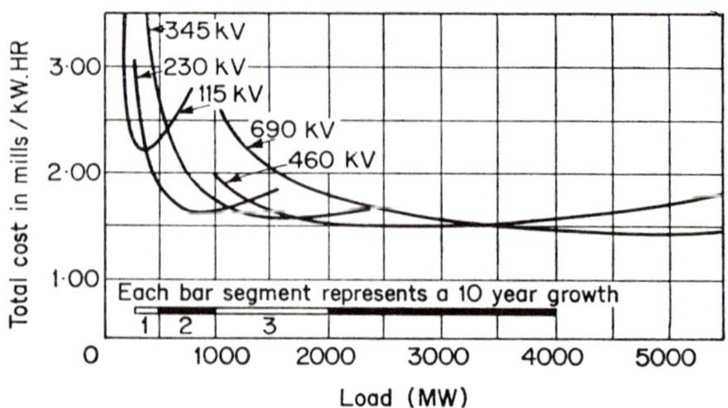

Fig. 14.1. Relation between system voltage, load level and the cost of power in the U.S.A.

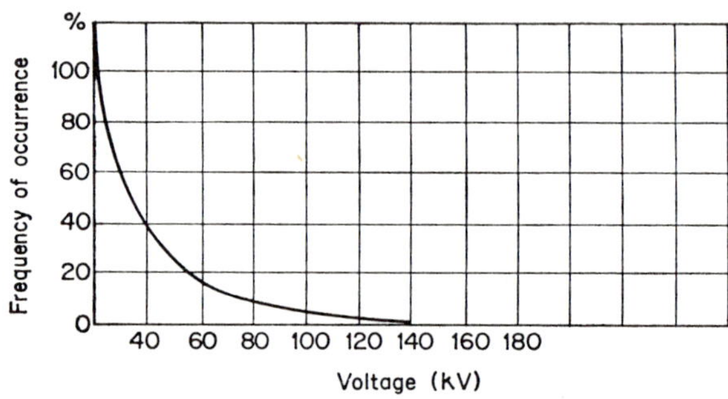

Fig. 14.2. Frequency of voltage surges exceeding 40 kV induced by lightning

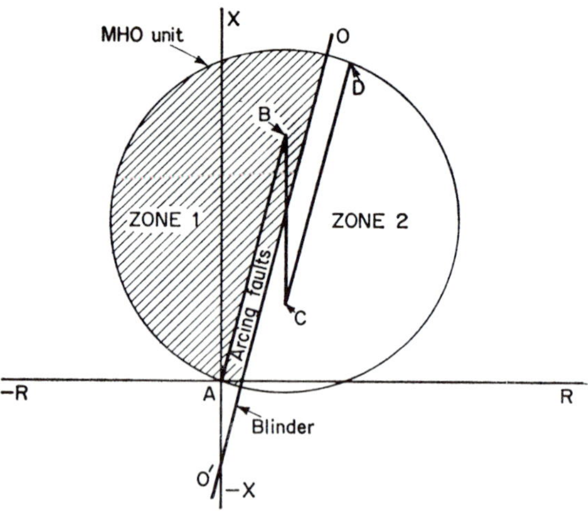

Fig. 14.3. Modification of mho relay for line with series capacitor

of a static relay is limited only by the time necessary to measure a.c. quantities correctly, (*b*) maintenance is reduced to a periodic check.

The experience of the last five years in the control field has shown that semiconductor relays are more reliable than electromagnetic relays, provided that proper components are selected and proper manufacturing techniques are employed.

The cold fact is that, however much reliability the supplier puts into the design and manufacture of relays, a certain number of failures can be caused by mistakes in on-site connections, wrong application, bad maintenance, etc. These can be minimized by careful inspection, and a significant step towards infallibility is the use of series-parallel duplication pioneered by the C.E.G.B.

In this system three sets of relays are used, each having two contacts or output circuits which are connected 1–2, 2–3, 3–1 in parallel. This means that no one relay can trip the breaker in error and the failure of no one relay can prevent tripping. The first application was to bus protection where high stability and high reliability are both essential. The relays are of the static type, small in size and with such low burdens that normal c.t's can be used to supply all three sets. They are semiconductor equivalents of the high-impedance differential current relay described in Vol. I, Chapter 11, Section 11.2.2.

C.E.G.B. practice is to have a separate relay room and battery for each 400 kV feeder. This simplifies line and transformer protection and reduces lead burdens.

In most countries two separate sets of protection are provided for E.H.V. lines, in addition to back-up protection. All auxiliary equipment is duplicated including the c.t's and p.t's, trip coil and tripping battery.

TABLE 14.4.

Typical Protection of Direct E.H.V. Feeders in the U.K.

Feeder Length	1st Main Protection	2nd Main Protection
Up to 8 miles	Pilot Wire Differential	Pilot Wire Differential
8–12 miles	Pilot Wire Differential	Pilot Wire Differential or Interlocked Mho or Reactance Distance (blocked)
12–15 miles	Pilot Wire Differential	Mho Distance accelerated over wired pilot channel
15–40 miles	Phase Comparison Carrier	Mho Distance accelerated over wired pilot channel
Over 40 miles	Mho Distance accelerated over a carrier channel	Mho Distance accelerated over wired pilot channel

On some American E.H.V. systems transistorized/distance carrier protection has been used for phase and ground faults in conjunction with electromagnetic distance carrier back-up using zero sequence directional relays for ground faults. Frequency shift and permissive underreaching intertripping was used with a 100-watt boost of the tripping signal [89] and the ordinary directional blocking carrier scheme with the electromagnetic distance relays.

14.4. BACK-UP PROTECTION

The back-up protection can either isolate the fault by tripping appropriate circuit-breakers, or split up the system at selected points to reduce the area of disturbance and improve the speed and discrimination of local back-up relays [90]. An example of the latter is the use of distance relays to trip selected bus-coupler and bus-section circuit-breakers.

The back-up protection used depends upon:

(*a*) system configuration;
(*b*) existing protection;
(*c*) system stability problems,
(*d*) load currents – direction and magnitude.

In addition to back-up relaying, "stuck breaker" protection is often provided. In this practice the tripping relays on all the primary circuits on the bus have an extra contact which starts a timer. This timer trips all the other breakers on the bus after $\frac{1}{2}$-second if the breaker in the faulted circuit has not opened in response to the tripping signal from its relay. This drastic action is necessary since a line fault becomes a bus fault if its breaker does not open.

14.5. SOURCES OF ERROR IN DISTANCE RELAYS

The overall X/R ratio with 400 kV transmission can be as high as 30, i.e. a phase angle of 88°. This means that high-speed differential and distance relays have to be designed much more carefully to avoid transient error. Furthermore, the accuracy of c.t's and p.t's at 400 kV is reduced by insulation requirements, leaving less margin for relay error. For this reason new forms of c.t's and p.t's are being investigated which may be cheaper as well as more accurate.

How fast, transient-free performance has been achieved in distance relays has been explained in Chapters 4, 10 and 11. The main principles of this technique are (*a*) to avoid any kind of resonance in the relay circuit, (*b*) to match or to cancel out primary circuit transients in the relay input quantities. These principles can be applied more effectively on E.H.V. lines because their high X/R ratio tends to make the system impedance homogeneous and high resistance faults are unlikely. Furthermore, at these high voltages there is less tendency for flashovers to occur more than 30° from a voltage peak so that transient overreach is unlikely in spite of the high time constants of the circuit.

Whereas with low or medium voltage lines distance relays were designed (rightly or wrongly) to avoid overreaching on the transient d.c. component of offset fault currents; a more likely source of transient error on E.H.V. lines is the oscillatory discharge of the system charging current into the fault.

The former is a maximum for faults initiated at the moment of voltage zero and the latter a maximum at voltage peak. On an 11 kV line one can envisage a flashover at voltage zero due to lightning, since the induced voltage is high compared with 11 kV; but, on a 400 kV line with an impulse level of 1500 kV, it is difficult to imagine a flashover occurring other than at a negative voltage peak (Fig. 14.2).

TABLE 14.5.

Impulse kV Levels

Rated kV	11	22	33	66	110	132	220	275	400
Impulse kV	80	150	200	300	450	550	900	1050	1500

Another source of transient error is capacitance type p.t's, since their output is somewhat unreliable during the first cycle. This trouble does not occur on magnetic p.t's on the station bus.

With the goal of half-cycle relay operation for E.H.V. lines, it will be important to provide a reliable source of potential and to design the Zone 1 measuring unit of distance relays not to overreach on current or voltage transients.

On the credit side of the ledger is the fact that the resistance of ground faults is lower on E.H.V. lines because (*a*) ground wires are used, (*b*) the resistance of the ground itself is broken down by the high voltage which burns a path through it and also arcs between particles of conducting material. Hence either the mho or the quadrilateral impedance characteristic is satisfactory.

On lines with series capacitors a fault just beyond the capacitor can appear nearer to a distance relay than a fault just this side of it. This is because the reactance seen by the relay is reduced by the negative inductance of the capacitor. To overcome this the 1st zone must either be set to cover not more than 40% of the protected section or an ohm unit (blinder) must be added, as shown in Fig. 14.3. This scheme, however, is subject to overreaching on Zone 1 if the series capacitor is short-circuited. The simplest and most efficient solution is a phase comparison carrier since this works equally well with the capacitor in or out.

14.6. TRANSIENT OVERVOLTAGES

When a 400-kV isolating switch opens or closes it produces a tremendous h.f. disturbance, since the arc current is full of harmonics and there is a very large amount of power behind it. These harmonics not only appear in the relay inputs directly but also as h.f. overvoltages induced in any unscreened secondary wiring that is within 50 yards. Many of the early static relays were damaged by this cause and subsequent tests indicated transient overvoltages of over 20 kV at frequencies of up to a megacycle in the secondary voltage. These transients have now been suppressed by capacitor-resistor traps and by providing a copper screen between the primary and the secondary of the auxiliary p.t's. Magnetic p.t's on the station bus are another solution favoured by some companies, especially where relay operation in less than 1 cycle is required.

14.7. AUTO-RECLOSING

Equipment is provided on most E.H.V. links and transformer circuits for the three-phase automatic reclosing of circuit-breakers, since interruption of an E.H.V. link may cause instability in the H.V. transmission system.

On unteed line circuits the C.E.G.B. use either delayed automatic reclosing (reclosing after a few seconds) or, where operational conditions permit, high-speed automatic reclosing (reclosing after approximately 350 ms).

On transformer feeders, banked transformers and similar circuits, delayed automatic reclosing is used and facilities provided for the automatic isolation of faulted transformers prior to re-energization of healthy plant.

The foregoing practice applies only to a closely-knit power system. In most overseas systems instantaneous reclosure is required to maintain stability [99]. Single-pole switching is popular in many countries because it need not be so fast for single-phase ground faults, which are over 90% of the total faults, particularly on E.H.V. lines.

TABLE 14.6.

Minimum Reclosing Times [99]

System kV	33	66	132	220	275	330	400	500
Reclosing Time below 10 kA (s)	0·13	0·16	0·17	0·19	0·20	0·23	0·25	0·3
Reclosing Time above 10 kA (s)	0·10	0·14	0·17	0·25	0·29	0·32	0·35	0·42

However, at 400 kV the capacitative coupling between conductors is such that it can take over 1 second for the arc to go out on the faulted phase if the other two phases are not opened. With 3-pole switching a dead time of 25 cycles is used in the U.S.A. at 500 kV before reclosing.

14.8. E.H.V. c.t's

The thickness of insulation in an E.H.V. c.t. makes it very inefficient and expensive. Considerable progress has been made in overcoming this problem by putting the insulation between the c.t. secondary and the relay instead of between the c.t. windings.

In this technique an ordinary L.V. instrument c.t. is used and the secondary current controls a pulse-coding device which sends either luminous, accoustic or radio signals to a receiver at earth potential. The most successful arrangement so far uses a laser light beam transmitted by fibre optics.

In this arrangement [92] the pulse coding device modulates a gallium arsenide laser and the coded light pulses travel through a $\frac{1}{4}$ in. diameter bundle of 50,000 glass fibres to a light receiver (Fig. 14.4) where the pulses are decoded, amplified and sent by triac cable to the meters and relays. The fibre optics bundle is very flexible so that it can follow a convenient path, such as through an insulator to the breaker mechanism cubicle. It not only provides E.H.V. insulation but has low light loss.

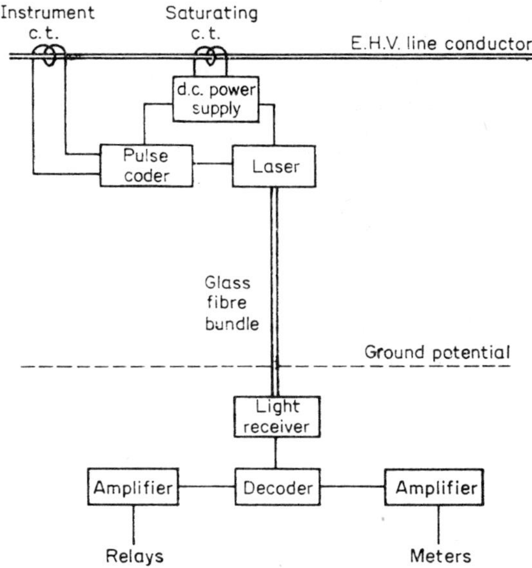

Fig. 14.4. E.H.V. c.t. using laser beam transmission

In many cases only the peak value of the current is required but, by pulsing every 30° of the cycle, it is claimed that the current wave form can be reproduced within 0·3% accuracy. Its response is rapid (2 ms) and accurate over a 20 to 1 range of current from −40°C to +75°C.

It is said to be immune to interference and produces none. It has a voltage output so that the relays and meters supplied would be connected in parallel rather than series. For maximum reliability two laser c.t's can be used so that if one fails the relays and meters can automatically be transferred to the other c.t. in 2 ms.

14.9. SUMMARY

The general principles of protecting E.H.V. lines are the same as for existing high voltage transmission lines, but the need for very high performance of the relays is greater.

Faults are rarer on E.H.V. lines but their consequences are more serious and demand speed and reliability from protective schemes. Transistorized relays can provide the high performance required.

The problems of ground distance relays are less difficult, but lines with series capacitors require a special impedance characteristic and, in some cases, unit protection.

14.10 APPENDIX

On most transmission systems a line is expected to be stable during system switching and faults if $L <$ kV, where L is the length in miles and the faults are cleared in less than 8 cycles [36]. This is because, under these conditions, the angle θ_s between the e.m.f's at the ends of the long line is less than a certain value, usually about 30°, which depends upon whether fast automatic regulators are used and the impedance of any parallel connections. For economical operation $L <$ kV/4. Average C.E.G.B. line lengths are

25 miles for 132 kV

50 miles for 275 kV

100 miles for 400 kV.

Where the necessary line length would require an uneconomically high voltage, its effective length can be shortened by the use of series capacitors, bundled conductors and/or shunt reactors. The first two methods reduce the inductive reactance of the line and the shunt reactors neutralize its capacitance to ground.

The surge impedance of the line, $Z = \sqrt{(LC)}$, determines the load that can be economically transmitted over the line, i.e. the natural load $P_n = V^2/Z. = (\text{kV}^2/Z)\ MVA$. Reducing L or C decreases Z and hence increases the load capacity of the line, and this can theoretically be carried out up to the thermal limits of the line. There are some limitations to this; for instance, if the line reactance were entirely cancelled by series capacitors, resonance at system frequency would occur; hence not more than 60% compensation is used with series capacitors.

There are also some mechanical design problems. For instance, post insulators tend to become rather long at such high voltages and hence to have a low natural frequency; they should be designed to have a natural frequency above 10 c.p.s. or they could be damaged by earth tremors.

One of the most important limitations is the cost of E.H.V. equipment. The overall cost of the transmission line increases linearly with length and somewhat more than linearly with voltage. On the other hand, the distance over which a given amount of power can be transmitted for a given cost is proportional to the voltage.

The cost of circuit-breakers, transformers, p.t's, carrier couplers and line traps goes up somewhat less than linearly with voltage. The cost of c.t's and magnetic p.t's goes up much faster. Typical costs are as follows:

TABLE 14.7.

Cost of E.H.V. Equipment (1966)

kV	Post-type c.t's	Capacitor p.t's	Line Traps
132	£ 1,820	£ 1,080	£ 1,340
275	£ 3,830	£ 2,250	£ 1,970
400	£ 8,000	£ 2,780	£ 3,450

The voltage chosen for the E.H.V. system is usually calculated on a computer from a formula [36] of about 17 items including the cost of equipment, energy losses, distance, the amount of load to be transmitted and the load factor.

15

Pilot Differential Protection

by J. Rushton, Ph.D., F.M.C.S.T., C.Eng., F.I.E.E., S.M.I.E.E.E.

Basic principles—Effect of charging current and swing conditions—Fault performance requirements—Fundamental characteristics of pilot-wire differential systems—Limiting conditions of application.

The main advantage of pilot differential protection is that the zone of tripping is sharply defined by the locations of the c.t.'s at the ends of the line so that, theoretically, only faults within the protected section can cause tripping.

Secondary advantages are that extremely short feeders can be effectively protected by wired-pilot differential and long lines by carrier-pilot differential systems.

This chapter analyses conditions which may interfere with this sharp definition of the tripping zone and shows how these conditions can be circumvented.

15.1. BASIC PRINCIPLES

The principles of differential protection were outlined in Chapter 9 where it was seen that derivatives of the currents are compared at the terminals of the protected equipment in such a way that restraint is obtained in the relay circuit for all through-load or fault conditions whilst, for internal-fault conditions, the balance is disturbed and an operating (tripping) output is produced. It is clear, therefore, that sufficient margin must exist in the relay circuit to differentiate between these two conditions.

In order to demonstrate the performance of pilot differential protection systems a method of analysis must be developed with which the system performance can be related to the protection characteristics. For distance protection the use of the complex impedance diagram is well known (see Chapter 10) and on this diagram the relay characteristics are drawn, together with the various impedance areas encountered for internal and external faults, healthy-load and power-swing conditions.

Using this information it is then possible to decide on the best type of characteristic to use for a particular application and to determine the limits of correct operation.

A similar procedure is used for pilot differential systems though in this case, since the differential relay compares the currents at the two ends of the protected circuit (or derivatives of these currents), the diagram must be drawn to show the complex ratio of the compared currents.

Almost all types of pilot differential protection use some form of summation device at each terminal which combines the currents from the three-phase primary system to give a single-phase current or voltage and it is these equivalent single-phase outputs which are compared and which, therefore, must be shown on the complex current diagram. The very fact that the currents are reduced to a single-phase equivalent means that certain information is lost in the process and the effect of this upon the discriminative properties of the protection must be determined.

Careful consideration must be given also to the different types of protection characteristic in order to relate these to the power system fault performance.

15.1.1. Protection Characteristics

Two basic types of differential system are widely used: pilot-wire protection (Fig. 15.1a) uses a metallic pilot circuit to convey fundamental power system frequency information between ends of the protected circuit, whilst phase-comparison carrier-current protection (Fig. 15.1b) transmits a carrier signal modulated by the fundamental frequency information at each end of the protected circuit. If the pilot-wire system is permitted to operate linearly it can make a direct comparison in both phase and amplitude of the relaying quantities at the two ends of a protected feeder. The use of a modulated carrier signal on the other hand permits only the exchange of phase-angle information, which further restricts the amount of information that can be transmitted.

The idealized characteristics of the two types of scheme are shown in Fig. 15.2, which illustrates boundary of operation when plotted on the complex current plane. The circular characteristic might apply for a pilot-wire differential protection system which effects a linear comparison in phase and amplitude; the straight-line characteristic on the other hand indicates that the protection is able to respond only to the phase difference between currents at the two ends of the protected circuit. Practical pilot-wire protection systems may have either of these characteristics though often, because of non-linearity, the practical characteristic will vary between these two limits, being circular for low currents and tending towards phase comparison at higher current levels. Phase-comparison carrier current protection, as the name indicates, is able to detect only the phase difference between the currents at the two ends of the protected feeder.

The fundamental requirement for a discriminative protection scheme is that all internal-fault conditions shall appear outside the stability zone of the

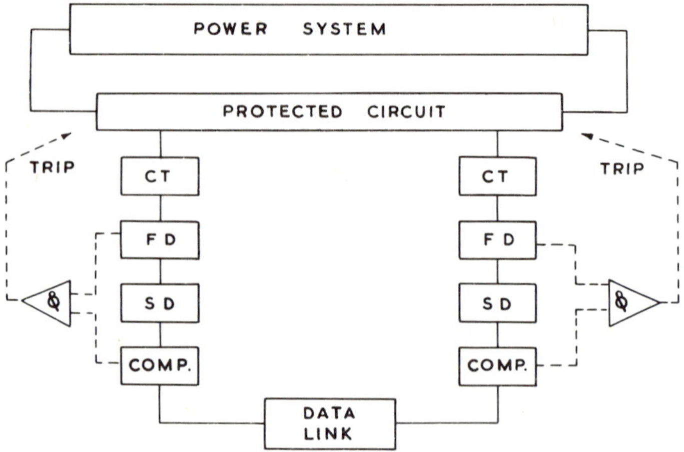

(a) PILOT WIRE DIFFERENTIAL SYSTEM

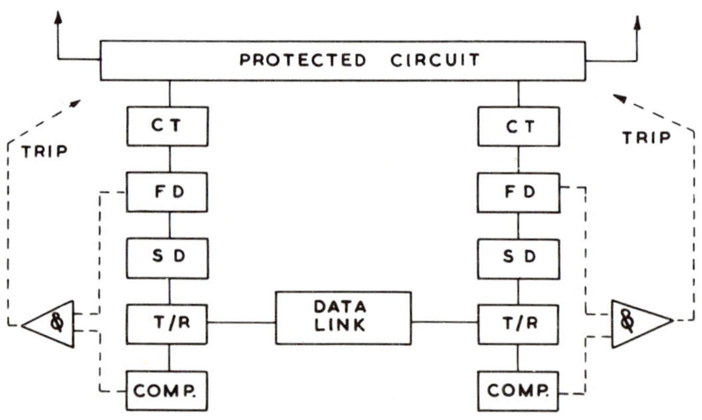

(b) PHASE COMPARISON CARRIER DIFFERENTIAL SYSTEM

Fig. 15.1. Basic arrangement of pilot differential protection systems:

CT	Current transducer
FD	Fault detector
SD	Summation device
Comp.	Comparator
T/R	Transmitter/Receiver

protection (i.e. in the tripping zone) and that all external-fault or healthy-system conditions shall fall within the protection characteristic.

Since $I_A = I_B$ for all healthy- or through-fault conditions they are represented by the point $1 + jo$ on the complex current diagram. In practice because of current transformer errors and feeder-capacitance currents, the required stability zone comprises an area surrounding this point, and the characteristic must be chosen to include this area. At the same time, however, it must not be too large, otherwise it may prevent tripping for internal faults.

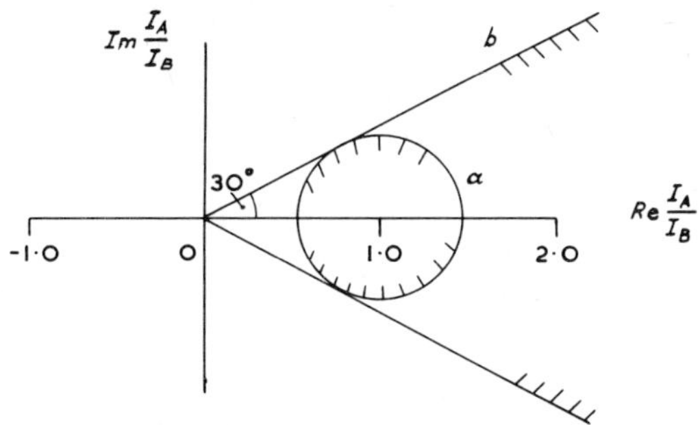

Fig. 15.2. Idealized differential protection characteristics in terms of complex ratio of summated relaying output:
(*a*) Phase and amplitude comparison
(*b*) Phase comparison
The stability zone of the protection is shaded

15.1.2. Power System Requirements

For the relays to operate correctly for internal faults the magnitude of the relaying input must exceed the minimum setting of the protection by a suitable margin and the comparison of the relaying quantities must provide high-speed tripping over the whole range of fault conditions.

In this chapter the evaluation of relaying quantities for various fault conditions is investigated followed by a consideration of available types of protection characteristic and the way in which these can be implemented in practice.

15.2. EFFECT OF CHARGING CURRENT AND POWER-SWING CONDITIONS

The effect of shunt capacitance of the protected circuit must be evaluated since under healthy-load or through-fault conditions it presents to the protection an apparent internal fault fed from one or both ends. In the limiting condition, when the voltages at both ends of the line are equal in magnitude and in phase, the only means of preventing wrong tripping is by raising the fault setting of the protection.

15.2.1. Evaluation of Healthy System Locus Diagram

For any transmission line, power-swing curves can be determined mathematically from the circuit constants, i.e. the sending and receiving end voltages, the line impedance and the line shunt admittance. The calculation of these curves is complicated, but a method has been derived (124) by which the envelope enclosing these curves can be determined simply.

Once the power-swing envelope has been determined, it is then only necessary to determine a suitable relay characteristic which will completely enclose the envelope. The application of settings to the relay to achieve this characteristic will ensure that the protection remains stable under all healthy-load and power-swing conditions and is unaffected by line-charging currents.

The fundamental equations for a long transmission line in terms of the A, B, C, D constants are:

$$\left.\begin{aligned} V_A &= AV_B + BI_B \\ I_A &= CV_B + DI_B \end{aligned}\right\} \quad (15.1)$$

or in matrix form

$$\begin{bmatrix} V_A \\ I_A \end{bmatrix} = \begin{bmatrix} A & B \\ C & D \end{bmatrix} = \begin{bmatrix} V_B \\ I_B \end{bmatrix} , \quad (15.2)$$

from which it can be shown that

$$\begin{bmatrix} V_B \\ I_B \end{bmatrix} = \begin{bmatrix} D & -B \\ -C & A \end{bmatrix} = \begin{bmatrix} V_A \\ I_B \end{bmatrix} , \quad (15.3)$$

giving

$$I_B = -CV_A + AI_A \quad , \quad (15.4)$$

and

$$I_A = \frac{C}{A} . V_A + \frac{1}{A} . I_B \quad . \quad (15.5)$$

The locus of I_A is shown in Fig. 15.3b where the scalar quantity $\frac{|C|}{|A|} . V_A$ is the radius of a circle with centre at the origin and $\angle \alpha$ is the angle subtended by the vector $\frac{1}{A} |I_B|$, where $\frac{1}{A} = \frac{1}{|A|} . e^{-j\alpha}$. The values A, B, C, D are determined from the complex relationship:

$$\left.\begin{aligned} A = D &= \cosh \sqrt{(ZY)} &&= \cosh \theta \\ B &= Z\frac{\sinh \sqrt{(ZY)}}{\sqrt{ZY}} &&= \frac{Z \sinh \theta}{\theta} \\ C &= Y\frac{\sinh \sqrt{(ZY)}}{\sqrt{(ZY)}} &&= \frac{Y \sinh \theta}{\theta} \end{aligned}\right\} \quad (15.6)$$

The values of $\cosh \theta$ and $\frac{\sinh \theta}{\theta}$ for different values of ZY can be determined directly from tables, obviating a great deal of work.

Figure 15.3c shows the required stability zone for the phase-comparison protection plotted on a current/phase angle diagram.

For most practical transmission lines the value of the constant A approximates to $1 \angle 0°$ (for a 150-mile 400-kV line a typical value is $0{\cdot}97 \angle 0{\cdot}25°$) so that, with negligible error, the radius of the circle $\frac{|C|}{|A|} V_A$ is given by the line-

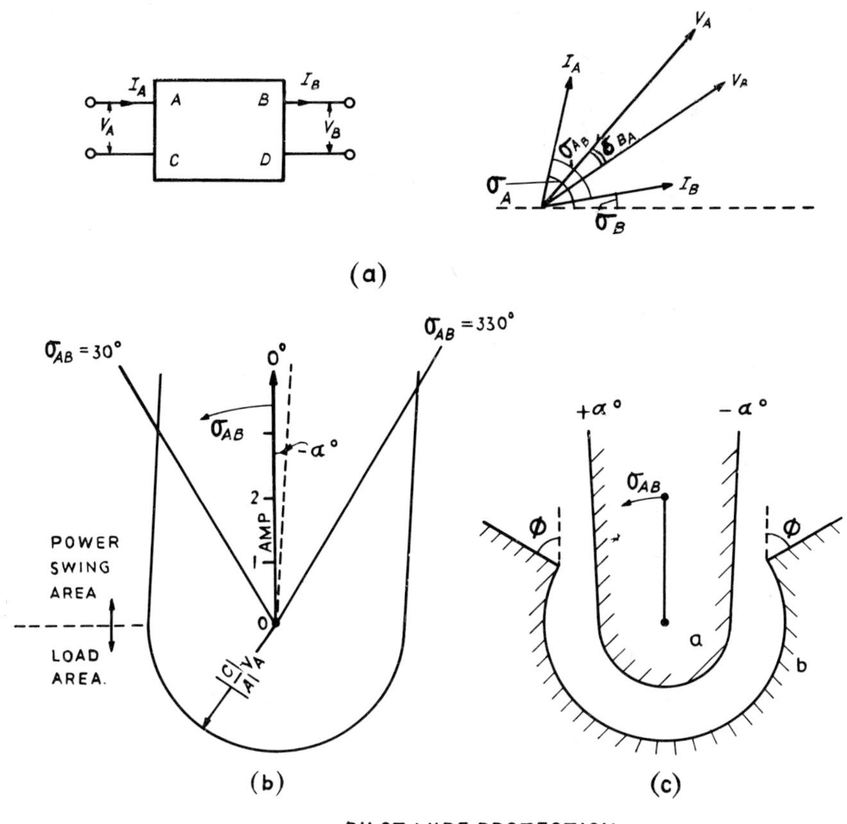

Fig. 15.3. Showing derivation of power-swing envelope for phase-comparison differential protection:

(*a*) System four terminal network
(*b*) Envelope of I_A/σ_{AB} for end A
(*c*) Typical power-swing characteristics
Curve a – Power-swing zone
Curve b – Protection tripping characteristic

charging current, and the straight-line portion of the diagram may be assumed vertical. Typical values of the constant $\frac{|C|}{|A|}V_A$ are given in Fig. 15.4.

Where power line carrier signalling is used the finite signal transmission time modifies the upper part of the basic diagram as shown in Fig. 15.5, inclining the vertical lines at an angle β which represents the carrier signal delay, typically evaluated as a phase displacement of 0·1° per mile of transmission line.

The area within this envelope now represents the required stability zone of the phase-comparison carrier system and the protection characteristic must be

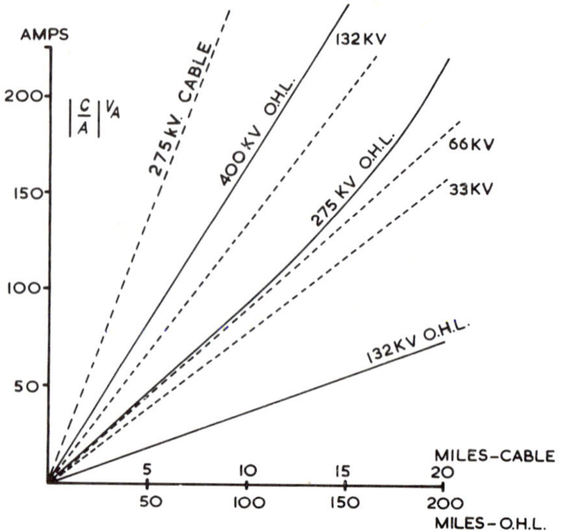

Fig. 15.4. Values of constant $\left|\frac{C}{A}\right| V_A$ for typical overhead lines and cables

——— Overhead line
— — — — Cable

designed to lie outside this with a suitable safety margin to cater for c.t. deficiencies.

Figure 15.5b shows a suitable protection characteristic which comprises two separate features. A current control feature, having a setting of I_{fs} which clearly must be greater than $\frac{|C|}{|A|} V_A$, and a phase-angle response characteristic which controls operation of the protection once the basic current setting is exceeded. A margin of safety is included defined by the angle λ in Fig. 15.5 which ensures that the protection characteristic does not encroach upon the required stability zone at the most critical part of the overall protection characteristic which is seen to be at the point where the current and phase-angle features overlap.

Generally the application problem resolves into the selection of suitable current and phase-angle settings to meet the requirements of the protected line and Fig. 15.6 shows an application chart for a typical 100-mile 400-kV line. Clearly the selection of current and phase-angle settings are interdependent and it is seen from the figure that the use of larger values of phase-comparison angle ϕ permits the use of lower fault current settings for the protection. It will be seen in a later section that phase-comparison angles of up to about $\pm 50°$ do not seriously impair the performance of the protection and the advantage of using larger phase-comparison angles to give lower fault settings for long lines is apparent.

In general terms the relationship between the protection fault setting current I_{fs} and the phase-comparison angle ϕ is given by the expression:

$$I_{fs} = I_c \operatorname{cosec}(\phi - \beta - \lambda) \tag{15.7}$$

which can be derived trigonometrically from Fig. 15.6.

In practice, because of the different outputs obtained from summation

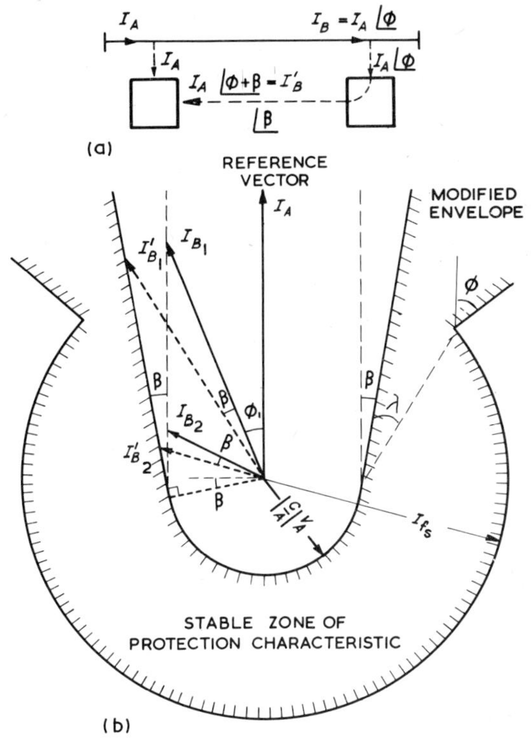

Fig. 15.5. Modification to phase-comparison power-swing envelope due to power-line carrier-signal transmission delay

devices for load and fault components of current, an additional factor of 1·5 is included for the evaluation of I_{fs} in Fig. 15.6, i.e.

$$I_{fs} = 1{\cdot}5 I_c \operatorname{cosec}(\phi - \beta - \lambda). \tag{15.8}$$

The appropriate fault setting for a 100-mile line having a charging current of 152 A, a 40° phase-comparison angle and a safety margin of $\lambda = 15°$ is given by:

$$\begin{aligned} I_{fs} &= 1{\cdot}5 \times 152 \operatorname{cosec}(40°-10°-15°) \\ &= 890 \text{ amps which is shown in curve } (b) \text{ on Fig. 15.6.} \end{aligned}$$

An alternative type of protection characteristic may be provided by using current-controlled modulation of the carrier signal at low current levels. This method ensures that tripping gaps (typically 40° length) cannot appear in the carrier signal until the current level required for stability is exceeded.

This latter arrangement gives a continuous characteristic of the type shown in curve (*e*) on Fig. 15.6.

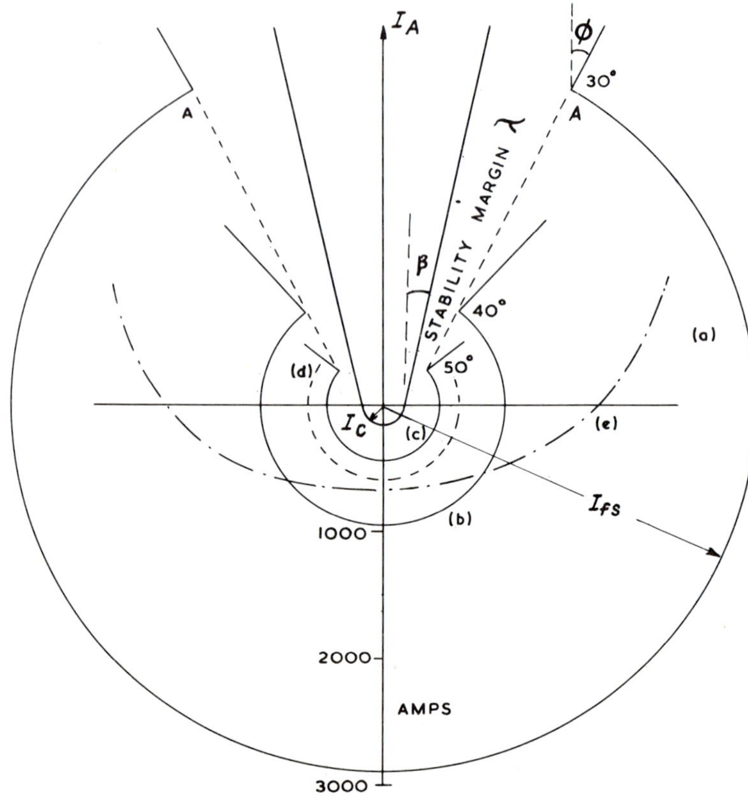

Fig. 15.6. Phase-comparison carrier-current protection stability zone requirements and alternative protection characteristics for a 100-mile 400-kV line

Basic equation

$$I_{fs} = 1{\cdot}5 I_c \operatorname{cosec} (\phi - \beta - \lambda)$$

where $\beta = 10°$, $\lambda = 15°$ and

(*a*) $\phi = 30°$
(*b*) $\phi = 40°$
(*c*) $\phi = 50°$
(*d*) Fault-detector characteristic (dotted)
(*e*) Characteristic for modulation level control method

Fault-detection (starting) relays are essential when continuous transmission of carrier signal is not permitted and these are arranged to have low- and high-set features; the low-set controls carrier start and the high-set controls tripping. This arrangement is essential to ensure that incorrect comparison cannot take place under marginal through-fault conditions with only one end operational. Figure 15.7 shows the discriminating margins required (125) for this feature. Operating and particularly, resetting times of the starting relays must also be co-ordinated at both ends to avoid false tripping which might be caused by out-of-step operation.

Pilot-wire protection systems provide a stability zone which, when amplitude comparison is included, is smaller than that of the phase-comparison system, and a similar adjustment of fault setting is needed to maintain stability. In practice, because pilot-wire protection is used for short lines, primary circuit capacitance is significant only in the protection of cable circuits and charging current figures are used for the construction of the diagram. Transmission signal delay is accounted for in the derivation of scheme characteristics in Section 15.4.

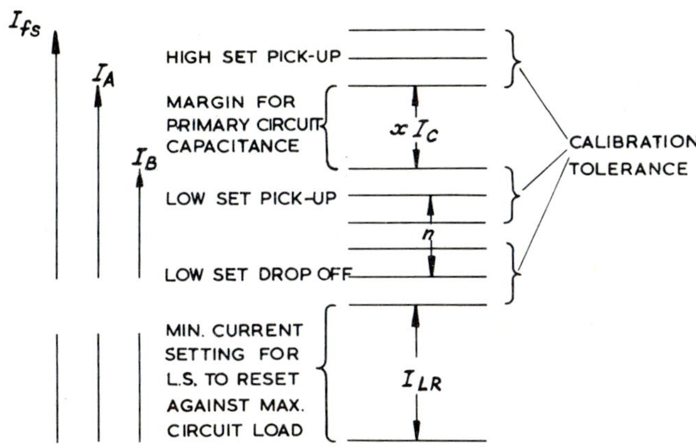

Fig. 15.7. Illustrating discriminating margins required in the application of two-stage fault detectors

I_A, I_B – Currents at ends *A* and *B* for marginal through-fault conditions (i.e. operation of *HS* at *A* must not occur without corresponding operation of *LS* at *B*)

I_{fs} – Protection fault current setting

15.2.2. Limits of Application

The need for careful selection of fault settings to overcome the effects of primary capacitance currents imposes a minimum setting requirement on the protection. This is illustrated in Fig. 15.8 which is a general chart derived from Equation 15.8. It is seen that the use of larger phase-comparison angles permits the use of lower fault settings and this may be especially useful for the protection of long lines. From the investigations into fault performance in Section 15.3 it is seen that the selection of phase-comparison angle is not a critical parameter except for extreme system conditions and that, for all practical purposes, a 50° tripping angle will give satisfactory fault per formance.

The minimum fault settings derived from Fig. 15.8 must be investigated to determine their ability to provide acceptable fault performance under low fault infeed conditions.

The importance of high-speed fault clearance may exert some influence upon the particular criteria chosen for assessment.

If, for example, high-speed fault clearance is essential, then the protection should trip simultaneously at each end. In this event the effect of pre-fault load current must be considered. The settings shown in Fig. 15.7 refer to zero pre-fault load but where, as is often the case, an "impulse" starting network is used, its operating current is given by $I_L + I_{FS}$, where I_L is the pre-fault load current, and fault settings may therefore be quite high.

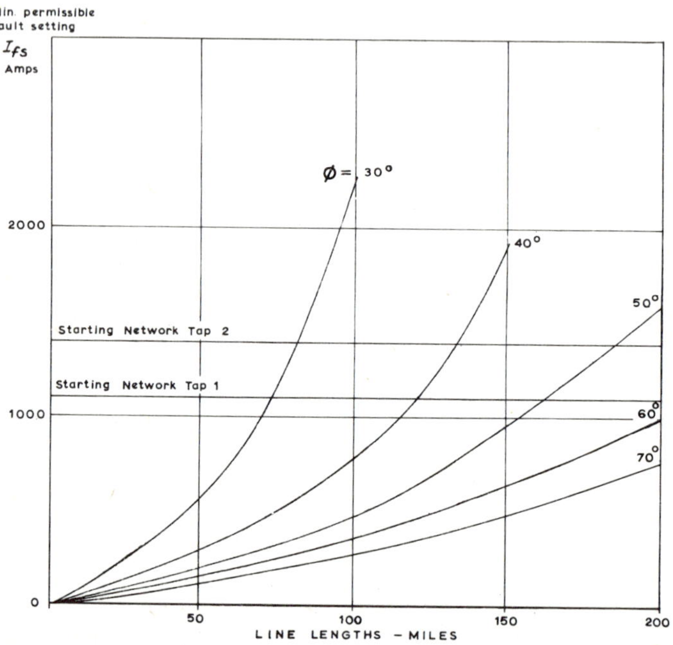

Fig. 15.8. Fault-setting limits for practical phase-comparison carrier-current protection applied to a 400-kV line

Basic equation

$$I_{fs} = 1{\cdot}5 I_c \operatorname{cosec}(\phi - \beta - \lambda)$$
$$I_c = 1{\cdot}32\ A/\text{mile},\ \lambda = 15°,\ \beta = 0{\cdot}1° \text{ per mile}$$

In many other applications the sequential clearance of a low fault infeed end may be acceptable, in which case the settings of Fig. 15.8 apply directly. Thus lines which may be regarded as unprotectable where simultaneous tripping is required, may be quite adequately protected on the basis of sequential tripping.

The successful application may sometimes be extended in situations where sequential tripping is possible by the addition of high-speed direct intertripping between ends, since this will usually minimize the time required for the second circuit breaker to trip.

15.2.3. Stability Requirements

Stability under high-current through-fault conditions depends wholly upon the performance of the current transformers, since unequal saturation at the two ends will result in a movement of the stability point away from the $1 + jo$ location on the I_A/I_B diagram. Excessive saturation cannot be tolerated and acceptable linearity of output must be maintained, taking account of the maximum values of d.c. offset likely to be encountered in practice. In consequence of the relationship between c.t. performance and protection characteristic it is necessary to determine c.t. requirements based upon the results of exhaustive heavy current tests, and from these to derive an expression which includes maximum fault current, system X/R ratio, relay, c.t. and lead burdens, which forms the minimum requirements for a c.t. specification.

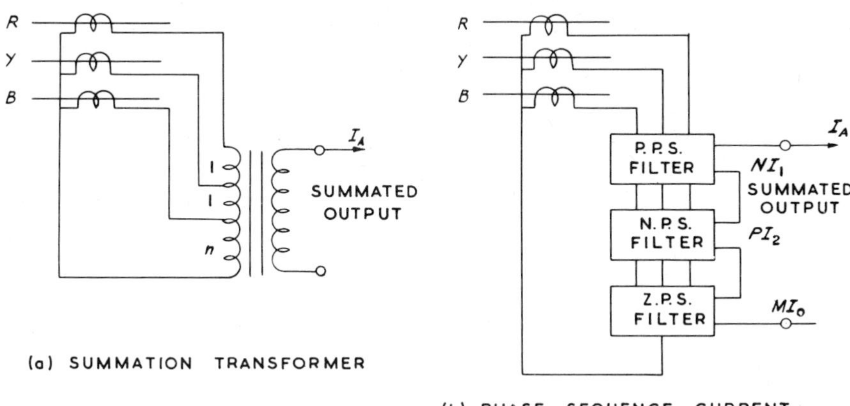

Fig. 15.9. Basic arrangement of summation devices

15.3. FAULT PERFORMANCE REQUIREMENTS

15.3.1. Minimum Outputs from Summation Devices

The problems involved in the use of summation devices have previously been mentioned and one effect of the reduction from three- to a single-phase quantity is that reduced outputs may be encountered for certain types of fault.

The basic schematic of two widely used types of summation device is shown in Fig. 15.9; the summation transformer and the phase-sequence current-segregating network. The output from the summation transformer is given by:

$$I_S = (n + 2)I_R + (n + 1)I_Y + nI_B, \tag{15.9}$$

which can be written in terms of symmetrical components as:

$$I_S = 3(1 + n)I_0 + (2 + a^2)I_1 + (2 + a)I_2. \tag{15.10}$$

In general terms the output of any kind of summation device can be expressed in the form:

$$I_S = MI_0 + NI_1 + PI_2 \tag{15.11}$$

where the values of M, N and P depend upon the constants of the summation device and the phases involved in the fault since, in most cases, it is convenient to evaluate fault currents once only using R phase as reference, and to evaluate the effects of faults in other phases by using the 120° operator to define the output from the summation device. Table 15.1 gives values of the constants M, N and P for the summation transformer and sequence network shown in Fig. 15.9.

TABLE 15.1

Values of Constants M, N *and* P *for Summation Devices*

Type of fault	Summation transformer			Phase-sequence network		
	M	N	P	M	N	P
R–E, Y–B, Y–B–E etc.	$3(1 + n)$	$2 + a^2$	$2 + a$	K_M	K_N	K_P
Y–E, B–R, B–R–E etc.	$3(1 + n)$	$1 + 2a$	$1 + 2a^2$	K_M	aK_N	a^2K_P
B–E, R–Y, R–Y–E	$3(1 + n)$	$2a^2 + a$	$2a + a^2$	K_M	a^2K_N	aK_P

In order to evaluate the outputs from summation devices under various fault conditions the appropriate constraints must be applied, i.e. for

$$\left.\begin{array}{ll} \text{single-phase-to-earth fault} & I_1 = I_2 = I_0 \\ \text{phase-to-phase fault} & I_1 = -I_2,\ I_0 = 0 \\ \text{phase-phase-earth fault} & I_1 = -(I_2 + I_0) \\ \text{three-phase fault} & I_0 = I_2 = 0 \end{array}\right\} . \tag{15.12}$$

Thus for the single-phase-to-earth fault

$$I_S = I_1\,(M + N + P) \tag{15.13}$$

For phase to phase

$$I_S = I_1\,(N - P) \tag{15.14}$$

For two phase to earth

$$I_S = I_1\,(N - P) + I_0\,(M - P) \tag{15.15}$$

and since

$$I_0 = \frac{Z_2}{Z_2 + Z_0}\,I_1$$

$$I_S = I_1\left[(N - P) + (M - P)\frac{Z_2}{Z_2 + Z_0}\right] \tag{15.16}$$

For the three-phase fault

$$I_S = I_1 N \tag{15.17}$$

TABLE 15.2

Output Ratios for Summation Devices (single-infeed fault conditions)

Fault type	$I_{sf}/I_{s3\phi}$
Single phase to earth	$\dfrac{M+N+P}{N(2+K)}$
Phase to phase	$\dfrac{N-P}{2N}$
Two phase to earth	$\dfrac{(N-P)K+N+M-2P}{N(1+2K)}$

Table 15.2 summarizes the output for each type of fault expressed as a fraction of the output for a three-phase fault at the same location.

By substituting the appropriate constant M, N and P the relative outputs for different types of fault can be determined.

Figures 15.10 and 15.11 illustrate the relative output for two different types of summation device.

For the 1:1:2 summation transformer (Fig. 15.10) it is seen that RY, YB and RYE faults may each give lower outputs than for the three-phase fault

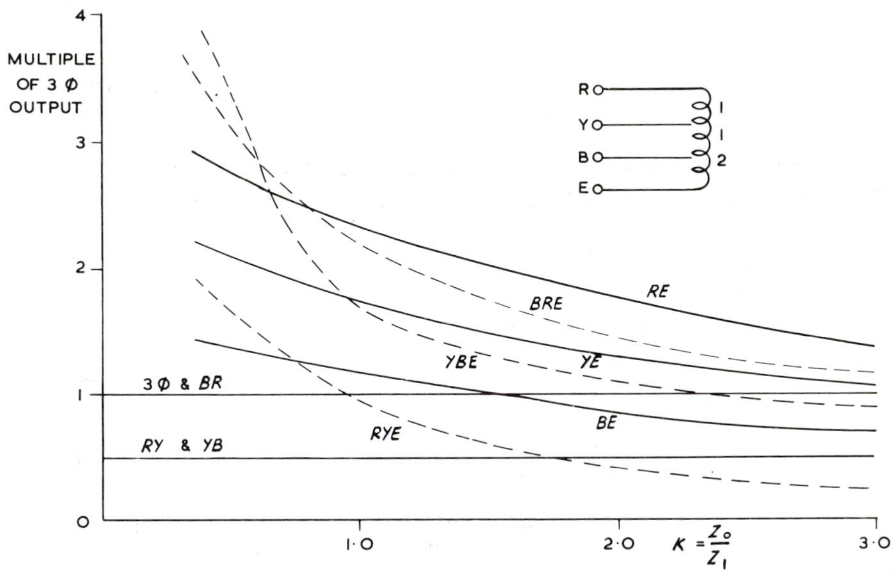

Fig. 15.10. Relative outputs for 1:1:2 summation transformer for different fault types – single-infeed fault condition
Three-phase fault current = E/Z_1

and that this may be particularly significant for the RYE fault where Z_0/Z_1 exceeds 2.0. On the other hand for the phase-sequence current-segregating network $(I_1 - 5I_2)$, the lowest output is generally obtained for the three-phase fault, and this may be used to determine the suitability of the protection to deal with particular system conditions.

By using this approach it is possible to ensure that the fault settings available on the protection are sufficiently low to give correct tripping for single-infeed internal faults under minimum plant conditions.

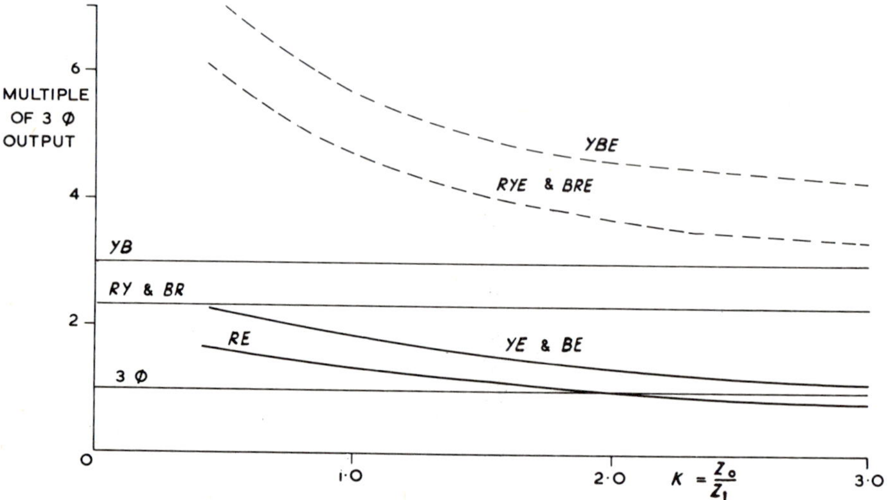

Fig. 15.11. Relative outputs for I_1-$5I_2$ sequence network for different fault types. Single-infeed fault condition

15.3.2. Fault Performance under Double-infeed Conditions (112)

The complex ratio of currents at the two ends of the protected line is given by

$$\frac{I_A}{I_B} \angle\phi = \frac{MI_{A0} + NI_{A1} + PI_{A2}}{MI_{B0} + NI_{B1} + PI_{B2}}, \tag{15.18}$$

which, from Fig. 15.12, may be written

$$\frac{I_A}{I_B} \angle\phi = \frac{M(I_{f0} + I_{g0}) + N(I_{f1} + I_{g1}) + P(I_{f2} + I_{g2})}{I_{g0} + NI_{g1} + PI_{g2}}$$

$$= \frac{MI_{f0} + NI_{f1} + PI_{f2}}{MI_{g0} + NI_{g1} + PI_{g2}} + 1. \tag{15.19}$$

By substitution for I_g from the general diagram of Fig. 15.12 and applying the appropriate constraints to I_f for the different types of fault the ratio of outputs in the complex current plane can be evaluated, and these are given in Table 15.3. Substitution of the appropriate circuit constants will define the performance of the protection for different fault types.

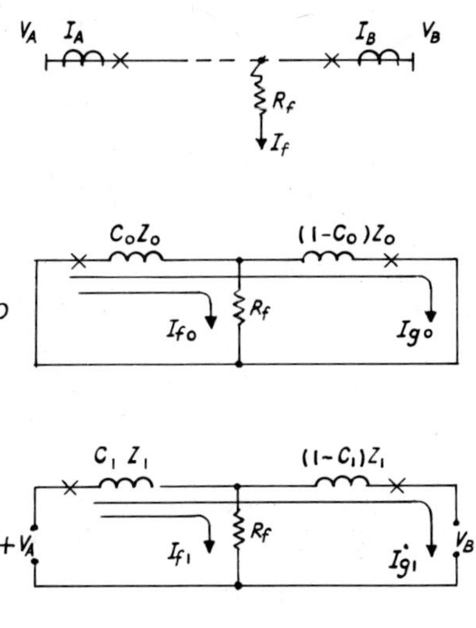

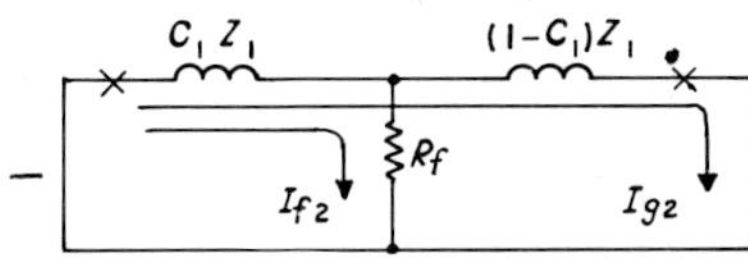

Fig. 15.12. Phase-sequence networks for balanced fault condition for condition $V_A = V_B$ (i.e. zero pre-fault load current)

TABLE 15.3

Summation Device Outputs when $V_A/V_B\angle\delta = 1\angle 0$ for double-infeed fault conditions

Fault type	$(I_A/I_B)\angle\Phi$
Single phase to earth	$1 - \left\{\dfrac{M + N + P}{C_0M + C_1(N + P)}\right\}$
Phase to phase	$1 - \dfrac{1}{C_1}$
Two-phase to earth	$1 - \left[\dfrac{M + N - 2P + (N - P)\left\{\left(\dfrac{C_0Z_0}{C_1Z_1}\right)\left(\dfrac{1 - C_0}{1 - C_1}\right)\right\}}{C_0M + C_1N - 2C_1P + (N - P)\left(\dfrac{C_0Z_0}{Z_1}\right)\left(\dfrac{1 - C_0}{1 - C_1}\right)}\right]$
Phase to earth with broken conductor	$1 - \left\{\dfrac{\left(\dfrac{Z_0}{Z_1} + \dfrac{1}{2}\right)(M + N + P)}{\left(C_0\dfrac{Z_0}{Z_1} - C_1\right)\left(M - \dfrac{N}{2} - \dfrac{P}{2}\right)}\right\}$

By substituting the appropriate constants M, N and P for the different types of summation device it is possible to determine the value of $I_A/I_B \angle\phi$ and to locate this on the complex diagram. It is apparent from Fig. 15.2 that, for internal-fault conditions, negative real values of the complex current ratio will give a correct relay response since they fall outside the stability characteristics.

On the other hand, if the summation device output appears along the positive real axis for internal-fault conditions, the possibility of failure to trip must be investigated.

From an examination of the various types of summation device using the expressions of Table 15.3 the particular combinations giving rise to negative outputs can be identified.

For the phase-sequence current-segregating network it is found that the most difficult case is the two-phase-to-earth fault but that correct outputs will be obtained for a network having an output

$$I_S = I_1 - KI_2,$$

where K is large (typically 5 or 6).

The summation transformer on the other hand may encounter difficulties under B–E fault conditions, for which case, by substituting the appropriate (B–E) expression for M, N and P in Table 15.1 in the single-phase-to-earth fault expression in Table 15.3.

$$\frac{I_A}{I_B} \angle\phi = 1 - \frac{n}{C_0(1 + n) - C_1}, \tag{15.20}$$

which is positive when

$$\frac{C_1}{C_0} > 1 + n.$$

This can happen where n is low (less than 3) and a B–E fault occurs close to one end of a line which has a high zero sequence infeed at that end.

The condition is illustrated in Fig. 15.13 which shows how healthy phase components of fault current at end A can give rise to a summated output which is in the reverse sense to the current in the faulted phase and which, in consequence, gives rise to a positive output in the complex current diagram, preventing tripping. The solution in this case is to increase the turn ratio of the summation transformer to 1:1:4 or greater which gives a positive output at end A and correct tripping.

15.3.3. Fault Loci when $V_A/V_B \angle\theta \neq 1\angle 0$

When the voltages at the two ends of the protected circuit are dissimilar, in either phase or magnitude, the location of the fault point will change on the diagram and, in general terms, a fault area can be drawn which describes the

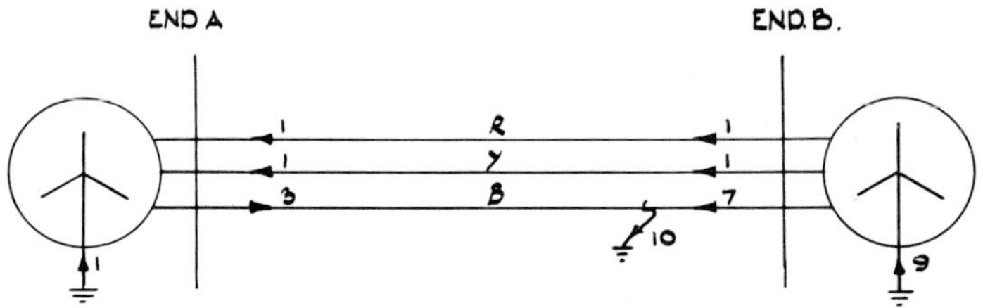

(a) Fault current distribution for B phase to earth fault conditions

R
Y
B
N

OUTPUT $\propto$ $4I_R + 3I_Y + 2I_B$ giving:-

FOR END A.

$(4 \times -1) + (3 \times -1) + (2 \times 3) = -1$

FOR END B.

$(4 \times 1) + (3 \times 1) + (2 \times 7) = +21$

Note. Output reversed at end A due to healthy phase currents which prevents operation of Phase Comparison Protection.

Fig. 15.13. Conditions for reversed output from 1:1:2 summation transformer

envelope of the locus of the fault point taking into account practical circuit-loading conditions.

The general expression for fault area of this kind is given

$$\frac{I_A}{I_B} \angle \phi = \frac{K_1 \frac{V_A}{V_B} + K_2}{K_3 \frac{V_A}{V_B} + K_4} \tag{15.2.1}$$

and reference 112 evaluates the values of these constants for the various fault types. Typical fault areas for two conditions of interest are shown in Figs. 15.14 and 15.15 [97]. In both figures it is seen that fault conditions can arise which encroach upon the stability area of the phase-comparison characteristic, indicating that particular care is needed in selecting relaying quantities for phase-comparison protection. Where the protection compares both phase and amplitude however, as indicated by the circular characteristic, encroachments along the positive real axis of the complex current diagram do not extend into the relay characteristic circle.

Using methods of this kind it is possible to determine the overall performance of the differential protection system based upon a correlation of power-system conditions and protective-relay characteristics. The problems involved in providing desirable relay characteristics are considered in the next section.

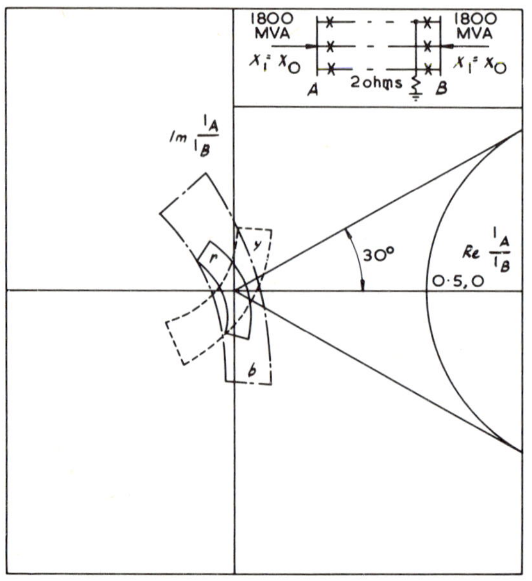

Twenty-five-mile 132 kV circuit
R phase-earth fault ————
Y phase-earth fault — — — —
B phase-earth fault —·—·—

Fig. 15.14. Fault areas for single-phase-to-earth open-conductor faults for a 1:1:3 summation transformer

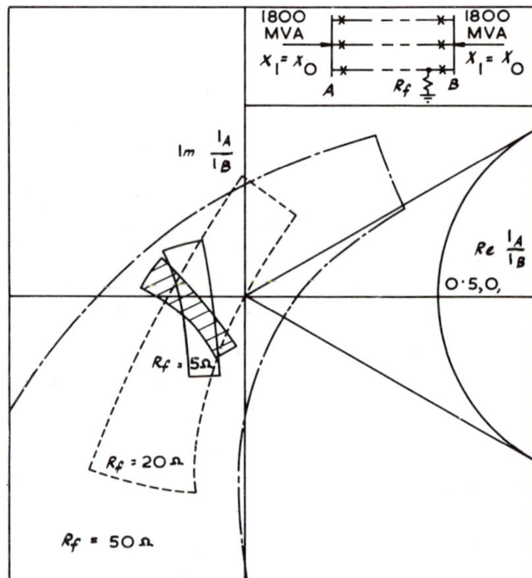

Twenty-five-mile 132 kV circuit
R_f = 5 ohm ————
R_f = 20 ohm — — — —
R_f = 50 ohm —·—·—
Zero fault resistance area is shaded

Fig. 15.15. Fault areas for *B* phase-to-earth fault for a 1:1:3 summation transformer showing effect of fault resistance

15.4. FUNDAMENTAL CHARACTERISTICS OF PILOT-WIRE DIFFERENTIAL SYSTEMS (69)

15.4.1. General Considerations

The objective to be met in the design of pilot-wire systems is basically that of realizing, in a practical sense, the types of characteristic previously considered.

Pilot-wire protection may use either the direct comparison of summated power frequency quantities over a metallic pilot circuit or alternatively some form of modulated audio frequency over a telephone-type circuit.

The use of linear direct power frequency comparison can provide a phase and amplitude comparison characteristic of the type discussed as long as the relay circuits are devised to include a method of compensation for the wide range of pilot-circuit characteristics which can be encountered in practice, and this method provides the simplest approach where a direct metallic pilot circuit interconnects the relays at the two ends.

Where the only available pilot interconnection is via an audio frequency telephone channel, and some form of signal modulation effects the comparison, the transmission of phase-angle information provides the most simple practical arrangement, and protections systems of this kind are similar in concept to the phase-comparison carrier-current protection system.

The fundamental design principles of power frequency comparison pilot-wire protection systems are discussed in this section. In particular the relationship between pilot-circuit impedance characteristics and comparator characteristics is developed.

Pilot-wire protection systems of this type may use either *opposed voltage* or *circulating current* methods and a duality exists between these two arrangements so that the analysis of one follows by duality from the other. In the following analysis the opposed voltage arrangement is considered in detail and it will be seen that the use of an admittance rather than an impedance diagram gives a simpler evaluation. For the circulating current system, analysis would proceed in terms of impedance in order to provide similar characteristics using dual-circuit arrangements.

The following procedure is used to determine the design and performance characteristics of a pilot-wire protection scheme.

(*a*) Prepare a chart in the axes of pilot-circuit conductance and susceptance from measured values of pilot-circuit constants.
(*b*) Specify the comparator characteristic requirements in the admittance plane.
(*c*) Design comparator control circuits to provide the required operating characteristic.
(*d*) Transform the characteristic from the admittance plane to the complex current ratio plane.
(*e*) Derive effective local and remote end characteristics to display the overall performance of the protection.

15.4.2. Basic Pilot-wire Differential System

The basic arrangement of an opposed voltage pilot-wire differential protection system is shown in Fig. 15.16, in which the single-phase relaying output signal is derived from the three-phase input at each end of the protected circuit. The single-phase output is derived in the form of a voltage (low-source impedance) and the voltages at the two ends are compared via the metallic pilot circuit.

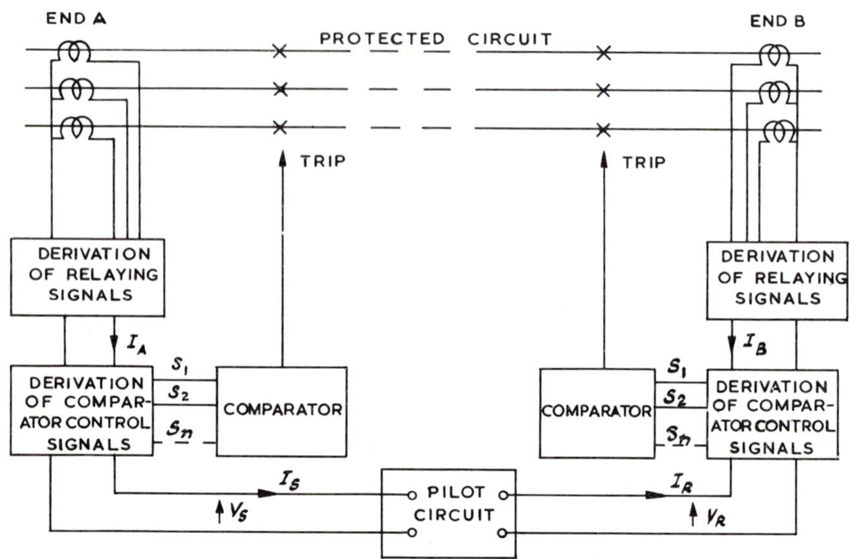

Fig. 15.16. Basic elements of pilot-wire protection system

The characteristic of the protection defines the boundary between operation and non-operation. It is determined by the ratio of the relaying signals, and the selected signal inputs to the comparator, and also by the basis on which the comparison is made.

The comparison between the applied quantities may be responsive to either the amplitude or the phase-angle difference between the signal inputs (i.e. either amplitude or phase comparison). The equivalence between amplitude and phase comparators having two inputs was demonstrated in Chapter 3 of Volume 1 and the type of comparator used is therefore not fundamentally important.

It is perhaps important, at this stage, to differentiate between phase and amplitude comparators as such and the overall characteristics of the protection scheme which may provide either phase and amplitude comparison or simply phase comparison. The type of comparator used merely influences the choice of its own signal inputs but the overall characteristics of the scheme depend mainly upon the method by which the relaying signal inputs are derived and particularly upon their linearity over a wide range of current levels.

15.4.3. Protection Scheme Characteristics

The protection characteristics are plotted in terms of the complex ratio of the summated outputs from the summation networks expressed in terms of the equivalent single-phase current.

Linear comparison can give a characteristic of the kind shown in Fig. 15.17a and characteristics of this kind were considered in previous sections in order to assess fault performance.

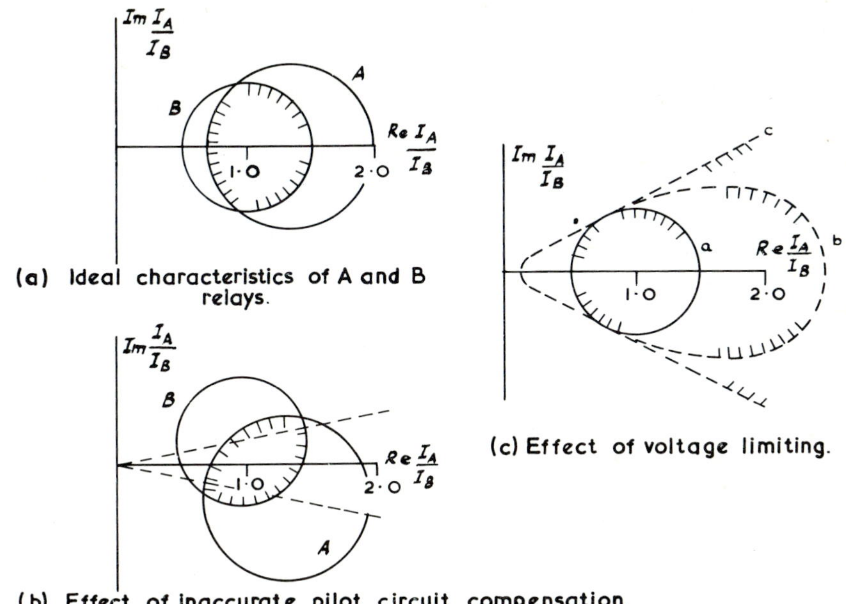

Fig. 15.17. Polar characteristics of pilot-wire differential relays. Overall stability zone is shaded.

In order to produce these characteristics, some form of compensating circuit is necessary to cater for the wide variation encountered in practical pilot circuits. Without effective compensation displaced characteristics of the kind shown in Fig. 15.17b are obtained which combine to give a smaller stability zone and larger areas for which sequential tripping takes place between ends for internal-fault conditions. Where the relaying signal inputs are non-linear, for example when the maximum permissible voltage between pilot cores is low and voltage limiters are used, the characteristic shape will vary as voltage limiting takes effect as shown in Fig. 15.17c and will eventually revert to phase comparison when the relaying signal voltage is fully limited and is, therefore, unable to respond to an increase in amplitude of the effective summated value of the three-phase primary currents.

It is apparent that the characteristics of the pilot circuit exert a great influence upon the design and performance of a pilot-wire protection scheme and the characteristics of practical pilot circuits must be examined to determine their effect.

This is most conveniently carried out by determining the variation in steady-state admittance measured at one end of the pilot circuit produced by variation of the relaying signal at the remote end.

The admittance locus is then related to the comparator characteristics which, of course, must be expressed in terms of admittance measurement.

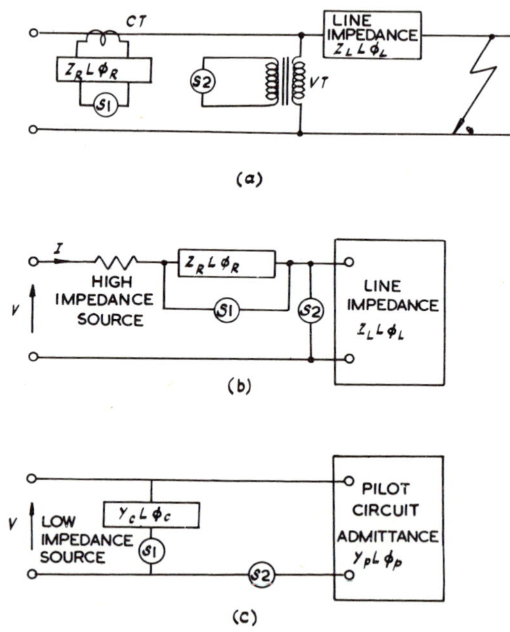

Fig. 15.18. The use of replica impedance for distance and opposed voltage pilot-wire protection illustrating concept of duality:
(*a*) Basic distance protection circuit
(*b*) Equivalent circuit for distance protection referred to the secondary side of c.t.s and v.t.s
(*c*) Opposed voltage pilot-wire protection

It is interesting to note that the admittance concept applied to opposed voltage pilot-wire protection is the circuit dual of impedance measurement applied to transmission lines. This is seen from Fig. 15.18.

15.4.4. Characteristics of Pilot Circuits

The following analysis assumes sinusoidal voltages applied to a linear pilot circuit.

A four-terminal network energized at its terminals S and R and having generalized constants A, B, C, D is considered (Fig. 15.19), these being related

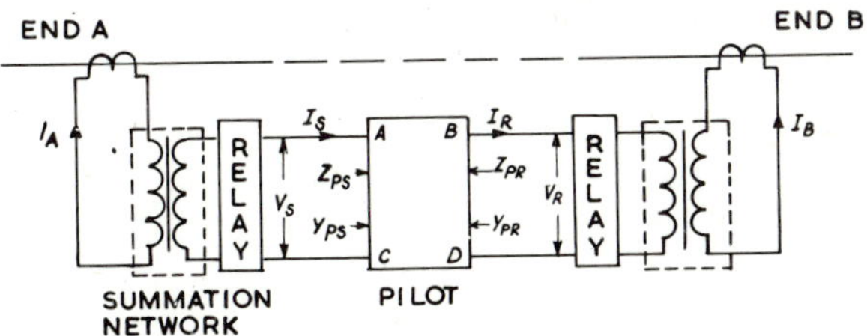

Fig. 15.19. Basic arrangement showing conventions used in analysis of overall performance of pilot-wire protection scheme

by Equation 15.22 where S is the sending, or local end and R is the receiving or far end,

$$\left.\begin{aligned} V_S &= AV_R + BI_R \\ I_S &= CV_S + BI_R \end{aligned}\right\} \qquad (15.22)$$

Taking the case where V_R is fixed and V_S varies in phase and magnitude, then $V_S = KV_R\angle\theta$ where K is the scalar ratio $\frac{|V_S|}{|V_R|}$ and $\angle\theta$ is the phase angle between the two voltages; then substituting for V_S we have

$$KV_R\angle\theta - AV_R = BI_R, \qquad (15.23)$$

from which

$$\frac{I_R}{V_R} = Y_{PR} = \frac{K\angle\theta - A}{B}, \qquad (15.24)$$

and, since the current at end R is reversed, then the admittance looking into the pilot circuit is given by

$$Y_{PR} = \frac{A}{B} - \frac{K\angle\theta}{B}. \qquad (15.25)$$

Similarly, for the remote end

$$Y_{PS} = \frac{D}{B} - \frac{1/K\angle\theta}{B}. \qquad (15.26)$$

It is apparent that the locus of Y_{PR} is a series of concentric circles of radius K/B with their centres displaced A/B from the origin.

In order to draw a practical locus diagram constants A and B must be

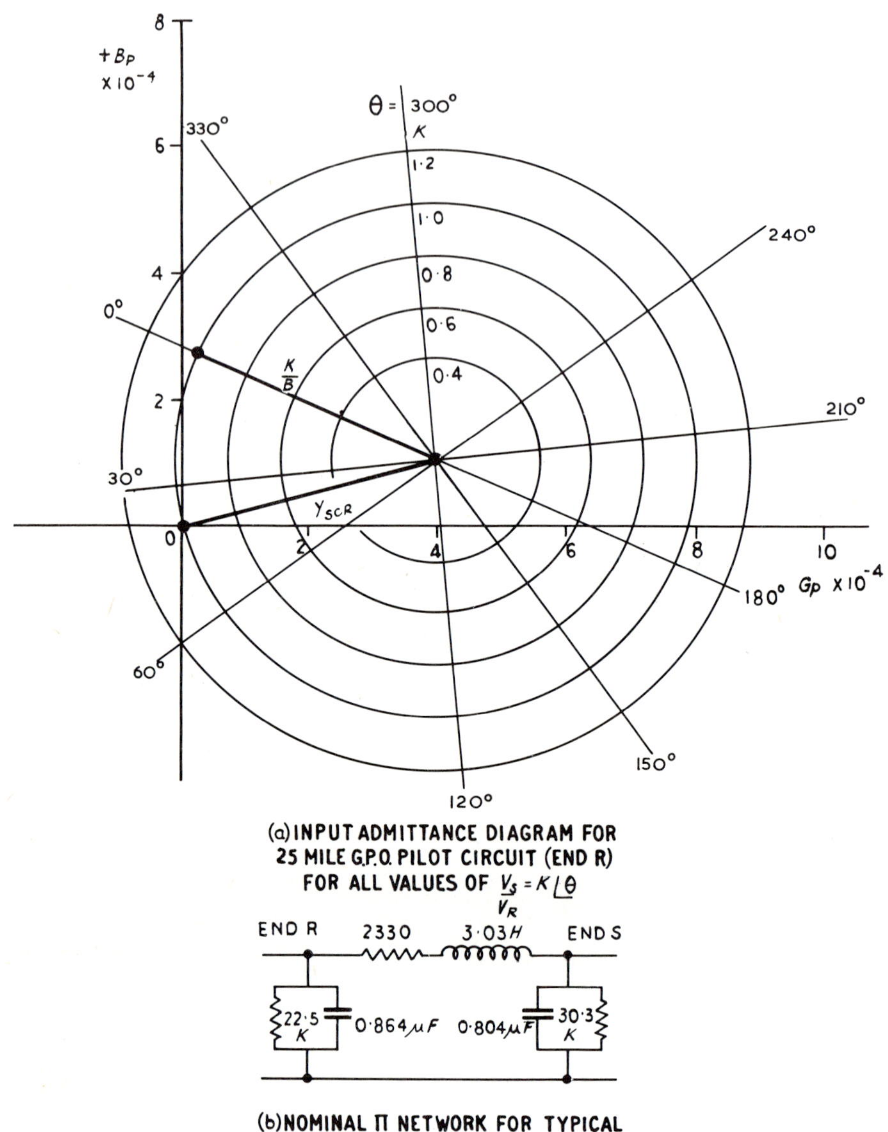

Fig. 15.20. Admittance locus diagram for twenty-five-mile loaded telephone pilot circuit

found, and this is most readily done from measurements of the open- and short-circuit admittances Y_{OC} and Y_{SC} of the pilot circuit from which

$$\frac{A}{B} = Y_{SCR}$$

(the short-circuit admittance viewed from end R) and

$$\frac{1}{B} = \sqrt{\{Y_{SCS}(Y_{SCR} - Y_{OCR})\}}.$$

The admittance locus for a practical pilot circuit is shown in Fig. 15.20.

When θ is zero and K is unity, the voltages at the two ends are in opposition and this describes the stability point for an opposed voltage scheme. With K at unity and $\theta = 180°$ the condition represented is that of an internal fault fed equally from both ends of the protected circuit. Other relationships between the voltages give rise to different admittance values on the diagram.

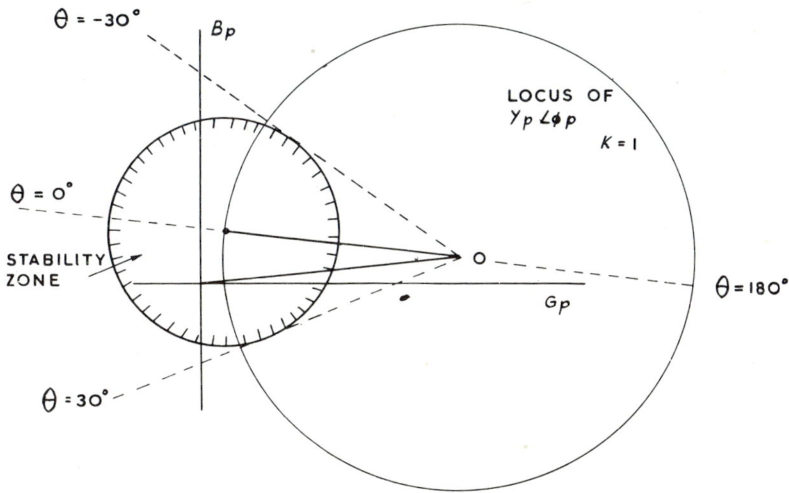

15.21. Circular protection characteristics on pilot-circuit input-admittance diagram

The phase and amplitude comparison circle used in previous sections for fault performance assessment is shown in the admittance plot of Fig. 15.21. A wide range of comparator characteristics is available, however, and these are considered in the next section.

15.5. ADMITTANCE CHARACTERISTICS OF TWO-INPUT COMPARATORS

A generalized two-input comparator circuit is shown in Fig. 15.22a which illustrates the various alternative signal inputs which can be derived and compared in combination.

It can be seen from Fig. 15.22 that the available signal inputs are as follows:

$$\left.\begin{aligned} S_1 &= Y_p\angle\theta_p - Y_a\angle\theta_a \\ S_2 &= Y_a\angle\theta_a \\ S_3 &= Y_p\angle\theta_p \\ S_4 &= Y_b\angle\theta_b \\ S_5 &= Y_p\angle\theta_p + Y_b\angle\theta_b \end{aligned}\right\} \quad (15.27)$$

Figure 15.22b to l shows the range of characteristics available from an amplitude comparator whilst Fig. 15.23a to j shows the characteristics obtained from a $\pm$ 90° phase comparator for similar signal inputs.

In each case the restraining (non-operation) zone of the protection characteristic is shaded. Interchanging operating and restraining inputs in the case of the amplitude comparator, or reversing the sign of one input for the phase comparator will result in the shaded area becoming the operating zone of the protection characteristic.

It is seen from Figs. 15.22 and 15.23 that it is possible to locate a straight-line or circle characteristic in any desired part of the complex admittance plane by suitable choice of comparator signal inputs and value of compensating impedance. Thus a wide choice of protection characteristic is available. This choice may be even further extended by the use of either asymmetrical phase-comparison angles or multi-input comparators (126), but generally it is found that the simple circle or straight-line characteristic provides adequate performance.

15.5.1. Complex Current Plane Characteristics

15.5.1.1. Transformation from Admittance Plane to Complete Current Ratio Plane

It is now possible to interpret in the plane of the complex ratio of relaying signals, the characteristics that have so far been given in the admittance plane. Using the input admittance chart of Fig. 15.20, the point 0 in Fig. 15.24a defines the condition where the end S voltage is zero and therefore defines the point where $I_A/I_B = 0$ since V_S is produced by I_A. The real axis of the current ratio plane must pass through the points at which the relaying signals are in phase and in antiphase, and the new plane is shown in Fig. 15.24b. Furthermore, where the system is operating in a linear mode, the ratio V_S/V_R is identical with the ratio I_A/I_B and so the protection characteristics may be shown directly in the plane of the complex ratio of relaying signals I_A and I_B, as derived from the current summation device. Although in Fig. 15.24a, negative values of the real part of the ratio V_S/V_R appear to the right of the origin, this is reversed in Fig. 15.24b in which positive values are shown to the right in the normal way.

15.5.1.2. Inverse Characteristics

The characteristics so far derived are those that obtain at each end of the protected circuit without giving consideration to the effect that the character-

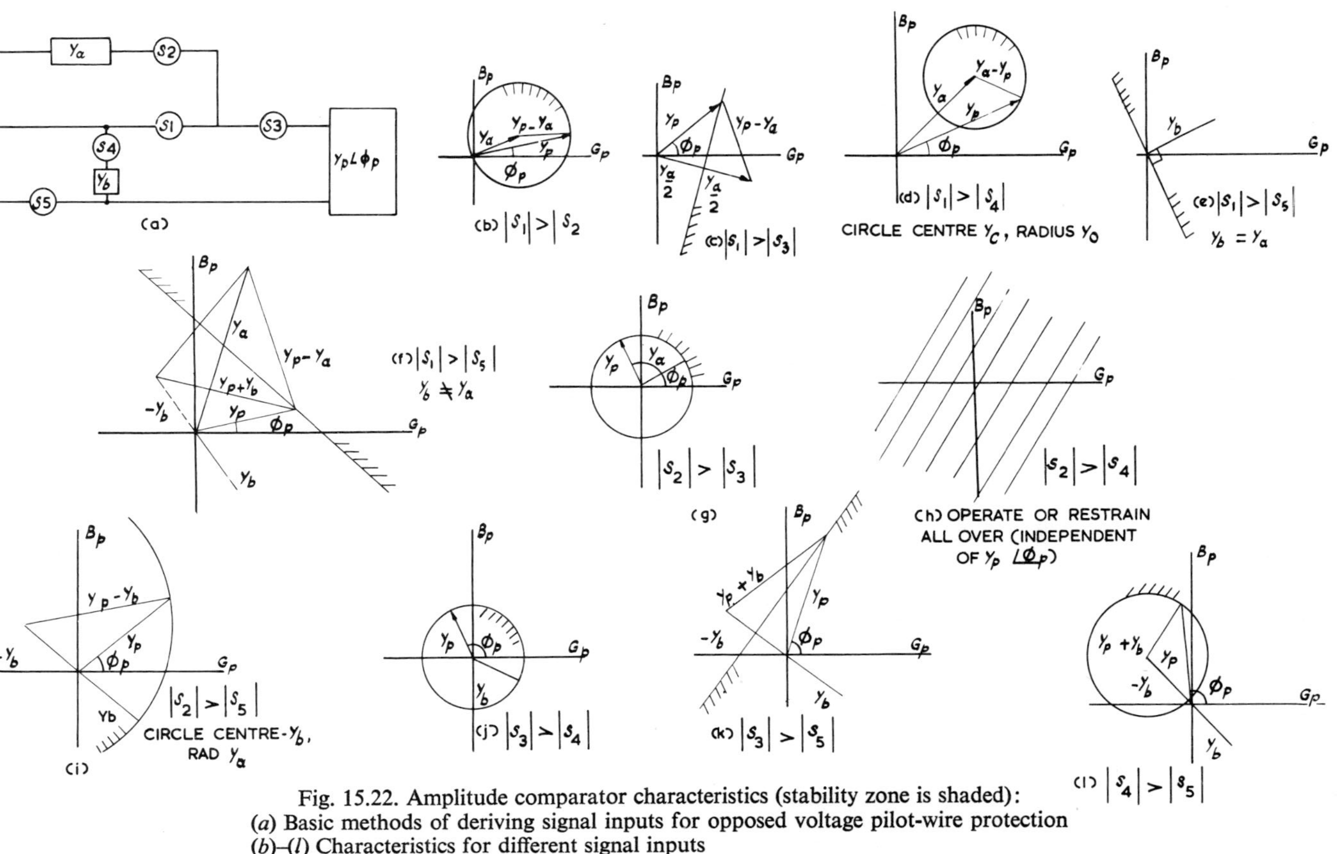

Fig. 15.22. Amplitude comparator characteristics (stability zone is shaded):
(*a*) Basic methods of deriving signal inputs for opposed voltage pilot-wire protection
(*b*)–(*l*) Characteristics for different signal inputs

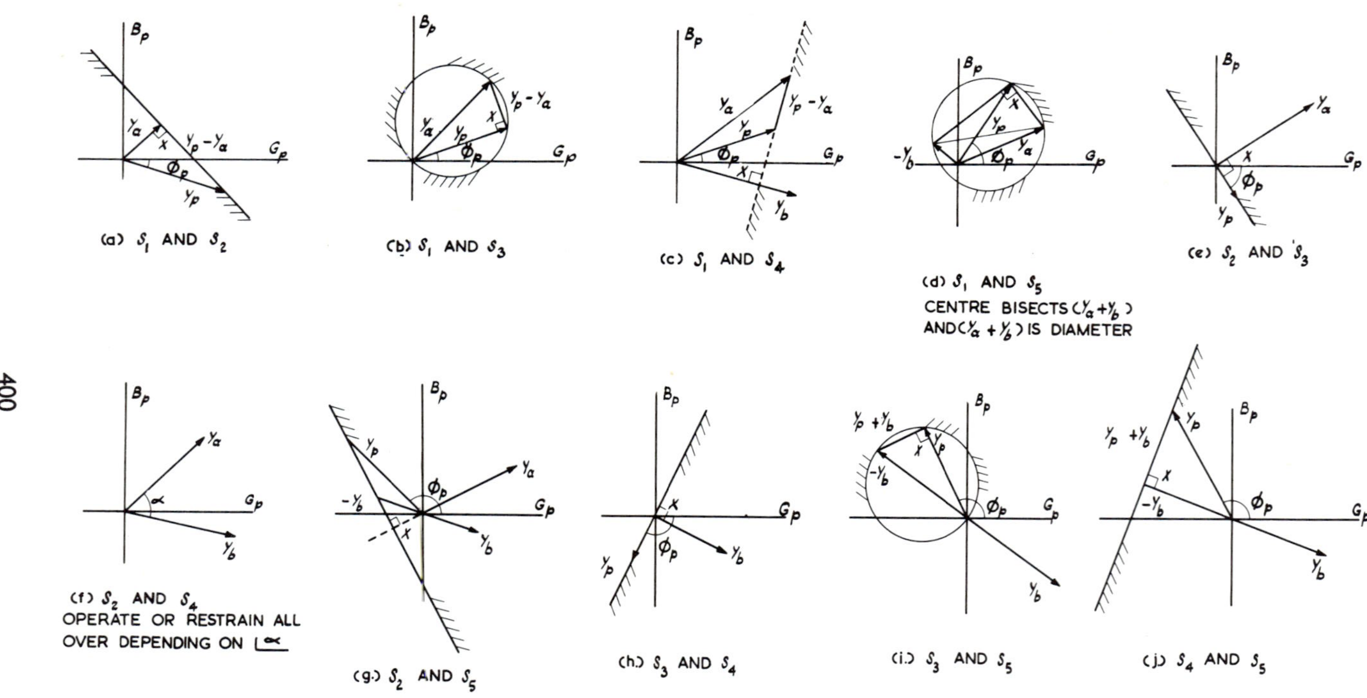

Fig. 15.23. Phase-angle comparator characteristics. Operation occurs when $\angle$ X > 90°. Stability zone is shaded: (*a*)–(*j*) Characteristics for different signal inputs

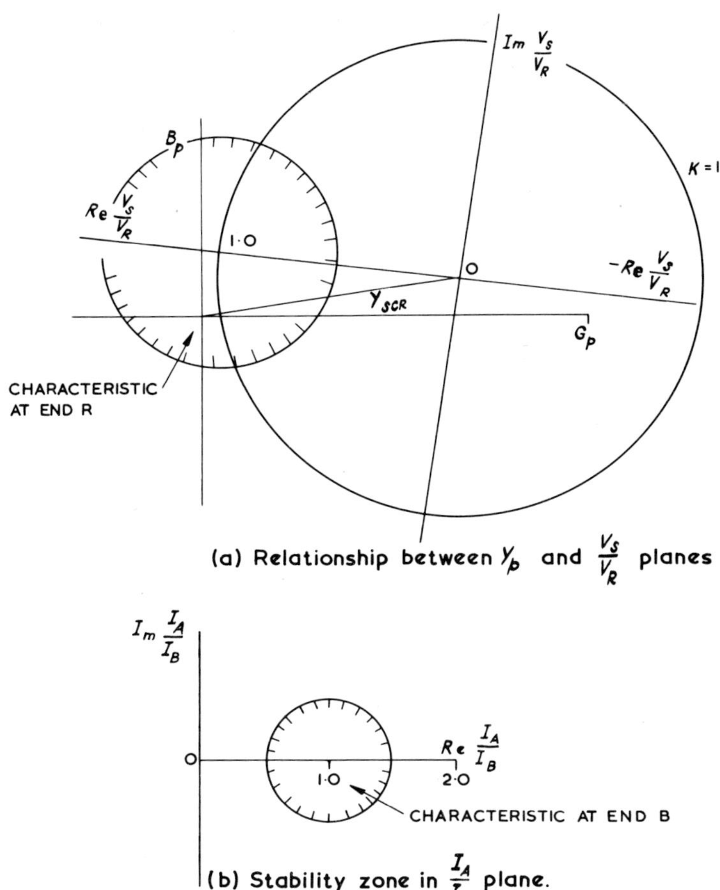

Fig. 15.24. Transition from admittance plane to complex plane of relaying signals

istic at one end has on the other. Now the characteristic at end 'A' may be shown directly on the I_B/I_A plane and similarly, that at end 'B' on the I_A/I_B plane. The characteristic at one end therefore appears as its inverse at the other, and it is the combination of the local and the referred remote end characteristics that define the resultant stability zone for the protection as a whole. In Fig. 15.25a is shown the case of a local end circular characteristic and the equivalent far end characteristic, where both characteristics are identical at their respective ends. For conditions completely outside the two characteristics both ends of the protection trip together. In the range that is covered by one characteristic only, the two ends trip sequentially. The common overlapping area to local and referred remote characteristics represents the conditions for which both ends restrain. Figure 15.25b shows the case where local and remote end circles are coincident.

Inevitably where pilot-wire schemes are applied over long pilot circuits, particularly where these employ low insulation telephone-type cores, some part of the protection characteristic is influenced by the action of the voltage limiter.

In the event of full limiting, the relay operates about the $V_S/V_R = 1$ point and tripping occurs where the $K = 1$ circle cuts the relay characteristic circle. The relay then operates at constant phase angle (see Fig. 15.17c) and is completely insensitive to magnitude difference. The characteristic is one of phase comparison and the characteristic angle is determined by the tangent drawn from the origin of the I_A/I_B diagram to the characteristic circles. Thus in the case of Fig. 24, the phase-comparison angle under voltage-limiting conditions is $\pm 30°$.

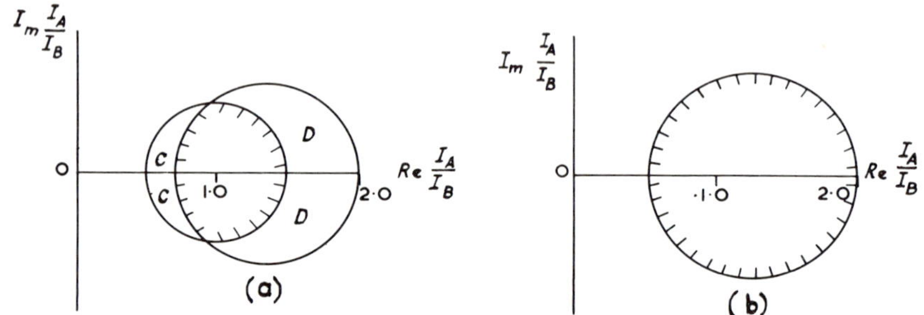

Fig. 15.25. Types of symmetrical protection characteristic:
(*a*) Local and remote end characteristics. Stability zone is shaded. *C* and *D* are areas of sequential tripping
(*b*) Self-dual characteristic. Centre 1·25, radius 0·75

15.5.2. Electrical Characteristics of Practical Pilot Circuits

The essential information required for the construction of the pilot-circuit input-admittance diagram can be obtained from measurements of open- and short-circuit impedances which provide the basic data for calculation of constants A/B, D/B and $1/B$.

Using these constants the input-admittance diagram can be drawn. Figure 15.26a shows the location of the $1/B$ phasor for practical pilot circuits and Fig. 26b shows the calculated values for homogeneous 20 lb./mile conductor circuits; the dotted lines of Fig. 15.26b illustrate the effect of distributed loading coils which are widely used in rented telephone circuits to facilitate speech transmission. It should be noted that the practical pilot circuits shown in Fig. 15.26a are all unsymmetrical (i.e. the input-admittance locus is different at the two ends). Clearly these practical pilot circuits are non-homogeneous and comprise various lengths of conductor of different cross-sectional area, with both loaded and unloaded sections within the total length.

The effect of the unsymmetrical admittance loci means that different values of compensating impedance are required for the relays at each end of the protected circuit in order to obtain the same protection characteristics.

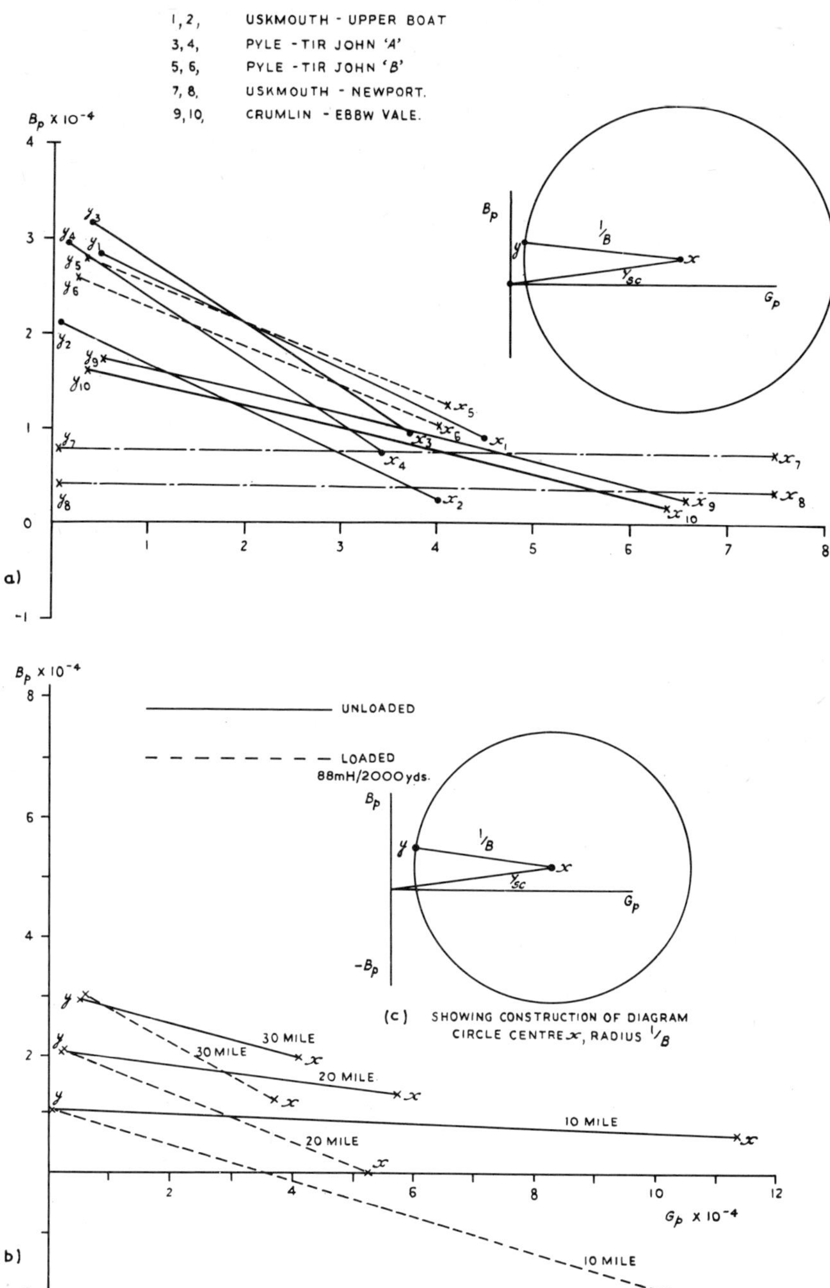

Fig. 15.26. Construction of input-admittance diagram from pilot-circuit constants:
(*a*) Some practical pilot-circuit constants. *Note:* Suffixes 1 and 2, 3 and 4, etc. denote different ends of the same pilot circuit
(*b*) Calculated pilot-circuit constants for homogeneous 20 lb/mile telephone circuits

15.5.3. Design Basis for Practical Schemes

A practical design will now be considered using an amplitude comparator and the pilot-circuit characteristics shown in Fig. 15.20a. In deriving the comparator circuit constants two alternative methods of compensation will be considered; the shunt compensated-impedance circuit of Fig. 15.27 and the bridge-compensation circuit of Fig. 15.28.

Both types of compensation can provide equivalent overall protection characteristics but, as is seen from (*c*) on the two diagrams, one set of characteristics is the inverse of the other. Using the opposed voltage connection it is clear that an inductive-compensation circuit is required in Fig. 15.27 and a capacitative-compensation circuit in Fig. 15.28.

15.5.3.1. Shunt Reactance Compensation Method

The characteristics obtainable are those shown in Fig. 15.26c, and these are obtained by adjustment of the pilot-compensating reactor Y_C and the restraint resistor Z_R in Fig. 15.27a. In practice the compensating reactor will usually have adjustable tappings which, of course, does not permit fine adjustments to be made.

In calculating the setting adjustments it follows from Fig. 15.27c that the circular characteristic (centre 1·0 radius 0·5) is obtained at end S on the V_S/V_R diagram where $V_S/V_R \propto I_A/I_B$.

The calculation of circuit parameters must be in terms of end R on the V_S/V_R diagram so that in this case, the values of compensation and restraint resistor must be evaluated from the inverse circle. The inverse circle has its centre at 1·375, 0 and radius 0·625.

When transformed to the admittance diagram this gives a circle radius of $\frac{0{\cdot}625}{B}$ displaced $\frac{A - 1{\cdot}375}{B}$ from the origin.

Thus for the practical pilot circuit of Fig. 15.20,

$$Y_C = \frac{1{\cdot}375 - A}{B} = (1{\cdot}06 - j3{\cdot}085) + 10^{-2} \text{ mho}$$

and

$$Z_C = 1100 + j3210 \text{ ohms}$$

the circle radius

$$K_R = \left|\frac{N_R}{N_0} Z_R\right| = \left|\frac{0{\cdot}625}{B}\right|,$$

where

$$|B| = 2520 \text{ ohms}$$

Thus

$$Z_R = 4030 \frac{N_R}{N_0},$$

where N_R/N_0 is the restraint/operating circuit turn ratio.

These settings give a circular characteristic corresponding to a $\pm 30°$ tripping angle, where $V_S/V_R = 1{\cdot}0$.

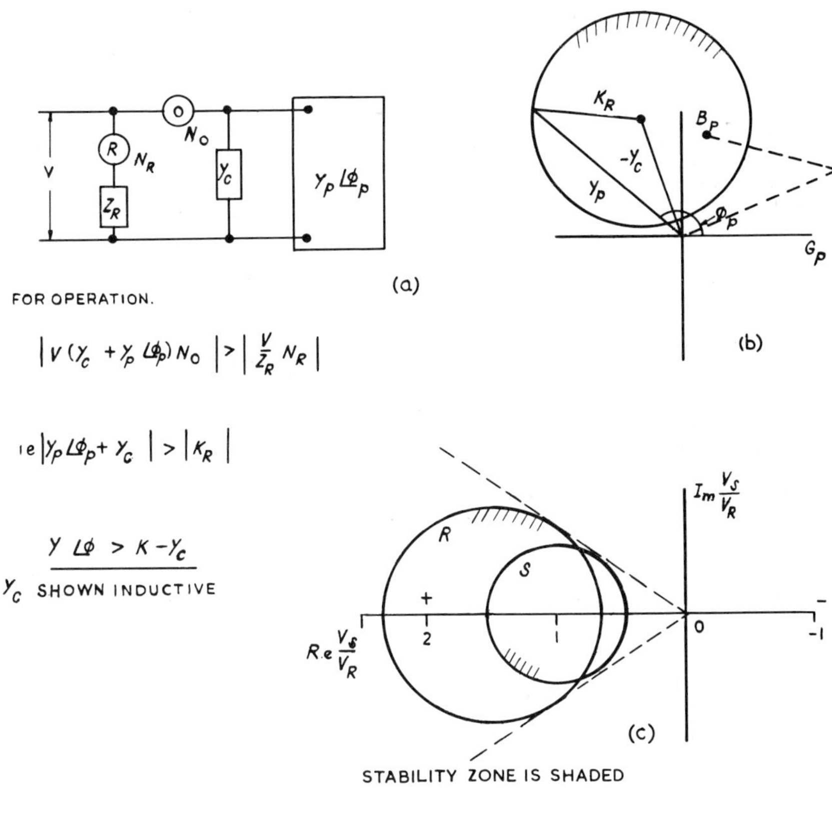

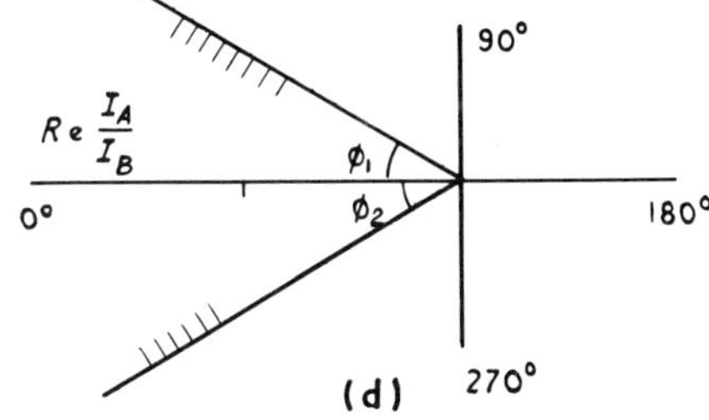

Fig. 15.27. Impedance-type comparator using shunt reactance compensation:
(*a*) Local diagram on admittance plane
(*b*) Relay characteristics on complex voltage plane
(*c*) Phase-comparison characteristics under voltage limiting conditions.

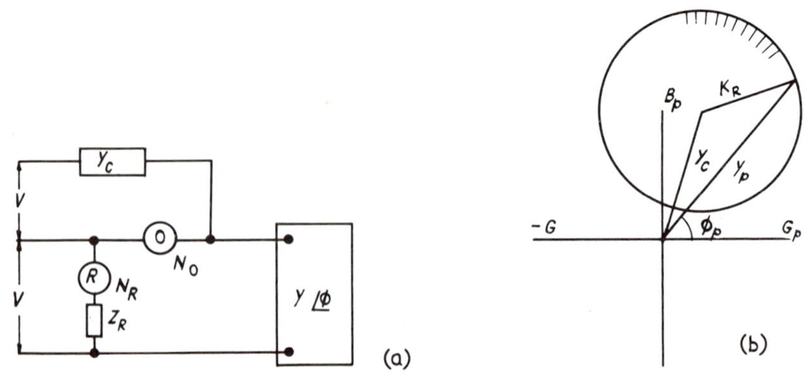

FOR OPERATION.

$$\left| V(Y_p \angle\phi_p - Y_C)\, N_O \right| > \left| \frac{V}{Z_R} N_R \right|$$

$$\left| Y\angle\phi - Y_C \right| > \left| \frac{N_R}{N_O} \frac{1}{Z_R} \right| = \left| K_R \right|$$

Y_C SHOWN CAPACITATIVE.

$I_m \frac{V_S}{V_R}$

$Re \frac{V_S}{V_R}$

S

R

2

1

-1

(c)

STABILITY ZONE IS SHADED

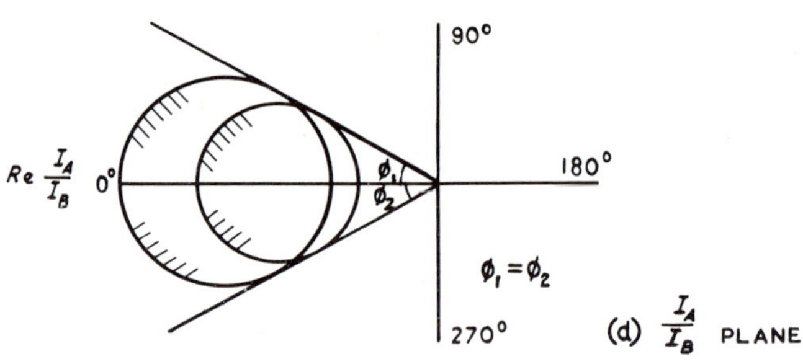

Fig. 15.28. Impedance-type comparator using capacitative bridge-circuit compensation:
(*a*) Derivation of relaying signals
(*b*) Locus diagram on admittance plane
(*c*) Relay characteristics on complex voltage plane
(*d*) Relay characteristics on complex current plane

In general terms and to a close approximation

$$Z_R = \frac{|B|\ N_R}{(1{\cdot}35 \sin \phi)N_0}$$

where $\pm\phi$ is the required tripping angle.

15.5.3.2. Bridge Circuit Compensation Method

The basic arrangement is shown in Fig. 15.28 and the adjustments are again of compensating impedance and restraint resistor. The required characteristic is obtained for end B on the I_A/I_B diagram corresponding to end R on the V_S/V_R diagram which requires a circle radius $K_R = \frac{N_0}{N_R} Z_R$, displaced Y_C from the origin.

For the pilot circuit of Fig. 15.20:

$$Y_C = \frac{A-1}{B} = (0{\cdot}3 + j2{\cdot}53) + 10^{-2} \text{ mho}$$

and

$$Z_C = 426 - j3900 \text{ ohms.}$$

For a $\pm 30°$ tripping angle

$$\left|\frac{N_0}{N_R} Z_R\right| = \frac{0{\cdot}5}{|B|}$$

from which

$$Z_R = \frac{|B|\ N_R}{0{\cdot}5\ N_0}$$

and for different tripping angles

$$Z_R = \frac{|B|\ N_R}{N_0 \sin \phi}.$$

15.6. PRACTICAL PILOT-WIRE PROTECTION SYSTEMS

In Great Britain, two types of protection schemes are in general use; the fundamental principles being similar for both private and rented telephone circuit applications though, of course, in the latter case, special design features are incorporated in the scheme to comply with additional requirements imposed by the telephone company which may limit the maximum voltage applied to the pilot circuit and specify high insulation between all protection equipment and the pilot circuits.

15.6.1. Opposed Voltage Protection (G.E.C.)

The opposed voltage principle is used with outputs provided for operating and restraining circuits of an amplitude comparator. The arrangement is shown in Fig. 15.29. Adequate discriminating margin is achieved by the use of a tapped pilot tuning reactor and the restraint resistor setting is adjustable.

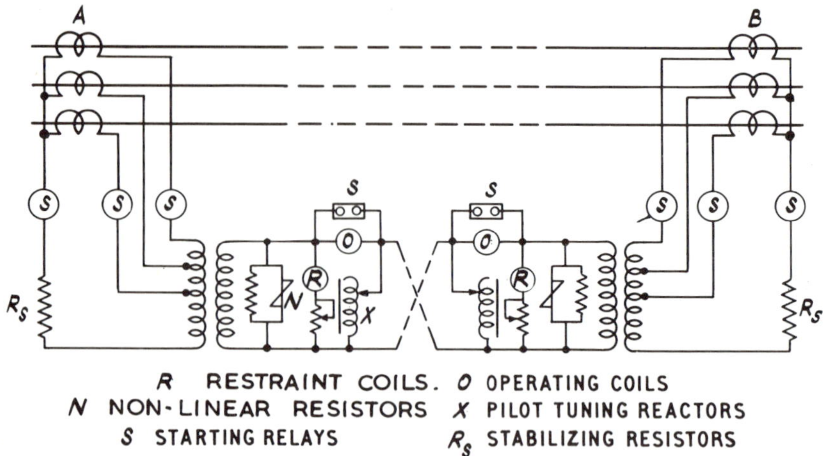

Fig. 15.29. Opposed voltage protection system (G.E.C.)

Three starting relays are used, two connected in the phases and the third in the c.t. residual circuit. The starting relay contacts control tripping and, in Fig. 15.29, also short-circuit the relay operating coil.

Low fault settings are obtained on the differential relays, and full advantage can be made of this where minimum fault currents are limited by using load-compensated (rate of change) starting.

Correct operation of the relay depends upon the setting of the pilot-tuning reactor and restraint resistor, particularly on long pilot circuits. Shorter pilot circuits (up to approximately 1,000 ohms loop resistance) do not usually require compensation, and a reasonable characteristic is provided by adjustment of restraint resistor.

The protection is suitable for pilot circuits up to 3,500 ohms loop resistance and 2·5 microfarad intercore capacitance, and these high values are often encountered on rented telephone-type circuits.

In the practical version of this protection scheme, the restraint-circuit resistor is shunted by a second harmonic (100 Hz) acceptor filter. This has been included particularly for transformer feeder applications where the high harmonic content of the magnetizing inrush currents during transformer switching can result in detuning of the pilot compensation. Transformer switching currents contain a high second harmonic component (this component is used for restraint on all power transformer differential protection schemes) and the 100 Hz tuned circuit provides an additional restraint under this condition.

15.6.2. Series Voltage System for Telephone-type Circuits – (Reyrolle)

The scheme uses the series voltage principle with parallel connected relays. Rectifiers are connected in series with the pilot circuit to obtain an artificial mid-point connection for the relay coil. The various components of the scheme are shown in Fig. 15.30.

The pilot loop contains a half-wave rectifier at each end shunted by resistance which are arranged so that their conducting directions are in opposition. The relays are connected across the pilot terminals in series with half-wave rectifiers.

Under external-fault conditions for the condition where the rectifier shunting resistors are each equal to the pilot-circuit resistance the equipotential point alternates between the two ends, each relay being at the electrical mid-

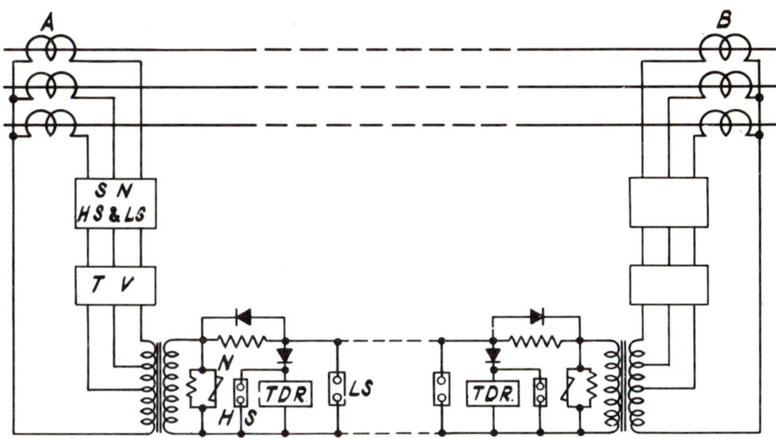

SN, LS & HS STARTING NETWORK, LOW SET AND HIGH SET RELAYS.
TV–TRANSDUCTOR RELAY VOLTAGE SUPPLY.
TDR–TRANSDUCTOR RELAY.

Fig. 15.30. Series voltage system for telephone-type circuits (Reyrolle HRPW)

point on alternate half-cycles. The series rectifier in the relay coil circuit prevents current from flowing in the relay circuit when it is not operating at the electrical mid-point.

In practice the rectifier shunt resistors are made greater than the pilot loop resistance so that both relays are polarized by reverse voltage for both half-cycles of the external-fault condition. This, in effect, provides the relay with a bias feature by displacing the relay connection beyond the mid-point in the stable direction.

The pilot replica impedance is purely resistive because the unidirectional nature of the applied pilot voltage under external-fault conditions, tends to reduce the effect of pilot capacitance currents. In practice, however, the effect of long pilot circuits is to produce an unsymmetrical tripping angle.

Under internal-fault conditions, only the relay at the infeeding end is energized. The relay current is unidirectional, being half-wave rectified and

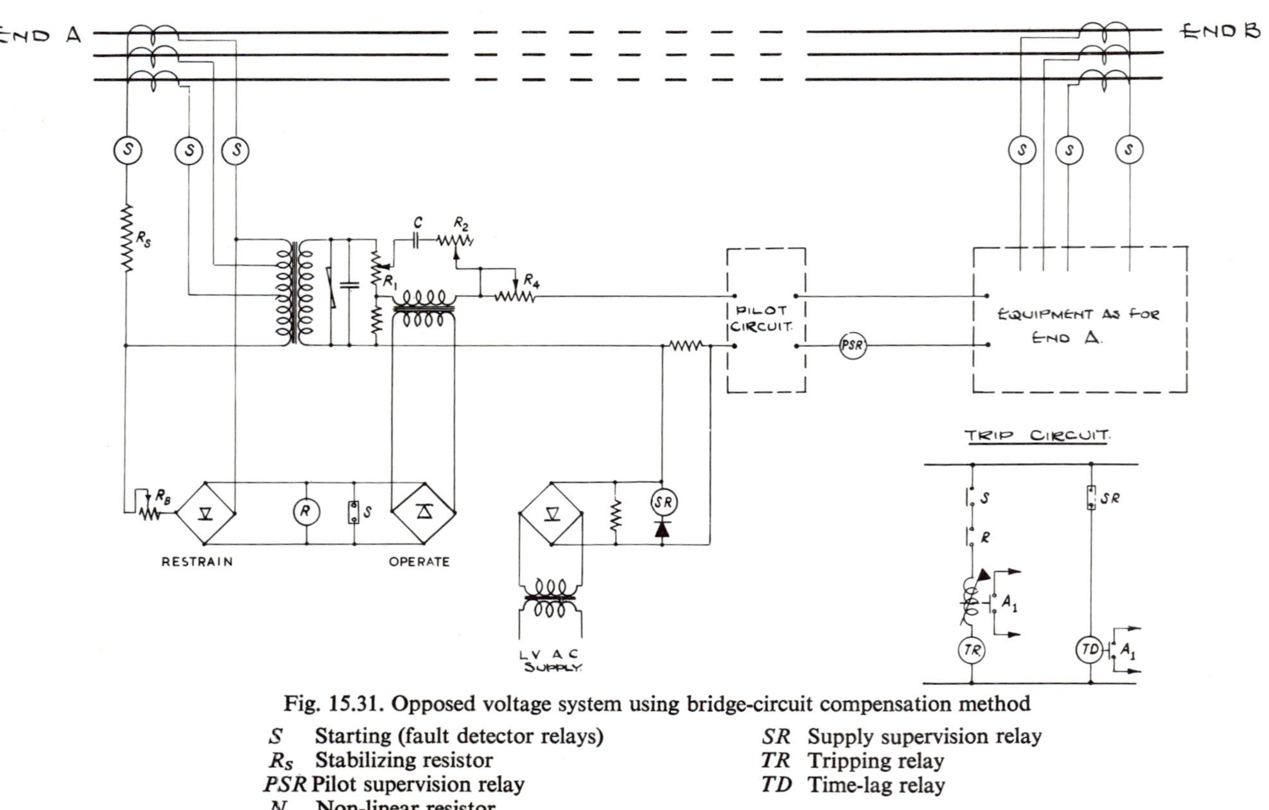

Fig. 15.31. Opposed voltage system using bridge-circuit compensation method

S	Starting (fault detector relays)	*SR*	Supply supervision relay
R_S	Stabilizing resistor	*TR*	Tripping relay
PSR	Pilot supervision relay	*TD*	Time-lag relay
N	Non-linear resistor		

used as control current for a transductor-type relay, which obtains its polarizing supply at constant voltage through a network energized by secondary fault current.

The starting equipment consists of three identical transformers, the outputs of which are connected to three rectifiers, the outputs of which in turn are paralleled and connected to the coils of two starting relays in series, low set and high set.

On the occurrence of a fault low-set contacts are opened, which remove a short-circuit and the supervisory supply, respectively, from the pilots. At the same time high-set contact is opened, which removes a short-circuit from the control winding of the transductor relay.

The scheme is designed for use over telephone-type circuits of up to 3,500 ohms resistance and 2 microfarad intercore capacitance.

15.6.3. Scheme Using Bridge Circuit Compensation (127)

Figure 15.31 illustrates schematically a design of an opposed voltage pilot-wire protection for use over rented telephone-type pilot circuits, using the principles of Section 15.4.8.2. The arrangement embodies fault-detector relays, continuous pilot supervision equipment and appropriate insulation levels to comply with the requirements laid down for the use of G.P.O. circuits for power-system signalling.

Adjustment of the protection characteristic is provided by two potentiometers R_1 and R_2. R_1 controls the amplitude of the injected compensation signal and R_2 its phase angle. The two adjustments, which are substantially independent, locate the centre of the protection characteristic circle and the setting of the restraint circuit resistor R_3 determines the radius of the characteristic circle. The setting up procedures required *in situ* are in this way simplified, and the adjustments provide precise control of the protection characteristic shape.

The amplitude-comparator circuit may comprise either a conventional two-rectifier arrangement with operating and restraint circuits energizing a sensitive relay, or some form of static amplitude comparator. An alternative arrangement is shown in Fig. 15.32 which uses a different arrangement of compensation circuits to achieve similar characteristics.

Schemes of this kind provide a more precise adjustment of compensating and restraint circuit impedances and this is particularly important since these adjustments, which determine the overall characteristics of the protection, must be made on site. Thus, apart from the more precise control of characteristics, these methods permit the use of accurate and efficient site commissioning methods to provide a practical realization of the idealized characteristics discussed in this chapter.

15.7. PILOT-WIRE PROTECTION FOR MULTIENDED CIRCUITS

A multiended circuit may be defined as one having more than two independent terminals capable of contributing fault current infeed at the end of con-

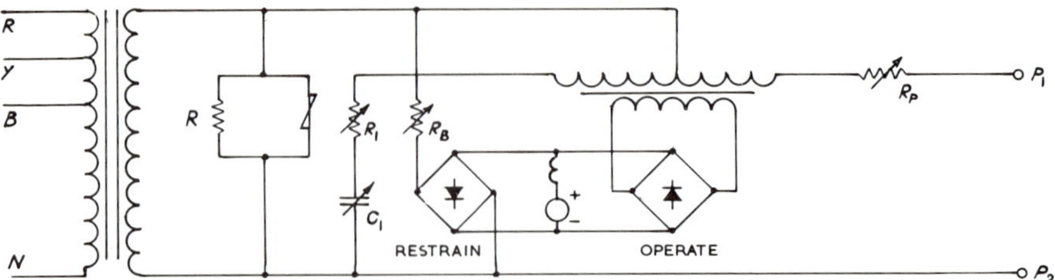

Fig. 15.32. Alternative form of opposed voltage protection using bridge compensation method (G.E.C.)

R_1C_1 Pilot-compensation circuit
R_B Bias setting resistor
R_P Pilot-circuit padding resistor
R Reference (current/voltage control) resistor
P_1, P_2 Input-to-pilot circuit

nection of the protection system. The definition includes all situations where the outputs from more than two sets of current transducers are required to be compared, and where one or more of the terminals is geographically remote.

Fundamentally, because of the large distance between ends, the techniques for ensuring stability discussed in Chapter 9 do not apply, and stability is wholly dependent upon the accuracy of current transformation.

The basic arrangement of opposed voltage protection for a multiended line is shown in Fig. 15.33 and the fundamental requirements for the successful

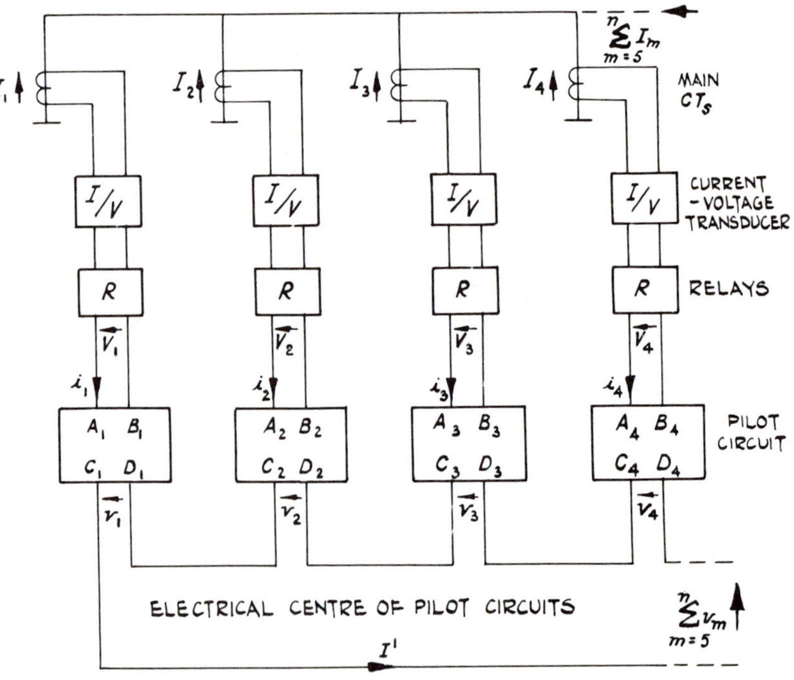

Fig. 15.33. Basic arrangement of opposed voltage protection for multi-terminal line

design are that (73) the pilot circuit must possess a well-defined electrical centre which provides a null point under all through-fault conditions. Since, in general, the individual limbs of the pilot circuit will be unbalanced, the pilot network must have artificial sections added so that the relevant electrical constants measured from each relaying end to the electrical centre are identical.

The input admittance then measured from one end is given by

$$Y_{P1} = \left[\frac{A}{B} - \frac{(n-1)}{nAB}\right] + \frac{K\angle\theta}{nAB}. \tag{15.25}$$

An input admittance chart drawn on this basis for 3 ends ($n = 3$) is shown in Fig. 15.34, where the centre is at

$$Y_{SC} - 2/3\,(Y_{OC} - Y_S)$$

and the radius is

$$1/3\,(Y_{OC} - Y_{SC}).$$

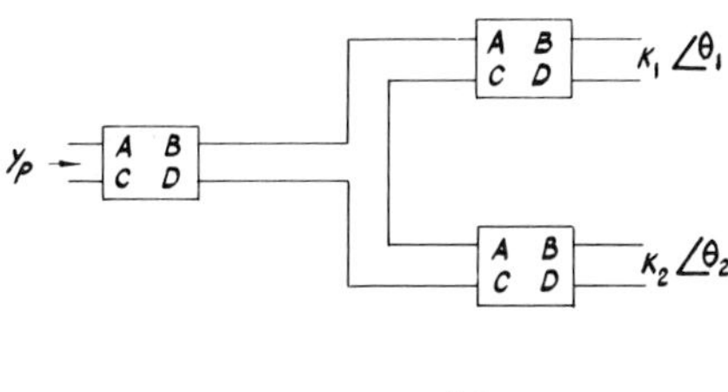

(a)

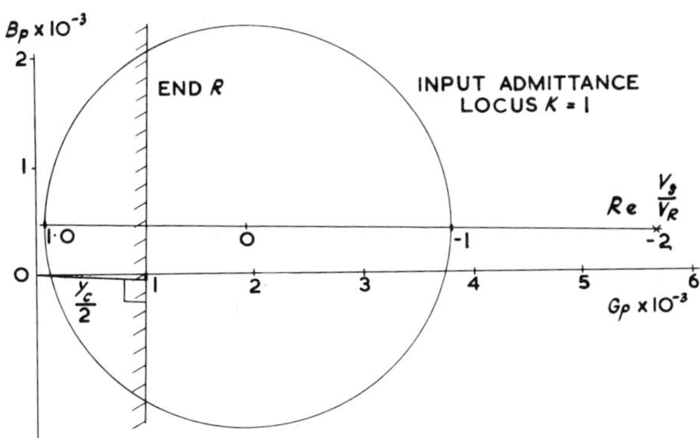

Fig. 15.34. Input-admittance diagram for three-ended pilot circuit:
(*a*) Circuit arrangement

$$\frac{I_A}{I_B} \propto \frac{V_S}{V_R} = (K_1\angle\theta_1 + K_2\angle\theta_2) = K\angle\theta$$

(*b*) Input-admittance diagram ($K = 1$) showing end B relay characteristic
Pilot circuit: $R = 18{\cdot}5$ ohm/mile: $C = 0{\cdot}18\ \mu F$/mile

Using this input-admittance chart, the comparator characteristic can be superimposed and the protection overall I_A/I_B performance determined.

The particular difficulty, of current transducer inaccuracy, remains however and has particular significance for an external fault where the fault current is fed largely through two ends with a small infeed at the third end. The low-infeed end is required to balance correctly against the two high-current ends which, with normal current transformers, must exhibit some unbalance.

The conventional circular characteristics of the previous section are found to be unsatisfactory because of this, and whilst straight-line characteristics can give considerable improvement (73) ultimately stability can only be maintained by the circulation of a d.c. biasing signal between ends through an additional pilot circuit. Such an arrangement is shown in Fig. 15.35. It is important to emphasize, however, that acceptable characteristics can only be obtained by creating an electrical centre for the pilot circuits and by providing some form of pilot compensation based on the measured admittance.

15.8. PHASE-COMPARISON CARRIER-CURRENT PROTECTION (125)

The fundamental principles of phase-comparison carrier-current protection are outlined in Section 8.11 of Volume 1 and the basic elements of the scheme are shown in Fig. 15.36. The modulated blocks of carrier signal which represent one half-wave of the relaying signal are compared on the principle that, under ideal through-fault conditions, a continuous carrier signal is applied to the comparator which is held inoperative. Dissimilarity between the phase angles of the compared quantities introduces gaps in the carrier signal which, for a predetermined gap length, produces tripping.

The phase-angle characteristic of a typical phase-comparison carrier-current protection system is adjustable between 30° and 50° and the correct

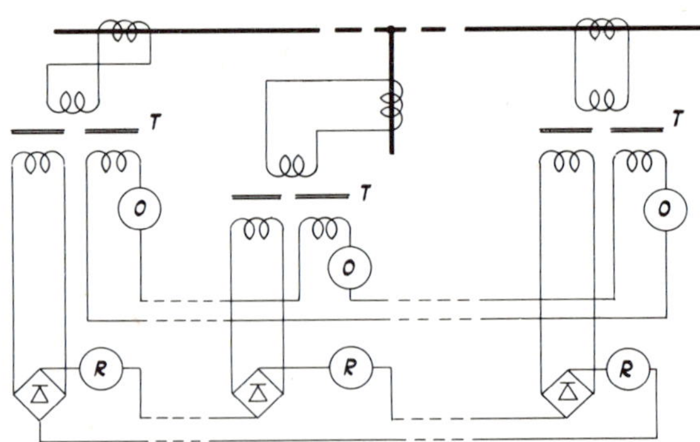

Fig. 15.35. Biased differential protection for a three-ended circuit using four pilot conductors

O Operating circuits of amplitude comparator connected to measure algebraic sum of currents at all three ends

R Restraint circuits connected to measure arithmetical sum of currents at all three ends

T Air-gap transformers (transactors)

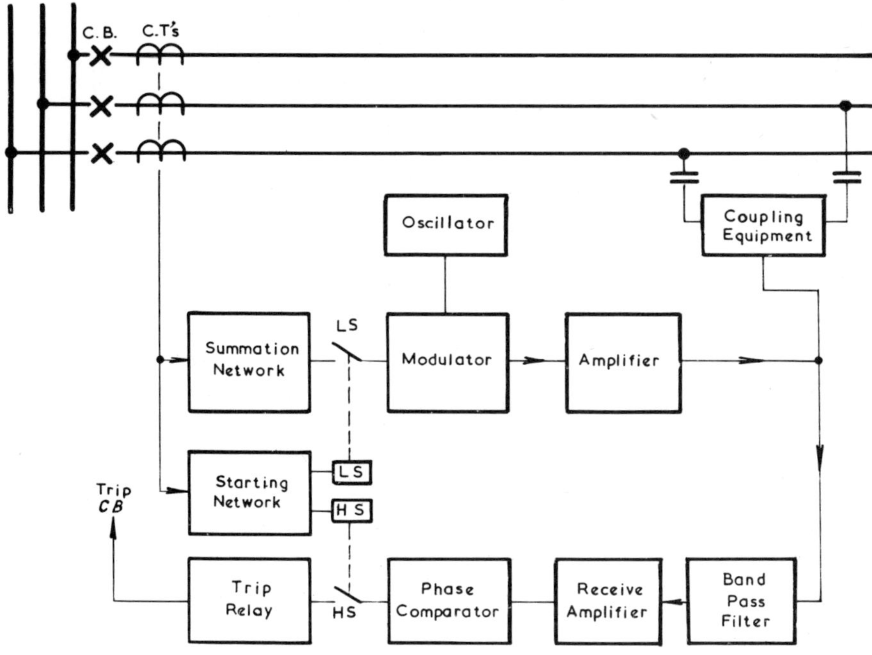

Fig. 15.36. Block diagram of phase-comparison carrier-current protection scheme

relationship between minimum fault settings and phase-comparison angle must be established by methods similar to those outlined in Section 15.2 taking account of the line-charging current and carrier signalling times.

The protection is controlled by a starting network which incorporates low- and high-set features to ensure that a carrier signal is transmitted only for fault conditions.

Modern protection systems use a negative phase sequence current network to provide starting for unbalanced faults and a positive phase sequence network having an impulse characteristic for three-phase faults. The compared signal utilizes mixed positive and negative sequence current signals, typically $I_1 - 5I_2$.

The limits of application of any particular practical scheme may depend upon one of three factors

(*a*) Attentuation of the carrier signal.
(*b*) Shunt capacitance of the protected circuit.
(*c*) Minimum fault infeed levels.

15.8.1. Attenuation of the Carrier Signal

The correct operation of a carrier system depends upon the reception of an adequate signal level at the comparator. This requires that the transmitter output shall be large enough to accommodate the attenuation of the coupling equipment and of the line whilst providing an input in excess of the receiver setting level.

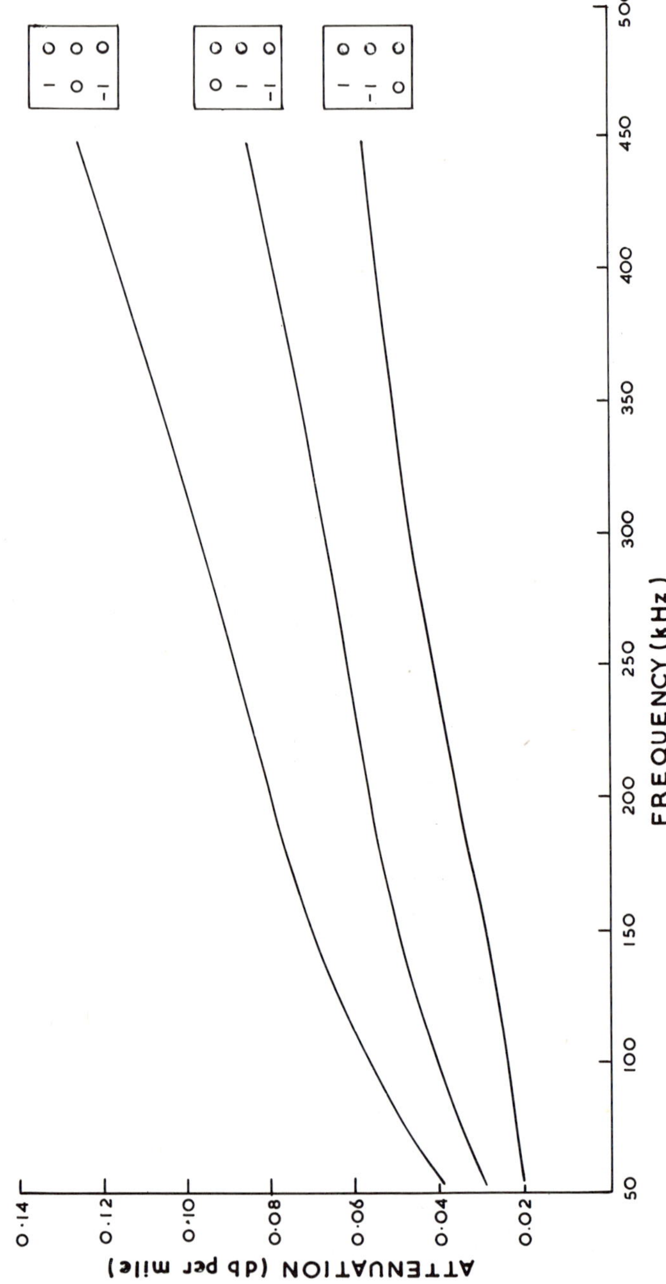

Fig. 15.37. Calculated power-line carrier attenuation on a 2 × 400 sq mm 275 kV overhead line (phase-to-phase coupling) Soil resistivity – 100 ohm metres

The insertion loss of the coupling equipment is constant for a particular design, but the line attenuation is dependent upon the carrier frequency, the length of line and its construction and method of carrier coupling, which may be either phase to earth or between phases.

Attenuation is also increased by line icing, which increases the conductor radiation loss, and allowance must be made for this in that ample margin must be available between normal received signal level and receiver setting.

The attenuation/frequency characteristic of an overhead line depends also upon the phases used for coupling of the carrier signal and Fig. 15.37 shows the variation of line-attenuation for a 2 × 400 sq mm. 275 kV line. The values given in the figure assume that the coupling phases are correctly terminated and the additional attenuation introduced by the coupling equipment is not included (128). Similar figures taken for 4 × 400 sq mm. 400 kV line do not differ greatly from those of Fig. 15.37.

Phase-to-earth carrier-signal injection is often used but in this case the line attenuation is higher (typically by about 30%) and is also more susceptible to interference by noise (125).

The complete carrier channel is shown schematically in Fig. 15.38. Typical

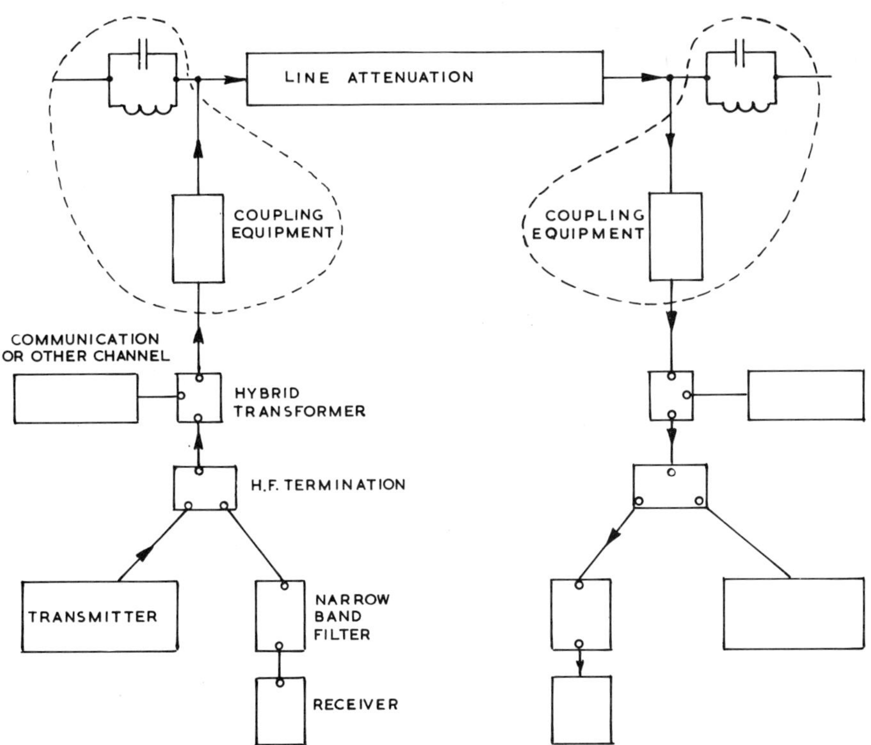

Fig. 15.38. Schematic of power-line carrier-current signalling channel

carrier equipments are designed for a 30 db insertion loss between transmitted signal and receiver setting. Allowing a 10–12 db margin (to cover exceptional losses due to icing, etc.), a total of 7 db for line traps and couplers at both ends and an additional 4 db for a separation filter since the total carrier channel will often include other signalling functions, e.g. intertripping, communication, etc., the permissible channel attenuation is 30–12–7–4 = 7 db.

Based on this figure the curve of Fig. 15.39 shows the maximum circuit length that can successfully permit carrier signalling for different carrier frequencies and modes of coupling.

The advantage of coupling to the top two conductors when long lines or high-carrier frequencies are involved is readily apparent.

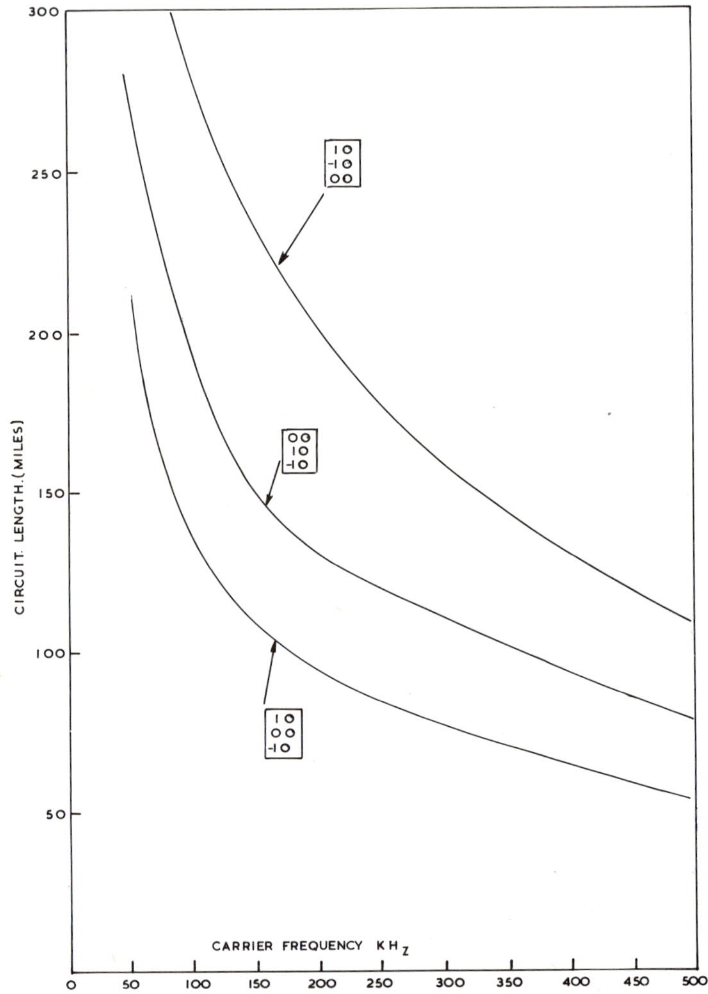

Fig. 15.39. Maximum permissible circuit length for reliable reception of power-line carrier signals

15.9. APPLICATION OF DIFFERENTIAL SYSTEMS

Generally, pilot differential systems are not so widely used as distance systems. There is ample justification for this where the time-graded protection provided by the basic multi-zone distance protection is acceptable.

Where high-speed fault clearance is essential, some kind of end-to-end communication must be used either by sending direct tripping or permissive tripping or blocking signals from distance protection. It is in these cases that differential systems provide a viable alternative. Their particular advantages over distance-measuring systems may be listed (129).

(*a*) Far higher values of fault resistance can be accommodated for internal faults. This is particularly advantageous for lines of insulated construction (i.e. without earth wires) and also for single-phase to earth faults where the faulted conductor touches ground in mid-span. In both cases high values of fault resistance may be encountered.

(*b*) When correctly adjusted to take account of the primary capacitance of the protected circuit, differential systems are unaffected by power swings and by high emergency load current levels.

(*c*) The protection equipment generally is simpler than distance protection and a failure of the communication signal will not inhibit high-speed tripping for internal faults, though, of course, incorrect operation may result on through faults.

Generally, then, pilot differential protection is particularly applicable to the protection of important lines and interconnectors where high-speed tripping is essential.

Pilot-wire systems are used extensively for the protection of cable networks up to the highest voltages as it is usually convenient to provide a multicore pilot laid with the main cable. Where reliable telephone circuits can be hired, the use of pilot-wire systems may be extended to overhead lines and, depending on economic appraisal, lines up to approximately twenty miles long may be protected.

Phase-comparison carrier systems are applied beyond this, or, where an existing carrier channel is available, up to a maximum line length of about 120 miles. Longer lines require a careful investigation of carrier signal attenuation characteristics to determine reliable performance.

On important interconnected EHV systems where the possibility of protection failure must be minimized to avoid widespread interruption, two independent protection systems are often used. Ideally these two systems should use different principles and different communication methods. The use of a pilot differential system and a high-speed distance measuring system will normally provide the best overall protection arrangement for these applications.

16

The Future

Miniaturization – Faster tripping – Communication channels – Use of computers in automatic protection – Other computer uses Fault location – Power from ground wires – New fields

Obviously the future of any technical field is difficult to predict because unforeseen new developments can change the course of trends which seem valid at the time of prediction. The following statements are based upon present trends and upon research being carried out by the C.E. G.B. In addition, attention has been called to some fields of investigation which have not yet been explored.

More new engineering tools appear each year. Transistors have revolutionised thinking in protective relay design. Electroluminescence offers an improved method of indicating relay operation. Lasers appear to offer a solution for the insulation problem of the E.H.V. current transformer. Semiconductor electronics has facilitated the development of very fast-responding generator excitation, regulation and kvar control and will probably be the means of relaying in less than a half-cycle.

16.1. MINIATURISATION

Semiconductor electronics has already reduced the size of many relays by 3 to 1. The use of integrated circuits instead of discrete components will cause further reduction in size.

The present static relays use modules on which discrete components are electrically connected by a printed circuit. In the 'thin film' circuit the passive components, such as resistors and small capacitors, are made by deposition of metal on a glass or ceramic surface; the semiconductors are of the standard type, soldered to the rest of the circuit.

Thin film circuits simplify trouble-shooting since it is easier to isolate a throw-away circuit unit than a defective component. They reduce the size of the equipment and the number of soldered connections to be made.

16.2. MICRO-MINIATURE CIRCUITS

In micro-miniature circuits there are no discrete components; all the components are integrated, including the semiconductors. With this technique an electronic circuit can be reduced to the size of a full stop; in fact as many as a thousand circuits can be put on a silicon chip the size of a penny. Saving in panel space is an obvious advantage of miniaturized relays but more important advantages of the micro-circuit are consistency, reliability, cheapness and speed of operation.

At present micro-circuits are made by chemical and lithographical processes under the control of an operator (usually female) with a binocular microscope. In a few years the operators will be replaced by automatic operation and, when that stage has been reached, ordinary printed circuits will be uneconomical except for small quantity special applications.

The reliability and consistency of these circuits will be valuable in protective relays, but their extreme speed of operation will offer no advantage until an accurate method of instantaneous measurement can be devised. At present this would appear to be possible only in differential current relays.

16.3. FAST CLEARING OF FAULTS

With the possibility of the development of fast vacuum-type breakers within a few years, pressure will be exerted upon protection engineers to produce even faster relays.

Because of the difficulty of correct measurement of derived quantities, like impedance, instantaneously, it is possible that digital computers will be used to achieve the necessary performance and at the same time to avoid error by automatic self-checking. Already in the U.S.A. computers have been programmed with the necessary signals for the isolation of a fault and restoration of service by appropriate switching.

One proposal for back-up protection is to have a single computer at headquarters and to feed it with information, such as currents and potentials from all pertinent points of the system. The location of any fault can be deduced by the computer from the fact that the ends of the faulted circuit carry more current than their neighbours and confirmed by the fact that their terminal buses have the lowest voltage values.

16.4. COMMUNICATION CHANNELS

In order to transmit the necessary information between protective equipments or to a computer, reliable and economical channels are required. Post Office channels are comparatively cheap, even for long distances but, unfortunately, are subject to interruption by P.O. personnel during maintenance and testing and to occasional failure due to flooding, digging operations and ground subsidence. P.O. overhead wires are vulnerable to lightning. Normal Post Office channels have a very limited bandwidth (3 kcs).

Carrier channels are expensive, except for very long overhead lines, and also rather limited in width and choice of frequency band. The quarter-wave length aerial will reduce the cost of carrier coupling and also eliminate line traps which, for 400 kV 4000 A lines, weigh 1100 lb and measure about $6 \times 5 \times 4$ feet and cost about £7800. Against this saving is the possibly higher cost of the aerial coupling. Typical performance values for a quarter-wave length aerial are 18% bandwidth, 1 db coupling loss and 35 db blocking efficiency. These are somewhat better than for a conventional capacitance coupling but are optimum values. On account of its more efficient blocking it can use the same frequency band more often and this may solve the problem of local spectrum space. Digital coding of the relay signal offers an alternative to variation of carrier frequency.

Another proposal for eliminating the line traps (which are not only very expensive but sometimes inefficient) is to use intra-bundle communication, the line traps being replaced by solid short-circuits at the ends of the bundled conductors and the communication channel being between insulated conductor wires. This method also would simplify the coupling equipment and eliminate it at T-junctions and underground sections.

Microwave links appear to be the most likely solution, especially in densely loaded areas such as exist in the U.K., Western Europe and the East coast of the U.S.A. Microwave transmission does not have to follow the path of the transmission line or cable and it is very economical if used for a large number of channels, as would be the case with complete automatic control of switching, regulation, load control, etc. Furthermore, it is much quicker and easier to adapt it to new system arrangements than is a power line carrier.

Microwave links offer a signal to noise ratio at least 20 db better than Post Office channels and only about 1% of the outage time. The wider bandwidth available gives a transmission time of about 0·8 ms per 100 miles compared with over 8 ms for the Post Office channel.

The tunnel diode has permitted the development of inexpensive and maintenance-free repeaters for transmission line towers, or other convenient eminences, which overcome the line-of-sight problem. The power consumption of these is so low that they can be supplied by solar cells. Also, unlike the H.F. and V.H.F. bands, a wide range of frequencies is available in the U.H.F. region, so that there is little restriction on the width or the number of channels available for the transmission of full information to the computer and the return of control signals.

16.5. OTHER COMPUTER USES

In addition to protection and system control, computers will be a powerful tool for:

(*a*) quickly determining the optimum power system arrangement;

(*b*) planning distribution systems and determining the location of shunt capacitors, etc. for optimum operatiom;

(*c*) quickly giving complete information on voltage conditions, etc. for different load flow programmes;

(*d*) determining the settings of relays, fuses and reclosers.

Better methods of computer load analysis, weather forecasting, etc. will permit more accurate load control and overload relay setting, taking into account the effect of sunshine, wind, rain, etc. upon the thermal capacity of the line.

It is probable that protection, automatic dispatching, regulation, etc. will all be integrated into one system of communication channels and computer control.

16.6. TESTING

At present, factory testing is a synthetic affair which consists largely of determining the threshold of operation of the relay and confirming that this threshold is not unacceptably changed by repeating the tests with fault conditions suddenly applied, usually at the improbable moment of voltage zero.

It should be possible with modern electronic equipment to analyse automatic oscillograms of actual fault conditions and automatically to apply the corresponding currents and voltages to the relays.

An alternative practical test is to cause a fault, not with the archaic use of a test circuit-breaker but by causing a flashover. The author's early method of doing this, with a bow shooting between conductors an arrow trailing 20 ft of thin iron wire, has been superseded by a pneumatic gun with accurate sights. This in turn will give place to the laser gun.

16.7. FAULT LOCATION

Line patrolling to locate a fault on an overhead line can be expensive in bad weather or rough country. Attempts have been made in the past to design into line relays, such as distance relays, a means for leaving an indication of the distance to the fault after tripping.

The time available for estimating the distance and leaving a record is the total clearing time of the fault which, nowadays, is only a few cycles and clearly insufficient for an electromagnetic movement.

With transistor circuitry very fast measurement and recording is possible and some distance relays already include a transistorised fault locator. In one arrangement a pulse signal is fed into the faulted phase through a phase selector relay and a timing device measures the time until the signal is reflected back from the fault and picked up by a receiver which stops the timer. The timer may consist of a capacitor which charges during the measured period and the time is obtained by measuring the voltage on the capacitor.

In another distance relay the inductance between the relay and the fault is measured by comparing the instantaneous voltage e with the rate of change of current di/dt at the moment when the current passes through zero; whence

$$L = \frac{e}{di/dt}.$$

16.8. POWER FROM GROUND WIRES

Considerable progress has been reported from Russia in the use of lightning shield wires of H.V. lines as a source of power for energising equipment in unattended substations and for supplying remote customers with domestic power [102]. It has been found that this use need not interfere with the original purpose of the wires, to deflect lightning from the conductors.

The potential induced on the wire is partly due to mutual induction from the conductor currents and partly due to electrostatic induction. The former is proportional to the length of wire; the latter is effective only if the wire is insulated through a spark gap to the tower; the usual practice is direct connection to each tower, but lightly insulated ground wires with spark-gaps have been ised in the U.S.A. and U.S.S.R.

16.9. UNTRIED FIELDS

As mentioned in Chapter 13, increased harmonics in the current or potential can be used for detecting faults and perhaps anticipating them. There are other properties of faults which could be utilized for fault confirmation, such as the intense light from the arc in a bus fault; also noise, radio energy, etc.

With the exception of the American K-dar relay and 8-pole induction cup directional relay, practically no attention has been paid to polyphase measurement, in spite of its obvious economy and the fact that most power system relays protect polyphase equipment.

It is clear that the future of protective relays still holds an interesting challenge to young engineers since the technique of automatic protection has by no means settled down to a predictable pattern; in fact the number of unsolved problems seems to increase each year.

References

[1] WIDERÖE R. Thyratron Tubes in Relay Practice, *Transactions A.I.E.E.*, **53**, 1934, pp. 1347–1353.

[2] NEUGEBAUER H. The Use of Rotating Coil Relays and Rectifiers in Protection, *Elektrotechnische Zeitschrift*, **71**, No. 15 August 1950.

[3] BIERMANNS O. Schnelldistanzrelais für Mittelspannungnetze, *A.E.G. Mitteilungen.*

[4] MACPHERSON R. H., WARRINGTON A. R. van C. and MCCONNELL A. J. Electronic Protective Relays, *Transactions A.I.E.E.*, **67**, Part III, 1948, p. 1702.

[5] SHERRIFF R. D. Reliability of Printed Circuit Connectors, *Electronic Industries* (*U.S.A.*) September and October 1960.

[6] SOROTKA V. T. Protective Relays based on the Hall Effect, *Elektrichestvo*, No. 11, 1958, pp. 68–71.

[7] SIROTA I. M. Galvanometric Directional Relays, *Elektrichestvo*, No. 4, 1959, pp. 38–43.

[8] LOVING (Jnr) J. D. Electronic Relay Developments, *Transactions A.I.E.E.*, **68**, Part I, 1949, p. 233.

[9] ADAMSON C. and WEDEPOHL L. M. Power System Protection with Particular Reference to the Application of Junction Transistors to Distance Relays, *Proceedings I.E.E.*, **103**, Part A, 1956, p. 379.

[10] BERGSETH F. R. An Electronic Distance Relay using a Phase-Comparison Principle, *Transactions A.I.E.E.*, **73**, 1954, p. 1276. Also: A Transistorised Distance Relay, Convention Paper, A.I.E.E. Summer Convention, San Franscisco, June 1956 (limited circulation).

[11] ADAMSON C. and WEDEPOHL L. M. A Dual Comparator Mho-type Distance Relay utilising Transistors, *Proceedings I.E.E.*, **103**, Part A, 1956, p. 509.

[12] FITZGERALD A. S. A Carrier Current Pilot System of Transmission Line Protection, *Transactions A.I.E.E.* **47**, No. 1, 1928, pp. 22–30.

[13] BRÄTEN J. L. and HOËL H. A New High-speed Distance Relay C.I.G.R.E. (Paris) 67, Paper No. 307, 1950.

[14] JARRATT T. J. Transistor SCR Firing Circuits, *Mullard Technical Communications*, **7**, No. 65, June 1963.

[15] DEWEY C. G., MATTHEWS C. A. and MORRIS W. C. Static Mho Distance and Pilot Relaying—Principles and Circuits, *Transactions, I.E.E.E.* Paper No. 63–112.

[16] HALMAN and HARRIS Voltage Surges in Relay Control Circuits, *Transactions A.I.E.E.*, **67**, 1948, pp. 1693–1701.

[17] A.I.E.E. Committee Report Insulation Level of Relay Control Circuits, *Transactions A.I.E.E.*, **68**, 1949, pp. 1255–1257.

[18] SONNEMANN W. K. Transient Voltages in Relay Control Circuits, *Transactions A.I.E.E.*, **80**, 1961, Part I, pp. 1155–1162.

[19] SEELEY H. T. Protection of Control Circuit Rectifiers against Surges from d.c. Coil Interruption, ibid. pp. 871–879.

[20] TRAVER O. C., AUCHINLOSS J. and BANCKER E. H. Pilot Protection by Power Directional Relays using Carrier Current, *G.E. Review*, **35**, No. 11, November 1932, pp. 566–570.

[21] WARRINGTON A. R. VAN C. Graphical Method for Estimating the Performance of Distance Relaying during Faults and Power Swings *Transactions A.I.E.E.* **68**, Part I, pp. 608–621.

[22] WARRINGTON A. R. VAN C. *Protective Relays: Their Theory and Practice* Vol. I. A book published by Chapman & Hall Ltd., London and John Wiley & Sons, New York.

[23] NEWCOMBE R. W. Electrical Protection of Large Generator Units. *The Electrical Journal*, May 22nd, 1959.

[24] DUMMER G. W. A. Component Reliability—A Survey, *British Communications and Electronics*, June 1963, pp. 434–437.

[25] WARRINGTON A. R. VAN C. Back-up Protection, C.I.G.R.E., Paris 1960, Report No. 334, Section III.

[26] CORDRAY R. E. and WARRINGTON A. R. VAN C. The Mho Carrier Relaying Scheme, *Transactions A.I.E.E.*, **63**, 1944, pp. 228–235. Disc. p. 434.

[27] RADKE G. E. A Method for Calculating Time-current Relay Settings by Digital Computer. I.E.E.E. Conference Paper 63–919, April 1963.

[28] KNABEL A. H. Simplify O.C. Relay Data for Computers, *Electrical World*, May 27, 1963.

[29] WARRINGTON A. R. VAN C. Application of the Ohm and Mho Principles to Protective Relays, *Transactions A.I.E.E.* **65**, June 1946, pp. 378–386. Disc. June Suplement 1946, pp. 490–491.

[30] YOUNG J. F. Use of Transistors in Industrial Timer Circuits, *Electronic Engineering*, June 1963, pp. 366–371.

[31] CORSON A. J. and ROWELL R. M. New Concepts in Systems for Electrical Mesaurement, *Electrical Engineering*, **82**, March 1963, pp. 214–219.

[32] CALECA V., HOROWITZ S. H., MCCONNELL A. J. and SEELEY H. T. Static Mho Distance and Pilot Relaying, *Transactions I.E.E.E.* August 1963, pp. 424–436.

[33] Mullard Booklet TP300, *Transistors for the Experimenter*.

[34] SCHWARTZ R. F. Introduction to Semiconductor Theory. *Electrical Manufacturing*, January 1959, pp. 107–130.

[35] MAPHAM N. The Rating of SCR's when switching into High Currents I.E.E.E. Conference Paper No. 63–498.

[36] JOHNSTON R. P. EHV Transmission Economics, *Allis-Chalmers Electrical Review* Fourth Quarter, 1961, pp. 26–28.

[37] BREWER R. and WYATT W. W. D. A Reliability Appraisal of Semiconductor Devices, *Proceedings I.E.E.* May 1959, Paper No. 2980 E.

[38] SUEKER K. Protecting Silicon Rectifiers from Transient Overloads, *Electro-technology*, November 1963.

[39] (General Electric Company of U.S.A. advertisement for d.c. power supply), *Electro-technology*, **71**, No. 3, March 1963, p. 210.

[40] PEACH N. Protective Relaying, *Power*, August 1961, pp. 67–90.

[41] MATHEWS P. and NELLIST B. D. Generalised Circle Diagrams and their Application to Protective Gear, *I.E.E.E. Transactions on Power Apparatus and Systems*, No. 2, February 1964, pp. 165–173.

[42] MATHEWS P. and NELLIST B. D. Transients in Distance Protection, *Proceedings I.E.E.*, **110**, No. 2, February 1963, pp. 407–418.

[43] DAVISON E. B. and WRIGHT A. Some Factors affecting the Accuracy of Distance-type Protective Equipment under Earth Fault Conditions, *Proceedings I.E.E.*, **110**, No. 7, July 1963.

[44] ATABEKOV G. I. *The Relay Protection of High Voltage Networks*, A book published by Pergamon Press, 1960.

[45] [Author not named] Schmitt Trigger Circuit, *Electrotechnology*, March 1964, pp. 132–133.

[46] SINCLAIR W. D. Thyristors and their Application, *Electrical Times*, August 27th, 1964, pp. 289–292.

[47] [Author not named] The Tecnetron, *Electronic Industries*, March 1958.

[48] I.E.E.E. Standard Test Procedure for Semiconductor Diodes, *I.E.E.E. Transactions on Electronic Devices*, ED-**11**, No. 8, August 1964, pp. 398–402.

[49] WEBER P. A New Distance Relay for Medium Voltage Networks *Brown-Boven Review*, April 1965, pp. 297–309.

[50] LYLE A. G. *Major Faults on Power Systems*, A book published by Chapman & Hall Ltd., London, 1952.

[51] GOLDE R. H. Lightning Surges on Overhead Distribution Lines caused by Direct and Indirect Lightning Strokes, *A.I.E.E. Transactions* (*Power Apparatus & Systems*) June 1959, pp. 437–447.

[52] YOUNG J. F. Visualising Symmetrical Components, *Electrical Review*, 15th February 1963, pp. 259–261.

[53] WRIGHT A. Saturable Current Transformers for Current Limiting, *Instrument Practice*, October 1962, pp. 1219–1224.

[54] WRIGHT A. Residual Flux in Current Transformer Cores, *Instrument Practice*, February 1963, pp. 160–163.

[55] PATRA DR. S. P. and BASU S. K. Transistorised Static Overcurrent Relays, *Electrical Times*, 14th November 1963.

[56] DALASTA D. Applying Modern Overcurrent Relay Time-current Curves, *Allis-Chalmers Review*, Third Quarter, 1963.

[57] LAMM U. The Transductor and its Applications *A.S.E.A. Journal*, **16**, No. 5, 1939.

[58] STAR J. Hall Effect Transducers, *Instrument and Control Systems*, April 1963, p. 113.

[59] STEIN L. B. A New Current-sensing Device, A.I.E.E. Paper, **62**, 239, April 1962.

[60] HARDER E. L., WENTZ E. C., SONNEMANN W. K. and KLEMMER E. H. Linear Couplers for Bus Protection *A.I.E.E. Transactions*, **61**, May 1942, pp. 241–248.

[61] HODGKISS J. W. The Behaviour of Current Transformers subjected to Transient Assymmetric Currents and the Effects on Associated Relays, C.I.G.R.E. Paper, No. 329, Paris 1960.

[62] NELLIST B. D. and MATHEWS P. The Design of Air-cored Toroids or Linear Couplers I.E.E. Paper No. 3921 M, June Y962.

[63] HAYWARD C. D. Prolonged Inrush Currents with Parallel Transformers affect Differential Relaying. *A.I.E. Transactions* **60**, 1941, pp. 1096–1101. Disc. pp. 1305–1312.

[64] SHARP R. L. and RICH W. E. A Static Relay for Generator Protection. Relay Conference paper at Illinois Institute of Technology, May 6th 1960.

[65] ROCKEFELLER G. P. *et al.* Magnetising Inrush Phenomena in Transformer Banks. *Transactions A.I.E.E.*, **77**, Power Apparatus and Systems, October 1958, p. 884.

[66] HARDER E. L., WENTZ E. C., SONNEMANN W. K. and KLEMMER E. H. Linear Couplers for Bus Protection. *Transactions A.I.E.E.*, **61**, May 1942, pp. 241–248, Disc. June Supplement 1942, p. 463.

[67] HARDER E. L. and BOSTWICK M. A. A Single-Element Differential Pilot-Wire Relay System, *Electrical Journal*, November 1938, pp. 443–448.

[68] NEHER J. H. and MCCONNELL A. J. An Improved A-C Pilot-Wire Relay. *Transactions A.I.E.E.*, **60**, January 1941, pp. 12–17.

[69] RUSHTON J. The Fundamental Characteristics of Pilot-Wire Differential Protection Systems, *Proceedings I.E.E.*, **108**, Part A, No. 41, Oct. 1961.

[70] GRINSHTEIN V. I. and RAKMANOV I. A. A Semiconductor Balanced Current Relay, *Elektrichestvo*, No. 10. 78–79, 1963.

[71] CHOWDHURI P. A Portable Fast-Response Transient Voltage Counter, *I.E.E.E. Transactions on Power Apparatus and Systems*, May 1965 pp. 417–422.

[72] WARRINGTON A. R. VAN C. Protective Relaying for Long Transmission Lines, *Transactions A.I.E.E.*, **62**, June 1943, pp. 261–268. Disc. June Supplement 1943, p. 427.

[73] RUSHTON J. Pilot-wire Differential Protection Characteristics for Multi-ended Circuits, *Proceedings I.E.E.*, **112**, No. 11, November 1965, pp. 2095–2102.

[74] PARTHASARATHY K. Static and Semi-static Conic Distance Relays. *Electrical Times*, 22nd July 1965, pp. 119–125.

[75] HUMPAGE W. D. and SABBERWAL S. P. Development in Phase-Comparison Techniques for Distance Protection *Proceedings I.E.E.* **112**, No. 7, July, 1965.

[76] UNGRAD H. Electronics in Protective Systems. *Brown Boveri Review*, **50**, No. 8, August, 1963, pp. 509–515.

[77] PFAFF C. J. R. and von BUZAY K. Circle Diagrams of Directional Impedance Relays. *Brown Boveri Review*, **49**, No. 5, May 1962.

[78] GIOT G. MARCHAL G. and VASQUEZ R. New Possibilities of Transistor Circuits for Distance Protection, C.I.G.R.E. Paper No. 309, Paris 1964.

[79] CHAPMAN P. C., HUMPAGE W. D. and RUSHTON J. The Discriminative Properties of Distance Protection, *Proceedings I.E.E.* **112**, July 1965 Parts I and II.

[80] BRATEN J. L. Desirable Shapes of Distance Relay Characteristics in Different Cases. Prelim. report given at CIGRE Study Committee No. 4 Meeting (Bucharest, May 1965).

[81] *Power System Transients*. A book edited by E. Openshaw Taylor, published by George Newnes Ltd., London.

[82] RODGERS H. E. and RIDGEWAY W. L. Reed Relays—Post Office Relays types 14 and 15. *Post Office Journal*, April 1965, pp. 46–49.

[83] BROWN G. H. Overload Protection of Industrial Motors, *Electrical Review*, 27th September 1963, pp. 5–6.

[84] DALZIEL C. F. Transistorised, Residual-Current Trip Device for Low-Voltage Circuit Breakers, A.I.E.E. Paper 62–1006, June 1962.

[85] VOSCRESENSKI A. A. Adjustment of Contact to Earth Signalling in Compensated Systems, *Electricheski Stantsii* (*Power Stations*), 1962, No. 2, pp. 90–91.

[86] GAFFORD B. N., DUESTERHOEFT W. C. and MOSHER C. C. Heating of Induction Motors on Unbalanced Voltages, *A.I.E.E. Transactions*, June 1959, pp. 282–295.

[87] HOBSON T. Get Complete Motor Protection, *Power*, January 1965.

[88] HAHN C. Line Protection Devices and Relays for Extra High Voltages. *Brown Boveri Review*, Jan/Feb. 1964, pp. 93–100.

[89] CHAMBERS F., HAMMER O. S. C. and EDWARDS L. Tennessee Valley Authority's 500 kV System—System Plans and Considerations. *Transactions I.E.E.E.*, Jan. 1966, Vol. PAS-85, No. 1, pp. 22–28.

[90] CLARKE S. A. and WHITTAKER J. C. The Choice and Application of Protective Systems for the 400 kV Transmission System, I.E.E. Conference on Design Criteria and Equipment for Transmisstion at 400 kV and Higher Voltage. Publication No. 15, Part I, pp. 25–27.

[91] SHOCKLEY W., SPARKS M. and TEAL G. K. p-n Junction Transistors *Phys. Review*, **83**, July 1951, pp. 151–162.

[92] PERRY E. R. Laser Measures Current, *Instruments and Control Systems*, **38**, July 1965, pp. 123–124.

[93] DALASTA D., FREE F. and DE SNOO A. P. An Improved Static Overcurrent Relay. *Transactions, I.E.E.E.*, Paper No. 63–218, January 1963.

[94] GERTSCH G. A. Capacitative Voltage Transformers and their Operation in Conjunction with System Protection Relays, C.I.G.R.E. 1960, Paper No. 318.

[95] STROM A. P. Long 60 Cycle Arcs, *Transactions A.I.E.E.* 1946, **65**, pp. 113–118.

[96] ADAMSON C. and TALKHAN E. A. Selection of Relaying Quantities for Differential Feeder Protection, *Proceedings I.E.E.*, **107**, Part A, No. 31, February 1960, pp. 37–47.

[97] RUSHTON J., LEWIS D. W. and HUMPAGE W. D. Fault Performance Analysis of Transmission Line Differential Protection Systems related to Their Polar Characteristic Requirements, *Proceedings I.E.E.*, **113**, No. 2, February 1966, pp. 315–324.

[98] STEPNOV T. V. Directional Protection from Short-Circuits to Earth, *Elektrichestvo* 1958, No. 8, pp. 75–79.

[99] MAIKOPAR A. S. Minimum Time of Automatic Reclosing, *Elektrichestvo*, 1959, No. 6, pp. 34–40.

[100] GRADY R. F. and CRAMP M. G. Direct Measurement of Generator Winding Temperature using a Miniature Transmitter, *Power Apparatus & Systems*, Vol. PAS/24 No. 11, November 1965, pp. 1073–1080.

[101] Symposium on Modern Power System Protection as practised in Europe. Account published in *Scientific Bulletin of Assoc. of Elec. Engineers*, Montefiore Electro-Technical Institute, September/October 1964, No. 5, pp. 341–399.

[102] MASON J. H. Discharge Detection and Mreasurements, *Proceedings I.E.E.* **112**, No. 7, July 1965, pp. 1407–1423.

[103] BOEHNE E. W. The Graphical Solution of Linear Circuits in the Steady State. *I.E.E.E. Transactions on Communications and Electronics*, July 1963, pp. 346–357.

[104] ELMORE W. A. and BLACKBURN J. L. Negative Sequence Directional Ground Relaying. I.E.E.E. Conference Paper 62–1072, July 1962.

[105] WARRINGTON A. R. VAN C. and DIENNE G. Recent Developments in the Transistorisation of Protective Relays and the Problems posed by their use in H.V. Systems. C.I.G.R.E. Paper No. 337, Part II, June 1966.

[106] PARTHASARATHY K. New Static 3-Step Distance Relay, *Proceedings I.E.E.*, **113**, No. 4, April 1966.

[107] PARTHASARATHY K. Three-system and Single-system Static Distance Relays, *Proceedings I.E.E.*, **113**, No. 4, April 1966.

[108] BAUDE J. The R-C Circuit with functionally variable Time-Constant and its Application to Static Protective Relays. A.I.E.E. Conference Paper No. 61–121.

[109] (Author not Named) Current Measurement at a Distance. *Electrical Times*, 10th July, 1964.

[110] ERDMAN H. G. and CALHOUN H. J. Improved Protection of Sub-Transmission Lines with Static Time-Distance Relays Minutes of Pennsylvania Electric Association Meeting, May 1966.

[111] SONNEMAN W. K. A New Static Time-Distance Relay. A paper presented at Georgia Institute of Technology Protective Relaying Conference, 7th May 1964.

[112] RUSHTON J. and JONES D. Generalized Locus Charts for the evaluation of Differential Feeder Protection. *Proceedings I.E.E.* **113** (7), July 1966, pp. 1194–1206.

[113] HAMILTON F. L. The Protection of Feeders by Pilot-Wire Systems. *Reyrolle Review*, No. 174, December 1959.

[114] ADAMSON C. Electronic Protection of Power Systems. *Electrical Times*, June 20, July 25, Oct. 3, Nov. 7, 1957; Feb. 27, March 6, 1958.

[115] HAUG H. and FORSTER M. Electronic Busbar Protection. *Brown Boveri Review*, **53**, No. 4, April 1966.

[116] REYROLLE Pamphlet No. 1297 Type H Distance Protection.

[117] *I.E.E. Colloquium Digest No. 1968/19*, Some Present-day Protection Problems, 28 May 1968.

[118] WARRINGTON A. R. VAN C. The History of Electronic Protection. *The Electrical and Electronics Technical Engineer*, Jan. 1967, Vol. 1, No. 3.

[119] RYDER, C., RUSHTON J., and PIERCE, F. M. A moving Coil Relay applied to Modern High-Speed Protective Systems, *Proceedings I.E.E.*, **100**, Part II, 1953 p. 261.

[120] SKUDERNA J. E. Mathematical Basis for a Protective Relay with Conic Pick-up Characteristics, A.I.E.E., Paper 62–20 Power Apparatus and Systems April 1962 pp 81–87.

[121] DURKIN, C. J. A Load Management System using Rate of Change of Frequency, Relays, *Pennsylvania Electric Association* Minutes, May 1970.

[122] NAYLOR, J. H. and NOBLE, J. O. Low Frequency Load Shedding *I.E.E. Colloquium Digest*, 1968/19.

[123] RUSHTON, J. and HUMPAGE, W. D. Power System Studies for the determination of Distance Protection Performance, *Proc. I.E.E.* **119**, 6, June 1972.

[124] DONALDSON, G. W. The application of phase comparison principles to power swing conditions on long high voltage transmission lines. *Proc. I.E.E.*, 1090S, February 1951 (98, Part II, p. 47).

[125] HAMILTON, F. L. *Power System Protection*, Vol. 2 (a book). Chapter 8, Feeder Protection: pilot wire and carrier current systems. Macdonald 1969.

[126] HUMPAGE, W. D., RUSHTON, J. and STEVENSON, P. D. Differential pilot wire protection systems using phase comparators. *Proc. I.E.E.*, 5038P, July 1966 (113, No. 7, p. 1183).

[127] RUSHTON, J. A new pilot wire protection scheme. *Electrical Times*, 24 October 1963, p. 605.

[128] WEDEPOHL, L. M. and WASLEY, R. G. Propagation of carrier signals in homogeneous, non-homogeneous and mixed multi-conductor systems. *Proc. I.E.E.*, 3433P, January 1968 (115, No. 1, p. 179).

[129] HUMPAGE, W. D. and RUSHTON, J. An evaluation of the comparative performance of distance and differential feeder protection systems. *I.E.E. Colloquium Digest*, 1968/19 – Some present-day protection problems, 267.

[130] RAO, T. S. M. Behaviour of a Rectifier Bridge as an Amplitude Comparator, Proc. *I.E.E.* (India), **XLV**, 10, Pt. EL5 June 1965.

Index